Physics of Shock and Impact, Volume 2

Materials and shock response

Physics of Shock and Impact, Volume 2

Materials and shock response

Dennis Grady
Applied Research Associates, Albuquerque, New Mexico

IOP Publishing, Bristol, UK

ISBN 978-0-7503-1257-8 (ebook)
ISBN 978-0-7503-1258-5 (print)
ISBN 978-0-7503-1259-2 (mobi)

DOI 10.1088/978-0-7503-1257-8

Version: 20171201

IOP Expanding Physics
ISSN 2053-2563 (online)
ISSN 2054-7315 (print)

British Library Cataloguing-in-Publication Data: A catalogue record for this book is available from the British Library.

Published by IOP Publishing, wholly owned by The Institute of Physics, London

IOP Publishing, Temple Circus, Temple Way, Bristol, BS1 6HG, UK

US Office: IOP Publishing, Inc., 190 North Independence Mall West, Suite 601, Philadelphia, PA 19106, USA

Contents

Preface xi

Introduction xii

Author biography xvi

6 Shock compression of ceramics with microstructure 6-1

6.1 Introduction 6-1
6.2 Classic mixture theory of shock compression 6-4
6.2.1 Shock compressibility of silicon carbide and silicon mixture ceramics 6-4
6.2.2 Shock compression of metal–ceramic composites 6-9
6.3 Shock compression of materials with microstructure strength 6-12
6.3.1 Objectives of the dynamic mixture model 6-14
6.3.2 Strength and EOS decomposition 6-15
6.3.3 Elastic and pressure-equilibrated mixture compressibility 6-15
6.3.4 Component compressibility relations 6-18
6.3.5 Mixture compressibility 6-19
6.3.6 Evolution of the compaction parameter 6-20
6.4 Elasticity and strength of ceramics with microstructure 6-22
6.4.1 Elasticity and microstructure 6-22
6.4.2 The Steinberg–Hugoniot elastic limit 6-23
6.5 Application to ceramics with microstructure 6-25
6.5.1 Commercial alumina ceramics 6-25
6.5.2 Boron carbide ceramics and shock-induced phase transformation 6-28
6.5.3 Silicon carbide ceramic with microstructure 6-41
6.5.4 Tungsten carbide ceramic and composition effects 6-50
6.6 Shock strength in silicon carbide and titanium diboride 6-55
6.6.1 Silicon carbide ceramics 6-56
6.6.2 Titanium diboride ceramics 6-58
6.6.3 Microstructure features of the shock strength 6-60
6.6.4 Failure deformation mechanisms in shock compression 6-63
6.7 Strength and failure models 6-65

6.7.1 Observations of failure wave phenomena 6-65
6.7.2 A meso-kinetic model of delayed failure 6-67
6.8 Assessment of experimental data and modeling of ceramics with microstructure 6-80
References 6-83

7 Dynamic compaction of crushable solids 7-1

7.1 Introduction 7-2
7.2 Spherical symmetry shock compaction of porous media 7-3
7.2.1 Shock propagation solution 7-3
7.2.2 Flow and pressure field solution 7-5
7.2.3 Attenuation properties in spherical shocks 7-6
7.3 Kompaneets shocks in porous media with other symmetries 7-7
7.3.1 Attenuation properties in other geometries 7-8
7.3.2 Thermo-mechanics of shock compaction 7-9
7.3.3 The crush model in distended media 7-11
7.4 Compaction in planar geometries 7-12
7.4.1 Variable compaction mass 7-13
7.4.2 Plate impact solutions 7-14
7.4.3 Constitutive compaction relations 7-15
7.4.4 Application of the incompressible-crush model to plate impact on porous media 7-18
7.4.5 Non-dimensional solutions 7-20
7.4.6 Power-law properties of the limiting solution 7-22
7.4.7 P-lambda compaction 7-24
7.4.8 Non-dimensional solution for p-lambda compaction 7-26
7.4.9 Explosive compaction of porous media 7-26
7.5 Compaction resistance in non-planar geometries 7-31
7.5.1 Kompaneets spherical geometry with compaction resistance 7-31
7.5.2 Compaction of a metal tube with porous media 7-33
7.6 Physics-based compaction models 7-37
7.6.1 Carroll and Holt porous compaction model 7-37
7.6.2 Continuum compaction models with microstructure 7-43
7.6.3 The p-lambda porous compaction model 7-51
7.7 Compaction with elastic recovery 7-55
7.7.1 Governing equations 7-56
7.7.2 Boundary and initial conditions 7-57
7.7.3 Elastic solution 7-57

7.7.4 Application of the boundary conditions 7-58
7.7.5 Rigid and elastic comparisons 7-59
7.7.6 Further elastic-compaction solutions 7-61
7.8 Dissipation and heating in shock compaction 7-61
7.8.1 Shock equation of state for distended solids 7-62
7.8.2 Temperature and shock compaction 7-70
7.9 Dynamic compaction and dissipation in distended solids 7-73
7.9.1 Compaction of the distended ductile solid 7-74
7.9.2 Compaction of the distended brittle solid 7-77
7.9.3 Dynamic compaction in crushable mixtures 7-86
References 7-96

8 The nature of structured shock waves 8-1

8.1 Introduction 8-2
8.2 Steady wave profiles and the fourth-power law 8-4
8.2.1 Experimental evidence for fourth-power dependence in the steady wave shock 8-4
8.2.2 Non-steady waves 8-6
8.3 Origin of the fourth-power law 8-8
8.3.1 Precursor attenuation in quartz 8-9
8.3.2 Steady waves in aluminum 8-10
8.3.3 Equivalence of the fourth-power relation and the shock invariant 8-10
8.4 Dissipative action and invariance in the structured shock wave 8-12
8.4.1 Action principles and implications for the shock wave 8-12
8.4.2 Correlation dynamics and length scales 8-17
8.4.3 Langevin dynamics and the dissipative action 8-20
8.4.4 Further applications in the shock transition 8-27
8.5 Viscosity and the structuring of shock waves 8-30
8.5.1 Viscosity of steady shock waves 8-31
8.5.2 Mechanisms of shock viscosity 8-32
8.5.3 Wave scattering and shock viscosity 8-33
8.6 Structured waves in porous and composite materials 8-34
8.6.1 Structured shock waves in granular and porous solids 8-35
8.6.2 Structured shock waves in composite solids 8-41
8.7 Physics of structured shock waves 8-45
References 8-47

9 Kinetics of shocks in brittle solids 9-1

9.1 Introduction 9-1
9.2 Shock compression of brittle solids 9-2
9.2.1 Nature of the shock wave 9-2
9.2.2 Experimental shock wave profiles 9-4
9.2.3 Shock- versus particle-velocity characteristics 9-7
9.2.4 Phase transformation under shock compression 9-8
9.2.5 Viscosity of brittle solids and the steady shock process 9-13
9.2.6 Hugoniot elastic limit and the shock wave elastic precursor 9-19
9.2.7 Off-Hugoniot measurements 9-22
9.3 Shock compression of dolomite rock 9-27
9.3.1 Experimental methods and materials 9-27
9.3.2 Features of the shock wave profiles 9-28
9.3.3 Ramp wave compression of dolomite rock 9-32
9.4 Rate dependence in brittle deformation and failure 9-35
9.4.1 Observations of rate dependence 9-36
9.4.2 Analysis of strain-rate-dependent deformation and failure 9-38
9.4.3 Steady elastic shock waves in dolomite rock 9-44
9.4.4 Strength at Hopkinson bar strain rates 9-48
9.4.5 Intermediate strain rates (10^4 s^{-1} to 10^5 s^{-1}) 9-52
9.4.6 Strain-rate dependence in brittle materials 9-55
9.5 Testing and failure kinetics of rocks and ceramics 9-58
9.5.1 Static testing 9-59
9.5.2 Shock wave testing 9-61
9.5.3 Intermediate rate testing 9-63
9.5.4 Kinetic effects in brittle solids 9-65
References 9-78

10 The shock wave equation of state 10-1

10.1 Introduction 10-1
10.2 Equation-of-state compressibility of a solid 10-2
10.2.1 Some elementary equations of state 10-3
10.2.2 Mie–Gruneisen equation of state 10-6
10.2.3 Enthalpy-based equations of state 10-23
10.3 Universal equations of state 10-32
10.3.1 Murnaghan equation of state 10-33
10.3.2 Linear shock-velocity–particle-velocity equation of state 10-34

10.3.3 Intermolecular potential equations of state 10-36
10.3.4 Finite-strain equations of state 10-42
10.3.5 Further universal equations of state 10-43
10.3.6 Linear-Z equation of state 10-49
10.3.7 Woolfolk Hugoniot equation of state and modifications 10-56
10.3.8 Assessment of universal equations of state 10-59
10.4 Temperatures under shock compression 10-60
10.4.1 Thermodynamics and temperature 10-60
10.4.2 Temperature on the isentrope 10-61
10.4.3 Temperature on the Hugoniot 10-62
10.4.4 Hugoniot temperature calculations 10-64
10.4.5 Melting on the Hugoniot 10-65
10.4.6 Plastic dissipation contribution to the shock temperature 10-67
10.4.7 The porous Hugoniot 10-68
10.4.8 Post-shock residual temperatures 10-70
10.4.9 Enthalpy equation of state and temperature on the Hugoniot 10-71
10.5 Additive mixture theory and experiment under intense shock compression 10-74
10.5.1 Shock equation of state through additive mixture models 10-75
10.5.2 Additive mixtures modeling through Gibbs energy 10-76
10.5.3 Mixture modeling through additive Hugoniot relations 10-77
10.5.4 Additive Hugoniot mixing at the molecular level 10-86
10.5.5 Dynamic isentrope from shock compression of mixtures 10-87
References 10-101

11 Shock wave analysis and applications 11-1

11.1 Introduction 11-1
11.2 Analysis in shock wave applications 11-2
11.2.1 Wave interaction analysis 11-2
11.2.2 Distance–time plots in shock analysis 11-7
11.3 Shock equilibration in a gage plane 11-11
11.3.1 Stress wave analysis 11-11
11.3.2 Stress wave attenuation from planar defects 11-15
11.3.3 Application to multiple planar defects 11-16
11.4 Application to stress pulse attenuation in oil shale 11-19
11.4.1 Elastic scattering and stress pulse attenuation 11-20
11.4.2 Viscous attenuation of a stress pulse 11-21

11.5 Shock and release in concrete 11-23

11.5.1 Reverberation shock decompression experiment 11-23

11.5.2 Shock decompression measurements for concrete 11-25

11.5.3 Decompression strain recovery 11-28

11.6 Shock velocity versus particle velocity 11-29

11.7 Elastic-plastic waves in uranium alloys 11-36

11.7.1 Uranium 6% niobium (U6Nb) 11-37

11.7.2 Impedance analysis of the uranium 6% niobium (U6Nb) wave profile data 11-39

11.7.3 Alpha uranium 11-42

11.7.4 Uranium 3/4% titanium (U3/4Ti) 11-44

11.7.5 Comparative behavior of uranium and alloys 11-44

11.8 Lagrangian analysis 11-48

11.8.1 Basic equations 11-50

11.8.2 Phase-velocity analysis 11-51

11.8.3 Path-derivative analysis 11-52

11.8.4 Application of the phase-velocity analysis 11-54

11.8.5 Application of the path-derivative analysis 11-60

11.8.6 Lagrangian analysis in spherical symmetry 11-65

11.8.7 Lagrangian analysis methods 11-70

References 11-71

Preface

This set of two books deals with the shock physics of solids, with an emphasis on impact applications that are brought about by any of a variety of intense-energy stimulations. A number of topics in shock physics are addressed in some depth. This book does not come close to exhausting the topic of shock physics, however. Rather, emphasis falls heavily on a selection of topics that have been the focus of my attention over the past several decades. The book covers the physical, material, mechanical and thermodynamic fundamentals of shock physics where appropriate. Also addressed are current research topics in shock physics, a number of which have yet to achieve a mature status of understanding. The various chapters also include a range of analytic solutions that address topics in the many applications involving shock physics. The solutions are tutorial in nature but are, in many cases, useful complements to comparable computational solutions, in that the underlying shock physics is usually more clearly revealed. In a number of cases unique shock wave and impact data are used to supplement the presented topic. The source of these data spans the three to four decades over which substantial strides in the development of shock wave instrumentation have been made. The shock data presented reflect such advances.

At this stage the book is a work in progress. The wonders of present day computer word processing allow, with some effort, completion of the text at intermediate points along the journey. My intent is to eventually supplement current chapters with further topics that have as yet to be prepared, and to add several further chapters. At periodic later stages the plan is to update the book when sufficient further progress has been made. Should personal energy and motivation stall along the path, the present organization will leave some product of the efforts expended.

Introduction

The title of this set of two books—*Physics of Shock and Impact*—has a noticeable redundancy. After all, impacts of bodies onto other bodies produce shock waves. And it is the physics of shock waves and the consequences of shock waves that are of paramount interest in this text. Other intense-energy impulsive loading events are the cause of similar shock waves. The sudden deposition of high-energy radiation, or the rapid release of chemical energy in the detonation of an explosive, are equally common events leading to strong shock waves.

Nevertheless, it is the various impact applications in practice over the past few decades which have provided the motivation for much of the analysis and physics modeling that is offered in the following chapters. Many shock physics issues are raised by the numerous terminal ballistics effects encountered in conventional armor and anti-armor applications. Most of the efforts on the strength and equation-of-state (EOS) modeling of ceramics have been driven by armor applications. Adiabatic shear failure in metals is clearly a consequence of conventional terminal ballistics. Hypervelocity impacts related to space station protection against meteoroid and orbital debris opens a broader arena of impact shock physics. Intense-energy fragmentation, a multiphase EOS and shock-induced vaporization, are all consequences of a hypervelocity impact event.

Other program applications, however, also provided impetus for the topics addressed here. The brief energy shortage in the early 1970s motivated efforts to find methods for effectively fracturing oil shale with explosives for the purposes of *in situ* retorting and oil extraction. Within the same time frame, concerns with the non-nuclear detonation of nuclear weapons due to accident or terrorist circumstances led to studies of the intense-energy breakup and dispersal of component nuclear materials. Both programs stimulated more fundamental investigations of dynamic fragmentation—the first focusing on more brittle materials such as hard rock and concrete, and the second on the spall and instability breakup of ductile solids and liquid metals.

Efforts to more accurately characterize the shock EOS and the strength properties of silicate and carbonate rocks for the purposes of nuclear and conventional explosive energy coupling and ground shock analysis, uncovered evidence for intense heterogeneous deformation on the microscale in the shock compression event. These observations led to some of the first endeavors to assess the characteristics of thermoplastic shear localization in the dynamic compression of solids. A similar program motivation drove more detailed investigations of the EOS and kinetics features of polymorphic phase transformation under strong shock compression. Both silicate and carbonate rocks and minerals were found to exhibit a rich range of phase transitions over the shock pressure ranges of interest. The rapidly improving time-resolved shock wave diagnostics over the same period, including piezo-resistance pressure gauges, electromagnetic particle-velocity gauges, and laser-velocity interferometry, allowed these shock wave effects to be examined in ever increasing detail.

The topics addressed in the present text have been partitioned over two volumes totaling eleven chapters. Briefly, these topics include the fundamentals of the shock EOS and shock wave propagation, fracture and fragmentation in shock and impact environments, the shock EOS and strength properties of brittle solids including ceramics and hard minerals, viscosity and structured shock waves, shock compaction of distended solids and adiabatic shear localization under high-strain-rate deformation.

Chapter 1, although far from exhaustive, covers a number of the fundamentals of shock wave physics. Acoustic wave propagation and the transition to large-amplitude nonlinear wave propagation is addressed. A description is provided of the respective Lagrangian, Eulerian and inertial representation of material position and motion. Classic nonlinear waves in solids, along with applications, are described, including the shock wave, simple waves, centered waves and steady structured waves. The governing equations of motion including the Hugoniot relations are developed, and solution techniques through the method of characteristics are described. The influences of material strength and viscosity are considered as they influence the structuring of shocks and large-amplitude waves. The spall wave with application to a number of material types is explored in some detail.

The dynamic fragmentation of condensed media subjected to intense-energy loading is covered in chapters 2–4. Chapter 2 focuses principally on the fragment size statistics of the resulting fragment distribution, and methods for predicting the governing distribution length scales, or equivalently, the characteristic fragment size. These topics are examined from both a historic perspective, and from that of more recent advances in theories of dynamic fragmentation. A wider range of applications in dynamic fragmentation is pursued in chapters 3 and 4, with the former leaning towards issues in physics while the latter focused on problems in engineering. Fragmentation due to both shock-induced spall and impact shatter is examined, along with specific spall and fragmentation criteria that have found application in shock and impact science. Instability fragmentation in ductile and liquid bodies subjected to intense distortional loading is pursued. Fragmentation in some special cases is described, including nuclear spallation, molecular dynamics simulation of intense-energy fragmentation and domain structure fragmentation in shocked ferromagnetic materials. Engineering models such as the Gurney theory and the equivalent Hugoniot theory have found profitable application in fragmentation. The models are relevant to space debris, exploding casings and ballistic impact.

Chapter 5 addresses the topic of thermoplastic shear localization in solids. This unique deformation phenomenon is of interest to the strong shock environment, as well as to the high-rate deformations achieved in ballistic impact and explosive loading. The chapter first explores the underlying physics of adiabatic shear and constraints imposed by the dimensional properties of the governing physics. Some calculations through stability analysis of the linearized equations of thermoplastic flow are described. The preponderance of attention, however, is focused on a two-temperature boundary layer model of the adiabatic shear localization process that finds inspiration from the seminal studies of Semenov, concerning thermal explosions in reactive materials, and of Mott, concerning plastic wave propagation in metals. The model identifies well-defined length scales associated with both the adiabatic shear-band width and the characteristic spacing of multiple shear bands. The time to localization and dissipation energy characteristic of the shear band also follow from the model. Topics briefly explored include turbulence in the wake of the shear band, dissipation contributions from elastic strain energy and concepts of an adiabatic shear-band toughness. Analysis and modeling are applied to a suite of metals for which appropriate thermomechanical properties are available.

Chapter 6, on the shock EOS and strength properties of ceramics, was motivated by programmatic opportunities to conduct extensive time-resolved shock profile measurements on this interesting class of materials in the early 1990s. The aim of the chapter is to pursue physics-based modeling and analysis for the purpose of ferreting out the complexities of the strong shock response of ceramics. The rich base of shock Hugoniot and shock wave profile data generated both by myself, and others, provides the backdrop for this effort. The lattice EOS, solid state phase transformation and high crystal strength intermix in ways that markedly complicate the interpretation of the measured data and the ability to achieve a clear physical understanding of the shock and high-rate response of ceramics. Boron carbide, despite several decades of continued study, still remains enigmatic.

The shock compaction of porous solids has received considerable attention since the seminal computational modeling efforts of Herrmann, and the resulting p-alpha crush model. Also notable are the even earlier geoscience-based analyses of Kompaneets on the propagation and attenuation of strong shocks through initially porous media. The self-similar features of shock propagation in crushable material that approximates extended geologic media parallels the better known solutions of Sedov and Taylor on the propagation of blast waves in air. In chapter 7 the solutions of Kompaneets are used as the starting point for exploring the analysis and modeling of shock compaction with an emphasis on brittle granular media. Both experimental data and crush models are examined that address the dynamic compaction of ceramic materials. The p-alpha model is discussed, along with extenuations introduced by Carroll and Holt. An alternative engineering crush model, the p-lambda model, which allows for the introduction of intermediate compliance materials such as water or polymeric media within the markedly stiffer porous material free space, is also explored.

The emergence of increasingly higher resolution shock wave profile measurement techniques during the 1970s revealed a rich time-resolved structure to the shock wave. The idealization assumed by many of a discontinuous shock wave was not observed. The assumption of such an idealized shock wave, in fact, confuses the physics of the material change of state through the shock wave. The material strength and phase transformation, of course, contribute a unique structure to the shock wave profile. Of further interest is the measurement of finite risetime through steady structured shock waves and consistent changes in that risetime with corresponding changes in the amplitude of the shock wave. This structure of the shock wave has been attributed to some form of solid state shock viscosity, although the responsible physics has not achieved consensus. In the early 1980s a systematic trend in this behavior over a reasonably wide range of materials was noted. This trend has subsequently become known as the fourth-power law. The history, and some of the developments, of the fourth-power law are described in chapter 8. Less well known was the proposition, during the same time frame, of a shock adiabatic invariant relating to the same shock wave structure. Whereas this shock invariant implies the fourth-power law of steady structured waves, the invariant has wider application to unsteady shock waves. The shock invariant, as well as issues of solid state shock viscosity, are also addressed in chapter 8.

In chapter 9 shock compression features unique to brittle solids are described, which have been garnered, after considerable thought and analysis, from time-resolved shock profile data. The chapter in part echoes some of the observations discussed in chapter 6 on ceramics. Other materials, analysis and modeling efforts enrich the description, however. The materials include natural minerals (quartz, calcite), rocks (dolomite, novaculite, limestone) and glass, among others. Again, the dynamic strength and phase transformation are observed to profoundly influence the shock wave structure in these materials. Dolomite, a rock formed of a natural carbonate mineral of the same name, has perhaps been the most extensively studied brittle material in terms of variations in both the impact amplitude and spatial propagation distance of experimentally measured time-resolved shock and release wave profiles. These data reveal unique kinetics characteristics in the dynamic strength of brittle solids that strongly hint at the fracture failure wave behavior later reported for similar brittle materials. The fundamentals of the fracture mechanics related to the shock failure of brittle solids are discussed in this chapter, as are descriptive models based on the governing fracture mechanics. Underlying causes of time and strain-rate-dependent strength and failure in brittle solids are revealed through such physics-based models.

Chapter 10 provides a more extensive exploration of the thermodynamic EOS of matter in the shock compression environment. The EOS of gases are briefly examined with emphasis on the shock Hugoniot and wave propagation. Several EOS, including the Mie–Gruneisen EOS, which have found frequent use in shock wave applications are covered. Also described is the less commonly used enthalpy-based Rice–Walsh EOS and its application in the EOS description of intense shocks in highly distended media.

The concept of a universal compression EOS commonly used as the reference compression path for more complete EOS descriptions of solids is examined. EOSs that have been identified with Murnaghan, Birch, Vinet, Holzapfel and others are contrasted and compared with material compression extracted from shock Hugoniot data. The applications in chapter 10 also explore the extreme shock compression of highly distended solids and the EOS of mixtures.

Lastly, chapter 11 focuses on methods of shock wave analysis. Useful tools for the interpretation of wave interaction in experiments and applications, including pressure–particle-velocity plots and distance–time plots are surveyed. Methods of Lagrangian analysis are described that were developed to extract dynamic constitutive properties from time-resolved unsteady experimental shock wave profile data. The shock wave analysis tools described are applied to several specific applications and identified with related experimental shock wave data.

Author biography

Dennis Grady

Dennis Grady is currently an Associate and Principal Scientist with the South West Division of Applied Research Associates headquartered in Albuquerque, New Mexico. He performed Military service from 1958 to 1963, in the U.S. Navy Submarine Service. He received his BS in physics and mathematics from Lewis and Clark College in 1967, and a PhD in physics from Washington State University in 1971. Following three years on the research staff of Poulter Laboratory at SRI International he joined Sandia National Laboratories. He retired from Sandia Laboratories as a Distinguished Member of the Technical Staff in 1996 and joined the South West Division of Applied Research Associates the same year. He has been involved with the experimental measurement and theoretical description of condensed matter under the extreme pressure and temperature stimulus of shock and high-velocity impact for over 40 years. He has published over 300 technical papers and reports on a range of materials and applications issues in the intense shock environment including experimental methods, electric and magnetic effects, phase transformation, high-pressure equation of state, transient strength, energetic reaction, hypervelocity impact, and dynamic fragmentation. He is a member or past member of the American Physical Society, International Hypervelocity Society, American Geophysical Union, International DYMAT Association, and has received honors from, and served various official roles in, these societies.

IOP Publishing

Physics of Shock and Impact, Volume 2

Materials and shock response

Dennis Grady

Chapter 6

Shock compression of ceramics with microstructure

The high crystal strength of many metal–nonmetal compounds accentuates the importance of material composition and microstructure in the equation-of-state and dynamic-strength response of ceramics in a shock wave environment. This feature of ceramics motivates the experimental assessment, analysis and theoretical modeling undertaken in the chapter, exploring the underlying physics governing compressibility—including polymorphic phase transformation and strength of ceramic materials—when subjected to shock wave compression. Of particular concern is the influence of composition or microstructure, such as the modest amounts of porosity, additives, or second phases common in the production of ceramics. Substantial shock wave data are examined in this study, some of which are not readily available elsewhere. A theoretical model of the material response is also described that accounts for composition and microstructure effects in the shock compression process. The theoretical modeling is in turn used in the assessment and physical interpretation of the experimental shock wave data. The discussion focuses on the ceramics of aluminum oxide, boron carbide, silicon carbide, tungsten carbide, titanium diboride and zirconium dioxide, among others.

6.1 Introduction

Ceramics prepared from metal–nonmetal compounds exhibit a rich array of material responses when subjected to shock wave and high-strain-rate loading (Grady 1998). Early Hugoniot measurements conducted at the Los Alamos National Laboratory (LANL; McQueen *et al* 1970) and the Lawrence Livermore National Laboratory (LLNL; Gust *et al* 1973) hinted at exceptional dynamic-strength characteristics and a propensity for polymorphic phase transformation under shock compression. In the intervening years, time-resolved wave profile measurements undertaken on ceramic materials at a number of laboratories have

doi:10.1088/978-0-7503-1257-8ch6

provided ever increasing detail on strength and equation-of-state (EOS) properties not commonly observed in other classes of materials. Shock compression and release profiles exhibit a multiple wave structure associated with both strength and phase transformation, along with precursor wave features unique to ceramic failure (e.g. Grady 1992a, 1992b, 1994e, 1995a, Kipp and Grady 1994). Shock profile measurements of ceramics subjected to complex load paths, including shock–reshock and shock–release, reveal dynamic failure characteristics and shock stress and strength properties inconsistent with more common elastic–plastic models (e.g. Vogler *et al* 2004, Vogler *et al* 2006). Wave profile measurements from experiments with impact obliquity show an unusual sensitivity of the dynamic tensile strength to imposed shear stress (Dandekar 2001). Wave profile measurements on brittle solids in the investigation of dynamic failure kinetics demonstrate time-dependent failure in ceramic materials (e.g. Kanel *et al* 1992, 2001).

Further, a number of studies of shock wave properties have demonstrated that a given ceramic is not a unique material, but is itself a class of materials. Silicon carbide ceramic is representative, in that enough investigations have been performed on materials from different suppliers, and on material from the same supplier but subjected to different preparation methods, to show that both the dynamic strength and EOS depend on details of the material's microstructure and ceramic composition. Aluminum oxide ceramic is a further example. In this ceramic both the porosity and glass-phase content are controlled to achieve properties consistent with specific applications. Such variations in material composition and microstructure markedly affect the shock wave and high-rate loading response of the ceramic.

At the heart of many of the shock compression investigations that have been performed, is the development of an understanding of the thermo-mechanical response of ceramics subjected to the intense shock and high-rate environment of the ballistic event. Testing in some applications is as simple as determining the ballistic resistance of a ceramic component in an engineering model. Today, however, knowledge of the material response is more frequently directed toward the development of computational models that faithfully described the response throughout the range of dynamic stresses and deformations that ceramics are subjected to in the ballistic application.

The present chapter has several objectives. Selected shock wave data for ceramics are examined with the objective of providing an understanding of the dynamic EOS and strength features of ceramics in a shock and high-rate event. One emphasis is directed primarily at the EOS or volumetric response of the ceramic, which here includes both mechanical compliance, such as pore compaction, as well as lattice compressibility. Where relevant, strength issues are also addressed. Of concern are the effects of ceramic microstructure and ceramic composition on the EOS response of the material. The shock response of aluminum oxide ceramic with both the glass phase and porosity is examined. Shock data for silicon carbide ceramic of differing composition are examined, including reaction-sintered silicon carbide with appreciable fractions of free silicon. Shock wave data on boron carbide and tungsten carbide ceramic as well as ceramic–metal composites are examined with the same objectives.

The shock wave data that are studied in this effort were selected to aid in uncovering some of the physics that have complicated the material response modeling of ceramics. These include shock Hugoniot data from the seminal studies of McQueen *et al* (1970) and Gust *et al* (1973), as well as more recent time-resolved shock compression data. Key issues of concern include phase transformation under shock wave loading, as well as the effects of additional microstructure or compositional phases, such as entrained porosity or disparate-compliance constituent materials such as glass or free silicon.

Theoretical modeling of the shock wave EOS response is also undertaken in an effort to provide analytic physics-based descriptions of the dynamics of material response. Classic shock wave mixture methods (e.g. Altshuler and Sharipdzhanov 1971, Duvall and Taylor 1971) are applied to the Hugoniot data of Gust *et al* (1973) on a silicon carbide ceramic now known to be a mixture of both silicon carbide and free silicon. The mixture method is also compared with shock wave data on several metal–ceramic composites. This classic mixture approach is shown to successfully describe key features of the shock wave EOS behavior. The approach fails to capture other important features of the shock compression process relating to the strength and inelastic dissipation associated with the composition and microstructure properties of the material.

Next, a more recent theoretical approach to modeling the shock wave response of ceramic mixtures is described (Grady *et al* 2000). This model provides continuum physics-based analytic relations for the shock EOS response of ceramics, which include compositional features affecting microstructure strength and compliance. These features can include porosity and highly compliant second phases such as free silicon or glass. The model is also shown to be appropriate to polymorphic phase transformation under shock compression where microstructure strength leads to spread in transformation pressure and hysteresis in the compression and decompression transformation cycle.

The later model is applied to the analysis and interpretation of the shock wave data on ceramics. This model provides a better appreciation of observed shock wave response and provides a beginning for understanding the trends in the constitutive response brought about by changes in ceramic microstructure and composition. These efforts are applied to ceramics of aluminum oxide, boron carbide, silicon carbide and tungsten carbide.

The observations of complexities in the compression precursor wave structure for silicon carbide and titanium diboride ceramics point to unusual features in the shock strength and onset of inelastic shear in these materials. These strength and deformation features, as extracted from experimental shock wave profiles, are described. An underlying cause is described based on elastic microstructure disparities.

Lastly, a theoretical analysis of the strength and kinetics of failure in brittle solids is pursued. Included in this investigation is the observation of delayed failure, or failure waves, which appears to be a unique feature of shock phenomena in brittle solids. This analysis describes a concept of a time-dependent equilibration of stress in materials with microstructure that can occur under shock or high-strain-rate loading. When combined with the notion of brittle failure in a heterogeneous material, a sensible model for failure kinetics in brittle solids emerges. The underlying microdynamics explored hint at even deeper roots embedded in irreversible statistical mechanics.

6.2 Classic mixture theory of shock compression

The early shock Hugoniot data for silicon carbide provided by Gust *et al* (1973) at LLNL, and also probably those of McQueen *et al* (1970), were provided from material that had significant amounts of free silicon (10%–15%). This fact was been verified by Cline (2003) and is discussed in more detail in later sections of this chapter. This feature of the early silicon carbide ceramic can account for the inconsistencies among the seminal shock Hugoniot data (McQueen *et al* 1970, Gust *et al* 1973) and more recent shock compression data for silicon carbide.

Silicon undergoes a first-order phase transformation with a substantial volume contraction under shock compression (Gust and Royce 1971, Goto *et al* 1982). The shock wave structure resulting from phase transformation in silicon is considered in chapter 1. The same phase transition is experienced in the shock compression of silicon carbide with a free second-phase component of silicon. The extent of the effect of the phase transformation on shock compression is proportional to the fractional amount of silicon in the ceramic.

Ceramics of silicon carbide and free silicon are also relatively easy to prepare through reaction-bonding (RB) techniques. The fractional amount of silicon can be varied over sensible limits. Thus, SiC–Si mixture ceramics offer unique materials for testing the influence of phase transformation on the mechanics of high-velocity impact and the ballistic performance of ceramics.

Here a consistent framework is described for modeling the response of a silicon carbide and silicon mixture ceramic with arbitrary mass fractions of the two components. The classic theoretical methods of shock mixtures (Dremin and Karpukhin 1960, Altshuler and Sharipdzhanov 1971, Duvall and Taylor 1971) are applied in this effort. The model addresses the issues of dynamic strength and phase transition, as well as the lattice compressibility of the mixture. In later sections a more extensive approach to modeling the shock compressibility of ceramic mixtures that accounts for microstructure features not addressed in the early theories is applied to the same silicon carbide and silicon ceramic, and contrasted with the present historic methods. The classic theory of mixtures is revisited in chapter 10, on the shock EOS of solids.

Shock compression data on composites of metal and ceramic provide further testing of the classic additive shock mixture theory. Hugoniot and release wave data on mixtures of copper with silicon carbide, boron carbide and titanium diboride (Grady 1994b, 1994c) are reasonably well described by the theory, but again, details of the comparison suggest the need for a more encompassing theory.

6.2.1 Shock compressibility of silicon carbide and silicon mixture ceramics

The measured shock Hugoniot curves for three related ceramic materials are shown in figure 6.1. The Hugoniot for monolithic silicon is provided by the shock wave data of both Gust and Royce (1971) and Goto *et al* (1982) on single-crystal silicon. The Hugoniot for monolithic silicon carbide is that provided by Sekine and Kobayashi (1997) on near-full-density α-SiC ceramic. More recent shock Hugoniot data on a different monolithic silicon carbide ceramic (Vogler *et al* 2006) spanning the higher pressure data of Sekine and Kobayashi suggest a modestly more compliant Hugoniot response within this range. Somewhat lower-pressure shock Hugoniot data for fully α-phase SiC–B and SiC–N ceramic provided by Cercom Company are also shown in figure 1.1 (Grady 2000). These latter data are reasonably consistent with the rather sparse data of Sekine and Kobayashi in this range. The shock Hugoniot data of Gust *et al* (1973) is for a ceramic identified as KT-silicon carbide with a content of free silicon. The specific ratio of silicon carbide to free silicon shown in figure 6.1 was determined through both discussions with Cline (2003), and through consistency with the mixture compressibility of the component materials.

The presentation of the shock Hugoniot data on the left emphasizes the disparate initial densities of the component materials and the marked phase transformation occurring in the silicon over the 10–25 GPa shock pressure range. The same data are plotted on the right as a function of volume compression. The separate Hugoniot data for monolithic silicon, monolithic silicon carbide, and the silicon carbide and silicon mixture (KT-SiC) of Gust *et al* (1973) are identified by the dashed curves in the figure. The large difference in compressibility between SiC and Si is a consequence of both the large volume–strain phase transformation in silicon and the more compliant lattice compressibility. The solid curve is the hydrodynamic compression EOS for silicon carbide determined by Holmquist and Johnson (2002), principally from static compression data (Bassett *et al* 1993). In the present mixture development the EOS of Holmquist and Johnson is used to describe the volume compression of the SiC component of the mixture. The Holmquist–Johnson relation

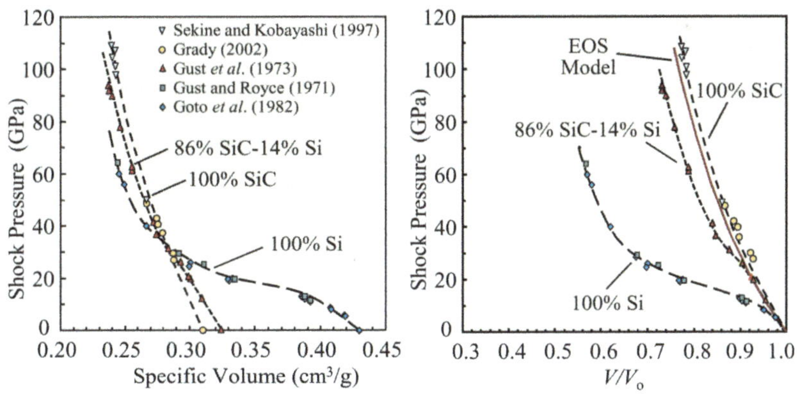

Figure 6.1. Shock Hugoniot data for full-density silicon carbide ceramic, single-crystal silicon, and a silicon carbide ceramic that is a mixture of silicon carbide and silicon in approximately the ratio shown. In the right panel, the hydrodynamic compression EOS model of Holmquist and Johnson (2002) for monolithic silicon carbide is compared to the Hugoniot data.

is fitted to a linear shock-velocity versus particle-velocity compression relation of the form

$$P = \frac{K\varepsilon}{(1 - S\varepsilon)^2}, \tag{6.1}$$

for the present analysis, where the volume strain $\varepsilon = 1 - \rho_0/\rho$. This alternative analytic representation falls within the thickness of the EOS curve shown in figure 6.1. Values for K and S are provided in the subsequent section.

6.2.1.1 Elasticity and strength properties

The volumetric response of the ceramic is provided by the material EOS. The elastic non-hydro response of the ceramic mixture, assumed to be isotropic, is characterized by specification of the elastic Poisson's ratio. Measured values of Poisson's ratio as assessed from ultrasonic data are provided on the left in figure 6.2 as a function of the silicon mass fraction for both ceramic and single-crystal silicon. Data are shown for Cercom SiC–B, a full-density SiC ceramic from Eagle Picher Company (Grady and Moody 1996), and the KT-SiC ceramic mixture tested by Gust *et al* (1973). Single-crystal samples of silicon in three crystal orientations were tested in the shock Hugoniot study of Gust and Royce (1971) and effective Poisson's ratios are also plotted in figure 6.2. Within the scatter of the data a Poisson's ratio of 0.18–0.20 for any mixture of silicon carbide and silicon is adequate.

The dynamic von Mises yield stress Y has been calculated from the shock Hugoniot elastic limit (HEL) data, described in chapter 1, for selected silicon carbide ceramics and for single-crystal silicon. These strength data are also plotted in figure 6.2. An additional ceramic, a Cercom RB silicon carbide (Grady and Moody 1996), with several percent free silicon, is also included in the plot. Although significant scatter is observed in the ceramic strength data, a linear representation captures the essence of the data (Grady 2005). The strength model ranges from $Y = 10$ GPa for 100% silicon carbide to $Y = 5$ GPa for 100% silicon.

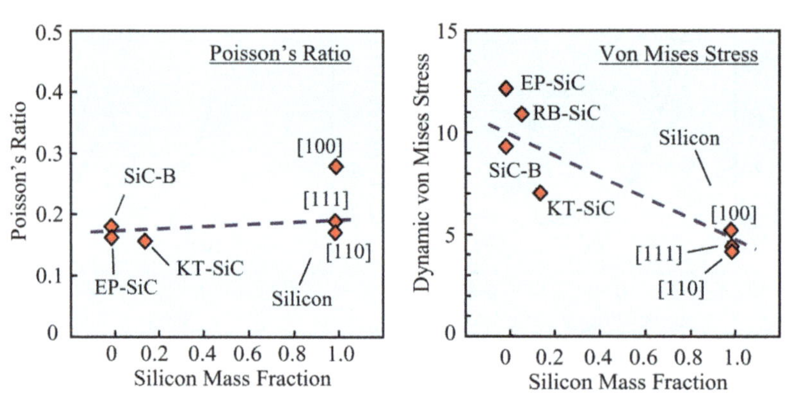

Figure 6.2. Elastic Poisson's ratio (left) and dynamic von Mises strength (right) from HEL data for selected silicon carbide ceramics and crystalline silicon.

The pressure dependence of the bulk modulus for monolithic silicon carbide, and both the low- and high-pressure phases of silicon, is shown on the left in figure 6.3. The plot illustrates the markedly higher compliance of the silicon over the entire pressure range. Additionally, the compliance of the silicon is compounded by further volume contraction during the pressure-dependent phase transformation. The pressure range of this transition on the shock Hugoniot is identified in the plot on the left in figure 6.3. Using the available shock wave and high-pressure static data for silicon carbide and silicon, analytic EOSs are constructed for the pressure (mean stress) versus volume compression response for silicon carbide, and both the low-pressure and the high-pressure phases of silicon. The analytic form, in each case, is that resulting from a linear shock-velocity versus particle-velocity representation of the pressure versus shock compression, $U = C_o + Su$. This relation is provided in equation (6.1) where the initial bulk modulus is $K = \rho_o C_o^2$, with ρ_o the initial density of the material. The EOS for monolithic silicon carbide is based on the compression relation of Holmquist and Johnson (2002). That of the low- and high-pressure phases of silicon is determined from the appropriate range of the Hugoniot compression data of Gust and Royce (1971) and Goto *et al* (1982).

The corresponding mixture EOSs are also of the linear shock-velocity versus particle-velocity form with parameters determined, in the spirit of the Altshuler and Sharipdzhanov (1971) mixture model, through a pressure-equilibrated averaging of the component EOSs. This approach results in the expressions

$$\rho_{io} = \left(\frac{1-\mu}{\rho_o^{SC}} + \frac{\mu}{\rho_{io}^{SI}}\right)^{-1}, \; K_i = \left(\frac{\rho_{io}}{\rho_o^{SC}}\frac{1-\mu}{K^{SC}} + \frac{\rho_{io}}{\rho_{io}^{SI}}\frac{\mu}{K_i^{SI}}\right)^{-1}, \qquad (6.2)$$
$$S_i = (1-\mu)S^{SC} + \mu S_i^{SI},$$

for the mixture density, the mixture bulk modulus and the mixture shock versus particle-velocity slope, respectively. In these relations μ is the mass fraction of silicon

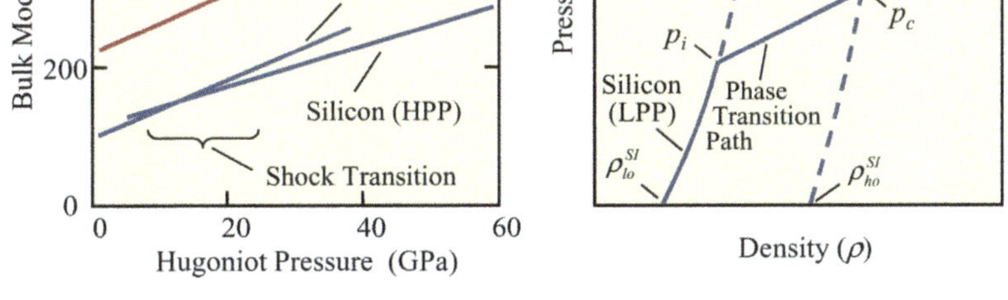

Figure 6.3. Pressure dependence of the bulk modulus for silicon carbide and for both the low- and high-pressure phases of silicon (left). Silicon compressibility including phase transformation is illustrated on the right.

in the mixture ceramic. The subscript i is either l or h depending on whether a mixture with low-pressure silicon or high-pressure silicon is required.

All densities in (6.2) refer to reference (initial) densities. For example, the mixture density ρ_{lo} in (6.2) corresponds to a mix of silicon with initial low-pressure phase density ρ_{lo}^{SI} and fractional mass μ, with silicon carbide at initial density ρ_{o}^{SC} and fractional mass $1 - \mu$. The model low- and high-pressure phase properties for any ceramic mixture of silicon carbide and silicon are shown in figure 6.4.

6.2.1.2 Mixture phase transformation

The mixture EOS for the silicon carbide and silicon mixture ceramic is completed by description of the path shown on the right-hand side of figure 6.3 through which the pressure-induced phase transformation in the silicon provides for the transition from the low-pressure mixture compressibility (LPP in figure 6.3) to the high-pressure mixture compressibility (HPP). The transformation is described by a linear path starting at a transformation-initiation pressure p_i on the low-pressure compression curve, and finishing at a transformation-completion pressure p_c on the high-pressure compression curve. The Hugoniot data for silicon (Gust and Royce 1971, Goto *et al* 1982) and for the silicon carbide–silicon ceramic tested by Gust *et al* (1973) are used to determine these transition pressure properties for any mass fraction of silicon in the mixture ceramic (Grady 2005).

Hugoniot data and the analytic forms selected to represent the data for silicon are shown in figure 6.5. The mean stress curve p_m is selected so that the Hugoniot pressure, $p_h = p_m + 2Y/3$, provides an optimum fit to the experimental Hugoniot data, assuming a constant dynamic yield stress of $Y = 5$ GPa. Initiation of transformation is provided by $p_i = 5.6$ GPa at a volume compression of 0.95. Transformation completion is at $p_c = 26$ GPa.

The corresponding Hugoniot data for the silicon carbide ceramic (assumed as 86% SiC and 14% Si carbide) tested by Gust *et al* (1973) is shown on the right in figure 6.5. The mixture model analytic pressure (mean stress) and Hugoniot stress EOS curves are compared with the Hugoniot data. A yield stress of 9.3 GPa based on the strength data provided in figure 6.5 is used to determine the mean stress versus Hugoniot stress offset. Onset of transformation at a mean stress of 10 GPa, and at a volume compression of 0.95, is consistent with the data. Transition completion is

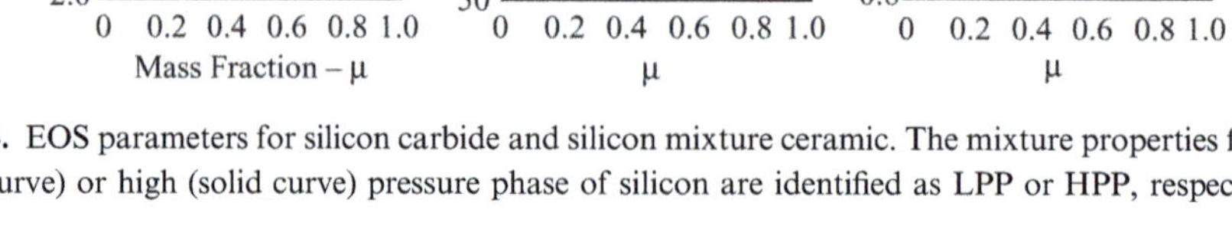

Figure 6.4. EOS parameters for silicon carbide and silicon mixture ceramic. The mixture properties for the low (dashed curve) or high (solid curve) pressure phase of silicon are identified as LPP or HPP, respectively.

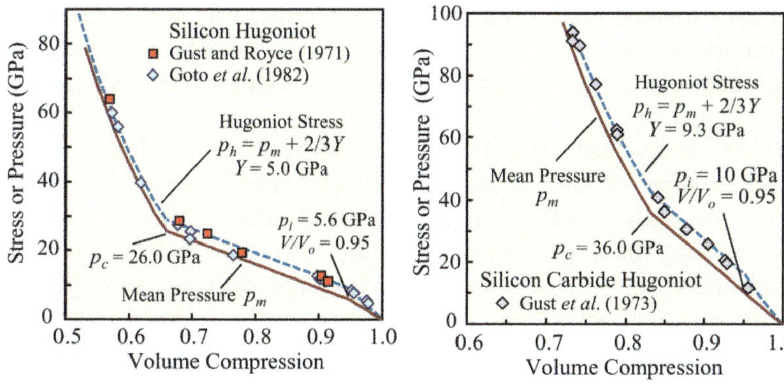

Figure 6.5. Mean stress and Hugoniot compression curves for silicon (left), and for 86% SiC and 14% silicon carbide ceramic (right), based on additive shock mixture model.

achieved at a mean stress of 36 GPa. Based on the present analysis for the shock compressibility of silicon and the silicon carbide ceramic of Gust *et al* (1973) that was determined to be a mixture of approximately 86% SiC and 14% Si, the phase transformation pressure range for any mix of silicon carbide and silicon is provided through the additive mixture relations.

In summary, a constitutive model for a near-full-density mixture of silicon carbide and silicon, including EOS, strength and phase transformation, is developed and explored based on the classic additive mixture methods of Dremin and Karpukhin (1960), Altshuler and Sharipdzhanov (1971) and Duvall and Taylor (1971), and used by numerous later workers. The model is based on shock Hugoniot data available for silicon and silicon carbide, and for a silicon carbide ceramic that is a mixture of the two components. Dynamic-strength and elasticity data are also incorporated. The model is a reasonable first-order description of the dynamic compressibility of the ceramic mixture and is readily implemented into production wave codes. This continuum modeling method has inherent physical flaws, however, that are of concern in more detailed dynamic constitutive modeling of ceramic mixtures. These issues are addressed in later sections, and uniquely different mixture modeling approaches are described to address the difficulties of the earlier mixture methods.

6.2.2 Shock compression of metal–ceramic composites

A further test of the classic shock compression mixture models is provided through shock wave experiments conducted on mixtures of metal and ceramic (Grady 1994b, 1994c). In this study, composite samples of copper and ceramic were prepared by mixing powders of copper with powders of boron carbide, silicon carbide or titanium diboride in ratios of 1:1 by volume. Samples were compressed to near-theoretical density (porosity 0.9%–2.4%). Uniaxial strain shock compression and release experiments were performed on the metal–ceramic composite specimens to Hugoniot pressures in excess of 30 GPa. The original purpose of the study was to investigate the accuracy of isentropic compression curves for monolithic ceramics determined from extrapolation of ultrasonic compression data obtained from hydrostatic pressure test methods up to 2 GPa. Copper was introduced to eliminate, or at least minimize, the deviator stress achieved at the shock state. Pressure–volume compression of the ceramic is determined by subtracting the known compression of copper.

Here the logic is inverted. An extrapolation of the ultrasonic data is assumed to be a reasonable description of the component isentropic compression curve for the ceramics, and the analysis will test application of the classic additive mixture theory. Consistency among the experimental data and mixture model, as will be shown, sensibly supports either line of reasoning. Measured densities and ultrasonic wave speeds for the metal–ceramic composites are provided in table 6.1.

The additive mixture EOS is developed for the copper and ceramic composites based on Hugoniot data for copper (Marsh 1980) and extrapolated isentropic compression curves for the monolithic ceramics B_4C, SiC and TiB_2 based on pressure-dependent ultrasonic data (Manghnani 1993). The specific volume of the mixture at pressure p is provided by the additive mixture relation

$$v(p) = \mu v_c(p) + (1 - \mu)v_m(p), \tag{6.3}$$

where $v_c(p)$ and $v_m(p)$ are specific volumes of the ceramic and metal, respectively, and μ is the mass fraction of the ceramic in the mixture. Within the range of the Hugoniot data, phase transformation of any of the components is not an issue. Strength issues in the shock compressed mixture are assumed to be mitigated by the copper component and ignored in this development. It is assumed that porosity is fully eliminated in the strong shock compression process. Compressibility of the individual ceramic and metal components is provided through a linear shock-velocity versus particle-velocity relation, $U = C_{io} + S_i u$, in the form for the component specific volume dependence on the equilibrated pressure,

$$v_i(p) = v_{io} - v_{io}\frac{\alpha_i}{S_i}\left(1 - \sqrt{1 - 1/\alpha_i^2}\right), \tag{6.4}$$

where $\alpha_i(p) = 1 + \rho_{io}C_{io}^2/2S_i p$. In equation (6.4) $v_i(p)$ is either $v_c(p)$ or $v_m(p)$ while ρ_{io}, C_{io} and S_i are the crystal density and shock-velocity versus particle-velocity parameters for the corresponding metal or ceramic material. The initial component specific volumes are $v_{io} = 1/\rho_{io}$. These properties for the component materials as determined from Hugoniot data (copper) and ultrasonic data (ceramic) are provided in table 6.2.

Table 6.1. Metal–ceramic composite properties. Table reprinted from Grady (1994b) with permission.

Sample	Density (g cm^{-3})	Calculated porosity	Longitudinal velocity (km s^{-1})	Shear velocity (km s^{-1})
SiC/Cu	5.96	1.9%	5.26	3.20
TiB_2/Cu	6.56	2.4%	6.60	3.82
B_4C/Cu	5.67	0.9%	5.50	3.49

Table 6.2. Linear shock-velocity versus particle-velocity properties. Table reprinted from Grady (1994b) with permission.

Property	Copper[a]	SiC[b]	TiB_2[b]	B_4C[b]
ρ_{io} (g cm^{-3})	8.93	3.22	4.51	2.52
C_{io} (km s^{-1})	3.94	8.31	6.71	9.64
S_i	1.49	1.44	1.19	1.40

[a] Marsh (1980)
[b] Manghnani (1993)

A linear shock-velocity versus particle-velocity representation of the compression has been shown by Jeanloz (1989) to be nearly equivalent to a finite strain Birch–Murnaghan EOS and is analytically convenient in the present analysis. Analytic EOSs with application to the shock compression environment are pursued further in chapter 10. The properties for copper are from Marsh (1980). The properties for ceramic are calculated from ultrasonic measurements of longitudinal and shear velocities under static pressure to 2.0 GPa (Manghnani 1993). The ultrasonic bulk modulus provides C_{io} through $K_i = \rho_{io} C_{io}^2$ while $S_i = (K_i' + 1)/4$, where K_i' is the pressure derivative of the ultrasonic bulk modulus.

The additive mixture relation in equation (6.3) provides the pressure versus specific volume relation for the metal–ceramic composite. Particle velocity and shock velocity at the shock compression state of the composite are calculated from the mixture model through the Hugoniot conservation relations

$$u = \sqrt{p(v_{oo} - v(p))}, \qquad U = \frac{1}{\rho_{oo}} \sqrt{\frac{p}{v_{oo} - v(p)}}, \tag{6.5}$$

where the initial mixture density $\rho_{oo} \leqslant \rho_o$, or specific volume $v_{oo} = 1/\rho_{oo}$, accounts for the small level of porosity in the metal–ceramic composite.

Hugoniot pressure versus particle-velocity data for the three composites are plotted on the left in figure 6.6. Solid curves are calculated with the additive EOS through equations (6.3), (6.4) and (6.5), assuming zero initial porosity in the samples, $\rho_{oo} = \rho_o$, in equation (6.5). The upper curve is for B_4C and the lower for TiB_2. The dashed curves are calculated using the initial porosities provided in table 6.1 and illustrate the magnitude of the correction in accounting for porosity.

The calculated curves are in sensible agreement with the experimental data. Scatter in the data are a result of both instrumentation uncertainties and sample preparation variations. Calculated curves from the additive mixture relations and experimental data in the shock-velocity versus particle-velocity plane are shown on the right in figure 6.6. Again, solid curves assume zero initial porosity whereas the dashed curves account for the initial porosities provided in table 6.1. Solid curves are in ascending order for TiB_2, SiC and B_4C, respectively. The corresponding dashed curves accounting for porosity are identified in the figure. Sensible correspondence between the additive mixture theory and experiment is also observed in this representation.

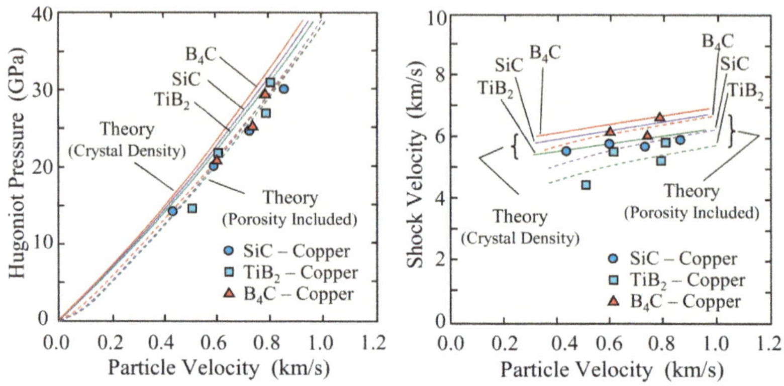

Figure 6.6. Experimental Hugoniot states for metal and ceramic mixtures. Predicted curves are based on the additive mixture theory (Grady 1994b, 1994c).

The calculated Hugoniot compression curves and corresponding experimental Hugoniot states are compared on the left in figure 6.7. The shock compression Hugoniot of copper is shown for comparison and initial distended (porous) densities of the composites are identified on the base line. Again, reasonable agreement between experiment and the classic additive mixture theory is observed.

Release wave measurements from Hugoniot states of the shock compressed mixtures reveal clear elastic-plastic wave profile signatures allowing for sensible experimental assessment of the bulk sound speed velocity at the Hugoniot state (Grady 1994b, 1994c). These bulk sound speed data are plotted in figure 6.7 and, although exhibiting considerable scatter, are at odds with predictions of the additive mixture theory. This comparison suggests that the additive mixture theory may be less reliable when comparisons are with higher order derivatives of the predicted EOS. Again, as with the silicon carbide and silicon mixture in the previous section, disparities are a consequence of small differences in component compressibility and concomitant dissipation not accounted for by the classic additive mixture model. An extended mixture theory that accounts for such differences is introduced in the next section.

6.3 Shock compression of materials with microstructure strength

The constitutive response of solids is commonly decomposed into the EOS (volumetric) and strength (deviator) properties. Microstructure properties including grain size, porosity and multiple phases play an important role in the deviator strength of ceramics. Not as commonly recognized is that the same microstructure properties can also influence the mechanical EOS (volumetric) response of ceramics leading to both volumetric hysteresis and dissipation.

In the present section, methods are described for modeling this volumetric constitutive response of ceramics with microstructure strength. Essential features of this model have been presented previously with application to geomaterials (Grady *et al* 2000), ceramic powders (Grady 2001) and porous polymer foams

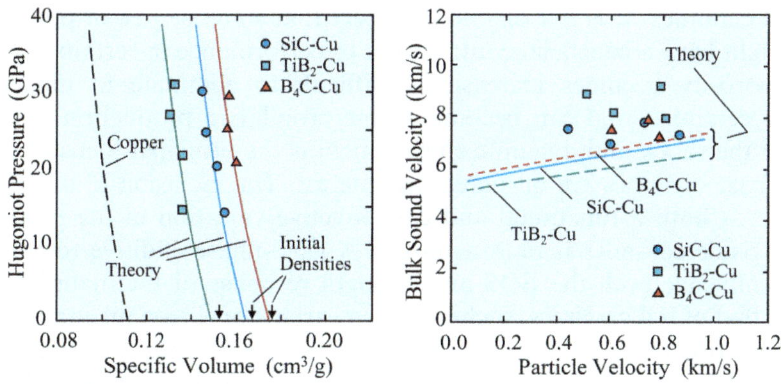

Figure 6.7. Experimental and additive mixture model pressure versus specific volume Hugoniot states (left) and bulk sound speed from release wave measurements (right) for metal–ceramic composites (Grady 1994b, 1994c).

(Grady and Winfree 2001). The model has been implemented in the CTH shock physics code (Bell *et al* 2000) and identified as the p-lambda model.

Engineering applications of ceramic materials require the use of a wide range of ceramics. Issues such as cost, manufacturability and functionality among others play a role in the selection of ceramics for applications. Such ceramics may be less than ideal and may not respond in an application in the same way as the ideal high-quality ceramics that are commonly selected for material properties testing and computational model validation. One example might be the high-quality silicon carbide ceramic identified as the SiC–N ceramic produced by Cercom Company (discussed later in this section), in contrast to a number of available lesser quality silicon carbide ceramics. Assessments of the response of such other-than-ideal ceramics should be the goal of computational codes and the underlying computational models in practical engineering applications.

One objective of the present model is to provide an adequate dynamic material response description of less-than-ideal ceramics based on limited experimental material property data. Non-ideal ceramics can contain a void volume (porosity), significant additives, secondary phases (such as glassy phases), as well as a range of other microstructure disparities. It is prohibitive to experimentally determine model properties for all ceramics spanning the range of microstructure complexities that can be produced. Hence, methods for adequately estimating the material response from limited data are paramount. This approach provides a sensible model for ceramics that addresses the issues of volumetric microstructure strength. More importantly, the approach provides a method for determining constitutive functional forms to account for the effects of the variations in microstructure on the constitutive response, making use of commonly available shock and high-pressure data.

Porosity in ceramics is perhaps the foremost microstructure feature that can have wide variation and is not usually adequately accounted for in computational models used in the evaluation of shock and high-strain-rate performance. Even a few percent or less of porosity can significantly influence both the EOS and the dynamic strength of ceramics. It is reasonable to expect that some degree of porosity in the ceramic might have a beneficial contribution to performance in certain applications. Further, porosity becomes increasingly difficult to eliminate in higher quality monolithic ceramics, and can become a cost prohibitive production requirement. Porosity in the shock and dynamic environment is the principal focus of chapter 7.

Multi-phase ceramics are of increasing interest. The inclusion of an appreciable glass phase as both a functional and cost-effective variation in the production of aluminum oxide ceramics is representative. A glass-phase additive to ceramic will markedly influence both the EOS and strength response of the material. Another example is that of RB ceramics, such as silicon carbide or silicon nitride, in which the production process can yield a mixture of the carbide ceramic and free silicon. As shown in the previous section, the contributing components of ceramic and free silicon in silicon carbide ceramic can have starkly different compressibility and strength properties. A further example is liquid-phase sintered tungsten carbide in which, with the various metal additives, constitute a metal–ceramic mixture.

Central to the need for a mixture model that addresses microstructure complications is the extreme crystal strength of some of the metal–nonmetal compounds that provide the dominant component of ceramics. These compounds consist of elemental metals strongly bonded to lighter elements such as boron, carbon, nitrogen and oxygen. Such borides, carbides, nitrides and oxides exhibit strong resistance to plastic flow. This crystal strength is not a dynamic effect. A good correlation exists between static hardness data and shock wave HEL data (e.g. Grady 1994d). Gilman (1970) has demonstrated that energies close to the bond strength are necessary to move dislocations on the glide planes in such materials. Thus, the potential for wide variations in stress on the microscale in ceramics surpasses that of any other class of materials.

6.3.1 Objectives of the dynamic mixture model

The model pursued here is focused on the dynamic response of a mixture of materials (with or without porosity) with wide disparities in density, strength and compressibility. The model is intended to be robust for both computational and analysis applications. Above all, the model is intended to provide the first-order dynamics response of the mixture material with relatively few well-known and readily available, or readily obtainable, material properties. In this philosophy the model has similarities to the p-alpha model of Herrmann (1969).

Although some deviator strength issues will be pursued, the present effort focuses primarily on the volumetric response of the material. This dynamic compression response is referred to here as compaction in the sense that one or more of the constituents is a relatively stiff skeletal microstructure (the dominant ceramic in the present applications to ceramic materials) that will undergo irreversible deformation (compaction) because of differing compliances and strengths of the mixture components. A clear example is that of a porous ceramic (void or air is the second phase), which undergoes irreversible compaction when compressive pressures lead to local stresses that exceed the intrinsic crystal strength of the ceramic microstructure.

Equally relevant, however, would be a non-porous mixture of aluminum oxide and glass phase. Although higher pressures are required, the markedly larger compliance of the glass phase relative to the ceramic phase leads to comparable irreversible compaction under compression in order to accommodate the larger contraction of the glass. Similar behavior is expected in an RB ceramic with free silicon content.

6.3.2 Strength and EOS decomposition

Arbitrary continuum deformation of a material element subjected to complex loading involves both compression and distortion, and requires a full tensor description of the stresses and strains. Most production codes decompose the stress and strain tensors, respectively, into two invariants. This decomposition separates compression from distortion and pressure from deviator shear stresses. The decomposition provides the mean pressure and the effective shear stress

$$p = -\sigma_{ii}/3, \quad \sigma = \sqrt{\frac{3}{2}s_{ij}s_{ij}}, \tag{6.6}$$

from the stress tensor σ_{ij} and stress deviator tensor s_{ij}. The corresponding decomposition provides volume dilation and the effective shear strain

$$1 - v/v_o = \varepsilon_{ii}, \quad e = \sqrt{\frac{2}{3}e_{ij}e_{ij}}, \tag{6.7}$$

from the strain tensor ε_{ij} and strain deviator tensor e_{ij}.

The present development is focused on the volumetric mechanical EOS deformation of the ceramic mixture. At present shear strength is treated through separate models such as are included in the CTH shock physics library (Bell 2000), or through methods such as those described by Holmquist *et al* (2001).

6.3.3 Elastic and pressure-equilibrated mixture compressibility

Intrinsic to the model are two bounding compression curves (or surfaces in a full thermo-mechanical description). The first captures the initial elastic response of the mixture during which strength of the stiffer skeletal component or components supports the required stress gradients on the microscale without failure. Second, is the final limiting response of the mixture following breakdown of the skeletal structure and when microscale stress differences become small compared to the average pressure.

The initial elastic compression of the mixture assumes equal compressibility, or iso-compressibility, of the mixture constituents. This assumption leads to a superposition or volume average of the partial pressures of the mixture constituents

$$p(v,T) = \sum_i \nu_i p_i(v,T), \tag{6.8}$$

where p and v are the pressure and specific volume of the mixture, while ν_i and p_i are the volume fraction and partial pressure of the ith component, and summation is over i components in the mixture. Temperature T is included here to illustrate a fuller thermo-mechanical response when needed, although it will generally be ignored in the development. The average thermo-mechanical response is more readily captured through an effective Gruneisen parameter for the mixture. Equation (6.8) provides a sensible approximation to the initial elastic compressibility of the mixture before failure on the microscale and compaction begins. In the limit of linear compressibility equation (6.8) provides the familiar Voigt upper-bound iso-strain elastic compressibility of a mixture.

It should be noted that the volume fraction ν_i under arbitrary compression is not a property of the mixture. Under the assumption of iso-compressibility, however, ν_i is constant and is a property of the mixture equal to its initial value $\nu_i = \nu_{io} = V_{io}/V_o$, where V_{io} and V_o are initial volumes of component i and the mixture, respectively. It follows that the specific volumes of component i and the mixture, υ_i and υ respectively, are proportional through,

$$\mu_i \upsilon_i = \nu_i \upsilon, \tag{6.9}$$

Where μ_i is the mass fraction of component i. Thus $p_i(\upsilon_i)$ is appropriately expressed as $p_i(\upsilon)$ in equation (6.8).

The equilibrated lower-bound compression, or iso-pressure, curve is given by the superposition of the partial volumes

$$\upsilon(p,T) = \sum_i \mu_i \upsilon_i(p,T), \tag{6.10}$$

where μ_i and υ_i are the mass fraction and specific volume of component i, respectively. Again, summation is over the i mixture constituents. Equation (6.10) provides the limiting compressibility of the mixture when full microstructure failure has occurred and stress differences on the microscale become small relative to the average pressure. In the limit of linear compression, equation (6.10) provides the Ruess lower-bound iso-pressure elastic compressibility of a mixture.

The reference density for the iso-compression curve is calculated from

$$\rho_{oc} = \sum_i \nu_i \rho_{io}, \tag{6.11}$$

while that for the iso-pressure curve is provided by

$$\rho_{op} = \left(\sum_i \mu_i / \rho_{io} \right)^{-1}. \tag{6.12}$$

For an initially fully compact initial mixture (no porosity), the reference densities (or specific volumes) for the upper (iso-compression) and lower (iso-pressure) curves coincide. When the mixture includes an evacuated pore volume, the ambient reference densities differ. (A gas filled void, as opposed to an evacuated void, is an appropriate mixture constituent within the model framework.)

The iso-pressure representation of the mixture is quite sensible at pressures in which material strength on the microscale becomes small compared to the applied pressure and has been used quite successfully in a wide range of EOS applications. This assumption, for example, is implicit in the additive classic shock mixture methods used to model the silicon carbide and free silicon mixture and the metal–ceramic composites in the previous section.

The elastic iso-compression response is approximate and becomes less appropriate with increasing pressure. As pressure increases and material compaction ensues through microstructure failure, the model becomes increasingly less sensitive to the chosen iso-compression response as will be shown. When a more exacting description of the initial elastic response of the mixture is required, improved representations can be used, some of which are described in a later section. The compression curves for a two-component mixture are illustrated on the left in figure 6.8, showing the independent component compression curves and the corresponding iso-compression and iso-pressure compression curves for the mixture calculated from equations (6.8) and (6.10) (e.g., Grady, 2001; 2003).

Compression data are commonly plotted as a function of the compression strain $\varepsilon = 1 - \upsilon/\upsilon_o = 1 - \rho_o/\rho$ in the analysis of shock and high-pressure wave propagation. Equations (6.8) and (6.10), describing the iso-compression and the iso-pressure mixture relations, are particularly convenient when expressed with this functional dependence. Iso-compression translates to iso-strain compression for the components and immediately yields, ignoring temperature,

$$p(\varepsilon) = \sum_i \nu_i p_i(\varepsilon). \tag{6.13}$$

Equation (6.10), using $\upsilon_i(p) = \upsilon_{io}(1 - \varepsilon_i(p))$ and the expression in (6.9), provides for the iso-pressure relation

$$\varepsilon(p) = \sum_i \nu_i \varepsilon_i(p). \tag{6.14}$$

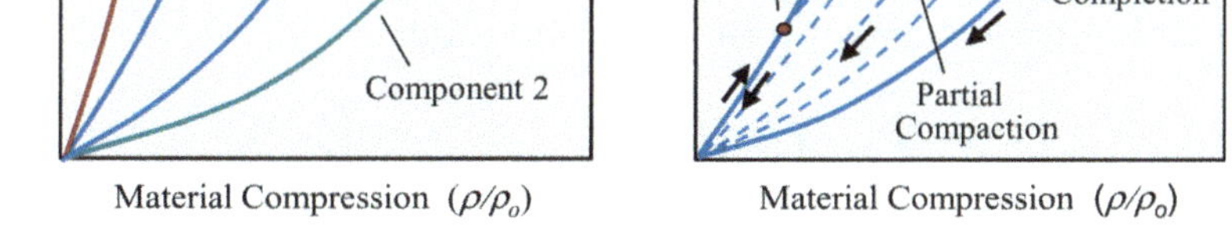

Figure 6.8. Component compression curves for a two-component mixture, and the respective upper-bound iso-compression and lower-bound iso-pressure mixture compression relations, are illustrated on the left. Compression and decompression paths as described by the two-state p-lambda compaction model are shown on the right.

In both equations (6.13) and (6.14) $\nu_i = \mu_i v_{io}/v_o$ is the initial volume fraction of mixture components.

6.3.4 Component compressibility relations

Application of the model requires analytic EOS compressibility expressions for the individual components within the mixture. The intent is to use physics-based relations for which material constants are familiar and are readily available or reasonably estimated. Several EOS relations are satisfactory for the present development. (These relations are introduced in chapter 1 and are discussed in further detail along with other relations in chapter 10.) The linear shock-velocity versus particle-velocity behavior yields the compression relation

$$p(v) = \frac{\rho_o C_o^2(1 - v/v_o)}{[1 - S(1 - v/v_o)]^2}. \tag{6.15}$$

Equation (6.15) provides the Hugoniot compression curve, and it follows from the Hugoniot conservation equations and the experimental observation that a large body of Hugoniot shock-velocity versus particle-velocity data plots linearly according to $U = C_o + Su$. Equation (6.15) is also appropriate for the analytic representation of isentropic or isothermal compression.

The Murnaghan EOS is a comparable analytic equation commonly used to extrapolate the isentropic or isothermal compressibility of solids using modest pressure ultrasonic data. The Murnaghan equation

$$p(v) = \frac{K_o}{K_o'}\left[\left(\frac{v}{v_o}\right)^{-K'_o} - 1\right], \tag{6.16}$$

follows from the assumption, sensibly demonstrated by a large body of materials, that the bulk modulus (isentropic or isothermal as necessary) exhibits a linear dependence on pressure, $K(p) = K_o + K_o'p$. Taylor expansion of the linear U–u relation in equation (6.15) reveals that the Murnaghan and linear U–u relations coincide to second-order if $K_o = \rho_o C_o^2$ and $K_o' = 4S - 1$. An analytic reference EOS identified as the linear-Z, or modified Murnaghan EOS, is finding increasing applications (Grady 2012). The model is based on the assumption of linearity of the inertial modulus $Z(p) = Z_o + Z_o'p$, where $Z = -dp/dv$. Comparative studies of analytic reference EOSs are pursued further in chapter 10.

A theoretically more acceptable EOS is the Birch finite-strain relation (Birch 1978), which has the functional form to second-order

$$p(v) = \frac{3}{2}K_o\left(\frac{v}{v_o}\right)^{-7/3}\left[1 - \left(\frac{v}{v_o}\right)^{2/3}\right]\left[1 + \frac{3}{4}\left(K_o' - 4\right)\left(\left(\frac{v}{v_o}\right)^{-2/3} - 1\right)\right]. \tag{6.17}$$

The Birch equation is commonly preferred to describe the compressibility of condensed matter. Again, the Birch and the linear U–u equations are second-order coincident if $K_o = \rho_o C_o^2$ and $K_o' = 4S - 1$ are satisfied (Jeanloz 1989).

A further satisfactory representation of the high-pressure compressibility of a single-phase component is the cubic expression used by Holmquist *et al* (2001),

$$p(\varepsilon) = \alpha_1\varepsilon + \alpha_2\varepsilon^2 + \alpha_3\varepsilon^3, \tag{6.18}$$

where the symbol $\varepsilon = 1 - v/v_o$ denotes the volume strain. Again, equivalence to equation (6.15) is achieved with $\alpha_1 = \rho_o C_o^2$, $\alpha_2 = 2S\rho_o C_o^2$ and, in a Mie–Gruneisen representation of the EOS, α_3 a calculable constant dependent on the material Gruneisen parameter.

Any of the several analytic compression curves provided are satisfactory for describing the compressibility of the individual components of a material mixture within reasonable limits of the lattice compression (see chapter 10). The selection of an analytic form will, in most cases, be dictated by the form and availability of material constants descriptive of the compressibility. Generally, the Hugoniot and the isentropic compression curves are second-order tangent, independent of the functional form (Duvall and Fowles 1963). Thus, a linear U–u description of the Hugoniot and either a Murnaghan or a Birch description of the isentrope is a theoretically acceptable analytic representation of the broader (energy- or temperature-dependent) thermo-mechanical compressibility. Third-order divergence of the Hugoniot and the isentrope constrain the functional volume dependence of the Gruneisen parameter.

6.3.5 Mixture compressibility

A two-state model of the mixture at intermediate states of compression-induced compaction is achieved through the additive superposition

$$v(p) = \lambda v_p(p) + (1 - \lambda)v_c(p). \tag{6.19}$$

In equation (6.19), $v_c(p)$ and $v_p(p)$ are the upper-bound iso-compression and lower-bound iso-pressure compressibility relations provided by the equations (6.8) and (6.10), respectively. The ability to analytically invert equation (6.8) is desirable and this is the case for the Murnaghan, the linear-Z and the linear shock-velocity versus particle-velocity compression relations. The parameter λ in equation (6.19) is defined over the interval $0 \leqslant \lambda \leqslant 1$ and is the mass fraction of the material achieving the lower-bound iso-pressure state at the pressure p. Thus, λ is the compaction parameter comparable to the distention parameter α in the p-alpha model. It is, in an abstract sense, the mass fraction of compacted material. The model views the compaction process as one in which the mixture is initially entirely within the upper-bound iso-compression state ($\lambda = 0$). As compression of the mixture proceeds, sites within the body undergo a compaction transition to the compacted lower-bound iso-pressure state. The compaction parameter is a monotonic increasing function of the compression, namely $d\lambda/dp \geqslant 0$. Compaction is concluded when the compaction transition fully subsumes the mixture and λ converges to unity; or when pressure ceases to increase. Upon decompression from intermediate states of compaction ($\lambda < 1$) the compaction parameter λ is assumed to be frozen at the maximum value achieved λ_{max}, and the mixture relation equation (6.19) at constant

λ_{max} provides the decompression pressure–volume relation. The compression-induced compaction and decompression responses from selected states of compaction are illustrated on the right in figure 6.8.

The physical behavior in an application most certainly differs from the two-state nature of the model. A more homogenous compaction process with local regions exhibiting intermediate states of compaction is more physically reasonable. The model, however, captures the average response of such intermediate compaction states of the mixture. Moreover, there is evidence that a two-state compaction process may, at least in some cases, approach physical reality. Compaction studies in some porous rocks (Olsson 1999) have shown that compaction initiates at sites of stress concentration and propagates as a compaction wave through the sample as the process progresses. This compaction phenomenon is considered further in chapter 7.

Experimental evidence demonstrates that accompanying shear deformation can influence compaction and compressibility. Shear deformation enhanced compaction and/or dilatancy is not addressed in the present p-lambda compaction model.

6.3.6 Evolution of the compaction parameter

Evolution of the compaction parameter λ as compression of the mixture proceeds remains to be described. In principal, any empirical compaction law governing λ could be implemented within the present model. Here, a physics-based approach based on statistical concepts of strength of materials is applied. A statistical hazard function $h(p)$ (Hahn and Shapiro 1967) is identified which provides the probability $h(p)dp$ that an element within a unit volume of the material will undergo compaction failure when the pressure is increased from p to $p + dp$. The hazard function could generally also depend on the rate of loading, the shear stress, the temperature or the extent of compaction. These dependencies are noted but are not pursued here.

The probability that some mass element within the unit volume will undergo compaction transition is then

$$d\lambda = (1 - \lambda)h(p)dp. \tag{6.20}$$

Integration provides the cumulative probability and the fraction of the mass,

$$\lambda(p) = 1 - e^{-\int h(p)dp}, \tag{6.21}$$

that has undergone compaction transformation at pressure p. Various functional forms have been pursued for the hazard function $h(p)$ in the statistical description of failure, including the power law attributed to Weibull, and an exponential law due to Gumbel (Hahn and Shapiro 1967). Although probably equally applicable for the present development, a two-parameter power-law hazard function,

$$h(p) = \frac{n}{\sigma}\left(\frac{p}{\sigma}\right)^{n-1}, \tag{6.22}$$

is assumed here for its modestly more convenient analytic properties. Equations (6.21) and (6.22) provide the transformation parameter

$$\lambda(p) = 1 - e^{-(p/\sigma)^n}. \tag{6.23}$$

The transformation parameter $\lambda(p)$ and its first derivative are illustrated in figure 6.9. In terms of a statistical interpretation, the scale parameter σ is identified as the expected value of the compaction pressure, $\langle p \rangle = \sigma$, while the shape parameter n provides a measure of the standard deviation through $\bar{s} \cong 1.28\sigma/n$. As n becomes large $\lambda(p)$ uniformly converges to a Heaviside function and its derivative, a Dirac delta function centered at the pressure $p = \sigma$. The effect of increasing n on the compaction transformation curve is illustrated in figure 6.9 (e.g., Grady 2001, 2003).

The current effort is focused on developing a model for compaction of material mixtures with special emphasis on mixtures with highly disparate initial densities, compressibilities and crystal strengths of the mixture components. Material property data are generally available for constructing the component compression curves and the mixture iso-compression and iso-pressure compressibility relations. At present, a definitive *a priori* theory has not been found that provides the compaction parameters σ and n. First-order estimates of the scale parameter σ can be made through consideration of the crystal strengths and relative amounts of the constituents. Compaction data are probably needed to constrain the shape parameter n. The work of Johnson (1975) directed toward developing a dynamic-strength criterion for mixtures incorporating an effective stress concept provides a useful starting point for developing a compaction model. An initial effort at a systematic calculation of σ is based on a critical excess elastic strain energy incurred on pressure loading (Grady 2001). This strain energy is calculated from the difference in the initial iso-compression and iso-pressure curves and leads to a criterion of $\sigma = (2K_c E_{\text{crit}}/\kappa)^{1/2}$, where $\kappa = 1 - K_p/K_c$, with K_c and K_p the initial iso-compression and iso-pressure bulk stiffness, and E_{crit} a critical elastic strain energy determined by the intrinsic crystal strengths of the mixture components. Whether such a criterion will correlate diverse compaction data is yet to be tested.

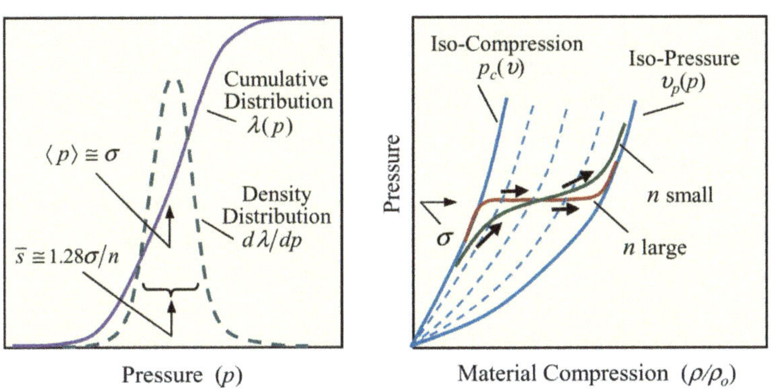

Figure 6.9. Statistical characteristics of the compaction transition parameter (left). Depicted on the right is the dependence of the compaction transformation path on the shape parameter n.

6.4 Elasticity and strength of ceramics with microstructure

The upper-bound iso-compression relation provided by equation (6.8), characterizing the initial elastic compressibility of the ceramic mixture model outlined in the previous section, is an adequate approximation for most applications when a better description is not available. For high-strength ceramics in which the initial elastic response supports substantial strain energy before compaction failure, improved modeling of the initial compressibility is sometimes warranted. Such improvements provide for better calculations of stress wave propagation, particularly with respect to the elastic precursor wave preceding onset of failure. When the material is a porous ceramic these improved relations also correlate with the onset of dynamic failure through a model developed by Steinberg (1975).

6.4.1 Elasticity and microstructure

A more general elastic iso-compression relation for a two-component ceramic mixture of the form

$$p(v) = \nu_1{}^b p_1(v) + \nu_2 p_2(v) \tag{6.24}$$

provides improved agreement with available data and is consistent with more detailed theories of elasticity. This relation would replace the expression provided in equation (6.8). Again, ν_1 and ν_2 are the initial constant volume fractions with ν_1 the volume fraction of the stiffer (ceramic) component, while ν_2 is that of the more compliant component. The second term on the right could accommodate more than one compliant component (for example water and air within a silica sand or a porous sandstone matrix). The exponent b is a material constant usually greater than one. For $b = 1$ equation (6.8) is recovered providing the Voigt upper-bound elastic compliance in the limit of linear compression.

From equation (6.24) the elastic bulk stiffness of the mixture is obtained,

$$K(v) = \nu_1{}^b K_1(v) + \nu_2 K_2(v). \tag{6.25}$$

If ν_2 corresponds to the evacuated void volume fraction then $K_2(v) = 0$ and

$$K(v) = \nu_1{}^b K_1(v) \tag{6.26}$$

provides the compressibility of the porous ceramic. In any case $K_2(v)$ for the compliant component is substantially less than $K_1(v)$and the second term on the right side of equation (6.25) is a correction to the mixture stiffness.

As noted, when $b = 1$, equation (6.26) provides the Voigt approximation to the elastic stiffness of a porous ceramic. For $b = 3$, it follows that

$$c = \nu_1 c_1 = (1 - \pi)c_1, \tag{6.27}$$

where c is the bulk sound speed of the porous ceramic, c_1 is that of the fully dense ceramic, and π is the porosity of the ceramic. This empirical relation for the elastic compression of a porous ceramic is occasionally assumed and usually provides a better approximation than does the Voigt relation (e.g. Roque *et al* 2000). If ultrasonic data as a function of porosity is available b is measured directly. The value of b has been related to pore geometry and the associated stress concentrating characteristics (Maitra and Phani 1994).

6.4.2 The Steinberg–Hugoniot elastic limit

The present development of a compaction model for a material mixture addresses only the volumetric part of the material constitutive relation. Deviator strength is modeled separately and is uncoupled, or at best weakly coupled, to the volumetric model. The simplest strength model is to assume a constant deviator von Mises strength Y. Although unphysical for the complex materials of interest, it does bound the deviator stresses and has been found to be sensible in many applications, such as with the commonly used p-alpha model (Herrmann 1969) for porous materials. Where warranted, more sensible strength-of-material models, such as described by Holmquist *et al* (2001), can be coupled with the volumetric mixture compaction model.

Strength properties for porous media are not commonly available and are difficult to estimate. It is not widely recognized that Steinberg (1975) developed a semi-empirical model for estimating the strength of a porous ceramic based on the HEL, or dynamic strength, of the fully dense material. This model can also extrapolate to other porosities if the dynamic strength at one porosity is known.

Steinberg proposed that the elastic volume change under uniaxial (shock wave) loading to failure is constant irrespective of the ceramic initial porosity. That is

$$v_o - v_{ho} = v_s - v_{hs}, \tag{6.28}$$

where v_s and v_o are the initial specific volume of the fully dense and the porous solid, respectively, while v_{hs} and v_{ho} are the corresponding specific volumes at the HEL of the fully dense and porous solid, respectively. The elastic volume strain at the HEL of the porous material can then be written in terms of that of the solid material according to

$$\varepsilon_{ho} = \frac{v_o - v_{ho}}{v_o} = \frac{v_s - v_{hs}}{v_o} = \frac{v_s}{v_o}\frac{v_s - v_{hs}}{v_s} = \frac{v_s}{v_o}\varepsilon_{hs}. \tag{6.29}$$

The volume fraction of solid in the porous material is $\nu = v_s/v_o$ and it follows that

$$\varepsilon_{ho} = \nu\varepsilon_{hs} = (1 - \pi)\varepsilon_{hs}, \tag{6.30}$$

where again $\pi = 1 - \nu$ is the pore volume fraction, or the porosity. Using equation (6.26) as descriptive of the longitudinal compressibility of a porous material one obtains

$$\sigma_{ho} = \nu^{b+1}\sigma_{hs} \tag{6.31}$$

for the Steinberg HEL strength of a porous ceramic (Grady 2003).

To test this the model ultrasonic data from the literature are shown in figure 6.10. Ultrasonic longitudinal and shear velocities as a function of the distention $\alpha = 1/\nu$ of the porous solid are for boron carbide (Brar *et al* 1992) and for aluminum oxide

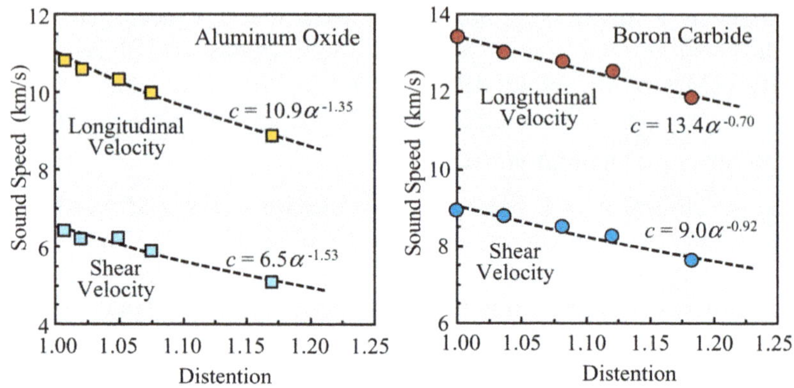

Figure 6.10. Ultrasonic longitudinal and shear wave speeds as a function of distension $\alpha = 1/\nu$ for porous aluminum oxide (left; Gust *et al* 1973) and boron carbide (right; Brar *et al* 1992) ceramics.

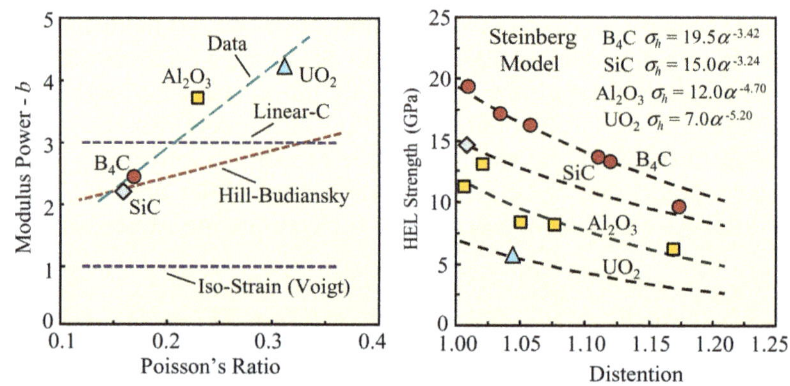

Figure 6.11. Ultrasonic distension modulus power for selected ceramics (references in the text) is shown on the left. Experimental HEL strength and predictions for porous ceramics based on the Steinberg (1975) model are provided on the right.

(Gust *et al* 1973). The coefficient b for the longitudinal elastic modulus in the form of equation (6.26) has been determined from the data for boron carbide and aluminum oxide, as well as for silicon carbide (Bartukowski and Spletzer 2001) and uranium dioxide (Roque *et al* 2000). The values for b are plotted as a function of the intrinsic elastic Poisson's ratio on the left in figure 6.11. These values are compared with the iso-strain Voigt theory (b = 1) and the linear c theory (b = 3), as well as with the more exacting theory of Hill (1965) and Budiansky (1965). The data are not adequately described by any of the three theories, although the dependence of the data on Poisson's ratio suggests a trend not unlike the Hill–Budiansky theory.

The experimental values of b are used to predict the dependence of the HEL on the porosity in figure 6.11 based on the Steinberg model from equation (6.31). The model provides a reasonable representation of the measured HELs for aluminum oxide (Gust *et al* 1973) and for boron carbide (Brar *et al* 1992), for which data at several porosities are available.

Agreement of the Steinberg model with the boron carbide data was previously noted by Brar *et al* (1992). Predictions are also shown for silicon carbide and uranium dioxide without supporting data. The physical mechanisms underlying the initial premise of the Steinberg model are not obvious and there is not sufficient data to indicate the range of validity of the model. Nonetheless, when nothing else is available and an estimate of the dynamic strength for a porous ceramic is needed, the model provides a reasonable starting point.

6.5 Application to ceramics with microstructure

In this section ceramics with evident microstructure issues, for which shock compression data are available, are examined in the light of the mixtures-with-microstructure modeling concepts that have been described in the previous sections. It is shown that these extensions of the mixture EOS theory offer modeling benefits not provided by the classical theory of mixtures that is applied to silicon carbide ceramics and metal–ceramic composites in the opening section. The model also provides a tool for analysis and physical interpretation of the experimental data. The section also displays a rich selection of shock data on ceramic materials.

6.5.1 Commercial alumina ceramics

Alumina, or aluminum oxide ceramic, is representative of how widely ranging manufacturing methods can result in engineering ceramics with vastly different mechanical strengths and EOS properties. The principal constituent, for which the ceramic is named, is crystalline α-Al_2O_3; namely the mineral corundum or the gemstone sapphire. Alumina ceramics are available that are high in purity and approach the theoretical density of 3970 kg m^{-3} of α-Al_2O_3. The Coors AD-999 ceramic is representative of a near-full-density reasonably high-purity alumina ceramic. Most alumina ceramics, however, are composites with significant fractions of other materials in addition to appreciable porosity. The Coors alumina ceramics such as AD-85, AD-90 and AD-995 are representative of materials in which a significant fraction of glassy phase material is introduced and a degree of porosity is retained in the manufacturing process.

An EOS description of these various alumina ceramics relevant to shock impact and ballistic penetration becomes challenging. Since EOS properties for the ceramic constituents as well as shock and penetration data on several of these Coors alumina ceramics are available, they provide appropriate materials for exploring multi-component EOS modeling. The alumina Coors AD-85 is included in the extensive seminal shock Hugoniot properties study of ceramics performed by Gust *et al* (1973). Shock profile measurements on AD-85 alumina were performed by Rosenberg *et al* (1985), while shock profile measurements on AD-999, AD-995 and an equivalent to AD-90 are provided in the report of Grady and Moody (1996).

6.5.1.1 Component compressibility

Coors AD-85 alumina is reported to be 15% by weight glass phase with the remainder aluminum oxide and about 5%–6% other additive oxides. Assuming average compressibility and density of the latter additives as approximated by that of aluminum oxide, it is sensible to describe the material as a ceramic of composition 15% by weight glass phase and 85% by weight aluminum oxide. Different ratios of these two components would adequately approximate the others of Coors series of glassy alumina ceramics. The Coors AD-85 alumina, for which Hugoniot data are available (Gust *et al* 1973), and for which wave profiles have been measured (Rosenberg *et al* 1985), is about 6%–7% porous. Other initial porosities for this ceramic are possible, and AD-85 alumina ceramics with porosities from a few percent to nearly 20% have been tested under shock wave loading (Bar-on *et al* 2002).

Compressibility properties for both monolithic aluminum oxide and glass, from which the component EOS curves can be constructed, are presented in figure 6.12. The pressure is plotted as a function of the specific volume, $\upsilon = 1/\rho$, in the left graph and as a function of compression strain, $\varepsilon = 1 - \rho_o/\rho$, in the right. Compressibility data for aluminum oxide is provided by the static compression measurements on α-Al_2O_3 to 12 GPa of Sato and Akimoto (1979). These data are extrapolated by the authors with a second-order Birch–Murnaghan equation yielding $K = 226$ GPa and $K' = 4(S = 1.25)$, and a third-order Birch–Murnaghan equation yielding $K = 239$ GPa and $K' = 0.9(S = 0.45)$. The second-order fit provides the higher curvature extrapolation in figure 6.12 (dashed curve) and is used in the present analysis. Both values for the initial bulk modulus from the static data are somewhat lower than the ultrasonic bulk modulus of $K = 250$ GPa. Extrapolation of ultrasonic data would yield a slightly stiffer compression of the aluminum oxide material. Disparities between Hugoniot and ultrasonic assessment of the compressibility of corundum is addressed in both chapter 1 and chapter 10.

A reasonable approximation for the glass phase within the commercial alumina is that of soda-lime glass for which Hugoniot data have been measured by Bourne and Rosenberg (1995) and by Grady and Chhabildas (1997). These shock Hugoniot data are also plotted in figure 6.12. The initial density for the Bourne and Rosenberg data is 2490 kg m^{-3} while that of the Grady and Chhabildas material is 2530 kg m^{-3}. The

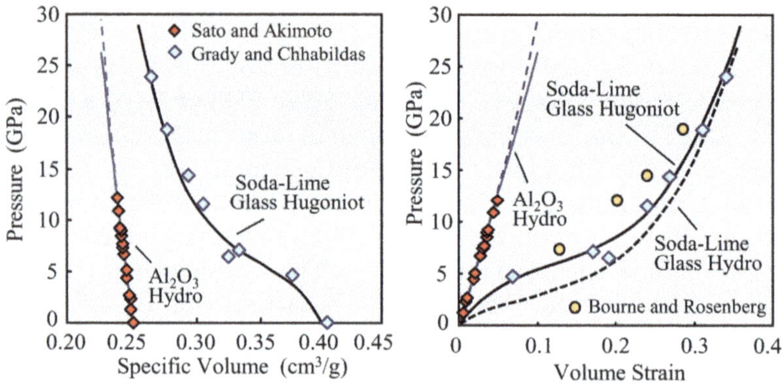

Figure 6.12. Experimental compression properties for crystalline Al_2O_3 (Sato and Akimoto 1979) and soda-lime glass (Grady and Chhabildas 1997, Bourne and Rosenberg 1995) and analytic representations of the compression curves.

data of Grady and Chhabildas are selected here over the somewhat stiffer soda-lime glass Hugoniot response of the Bourne and Rosenberg data to determine the model compressibility, because of the present author's familiarity with the test method and the data reduction. A cubic fit to the Hugoniot data is shown in figure 6.12. Strength is subtracted from the Hugoniot response curve by requiring that the initial compression slope be consistent with the initial ultrasonic bulk modulus of the glass, and the high-pressure compression meld with the Hugoniot where the strength offset is assumed to be negligible. The compression EOS parameters for α-Al_2O_3 and glass applied in the analysis are provided in table 6.3.

6.5.1.2 Mixture compressibility

The component compressibility determined from the compression relations provided in table 6.3 is reproduced as the solid curves in figure 6.13. Both the upper-bound iso-compression and the lower-bound iso-pressure curves for the p-lambda model representation of AD-85 alumina are also shown in the plot. These bounding mixture compression paths are calculated from equations (6.8) and (6.10) using the appropriate component compressibility and the relevant mass and volume fractions of the constituent components. The assumed mass fractions of aluminum oxide and glass for AD-85 alumina are 0.85 and 0.15, respectively. The corresponding volume fractions are 0.732 for aluminum oxide, 0.205 for glass and 0.063 for the void fraction or porosity. Shock Hugoniot data for AD-85 alumina provided by Gust *et al* (1973) to shock pressures in excess of 40 GPa are shown on the right in figure 6.13 and compared to the model curves.

The several HEL experimental points at approximately 6–7 GPa lie to the left of the iso-compression curve emphasizing the initial uniaxial strain elastic shock compression preceding the dynamic failure. The post-HEL Hugoniot states indicate progressing inelastic compaction of the alumina with the higher pressure (20–40 GPa) Hugoniot points lying close, but also to the left, of the calculated iso-pressure volumetric response of the mixture. Although this difference may identify some inadequacies in the mixture model, it is more likely that the discrepancy corresponds to retained strength of the ceramic at the shock state.

A p-lambda model compaction path is also shown with the data in figure 6.13. The model curve is based on the p-lambda superposition of the upper and lower-bound compression relations from equation (6.19) along with the analytic form of

Table 6.3. Compression model properties for component materials of commercial aluminum oxide (alumina).

Aluminum oxide	Glass
$\rho_o = 3.97$ g cm^{-3}	$\rho_o = 2.50$ g cm^{-3}
$C_o = 7.55$ km s^{-1}	$a_1 = 45$ GPa
$K = 226$ GPa	$a_2 = -270$ GPa
$K' = 4$	$a_3 = 1000$ GPa
$S = 1.25$	
$p = \rho_o C_o^2 \varepsilon/(1 - S\varepsilon)^2$	$p = a_1\varepsilon + a_2\varepsilon^2 + a_3\varepsilon^3$

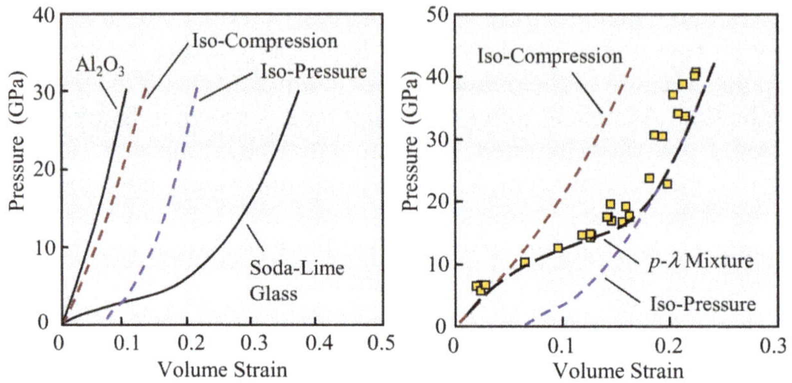

Figure 6.13. Mixture iso-compression and iso-pressure curves for 15 wt% glass, 85 wt% aluminum oxide and 6.7 volume percent porosity description of AD-85 alumina (left). Comparison of the model (Grady 2003) and Hugoniot data for AD-85 aluminum oxide ceramic (right; Gust and Royce 1971).

the compaction transition of equation (6.23). Compaction parameters of $\sigma = 14$ GPa and $n = 4.5$ are used and are based on a sensible best fit to the available Hugoniot data. The analytic and continuous compaction curve adequately characterizes the available data for AD-85 alumina and provides a theory for extrapolating to other mixes of aluminum oxide, glass and porosity in the alumina ceramic. Discrepancies with the data suggest the need for an accompanying strength model.

6.5.2 Boron carbide ceramics and shock-induced phase transformation

Boron carbide ceramic has been particularly problematic in attempts to develop adequate constitutive model descriptions for purposes of computational simulation of dynamic response (Grady 2015). Dynamic-strength properties of boron carbide ceramic differ uniquely from comparable ceramics, and are difficult to characterize. Further, boron carbide is suspected of undergoing polymorphic phase transformation within the shock wave compression process. These phase transformation features have been particularly elusive under experimental investigation, and consequently are also difficult to capture within an appropriate constitutive model.

Boron carbide ceramic is, however, an attractive candidate for ballistic applications. It is the lowest density and the highest strength ceramic among a suite of ceramics considered for such applications. The necessary physical and mechanical properties of boron carbide have been studied extensively and access into much of the literature is available in the report of Dandekar (2001). In particular, the Hugoniot and ultrasonic data integral to the development of shock wave and high-strain-rate constitutive response models are reviewed in the same report. In this section, two issues relating to high-pressure phase transformation in boron carbide ceramic in the dynamic environment are pursued.

First, although other methods are used to determine the high-pressure phase transformation properties in ceramic materials—diamond anvil cell (DAC) technology—shock compression techniques remain the standard when issues of dynamic transition pressure and compressibility are of paramount concern. Extraction of the required phase transformation properties remains elusive, however, and usually requires concerted effort by workers experienced in the field. The extensive shock Hugoniot data for boron carbide of McQueen *et al* (1970), provided in the LANL shock compendium (Marsh 1980), are representative. These data, joined with the necessary analysis, provide strong evidence for phase transformation in boron carbide ceramics, in addition to the dynamic compressibility properties accompanying the phase transformation.

Second, the model for compaction of mixtures outlined in section 6.3 is here extended to model phase transformation under shock compression, and applied to boron carbide ceramic. This undertaking is profitable for several reasons. Unless a tabular description of the constitutive response is used, analytic relations must be incorporated to specify functional behavior of material compression and deformation. Compression of a material undergoing phase transition, for example, is commonly described by a cubic polynomial to account for the material stiffening–softening–stiffening response through the transition region. The analytic model of Steinberg (1996) is representative where the experimental shock-velocity versus particle-velocity slope is expressed as a quadratic function of the Hugoniot strain $S = S_o + S_1\varepsilon + S_2\varepsilon^2$.

Empirical polynomials are approximations at best, pathological outside the range of the data fit, and, where wave propagation is involved, incur additional difficulties where sensitivity to the local derivative of the polynomial is critical. Extension of the mixture model naturally introduces physics-based functional forms with physical constants that are readily extracted from the data or sensibly estimated.

Further, a number of previous efforts have involved quite excellent modeling studies of phase transformation in solids, and the concomitant effects on shock wave propagation. Such efforts, however, have focused on phase transformation in thermo-mechanically ideal materials. These models do not capture the transformation pressure spread and hysteresis commonly observed in engineering materials and brought about by porosity, alloying, and a range of other material-processing features influencing the microscopic and mesoscopic character of phase transformation in the materials. Again such effects of microstructure heterogeneity are naturally reproduced through the present mixture modeling methods.

6.5.2.1 Phase transformation from shock Hugoniot data

Boron carbide subjected to high-rate shock wave compression exhibits dynamic strength, based on its HEL amplitude, that is one of the highest observed among the metal–nonmetal compounds. The strength of boron carbide ceramic subsequent to failure within the shock and post-shock domain is controversial, however, and considerable effort has been expended to achieve a better understanding of the underlying physics (Grady 1994a, 1994e, 2015, Vogler *et al* 2004). An informative study of the shock strength of boron carbide and silicon carbide through non-planar shock wave experiments is provided in the paper of Paris *et al* (2011). More is said on this topic in chapter 9.

The restricted loading path of the shock wave experiment complicates interpretation of strength and phase transformation in the shock compression event. In particular, phase transformation under shock compression with the accompanying change in specific volume and compressibility makes assessment of strength on the Hugoniot particularly difficult. The converse is also true. The occurrence of phase transformation in boron carbide within the shock and high-rate loading event is not yet definitive. Efforts have been made to assess high-pressure phase transformation in boron carbide through static DAC technology (Manghnani 2007). Vogler *et al* (2004) have suggested evidence of phase transformation in boron carbide from shock wave experiments. Phase transition under pressure has been explored through molecular dynamics analysis (e.g. Yan *et al* 2009).

The earlier studies and other available shock wave data are explored for evidence of shock-induced phase transformation (e.g. Grady 2010, 2015). The conclusion is that shock data alone suggest the occurrence of one phase transformation in the shock compression process. Shock data combined with extrapolation of modest-pressure ultrasonic data suggest the occurrence of a second transition under shock compression. Portions of these studies of boron carbide ceramic are highlighted in this section.

The principle shock wave data for boron carbide ceramic examined here are those of McQueen *et al* (1970) provided in the Los Alamos Compendium (Marsh 1980). Other shock data on boron carbide are available (e.g. Wilkins 1968, Gust and Royce 1971, Grady 1994a, Vogler *et al* 2004, Zhang *et al* 2006). The Los Alamos data, however, have unique features significant to the present analysis. Also, additional uncertainties inherent in addressing the different materials and experimental methods of the various authors are avoided. These further shock Hugoniot data are examined in Grady (2010, 2015).

Before proceeding, qualifying remarks concerning the available Hugoniot data are necessary. There is a marked disparity between the high-pressure Hugoniot compression data of Gust and Royce (1971) and of Vogler *et al* (2004) on one hand, and the data of McQueen *et al* (1970) and Grady (1994a) on the other. Whether these differences are due to manifestations of material strength, effects of phase transformation, uncertainties in initial material composition and chemistry, or difficulties with analysis and interpretation of the measured shock data is not known. Some of the issues have been discussed by Dandekar (2001). Additional comments concerning this disparity will be offered where relevant in the development that follows.

The McQueen *et al* (1970) data provide shock compression states for boron carbide over the pressure range of 15–110 GPa. These data encompass two starting densities for boron carbide of nominally 2400 kg m^{-3} and 1900 kg m^{-3}. Assuming reasonably pure material, the densities imply initial porosities of approximately 5% and 30%, respectively, based on a theoretical density of 2520 kg m^{-3}. (Although the ceramic composition was not reported, it is unlikely that any sintering additives would be lower density than the boron carbide. Appreciable higher density additives, and correspondingly higher calculated porosity, could account for discrepancies with the Gust and Royce data, but this seems unlikely.) Shock-velocity versus particle-velocity and pressure versus specific volume Hugoniot data for the boron carbide are shown in figure 6.14. In the test method of McQueen *et al* (1970) shock velocity and particle velocity are the measured quantities and are identified as the original data in the right-hand plot. There are two regimes associated with the two initial porosity levels. The scatter for the most part is due to sample-to-sample variations in density. Pressure and specific volume are calculated from the measured shock-velocity and particle-velocity data through the Hugoniot relations

$$p = \rho_{oo} U u, \qquad v = (1 - u/U)/\rho_{oo}, \tag{6.32}$$

with ρ_{oo} the initial distended sample density.

The calculated pressure versus specific volume Hugoniot for both densities is shown on the left in figure 6.14. Shock compaction of the porous ceramic leads to excessive dissipative heating—substantially more so for the nominal 30% porous material than for the nominal 5% porous material. The extent of heating, however, is not sufficient to separate the two sets of data in the pressure–volume plot through thermal expansion (Gruneisen) properties of the material. Where Hugoniot states for the two porosities overlap, pressure–volume values do not statistically differ. On the other hand, heating due to pore compaction is expected to markedly reduce residual strength at the shock state leaving material closer to a hydrodynamic state (unable to support significant stress deviator). Strength states on the Hugoniot for a near-full-density ceramic is pursued for boron carbide in the work of Vogler *et al* (2004) and Paris *et al* (2011). The lowest three Hugoniot points for the nominal 5% porous material appear to exhibit some strength on the Hugoniot, although it is known that the test method of McQueen *et al* (1970) is not reliable at the lower shock amplitudes where structured shock waves are probably present.

The last observation introduces an interesting point. Shock Hugoniot states achieved in near-full-density ceramic, which includes the work of Gust and Royce (1971), Grady (1994a) and Vogler *et al* (2004), do so through a structured shock wave. Hugoniot states are referenced from the HEL amplitude that is itself determined from the magnitude of a rather unusual, and probably unsteady, elastic

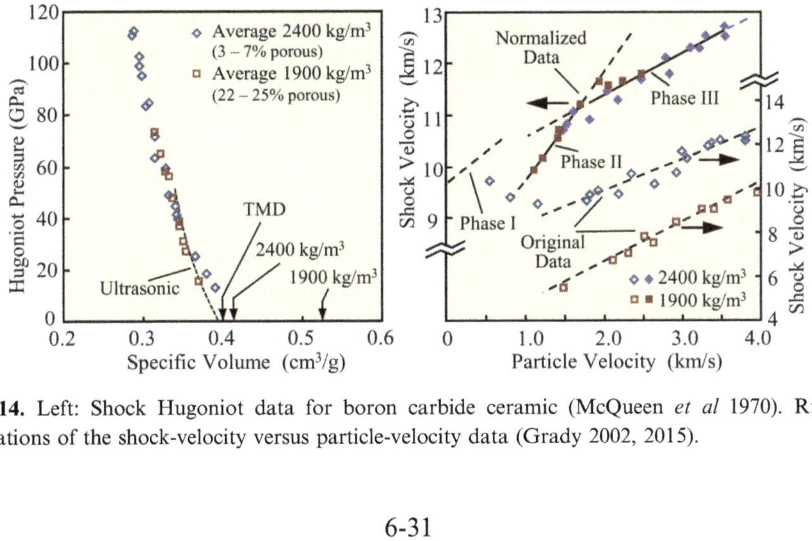

Figure 6.14. Left: Shock Hugoniot data for boron carbide ceramic (McQueen *et al* 1970). Right: Two representations of the shock-velocity versus particle-velocity data (Grady 2002, 2015).

precursor wave (Grady 1994a, Vogler *et al* 2004, Hayun *et al* 2009). With the probable exception of the three lower points mentioned previously, because of material porosity, any elastic precursor wave is overdriven and transition to the Hugoniot state for the McQueen *et al* (1970) data is through a single steady shock wave. In contrast, the elastic precursor for full-density boron carbide ceramic is not overdriven until quite high Hugoniot stress levels. This important difference, and associated complications in Hugoniot analysis, may relate to some of the disparity in the high-pressure Hugoniot data for boron carbide ceramics.

A synthesis of the two sets of the McQueen *et al* Hugoniot data is used to assess further high-pressure EOS features of boron carbide. The pressure versus specific volume Hugoniot data on the left in figure 6.14 are mapped back into the more revealing shock-velocity versus particle-velocity plane through the Hugoniot relations

$$U' = \sqrt{\frac{p}{\rho_o(1 - \rho_o v)}}, \qquad u' = \sqrt{\frac{p(1 - \rho_o v)}{\rho_o}}, \tag{6.33}$$

where $\rho_o = 2520$ kg m^{-3} is the theoretical density of boron carbide (Dandekar 2001). These effective shock-velocity versus particle-velocity states, shown as the solid symbols in figure 6.14, provide Hugoniot states for boron carbide referenced from the initial theoretical density. It is emphasized that this is an analysis method for providing a more sensitive display of the Hugoniot EOS data. These calculated shock-velocity–particle-velocity points would not be realized in any experiment. This analysis procedure is referred to as a normalization of the shock-velocity versus particle-velocity data to the solid or crystal density of the material.

With this normalization, the shock-velocity–particle-velocity data corrected to theoretical density in figure 6.14 display clear evidence for bi-linear behavior, indicative of a shock-induced phase change on the principal Hugoniot. The two linear curves shown with the normalized data in figure 6.14 are the result of a statistical analysis of the data to determine the linear shock versus particle-velocity constants for the regions of data identified as shock phase II and shock phase III in table 6.4 and in the figure. Various fits to the data provided a pressure intercept between 45 and 55 GPa (the shock transition pressure) with the optimum *U–u* parameters in table 6.4 intercepting closer to 45 GPa. The shock data is unable to distinguish between a second-order transition (no transition volume change) and a small first-order transition. The report of a shock transition in B_4C at 38 GPa with second-order transition characteristics (Zhang *et al* 2006) is possibly the same II–III transition identified here.

A plot of the McQueen *et al* (1970) Hugoniot data below 60 GPa is shown in figure 6.15. The overlying curve is the linear *U–u* Hugoniot provided by C_o and S for phase II in table 6.4. As argued previously, the curve should provide a sensible estimate of the hydrodynamic compressibility of boron carbide within the stated range. There are some additional arguments for this supposition. First, if there were a significant strength offset to the Hugoniot stress it would be highly fortuitous for

Table 6.4. Shock phase transformation properties of boron carbide (Grady 2010, 2015). Table reprinted from Grady (2010) with permission.

Phase	C_o(km s^{-1})	S	p_t(GPa)	Δv_t(cm^3 g^{-1})
B_4C-I	9.66	1.32	Ambient phase	Ambient phase
B_4C-II	7.70	2.15	25–35	~0.008
B_4C-III	9.80	0.80	45–55	<0.002

the stress offset for the markedly different starting density materials to be the same. This argument would also support full compaction of the pore volume by the initial shock wave, and further suggests that local shear stresses are small compared to the compressive shock pressure. Second, and more convincing, elevated temperatures at the Hugoniot states for the more porous materials are inconsistent with significant material strength. In the range of 40 GPa shock pressure, containing a cluster of both 1900 kg m^{-3} and 2400 kg m^{-3} material Hugoniot points, temperatures of the lower initial density material subjected to Hugoniot shock compression are readily estimated. Based on a constant specific heat of 961.6 J kg^{-1} K^{-1} (Dandekar 2001), an ambient average temperature rise due to shock heating is calculated to be about 2700 K. This level of temperature rise would approach the ambient melt temperature of boron carbide which is also about 2700 K (Dandekar 2001). The specific heat can be expected to increase with temperature and therefore the calculated temperature rise is likely an overestimate. Nonetheless, it is unlikely that such temperature states on the Hugoniot would support significant material strength. Further, due to low thermal conductivity, it is also unlikely that thermal equilibrium on the microscale is achieved, leading to locally hot deformation regions that would additionally reduce material strength on the Hugoniot.

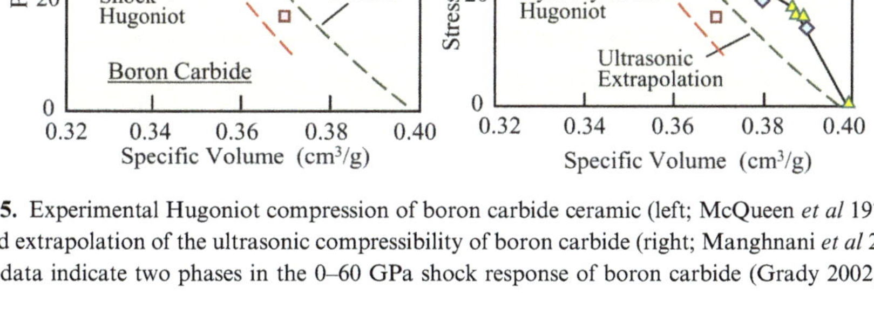

Figure 6.15. Experimental Hugoniot compression of boron carbide ceramic (left; McQueen *et al* 1970, Grady 1994a) and extrapolation of the ultrasonic compressibility of boron carbide (right; Manghnani *et al* 2000). The collective data indicate two phases in the 0–60 GPa shock response of boron carbide (Grady 2002, 2015).

There are complications. Although composition of the boron carbide tested in the LANL compendium is not known, sintered ceramics at that time (the 1960s) were typically prepared with 5–10 μm particle size powders. Thus, if particle size is $a = 10$ μm, then the shock compaction time is of order a/U (U is the shock velocity) and thermal diffusion time is of order a^2/χ (χ is the thermal diffusivity of approximately 25×10^{-6} m^2 s^{-1}). The thermal diffusion time is about three orders of magnitude lower, and would certainly not equilibrate temperature on length scales of order a. Shock compaction is, however, more complex. There is reason to believe that early in the shock rise particles are subjected to stress loads at strain rates of order $u/a \sim 10^8$ s^{-1} that shatter (fragment) the particle with very little dissipation. Estimates based on a fracture energy of boron carbide of order 100 J m^{-2} calculates a reduction in particle size of about one order of magnitude. Subsequent deformation on this granulated length scale ($\sim a/10$) accommodates the dissipation necessitated by the steady wave shock to the Hugoniot state. Thermal diffusion time is still one order of magnitude lower than the compaction time, but certainly closer to homogeneous. Microstructure details of the porous compaction in intense shock waves have not been adequately studied.

Manghnani *et al* (2000) have measured ultrasonic properties of boron carbide to 2.1 GPa and determined a bulk modulus of $K = 234.9$ GPa and the first pressure derivative $K' = 4.26$. These elastic constants are used to extrapolate compression to higher pressure with a Murnaghan EOS. The extrapolated compression curve is shown in figure 6.15. Shock parameters C_o and S calculated from the reported K and K' are provided in table 6.4 and correspond to the ambient boron carbide phase I. The corresponding U–u curve for phase I is also shown on the right in figure 6.14.

It should be noted that a fairly broad range of ultrasonic wave speeds for boron carbide has been reported (Dandekar 2001, Grady 2015). The reasons for this rather large disparity are currently unclear. It is suspected, however, that unusually low values are a consequence of microstructure effects (e.g. crack porosity) and do not fully represent the lattice compressibility of the material. The measurements of Manghnani *et al* (2000) were performed on superior quality material, provided the highest ultrasonic values, and are considered the closest to representing the lattice compressibility properties of boron carbide.

The volume offset of the hydrodynamic shock Hugoniot on the left in figure 6.15 and the extrapolated ultrasonic compression curve suggest a further phase transformation within the range of the data. It is unlikely that uncertainties in the Hugoniot data and the extrapolated ultrasonic compression curve could account for the 15%–20% volume strain offset.

The transformation pressure at which the possible phase change occurs is less certain. Shock profile studies (e.g. Grady 1994a, 2015) do not show clear evidence of a distinct transformation stress. The lowest pressure Hugoniot point of McQueen *et al* (1970) on 1900 kg m^{-3} material at about 18 GPa suggests that transformation from an ambient phase I to a shock phase II may have occurred under shock compression. Uncertainties in the single data point make this supposition speculative. More recent analysis of earlier shock profile measurements on boron carbide indicates wave profile anomalies above approximately 30 GPa shock pressure

(Grady 2002, 2015). Thus, the shock profile data may also indicate phase transformation from a B_4C-I phase (the ambient phase) to a B_4C-II phase (the shock phase II) at approximately 30 GPa, and suggest the range of 25–35 GPa provided in table 6.4.

Shock Hugoniot data of Grady (1994a) and the lowest three Hugoniot points for the 2400 kg m^{-3} material from McQueen *et al* (1970) are included on the right in figure 6.15. These data collectively illustrate that the Hugoniot strength of near-full-density boron carbide plays a role in this range of the data. The data provide a 15–20 GPa HEL for boron carbide ceramic and a Hugoniot strength offset to approximately 40 GPa. Agreement of the 40 GPa and 56 GPa Hugoniot states of Grady (1994a) with the hydrodynamic phase II–III Hugoniot from the McQueen *et al* (1970) data suggest loss of strength on the higher pressure Hugoniot for this boron carbide ceramic.

It is notable, however, that the Hugoniot data of Gust and Royce (1971) and of Vogler *et al* (2004) on near-full-density boron carbide ceramic are in sensible agreement with the McQueen *et al* (1970) and Grady (1994a) data up to Hugoniot pressures of about 30 GPa. Their data reside markedly higher (or to the right) above this pressure. A possible explanation is that dynamic strength and failure in the shock compression process have sensitivities to ceramic microstructure that are not yet understood. Such differences in behavior are hinted at by at least some of the precursor wave profiles measured in the study of Vogler *et al* (2004), which exhibit stable deformation hardening, unlike the marked stress relaxation observed in the wave profiles of Grady (1994a) and Hayun *et al* (2009). Further work on the strength issue in boron carbide ceramic is needed.

In summary, the consensus of the present analysis of the shock wave data is that boron carbide undergoes two polymorphic phase transformations under shock compression. Best estimates of the transformation pressures and volumes, along with hydrodynamic compressibility properties of the ambient phase and the two high-pressure phases, are provided in table 6.4. This conclusion is certainly arguable, but represents the best assessment of the data. An extended analysis of Hugoniot data for boron carbide is provided in later papers (Grady 2010, 2015). Boron carbide is and will probably remain one of the more enigmatic ceramics under investigation by the shock and ballistic community.

6.5.2.2 Analytic description of shock-induced phase transformation

The two-state p-lambda model previously used to describe the irreversible compaction of porous solids and heterogeneous mixtures is extended to the EOS descriptions of solids undergoing phase transformation under shock compression (Grady 2002). The model treats one, or several, phase transformations and addresses the spread in transformation pressure, and transformation hysteresis, brought about by the microscopic and mesoscopic heterogeneities of the materials. The previous analysis of shock Hugoniot data suggests that boron carbide ceramic undergoes two phase transformations within the pressure range of 25–55 GPa. Although the transformation pressures, and even the occurrence of these phase transformations, are still arguable, boron carbide is used here to illustrate application of the theory and model.

This model provides a description of the response of a material, or a mixture of materials, that undergo polymorphic phase transformation when subjected to large-amplitude shock wave loading. The model is intended to be robust for engineering and computer analysis purposes, and provides a satisfactory description of the material dissipation and compressibility under transient pressure loading using physics-based functions and experimentally available material constants.

Several features of the application of the model are noted. A full thermo-mechanical description of the phase transformation is not undertaken. Attention is instead focused on adiabatic mechanical compressibility through the phase transformation, which provides the first-order response of the materials under shock compression. Ideal homogeneous phase transformations are recognized to occur over a relatively narrow pressure range. For a number of reasons, such homogeneous transformation is not observed in materials with microstructure. In the model statistical concepts are applied to account for heterogeneous phase transformation. The formalism leads to analytically convenient physics-based relations which continuously spread the phase transformation over the requisite pressure range. A further feature commonly observed is hysteresis in the phase transformation cycle. That is, the pressure at which a transformation reverts back from the high-pressure phase on decompression is observed to be lower than the transformation pressure on compression. This hysteresis can be a consequence of thermally activated kinetics in the transformation process. More likely, in the present applications, hysteresis is a consequence of the heterogeneous microstructure leading to transformation hysteresis. In the present theory hysteresis is captured through the same statistical framework.

The stable compression of a monolithic substance exhibits familiar behavior until the current crystal lattice structure becomes thermodynamically unstable with respect to an alternative, lower-energy lattice structure. At this level of compression the substance may transform to a further stable lattice structure (equilibrium transformation), or a certain level of over compression may be required before transformation to the stable structure is achieved (nonequilibrium transformation). For many materials a multiplicity of such lattice structure transitions will occur over a sufficiently large compression range.

Without phase transformation, the compressibility of a material within any specific lattice structure exhibits predictable behavior. The present model requires EOS compressibility expressions for the individual phases of the material, and any of the physics-based expressions described in earlier sections or in later chapters are appropriate. The pressure versus specific volume relation emerging from linear shock velocity versus particle velocity is used here,

$$p(v) = \frac{\rho_o C_o^2(1 - v/v_o)}{[1 - S(1 - v/v_o)]^2}, \tag{6.34}$$

which inverts to

$$v(p) = v_o - v_o \frac{\alpha(p)}{S}\left(1 - \sqrt{1 - \alpha(p)^{-2}}\right), \tag{6.35}$$

where $\alpha(p) = 1 + \rho_o C_o^2/2Sp$.

As shock compression proceeds and phase transformation initiates, the material resides at successive states that are a mixture of the component phases. An additive superposition of the individual phases is assumed in the model,

$$v(p) = \sum_i \lambda_i v_i(p), \tag{6.36}$$

where the component-phase mass fractions λ_i satisfy

$$\sum_i \lambda_i = 1. \tag{6.37}$$

For a two-phase material, equation (6.36) simplifies to

$$v(p) = (1 - \lambda)v_1(p) + \lambda v_2(p), \tag{6.38}$$

where the transformation parameter λ, defined over the interval $0 \leqslant \lambda \leqslant 1$, is the mass fraction of the transformed, or the high-pressure, phase. A transformation involving more than two phases (such as the present boron carbide) also has application (Grady 2002).

It remains to characterize the evolution of the transformation parameter λ as compression and phase transformation proceeds. This facet of the model parallels the statistical physics-based development of the corresponding compaction parameter described earlier in the chapter. In a heterogeneous solid with microstructure, stress levels on the microscale will vary by about an average pressure. Such local stress fluctuations are a consequence of stress concentrations due to material crystal strength along with microscopic and mesoscopic features including crystalline anisotropy, defects, porosity, alloying elements, and inherent internal strain, among others. Phase transformation at some point in the material will occur only when the local stress state at that point achieves a critical level. As the average pressure increases, phase transformation will proceed until each and every point in the body attains the requisite transformation level. The spread in pressure necessary for transformation completion is determined by the unique heterogeneous character of the material.

A statistical hazard function $h(p)$ provides the chance $h(p)dp$ that an element will undergo phase transformation when the average pressure is increased by increment dp. More generally, the hazard function may also depend on the rate of loading, the temperature, and the past extent of transformation. Also, transformation kinetics may play a role in the dynamic phase transition process. In this development only a rate-independent, pressure-dependent hazard function is considered. Assuming, as before, a power-law hazard function, the analysis leads to an expression for the pressure-dependent transformation parameter,

$$\lambda(p) = 1 - e^{-(p/\sigma)^n}. \tag{6.39}$$

In this expression $\lambda(p)$ is the transformed mass fraction while σ and n are the statistical transformation properties of the material.

A uniform application of pressure to the boundaries of the sample translates to a complex stress state on the microscale within interior regions for a number of reasons. Inclusions of different compliance material or local regions of voids represent points of stress concentration that contribute to local variations in pressure and shear stress. The extent of local stress fluctuation depends on the density and intensity of such stress concentrators as well as the crystal strength of the material. Even for ideal polycrystalline material without pores or impurities, variations in local stress occur. A single crystallite subjected to a uniform hydrostatic pressure will undergo shape change as well as volume change due to crystal anisotropy, provided crystal symmetry is lower than cubic (e.g. hexagonal, tetragonal). Similarly, a single crystallite subjected to a uniform volume compression will sustain shear stresses as well as an average pressure state for the same reasons. Hence, crystal anisotropy leads to local stress fluctuations within a polycrystalline solid due to strain incompatibility within the randomly oriented crystal structure.

Local stress disparities within a solid with microstructure do not increase without bounds, but are limited by the crystal strength of the material. Crystal slip and deformation plasticity, or fracture, will limit the magnitude of local stress variations at some level of compression. Phase transformation at a point in a crystalline solid will depend on the state of pressure and shear stress at that point. It is reasonable to conclude that the statistical transformation parameters σ and n correspond, respectively, to the average local pressure at which phase transformation proceeds and to a measure of the dispersion in the local stress state responsible for the spread in transformation pressure. The parameter σ is sensibly estimated from the transformation pressure observed in the hydrostatic or shock compression of single-crystal samples. The parameter n is determined by the microstructure pressure dispersion properties and crystal strength of the material.

With some exceptions, a crystalline solid will exhibit pressure–volume hysteresis when cycled through a first-order solid state phase transformation. Perhaps a nearly defect-free single-crystal quasi-statically cycled through a displacive phase change comes closest to an ideal hysteresis-free phase change. Virtually all real solids exhibit hysteresis for the reasons described above. Pressure–volume hysteresis for initially monoclinic zirconium dioxide, which exhibits multiple pressure-induced phase transformations, is shown on the left in figure 6.16 (Leger *et al* 1993). These data are illustrative of transformation hysteresis on compression and decompression through the high-pressure phases. The extent of hysteresis will depend on both the material and the metallurgical makeup of the material, and can range from relatively minor to the full quenching of the high-pressure phase on decompression.

Most phase transformations are thermally activated. Thus kinetics in the thermal activation process can lead to transformation hysteresis. The silicon dioxide reconstructive tetrahedral-to-octahedral transformation in glass or crystalline quartz is representative. Theorists report (e.g. Sikka 1992) that pressures commensurate with mechanical instability (spinoidal) may be required before transformation is

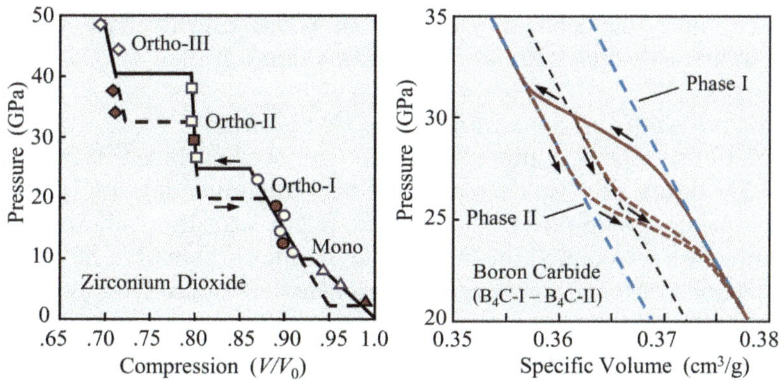

Figure 6.16. Left: Phase transformation during compression in zirconium dioxide; open symbols and solid lines are pressure–volume states on compression while solid symbols and dashed lines are the same on decompression (Leger *et al* 1993). Right: Analytic model of phase transformation compression and decompression paths in boron carbide ceramic with hysteresis and transformation pressure broadening.

achieved under very high-rate (shock) loading. Further, transformation hysteresis is brought about by the nature of pressure-induced phase transformation in a solid with microstructure. Anisotropic crystal elasticity and deformation incompatibilities with neighboring grains bring about stress concentrations and deformation plasticity through the transition process. The same phenomena are compounded when material contains disparate-compliance impurities and porosity. Both transformation kinetics and transformation incompatibility plasticity will contribute to hysteresis in the pressure-induced transformation cycle.

If, on compression, the pressure reaches a maximum pressure $p_{\max}$ then, on decompression, the following relations are used within the present model to characterize the reverse transformation,

$$\lambda_r(p) = \lambda_{\max}\frac{1 - e^{-(p/\sigma_r)^{n_r}}}{1 - e^{-(p_{\max}/\sigma_r)^{n_r}}}, \tag{6.40}$$

where

$$\lambda_{\max} = \lambda(p_{\max}) = 1 - e^{-(p_{\max}/\sigma)^n}. \tag{6.41}$$

For positive hysteresis $\sigma_r \leqslant \sigma$ is required. When $\sigma_r = \sigma$ and $n_r = n$ the transformation is reversible. When $\sigma_r = 0$, the transformation is fully quenched.

Compression and hysteretic decompression phase transformation paths based on properties for boron carbide are illustrated on the right in figure 6.16. In this example, an average transformation pressure on compression of $\sigma = 30$ GPa is based on a best estimate for the B4C-I to B4C-II transformation provided in table 6.4. A reverse transformation pressure of $\sigma_r = 25$ GPa on decompression is further assumed. Dispersion parameter values of $n = n_r = 25$ are also assumed. In one compression and decompression path a maximum pressure of 30 GPa is achieved. In the second path a maximum pressure of 35 GPa is reached. The analytic model very effectively captures the behavior observed in phase transformation within polycrystalline solids with physically acceptable functional forms.

6.5.2.3 Hydrodynamic compressibility of boron carbide

One final plot of the shock compression data for boron carbide ceramic is provided in figure 6.17. Based on the assessment of the Hugoniot data of McQueen *et al* (1970), and using the model incorporating phase transition compressibility, an analytic expression is obtained for the hydrodynamic compression of boron carbide, shown as the solid curve in figure 6.17. Arguments for hydrodynamic (no strength) behavior at the Hugoniot state are given in the earlier assessment of the shock wave data. Arguments are also made for three separate polymorphic phases over the range of the Hugoniot data (0–100 GPa). Independent compressibility of the three phases is shown by the dashed lines in the plot. The curve for B_4C-I is determined from low-pressure ultrasonic data (Manghnani *et al* 2000). Curves for B_4C-II and B_4C-III are determined from the McQueen *et al* (1970) Hugoniot data. Parameters for the analytic compression relation within the region of each phase are provided in table 6.4. Within the phase transformation domains identified in the figure, the transformation model previously described is used.

This analytic representation of the hydrodynamic compression of boron carbide requires considerably more material constants than, say, an empirical cubic relation, which could probably adequately fit the McQueen *et al* (1970) data within experimental uncertainty. The present analytic description has a stronger physical basis, however, and material constants have a more transparent origin in the underlying shock wave and static high-pressure experiments. Further, more confidence can be placed on derivatives of the analytic relation essential to wave propagation in the material.

Also included in this final plot are Hugoniot data for boron carbide ceramic from the study of Vogler *et al* (2004). There is reasonable agreement among all of the

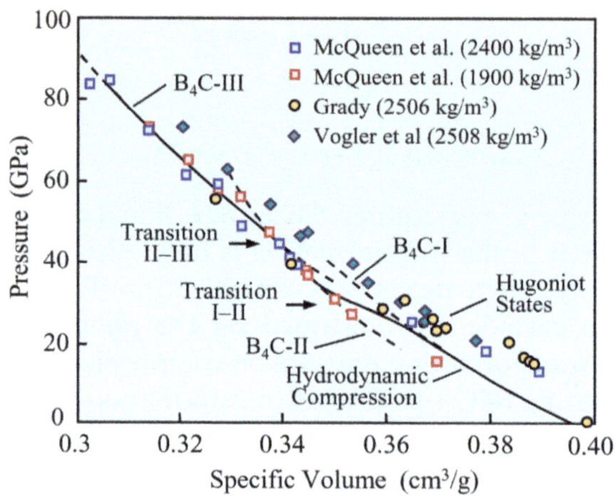

Figure 6.17. Experimental Hugoniot data and the model representation of the hydrodynamic Hugoniot compression of boron carbide.

higher density material Hugoniot states to shock pressures of about 25–30 GPa. At higher pressures the Hugoniot data of Vogler *et al* (2004) reside well above (or to the right of) the data of McQueen *et al* (1970). A tempting explanation is that strength on the Hugoniot accounts for the marked difference, and this may be. One difficulty with this explanation, however, is that the two Hugoniot points of Grady (1994a) in the neighborhood of 40 and 55 GPa, also on a high-quality near-full-density ceramic, are in agreement with the McQueen *et al* (1970) data and differ markedly from those of Vogler *et al* (2004). A fuller discussion of the unique shock properties of boron carbide and disparities among available shock Hugoniot data is provide in Grady (2015).

6.5.3 Silicon carbide ceramic with microstructure

Silicon carbide ceramics have received attention in applications where shock strength and ballistic resistance are important. Selected studies on silicon carbide ceramics have focused on such issues. Representative are the investigations of the ballistic resistance of silicon carbide reported in the study of Holmquist and Johnson (2005), and the work of Vogler *et al* (2006) which provides an extensive experimental study of the shock strength properties of silicon carbide ceramic.

Investigations of silicon carbide ceramic, in light of the various ceramic production and preparation methods, reveal wide-ranging variations in the microstructure of silicon carbide that potentially influence strength and ballistic performance. These can include sintering additives, the amount and type of secondary phases, silicon carbide polytypes, the size and shape of grain structure, the size and distribution of defects, and ceramic porosity. Accounting for these microstructure features, at least in part, in the computational models describing impact and ballistic resistance of silicon carbide ceramic is a daunting, but nonetheless necessary, undertaking confronting the engineers and scientists tasked with development of such models. The theoretical modeling described earlier in the present study is undertaken, at least in part, with this objective in mind.

Shock compression data obtained on silicon carbide ceramics that demonstrate the influence of ceramic microstructure on shock EOS behavior, and potentially ballistic performance, are investigated in this subsection. These data are also used where possible to further explore the efficacy of the p-lambda modeling methods to capture at least some of these microstructure effects. The modeling efforts pursued here contrasts with the classic mixture modeling of silicon carbide shock response presented earlier in the chapter.

First, shock Hugoniot data are examined on various initial density (and porosity) silicon carbide ceramics generated at LANL (McQueen *et al* 1970). The effects of initial density differences and porosity on shock EOS properties are assessed. Second, the Hugoniot response of reaction-sintered silicon carbide is treated as a mixture of silicon carbide and silicon within the framework of the p-lambda microstructure strength model and compared with the available Hugoniot data. Lastly, anomalies in the earlier LANL and LLNL Hugoniot data (Gust *et al* 1973) are investigated with the present microstructure strength model in light of the free silicon content in the silicon carbide used in the study.

Crystalline silicon carbide (SiC) is an interlayered tetrahedral coordinated solid with physical and chemical similarities to diamond (Sekine and Kobayashi 1997). Both the hexagonal α-SiC and the cubic β-SiC are available under standard pressure and temperature conditions. The crystal also occurs in a variety of polytypical forms (polytypes) relating to adjustments in the layer stacking sequence.

The compression properties of silicon carbide have been investigated through various methods. Static compression to approximately 100 GPa has been performed by several groups using DAC methods. Notable are the studies of Aleksandrov *et al* (1992) and Bassett *et al* (1993). Dandekar (2002) has surveyed the DAC compression data for silicon carbide. Ultrasonic longitudinal and shear wave speeds have been measured on selected high-quality silicon carbide ceramics and systematics of this ultrasonic data are reported by Bartkowski and Spletzer (2001). Shock Hugoniot data for silicon carbide ceramic are provided by McQueen *et al* (1970) and Gust *et al* (1973). The studies of Grady (1994a, 1995a, 2000a) and Vogler *et al* (2006) contribute further Hugoniot EOS data through structure shock wave measurements. More extensive dynamic compression properties including shock release, and shock–reshock and release, are provided from shock wave profile measurements (Grady 1994a, 2000a, Vogler *et al* 2006). Dandekar and Bartkowski (2001) have conducted detailed spall measurements on silicon carbide ceramics. The latter data are discussed in chapter 1. Dynamic strength through the measurement of both the longitudinal and transverse stress in impact generated shock waves has been determined by Feng *et al* (1996). Assessment of dynamic strength of silicon carbide at off-Hugoniot states is pursued in the divergent wave experiments of Paris *et al* (2011). The latter work are described in chapter 9. The data of Feng *et al* are examined in a later section.

6.5.3.1 Hugoniot properties of fully dense and distended silicon carbide

Phase transformation under shock loading is pursued through examination of the LANL high-pressure Hugoniot data of both the fully dense and porous boron carbide ceramics in the previous subsection. Here a similar examination of the near-full-density and distended silicon carbide ceramic Hugoniot data (McQueen *et al* 1970, Gust *et al* 1973) is undertaken. Where the data overlap, the LLNL data are consistent with the LANL data. The LLNL data are more appropriate for examining the effects of free silicon in the ceramic because of the more extensive coverage in the lower shock pressure range. The three silicon carbide materials tested by McQueen *et al* (1970) had average initial densities of 3.12 g cm^{-3}, 3.02 g cm^{-3} and 2.30 g cm^{-3}, respectively. The theoretical density of silicon carbide is approximately 3.22 g cm^{-3}. Consequently, all three ceramics tested had some degree of porosity along with possible lower density additives and free silicon or glass content. A chemical analysis of the ceramics was not provided. The higher density ceramic may be similar to the material studied by Gust *et al* (1973).

The McQueen *et al* (1970) Hugoniot data for silicon carbide are plotted in figure 6.18. Points on the abscissa identify the initial specific volumes of the respective ceramics. They are compared with the initial specific volume (theoretical density) from which the two continuous EOS compression curves for monolithic silicon carbide originate. These curves are, respectively, a fit to the DAC data of Bassett *et* al (1993) provided by Holmquist and Johnson (2002), and a best assessment cubic fit to more extensive data provided by Dandekar (2002).

Shock Hugoniot data for both of the higher density ceramics are observed to cross the hydrostatic EOS curves at about 40–50 GPa and reside at measurably higher density states at higher pressure. Gust *et al* (1973) have suggested a lower pressure (~24 GPa) phase transformation in silicon carbide under shock compression. This transformation has not been independently verified. An alternative explanation that sensibly clarifies the disparity between the shock data and DAC data relates to the higher compliance and phase transition in free silicon, as was explored in an earlier section and is addressed further in the present section.

Also at odds are measurable differences in the high-pressure Hugoniot curves of the two higher density silicon carbide ceramics on the right in figure 6.18. The small differences in starting density are not expected to sensibly separate the two Hugoniot curves through thermal considerations but, in any case, the trend is wrong. Comparisons of shock profile data with hydrostatic compression data (Grady 1994b, 1994c) show that a significant strength component to the Hugoniot stress is retained to at least 30 GPa. The three lower pressure points for the 3.12 g cm^{-3} material in the left-hand plot of figure 6.18 are representative of this strength offset. Differences in the two experimental Hugoniot curves at higher pressures may signify differing strengths on the Hugoniot of the two materials. A further explanation is differences in the free silicon content of the two ceramics.

Dissipation and thermal energy are expected to be extensive in the shock compression of the 2.30 g cm^{-3} starting density material, assuming that the reduced density reflects distention and microstructure porosity in the silicon carbide ceramic. Sensible agreement with the hydrodynamic compression curve at the lower range of

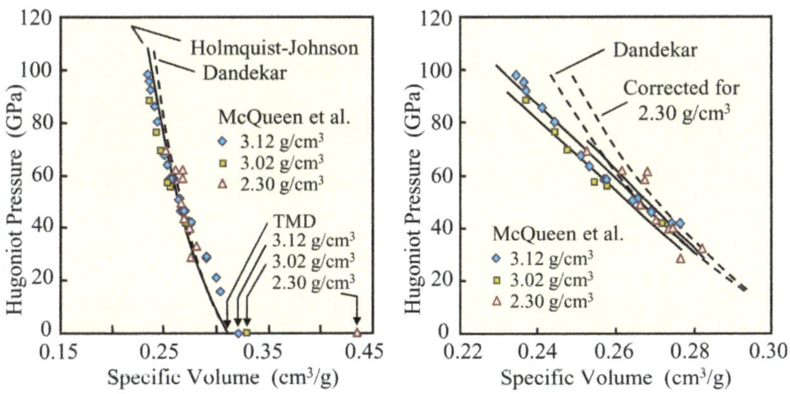

Figure 6.18. Left: Shock Hugoniot data for silicon carbide (McQueen *et al* 1970); compression curves based on DAC data are from Holmquist and Johnson (2002) and Dandekar (2002). Right: An expanded view compares the Dankekar compression curve and a thermal correction based on shock compression from an initial 2.30 g cm^{-3} density and a Gruneisen gamma of 0.5.

the data (~30–40 GPa) but flaring to lower density at the upper range of the data (~60–70 GPa) is consistent with expected thermal properties of the material.

Thermal corrections are illustrated in the right-hand plot of figure 6.18. The solid curves are sensible linear fits to the three Hugoniot data sets. The dashed curves are the Dandekar (2002) cubic representation of the compression of full-density silicon carbide (lower curve) and the thermal corrected compression Hugoniot curve for an initial distended density of 2.30 g cm^{-3} (upper curve), assuming a constant value for the Gruneisen parameter of 0.5. Larger or smaller values of gamma will correspondingly expand or contract the specific volume correction at the Hugoniot state.

6.5.3.2 Hugoniot properties of reaction-bonded silicon carbide

Silicon carbide ceramics prepared through RB technology are clear examples of materials for which EOS models incorporating microstructure features are warranted. Production processes provide for a controlled mix of silicon carbide and free silicon, in addition to microstructure porosity. Shock wave profile measurements have been conducted on one RB silicon carbide. Comparisons with the Cercom Company SiC-B series (SiC-B, SiC-N, SiC-C) silicon carbide ceramics (Grady and Moody 1996, Grady 2000a, 2002) reveal differing shock wave EOS features. Shock Hugoniot and HEL states for the various silicon carbide ceramics are shown on the left in figure 6.19.

The RB silicon carbide was also produced by Cercom Company. The material's mass density was 3.125 g cm^{-3} and was determined to be approximately 92% silicon carbide and 8% free silicon by weight without appreciable porosity. There are marked differences in the measured Hugoniot properties. The HEL of RB silicon carbide is higher (13–14 GPa) when compared with the SiC-B series (11–13 GPa). The subsequent Hugoniot states for the RB silicon carbide in the 25–40 GPa shock pressure range fall below corresponding states for the SiC-B ceramics and, when Hugoniot offset is compared with the hydrodynamic compression curve for monolithic silicon carbide, suggests significantly lower strength at the Hugoniot state. An earlier hot-pressed silicon carbide ceramic prepared by Eagle Picher

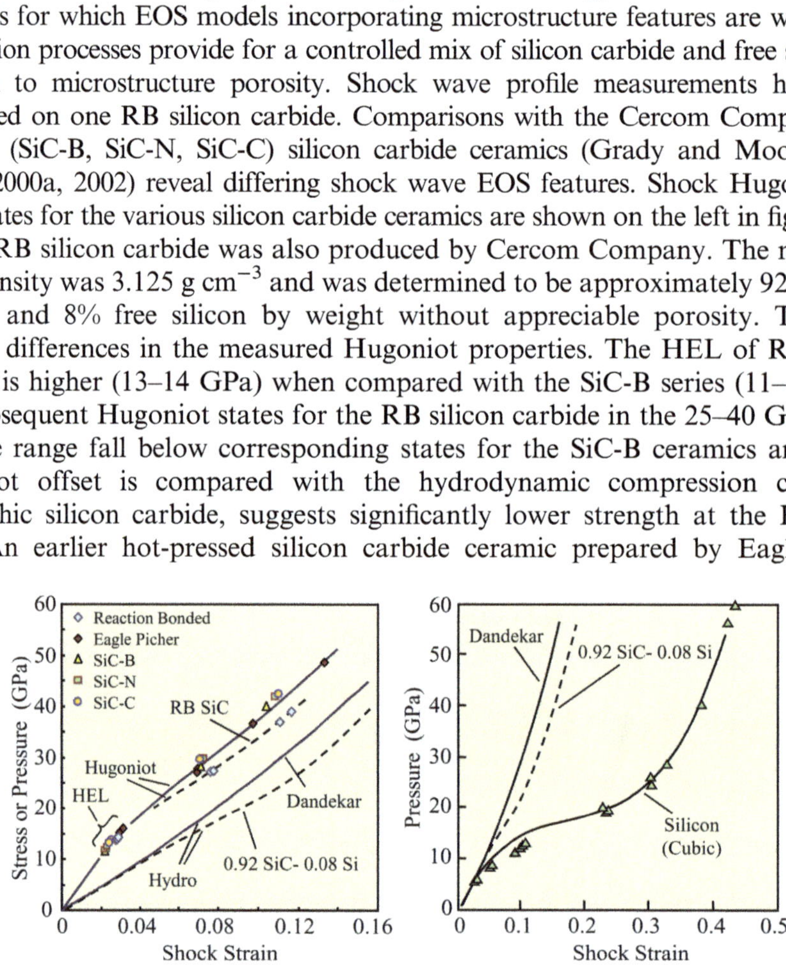

Figure 6.19. Left: Hugoniot compression of selected silicon carbide ceramics (Grady 2000a, 2002) is shown on the left; comparison is provided with calculated hydro-compression of silicon carbide and a silicon carbide and silicon mixture. Right: Hugoniot compression of silicon (Goto *et al* 1982) and a mixture representation of RB silicon carbide (Grady 2003).

Company reveals an HEL strength (15–16 GPa) in excess of all the Cercom ceramics. The variations in onset of failure among the various silicon carbide ceramics are believed to be a consequence of the different microstructures and their effect on the local stress concentrations precipitating failure.

Interferometer particle-velocity profiles from which strength and Hugoniot state properties are determined are shown in figure 6.20. HEL strength states are assessed from the first break in the slope of the precursor wave. Profiles for RB ceramic are adjusted (scaled) in time to account for modest differences in sample thicknesses. The experimental configuration for the measurements is illustrated in the later figure 6.30 where further details of the experimental method are described. Slower second wave speeds of the RB ceramic are consequence of the softer Hugoniot state of the material relative to the hot-pressed ceramic.

An EOS treatment of the Cercom RB silicon carbide ceramic as a mixture of silicon carbide and free silicon using the microstructure-based mixture model provides an improved comparison of the various shock wave Hugoniot data for the silicon carbide ceramics. A hydrodynamic compression of monolithic silicon carbide from Dandekar (2002) is shown on the right in figure 6.19. Compression of silicon is represented by a cubic fit to the Hugoniot data of Goto *et al* (1982), and is also shown in the same plot. The cubic representation of the silicon Hugoniot data is not an appropriate description of the compression in either the lower or the higher pressure range, but is adequate in the 20–60 GPa range of interest. A more appropriate representation of the silicon Hugoniot data is described in the next subsection. Hydro-compression of monolithic silicon carbide and silicon are provided by the properties and analytic relations in table 6.5.

The pressure-equilibrated compression curve for the Cercom RB silicon carbide calculated from the compaction model based on a 92% SiC and 8% Si composition by weight is displayed with the component compression curves on the right in figure 6.19. The same hydro-compression curves for monolithic silicon carbide and the RB

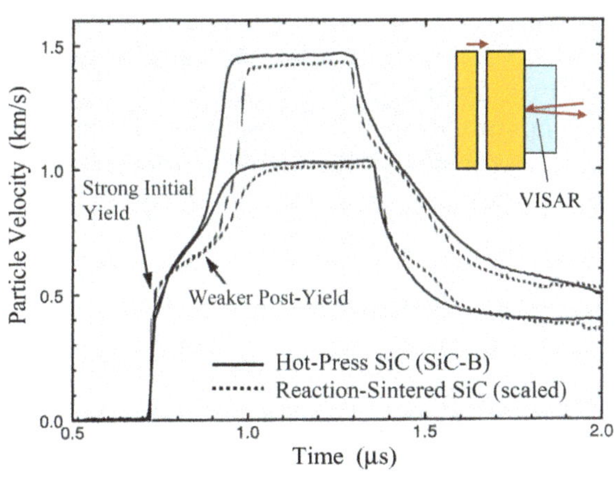

Figure 6.20. Interferometer particle-velocity profiles contrasting shock response of silicon carbide SiC-B and RB (sintered) SiC (Grady and Moody 1996). Maximum stress levels are approximately 28 GPa and 38 GPa for the two profile amplitudes.

Table 6.5. Compression properties for the component materials of silicon carbide and silicon.

Silicon carbide	Silicon
$\rho_o = 3.22$ g cm^{-3}	$\rho_o = 2.33$ g cm^{-3}
$C_o = 8.285$ km s^{-1}	$a_1 = 240$ GPa
$K = 221$ GPa	$a_2 = -1175$ GPa
$K' = 4.4$	$a_3 = 2184$ GPa
$S = 1.35$	
$p = \frac{\rho_o C_o{}^2 \varepsilon}{(1 - S\varepsilon)^2}$	$p = a_1\varepsilon + a_2\varepsilon^2 + a_3\varepsilon^3$

silicon carbide and silicon mixture are compared with the shock Hugoniot data in the left-hand plot. If the offset between the Hugoniot curve for the RB silicon carbide and the mixture hydrodynamic compression curve can be accepted as a measure of the strength on the Hugoniot, then the strength offset of RB silicon carbide is comparable to or somewhat larger than that of the SiC-B series ceramics.

Not addressed here is the corresponding iso-strain hydro-compression curve, which lies above the iso-pressure curve, and the complexities of compaction of the mixture matrix accounted for within the p-lambda microstructure model framework. These issues are pursued next.

6.5.3.3 Hugoniot response of silicon carbide and free silicon mixture

The shock compression properties of the silicon carbide Hugoniot data of Gust *et al* (1973) are considered in light of the free silicon content in this material, using the microstructure-based mixture model described earlier. This modeling method is contrasted with the classic mixture model applied to the same material in section 6.2 of this chapter. Carl Cline, a retired material scientist from LLNL, worked closely with Gust and co-workers during the performance of their seminal shock EOS work on ceramics. Discussions were held with Cline on this topic at the 2003 American Physical Society Topical Conference on Shock Compression. According to Cline (2003) there were only two suppliers of appropriate silicon carbide ceramic materials at that time—Carborundum Company and Norton Company. He worked closely with Carborundum and specifically Ken Taylor (note KT-silicon carbide) who developed the technology for preparing the KT-silicon carbide tested in the Gust *et al* (1973) study. To the best of my understanding (not being a ceramicist), it was an early version of reaction sintering that was only partially effective leaving large amounts of free silicon (10%–15%) and free carbon. He also stated that it was likely that McQueen and co-workers obtained their silicon carbide from the same source.

Based on this added understanding of the nature of the KT-silicon carbide, a further analysis of silicon carbide shock compressibility is undertaken (Grady 2003). Again the intent will be to assess the high-pressure (50–100 GPa) discrepancies between the Gust *et al* (1973) and McQueen *et al* (1970) Hugoniot data on one hand, and the Sekine and Kobayashi (1997) and more recent Vogler *et al* (2006) Hugoniot data on the other (see figure 6.1). The modeling also investigates the source of the approximately 24 GPa phase transformation reported by Gust *et al* (1973) for silicon carbide, but not observed in later shock Hugoniot measurements.

A fully dense homogeneous mixture of silicon carbide and silicon is assumed, although the actual mixture tested by Gust *et al* (1973) is probably more complex. Treatment of the individual silicon carbide and silicon components is modified, however, to better model the response over the full 0–100 GPa pressure range of interest. (The cubic expression for silicon used in the previous subsection is not appropriate.) For the monolithic silicon carbide the linear shock-velocity versus particle-velocity approximation to the cubic hydrodynamic compression relation of Dandekar (2002) will continue to be used. For compression of the silicon, independent expressions are used for the low- and the high-pressure phase, and the p-lambda model description for phase transformation described in the previous analysis of boron carbide ceramic. Experimental shock Hugoniot data for silicon (Gust and Royce 1971, Goto *et al* 1982) and the model representation of the compression are shown in figure 6.21.

Low-pressure compression of silicon is determined from the compressibility properties, $K_o = 97.9$ GPa and $K_o' = 4.2$, provided by Goto *et al* (1982). High-pressure compression is provided by a best fit to the four 40–65 GPa Hugoniot points. Compression through the silicon transition is determined through (6.39) with the transformation parameters of $\sigma = 23$ GPa and $n = 5$. Hugoniot compression properties are provided in table 6.6. The resulting compression curve for silicon through the phase transformation is shown with the data in figure 6.21. The objectives of this analysis differ from the classic mixture modeling of silicon carbide and silicon described earlier in the chapter, and issues of dynamic strength are ignored in the present calculations.

The appropriate fractions of material in the two-component silicon carbide and silicon ceramic tested by Gust *et al* (1973) is only approximately known. Through

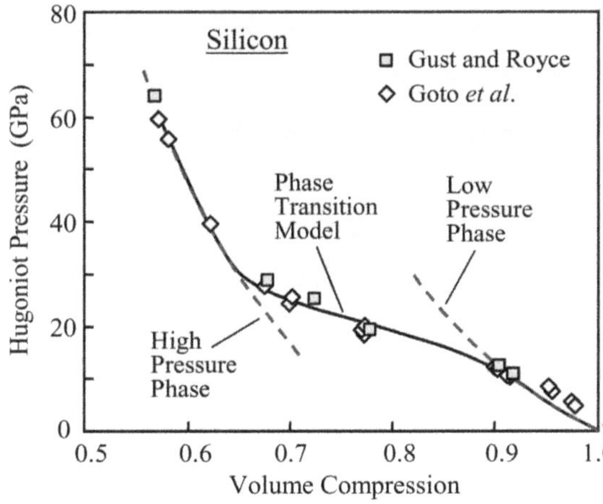

Figure 6.21. Pressure versus volume Hugoniot compression for silicon based on the p-lambda phase transformation model (Grady 2003). Silicon Hugoniot data are from Gust and Royce (1971) and Goto *et al* (1982).

Table 6.6. Compression properties for component materials of silicon carbide and silicon mixture.

	Silicon	
Silicon carbide	Low density	High density
$\rho_o = 3.22$ g cm^{-3}	$\rho_o = 2.33$ g cm^{-3}	$\rho_o = 2.95$ g cm^{-3}
$C_o = 8.28$ km s^{-1}	$C_o = 6.49$ km s^{-1}	$C_o = 6.19$ km s^{-1}
$K = 221$ GPa	$K = 97.9$ GPa	$K = 113$ GPa
$K' = 4.4$	$K' = 4.2$	$K' = 3$
$S = 1.35$	$S = 1.30$	$S = 1.00$

optimization of the model to the available data, a mixture of 86% silicon carbide and 14% silicon is determined, not inconsistent with the recollections of Cline (2003). This mixture provides an initial density of 3.056 g cm^{-3}, slightly lower than the 3.06–3.11 g cm^{-3} density range reported by Gust *et al*, but only by about 1%. The initial iso-compression modulus yields a bulk wave speed of 8.06 km s^{-1}, reasonably consistent with the ultrasonic bulk speed of 7.71 km s^{-1} measured by Gust *et al* (1973).

Pressure versus specific volume provided by the mixture model is shown in figure 6.22. Reference curves are provided by the compression relation for monolithic silicon carbide of Dandekar (2002) and the shock Hugoniot data of Gust *et al* (1973). Dashed curves provided the model iso-compression and iso-pressure compressibility of the mixture.

Compaction parameters of σ = 32 GPa and n = 6 provide the mixture compression curve that is compared with the Hugoniot data in figure 6.22. Agreement with the data through the phase transformation and up to about 60 GPa is reasonably good. The model compression fades to the right of the data at increasingly higher pressures. There are, however, no Hugoniot data for silicon above 65 GPa and silicon compression at higher pressures is based on extrapolation

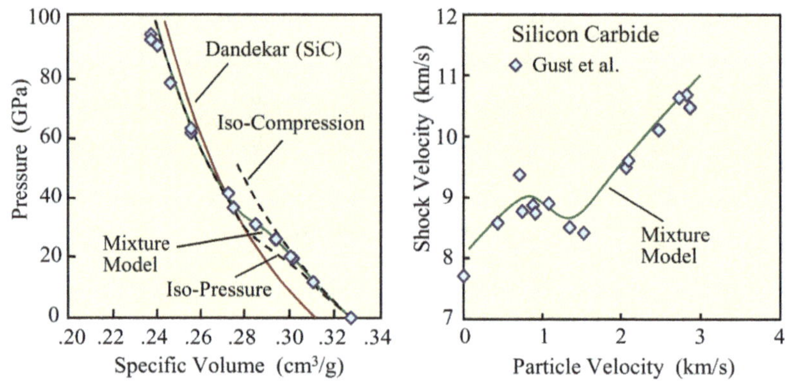

Figure 6.22. Pressure versus volume and shock-velocity versus particle-velocity Hugoniot compression for a mixture of 86% silicon carbide and 14% silicon by mass based on the p-lambda mixture model. Hugoniot data for KT-silicon carbide from Gust *et al* (1973).

of a fit to four Hugoniot points in the 40–65 GPa range. The comparison, nonetheless, offers a sensible explanation for the phase transformation and soft high-pressure Hugoniot anomalies of the Gust *et al* (1973) Hugoniot data for KT-silicon carbide as well as for the silicon carbide data of McQueen *et al* (1970).

The plot on the right of figure 6.22 provides a comparison of the mixture model compression with the shock Hugoniot data in the more sensitive shock-velocity versus particle-velocity plane. Again, the model agreement with the data is good. Better agreement at the higher pressure states is a consequence of this plot not reflecting the approximately 1% discrepancy in the mixture model initial density. The higher phase transformation pressure in the model compression, and sensibly also in the data, is a consequence of the differing compliances of the two components, with that of silicon being substantially higher. Thus, higher average pressures are needed to achieve local pressures in the silicon sufficient to initiate phase transformation. This feature of the analysis is a clear example of the effect of microstructure strength in the current model that is not possible with the classic mixture model. The issue is pursued further in chapter 7.

Lastly, model compression curves for full-density silicon carbide and for silicon mass fractions of 10%, 15% and 20% are shown in figure 6.23. This parametric study illustrates sensitivity to component variations and the potential of the present model. The analysis also provides support for the explanation that a free silicon component in the Gust *et al* (1973) silicon carbide material is responsible for both the reported discrepancies in Hugoniot compressibility and the phase transformation, but not observed in later studies. Namely, phase transformation under shock compression is in the silicon component, not the silicon carbide.

6.5.3.4 Metal–ceramic matrix composite

The shock wave properties of silicon carbide ceramic can be contrasted with a metal–ceramic mixture of predominantly silicon carbide compound. The material is identified as a Lanxide ceramic matrix composite (DIMOX 109). The mixture composition by weight is 66% silicon carbide, 28% aluminum oxide and 6%

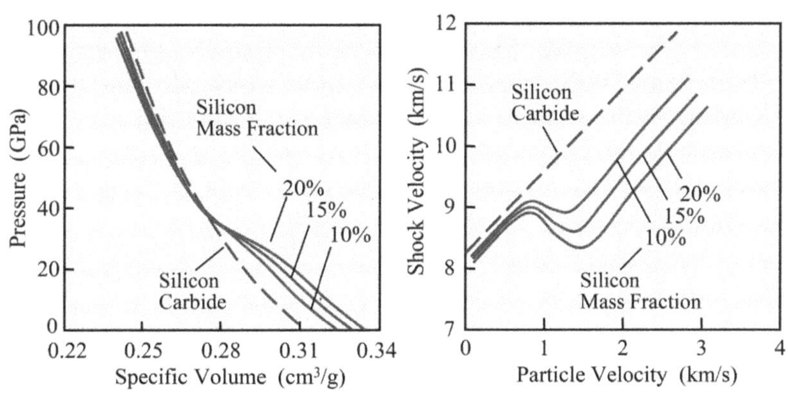

Figure 6.23. Hugoniot compression properties of silicon carbide and silicon mixtures for different levels of silicon content based on the p-lambda compaction model.

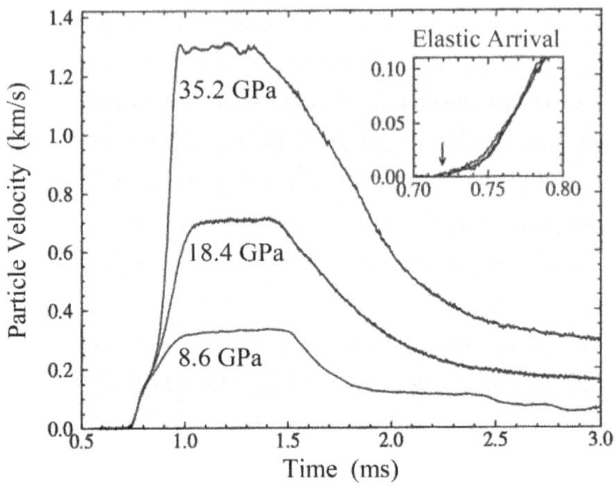

Figure 6.24. Interferometer particle-velocity profiles of DIMOX 109 metal–ceramic matrix composite for Hugoniot stress states of 8.6, 18.4 and 35.2 GPa (Grady and Crawford 1994).

aluminum metal. Shock wave profiles measured by velocity interferometry (see figure 6.30) on the material are shown in figure 6.24. Symmetric sample planar impact velocities of 0.51, 1.16 and 2.20 km s^{-1} achieved corresponding Hugoniot stress states of 8.6, 18.4 and 35.2 GPa. Matrix microstructure is observed to markedly disperse the elastic precursor wave. The break in the slope at approximately 0.13 km s^{-1} identifies an HEL strength of the material of approximately 3 GPa. The modest metal component notably reduces dynamic strength below the expected strengths of the component ceramics; approximately 15 GPa and 9 GPa for silicon carbide and aluminum oxide, respectively. The Hugoniot EOS response of the composite is reproduced with classic mixture theory discussed in section 6.2. A low-amplitude precursor arrival, shown in the inset in the figure, was measured to propagate at the longitudinal sound speed of silicon carbide.

6.5.4 Tungsten carbide ceramic and composition effects

Tungsten carbide ceramics are commonly produced with metal additives through liquid-phase sintering methods. Metal additives are typically in the range of 5%–10% by weight. Such composition complications can be expected to influence the shock wave EOS and strength properties of tungsten carbide ceramics.

In this subsection the effects of composition variations on the shock EOS and strength of tungsten carbide ceramic are explored through examination of available shock wave data. Modeling methods are also described to account for composition variations to the ceramic composite EOS through the present microstructure mixture model.

6.5.4.1 Tungsten carbide materials

Shock wave EOS properties are examined for several tungsten carbide ceramics. Again, the extensive source of shock Hugoniot data of McQueen *et al* (1970) is exploited. The material tested in their study was a near-full-density tungsten carbide

Table 6.7. Density and elastic properties of tungsten carbide ceramic. Reprinted with permission from Grady (1999), with permission from Elsevier.

Material	Density(g cm^{-3})	c_l(km s^{-1})	c_s(km s^{-1})	c_o(km s^{-1})	Bulk modulus(GPa)
LANL[1]	15.01	6.89	4.18	4.92	363.0
KM[2]	14.93	6.90	4.17	4.94	364.5
AP[3]	14.91	6.92	4.15	4.99	371.4
CC[4]	15.56	7.04	4.30	4.96	383.0

[1] Tungsten carbide (5wt% Co), LANL (McQueen *et al* 1970).
[2] Refers to K68 material obtained from Kennametal Company.
[3] Material extracted from a 14.5 mm (BS-41) AP round.
[4] High-purity tungsten carbide ceramic from Cercom Company.

ceramic with 5wt% Co. Shock data for two other pressureless liquid-phase-sintered tungsten carbide ceramics are also included in the comparison. One is produced by Kennametal Company, and contains 5.7wt% Co, 1.9wt% Ta, along with less than 0.3wt% Nb and Ti. Several Hugoniot measurements along with shock strength properties were determined for this material. The second tungsten carbide ceramic was extracted from armor piercing (AP) rounds. This tungsten carbide ceramic contained 3–4wt% Ni, 0.4–0.8wt% Fe and 0.05–0.2wt% Co. Only shock strength properties were determined for this latter material. Selected data analyses on the latter two materials have been previously reported (Grady 1995a, 1999, 2000b).

Lastly, Hugoniot and strength data are provided for a hot-pressed high-purity tungsten carbide produced by Cercom Company. The material is reported to be 97.2wt% WC and 2.8wt% W_2C, the latter resulting from the densification process. Theoretical densities for WC and W_2C are 15.7 g cm^{-3} and 17.2 g cm^{-3}, respectively. The material is also reported to have about 1% porosity. Shock strength and EOS analyses of the shock wave data on this ceramic have previously been reported by Dandekar and Grady (2002). Density and elastic properties for the tungsten carbide ceramics investigated are provided in table 6.7 (Grady 1999, 2009).

6.5.4.2 Hugoniot properties of tungsten carbide ceramics

Plots of experimental Hugoniot data for tungsten carbide ceramics are shown in figure 6.25. Shock-velocity versus particle-velocity states are provided on the left while the Hugoniot pressure versus particle-velocity representation is shown on the right. As is observed, neither shock velocity nor shock impedance is markedly affected by the variations in composition among the three ceramics. The high-purity Cercom ceramic is observed to be modestly stiffer in both plots, followed, respectively, by the Kennametal material and the ceramic tested by McQueen *et al* (1970).

Elastic wave speeds are compared with the shock-velocity data in the left-hand plot. HEL amplitudes for the Cercom ceramic (~6 GPa) and the Kennametal–ceramic (~4 GPa) are identified in both plots. Curves in either plot represent a sensible average of all of the experimental Hugoniot data. A further display of the same Hugoniot data is provided in figure 6.26 in both a Hugoniot pressure versus

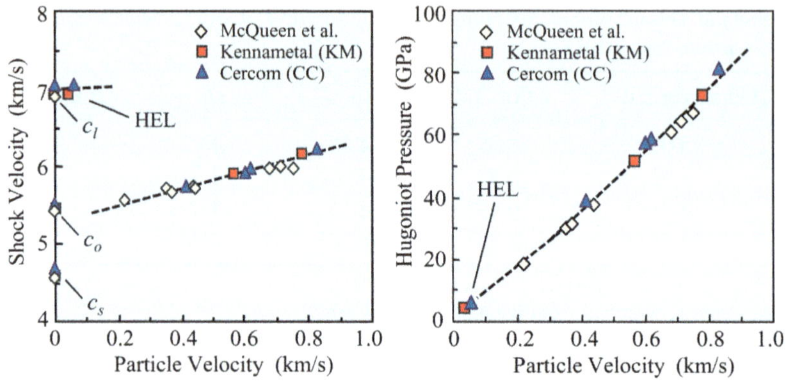

Figure 6.25. Experimental shock-velocity versus particle-velocity (left) and pressure versus particle-velocity (right) Hugoniot data for tungsten carbide ceramics (McQueen *et al* 1970, Grady 1995a, 2000b).

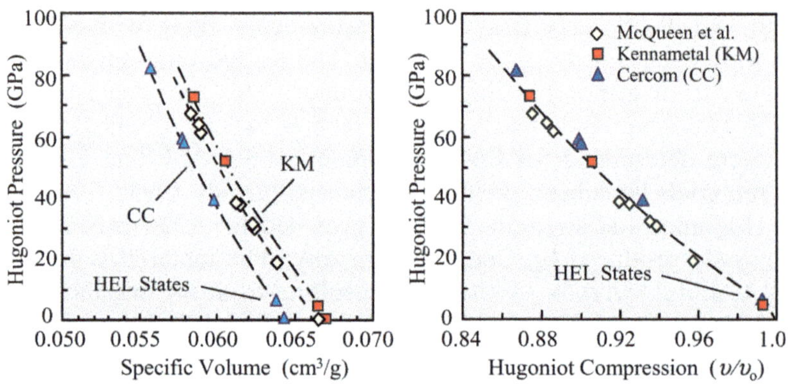

Figure 6.26. Experimental pressure versus specific volume (left) and pressure versus compression Hugoniot data (right) for tungsten carbide ceramics (McQueen *et al* 1970, Grady 1995a, 2000b).

specific volume and a Hugoniot pressure versus compression representation. Again, the Hugoniot compression representation reveals modest but increasing steepness of the Hugoniot curves for the McQueen *et al*, Kennametal and Cercom materials, respectively.

6.5.4.3 Hydrodynamic response

Through selective use of the very high-pressure Hugoniot data of McQueen *et al* (1970), a sensible hydrodynamic compression curve for monolithic tungsten carbide has been constructed (Dandekar and Grady 2001). A compression relation of the linear shock-velocity versus particle-velocity form

$$p = \frac{\rho_o C_o^2 (1 - \rho_o v)}{\left(1 - S(1 - \rho_o v)\right)^2}, \qquad (6.42)$$

with $\rho_o = 15.7$ g cm^{-3}, $C_o = 4.93$ km s^{-1} and $S = 1.31$, is determined from analysis of the shock data. It is recognized that such an analysis is subject to uncertainties due to unknown initial sample porosities and additives, along with possible subtle phase changes on the Hugoniot not recognized in the experimental data. Uncertainties due to these unknowns are difficult to bound.

In chapter 10, further assessment of the dynamic EOS of tungsten carbide is undertaken through a method of shock Hugoniot measurement on a mixture of the ceramic of interest with a low-strength material such as a ductile metal. The method assumes that deviator stresses in the assessed Hugoniot state are small and that compression of the ceramic approaches isentropic loading because of the several pressure steps incurred by the granular nature of the mixture. This test method yields an independent determination of the dynamic isentropic compression of tungsten carbide that is in good agreement of that assessed by Dandekar and Grady (2001) described by equation (6.42) through parameters provided in the adjoining text.

A description of the shock EOS response of any of the tungsten carbide ceramics can be generated through the monolithic hydrodynamic response of equation (6.42) and the methods for modeling compositional mixtures described in earlier sections. The Cercom ceramic, accounting for the small admixture of W_2C, has a theoretical density of ρ_o = 15.74 g cm^{-3}. The measured density of ρ_{oo} = 15.56 g cm^{-3} yields a calculated porosity of 1.1%; in good agreement with the independently reported porosity of about 1%. Although the corrections are modest in accounting for this small porosity difference, the p-lambda model methods can be applied to account for the EOS differences.

Compression curves for both the partial-volume averaged iso-pressure, and the partial-pressure averaged iso-compression EOS response are shown with the Hugoniot data for the Cercom ceramic in figure 6.27. The difference between the two curves is a consequence of the initial porosity. EOS compression is predicted to transition from the iso-compression to the iso-pressure curve as compaction of the

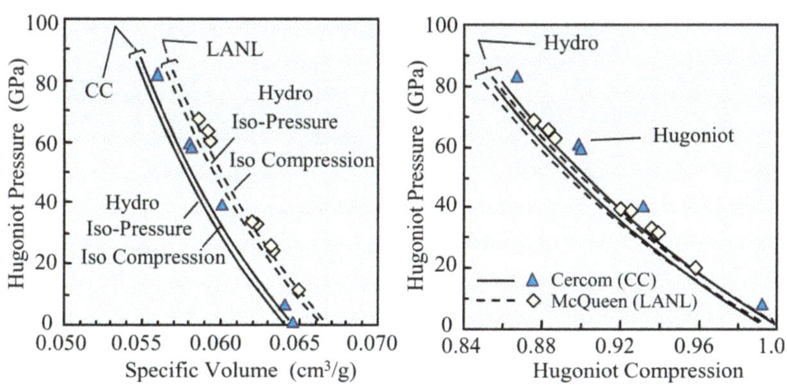

Figure 6.27. Left: Hugoniot compression of tungsten carbide ceramics and model predictions of the hydrodynamic EOS compression accounting for compositional differences (Grady 2009). Right: The curves plotted as a function of compression correspond to the same curves on the left. Differences between iso-compression and iso-pressure illustrate the model analysis accounting for small levels of porosity.

porosity takes place under shock compression. The compaction process probably requires several tens of GPa to achieve completion. Shock compaction experiments on tungsten carbide powder reveal that the material still retains approximately 8% porosity following shock compression to stresses in excess of 4 GPa (Vogler *et al* 2007). The p-lambda mixture model and porosity are addressed further in chapter 7.

Similar iso-pressure and iso-compression EOS curves are shown for the McQueen *et al* (1970) material assuming the reported composition of WC-5wt%Co. Calculated and measured densities provide a 0.7% discrepancy. The discrepancy is considered here to be due to initial porosity. Again, the principal difference between the two curves is due to the porosity.

For both materials the Hugoniot data resides above the model representation of the EOS hydrodynamic compression. Accepting the relation developed for the hydrodynamic response of the monolithic tungsten carbide, and appropriate application of the composite model, the discrepancy is attributed to strength (deviator stress) on the Hugoniot. The same conclusion was reached previously for the Cercom ceramic (Dandekar and Grady 2001). The strength offset for the Cercom material is of order 10 GPa. The offset for the McQueen material is somewhat less. The offset of the Cercom material is sensibly consistent with measured initial HEL amplitude and post-HEL hardening behavior determined form shock profile experiments (Dandekar and Grady 2001). See also the analysis in chapter 10 of shock data on a tungsten carbide and copper mixture where the issue of strength at Hugoniot states is clearly revealed.

6.5.4.4 Dynamic strength and composition

Shock wave strengths are influenced by compositional variations in tungsten carbide ceramic. Both the dynamic compressive strength (HEL) and dynamic tensile strength (spall stress) have been measured for several of the tungsten carbide ceramics reported in the previous subsection. Spall stress as a function of the shock wave pre-compression stress is plotted on the left in figure 6.28. Features of spall waves in tungsten carbide are considered further in chapter 1.

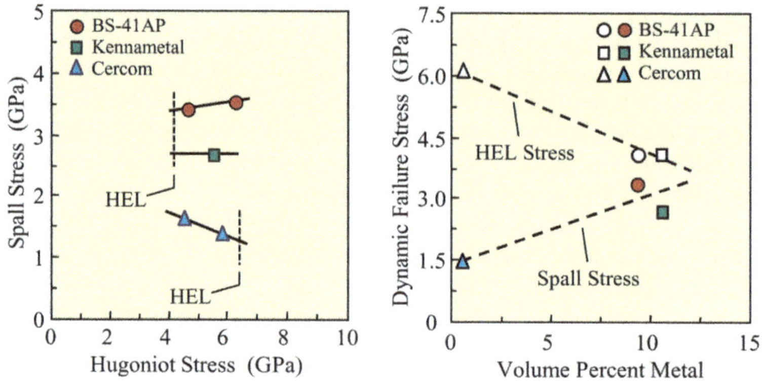

Figure 6.28. Tensile spall strengths (left) and compressive HEL strengths (right) for tungsten carbide ceramics showing dependence on shock pre-compression stress and metal–ceramic content.

Both spall stress and HEL are plotted on the right as a function of the metal content of the ceramic. Metal content increases the dynamic tensile strength and decreases the compressive shock strength, as observed in contrasting the two liquid-phase sintered ceramics with the hot-pressed Cercom tungsten carbide. Further, the strength dependence on both the specific metal content and ceramic preparation method is indicated by the marked difference in spall strength between the two liquid-phase sintered materials. Also significant is that spall stress measurements on the metal-phase tungsten carbide ceramics in the left-hand plot achieve shock pre-compression stress levels in excess of the HEL of the material. Thus, little or no fracture damage is indicated from deformation during shock compression. In contrast, reduction of spall strength of the Cercom material as pre-compression stress approaches the HEL may reflect some weakening of the ceramic due to compressive fracture damage. This trend has also been noted in the spall strength of selected silicon carbide ceramics (Dandekar and Bartkowski 2001).

6.6 Shock strength in silicon carbide and titanium diboride

The thrust of the analysis of shock wave data, and the theoretical modeling in the present chapter, has focused on the densification of ceramics (or any high-strength solid for that matter) when subjected to shock compression. The strength and the deformation of solids under the dynamic shear loading induced by the shock wave have largely not been addressed. Shear strength has only appeared in its role in resisting densification on the local scale in materials with disparate-compliance microstructure.

With some thought it is apparent that the same microstructure compliance issues can also contribute to the physics of the onset and evolution of shear deformation in brittle solids under shock compression. Although the issue of the dynamic strength of solids is not addressed here in detail, some relevant shock strength data are explored. The data examined are those of time-resolved elastic precursor profiles measured on several ceramics. Interrogation of the elastic precursor wave profile has become the gold standard in determining strength under shock compression—namely the HEL.

Two ceramics, silicon carbide and titanium diboride, exhibit complexities in the elastic precursor structure that make identification of dynamic strength from the elastic precursor problematic. In early investigation of these precursor wave features, an identification of several HEL amplitudes for the same material was proposed (Grady 1992a). Although the rationale for this added characterization of the shock strength can certainly be questioned, the notion of several HELs for a material is continued here. The reasons are partly for historic continuity, partly to aid the discussion of shock strength, and partly because it can be argued that the underlying shock physics is sensibly consistent with this interpretation. This section will start by presenting and discussing elastic precursor wave profile measurements on silicon carbide ceramics. Second, comparable wave profile data and interpretation are presented for titanium diboride ceramics. Lastly, this section explores a plausible discussion of how these shock strength features may relate to the microstructure characteristics of the material, a dominant theme of the present chapter.

6.6.1 Silicon carbide ceramics

Shock wave profile measurements with emphasis on the shock precursor structure have been obtained for several silicon carbide ceramics (Grady and Moody 1996, Grady 2000a). Measurements use the time-resolved velocity interferometer, or VISAR, method (Barker and Hollenbach 1972). Later investigations of the dynamic strength of silicon carbide through shock wave methods have been performed. These studies, which include the work of Vogler *et al* (2006) and Paris *et al* (2011), emphasized other methods in assessing dynamic strength. That of Vogler *et al* use shock-release and shock–reshock experiments to determine strength at the shock state. Experiments of Paris *et al* rely on divergent wave methods to assess dynamic failure states at reduced levels of confining pressure.

Features of the compressive precursor wave reveal complexity in the shock strength of the several silicon carbide ceramics investigated. The first is a hot-press silicon carbide ceramic provided by Eagle Picher Industries with a relatively broad grain size distribution (~2–16 μm) with a mean of about 8 μm. The density of the Eagle Picher material is 3.177 g cm^{-3}. The ultrasonic bulk modulus and Poisson's ratio are 213.1 GPa and 0.161, respectively. Shock properties of the Eagle Picher silicon carbide ceramic have been reported previously (Grady 1992a, 1993, 1994a).

The second silicon carbide ceramic was provided by Cercom Corporation and is actually a series of several ceramics. The baseline material is identified as Cercom SiC-B ceramic. This commercial grade hot-pressed ceramic uses aluminum nitride as a sintering aid (Shih 1998). The material has unique microstructure characteristics with impurities segregating in small, well-dispersed pockets along grain boundaries. These grain boundary features include much of the metallic impurities, which melt under sintering temperatures and are dispersed into the distributed grain boundary features through processes which are promoted by the sintering aids. The nominal grain size is 4 μm (Shih 1998). The second hot-pressed product is Cercom SiC–N ceramic. It is a refinement of SiC–B with improvements in powder homogenization and the addition of an organic binder. Carbon in the polymer binder removes oxygen and reduces the glassy SiO_2 grain boundary content. The SiC–N ceramic also has a nominal grain size of 4 μm (Shih 1998). The third hot-pressed ceramic product is Cercom SiC–C. Production of this ceramic generally follows the procedures for SiC–N, except that a 1 μm powder size is used. The SiC–B ceramic has a density of 3.15 g mm^{-3}, and an ultrasonic bulk modulus and Poisson's ratio of 226.5 GPa and 0.182, respectively. The density and elastic properties of the other Cercom ceramics are comparable.

Wave profiles for the Cercom SiC-B series of ceramics have also been reported previously (Grady and Moody 1996, Grady 2000a). Shock wave profiles for the Eagle Picher material are shown along with comparable profiles for the Cercom SiC-N ceramics in figure 6.29. The thickness of the samples is very close to 9 mm in each of the four tests. Velocity is monitored at the interface between a lithium fluoride

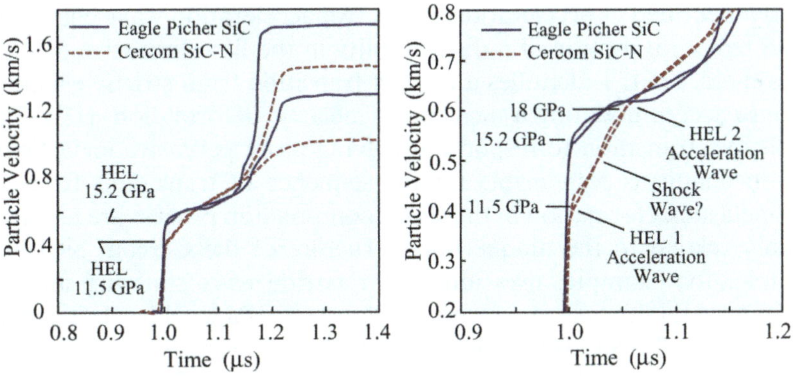

Figure 6.29. Shock compression wave profiles for Eagle Picher and Cercom SiC-N ceramics with emphasis on precursor wave structure (Grady and Moody 1996, Grady 2000a, Grady 2008).

window and the ceramic sample. The experimented geometry of the impact tests is illustrated in figure 6.30. The full compression shock wave profiles are shown on the left in figure 6.29 and details of the precursor wave are displayed on the right.

Previously reported HELs for both the Eagle Picher and the SiC-N ceramics are provided in the left-hand plot of the wave profiles. As such, the reported values would imply a markedly higher shock strength for the Eagle Picher ceramic than for the SiC-N ceramic. Such a straightforward interpretation of the data, however, ignores the strikingly different post-HEL complexities of the precursor waves of the two ceramics.

In a more detailed assessment of the precursor wave measured for the Cercom SiC-N ceramic, two HELs have been used to characterize the data—the first at about 11.5 GPa, and the second at about 18 GPa, as identified on the right in figure 6.29. At each of the two amplitudes, a distinct break in the slope of the velocity profile is observed. The wave features associated with HEL 1 and HEL 2 in the plot

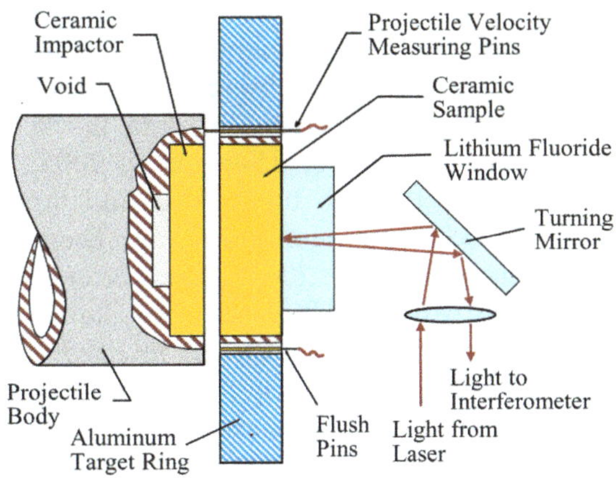

Figure 6.30. Experimental configuration for VISAR shock compression wave profile measurements on ceramic.

are properly identified as acceleration waves. An acceleration wave characterizes the spatial and temporal motion of a discontinuity in the first derivative of the velocity (and stress) field. HEL 1 identifies a distinct transition from strictly elastic response to a response accommodating a measure of inelastic deformation. HEL 2 identifies another distinct transition to response involving more pervasive inelastic deformation. The mechanisms responsible for the sequence of transitions from elastic to pervasive inelastic deformation in the shock compression process are not known, but are certainly related to the unique microstructure of the Cercom SiC-N ceramic. This ceramic, for example, uses aluminum nitride as a sintering aid. An HEL strength for an AlN ceramic at a somewhat lower 8 GPa level has been measured. It is noted that both the SiC-B and the SiC-C Cercom ceramics show shock structure qualitatively the same as that of the SiC-N ceramic (Grady 2000a). Also, with some visual imagination it is possible to discern a similar structure in the Eagle Picher ceramic—albeit considerably more subtle.

A portion of the precursor structure displayed in figure 6.29 for SiC-N ceramic is identified as a shock wave with a question mark. A shock wave is recognized as a discontinuity in the principal flow field, but the definition also encompasses a structure shock wave if the wave structure propagates unchanged in form—that is, a steady wave. The velocity field between the two acceleration waves would qualify as a structural shock wave if this region of the precursor profile propagated unchanged in form with increasing distance into the silicon carbide sample. This would require, for example, that the time duration between the two acceleration waves remain constant with propagation distance or the structure between the two acceleration waves evolve to a steady form as the waves separate. Whether or not this region is a structural shock wave has implications as to the mechanisms of inelastic deformation in the processes of failure in the shock compression event. The simple test, of course, in assessing the steady wave character of the region in question would be to measure wave profiles at several sample thicknesses. This was unfortunately not done.

6.6.2 Titanium diboride ceramics

A material with shock strength features comparable to silicon carbide is the ceramic titanium diboride. The strength from a high-pressure shock wave perspective has been examined for titanium diboride. Hugoniot data in excess of 100 GPa are available (Gust *et al* 1973, Marsh 1980, Zhang *et al* 2005). Static compression properties of titanium diboride through both DAC and large-volume cell x-ray diffraction studies, and ultrasonic measurements, have been determined to about 60 GPa (Amulele and Manghnani 2005, 2006). Shock strength is assessed through considerations of the offset between Hugoniot and static compression curves. Through this method titanium diboride ceramic is inferred to have an impressive shock wave strength, exhibiting approximately a 15 GPa offset at 60 GPa Hugoniot stress (Dandekar and Benfanti 1993). Bourne *et al* (2000) have inferred the propagation of a delayed failure wave in titanium diboride from longitudinal and transverse stress gauge measurements over a limited shock pressure range.

There are, however, unsettling features of these data that have not been fully reconciled. Shock Hugoniot data provide an anomalously high bulk sound speed (8.5–8.9 km s^{-1}) in comparison with the ultrasonic bulk velocity (6.9–7.3 km s^{-1}). Correspondingly, the slopes of the linear shock-velocity versus particle-velocity curve are low (0.72–0.85) when compared to the pressure derivative of the bulk modulus K_o' determined from ultrasonic or x-ray diffraction studies. There are reasons for concern with the available shock Hugoniot data. Titanium diboride ceramic has an unusually low Poisson's ratio and the elastic precursor wave is not overdriven in the Hugoniot measurements until stresses well in excess of 100 GPa are achieved. Over this range interface Hugoniot measurement methods on high-strength materials can be complicated by precursor wave interference (e.g. Grady and Kipp 1998, Alexander *et al* 2006).

This section, however, does not address shock strength inferred from high-pressure static and Hugoniot data. Rather, it is focused on shock strength features assessed from the shock wave precursor structure. Shock wave profile measurements with time-resolved VISAR techniques on selected titanium diboride ceramics, also using the configuration illustrated in figure 6.30, have revealed unique features in the profile structure relating to the shock strength of the material (Grady 1992a, 1992b, Grady and Moody 1996). Measurements have been made on titanium diboride ceramics provided by the companies Eagle Picher, Cercom and Ceradyne, as well as material provided by the Ernst-Mach Institute in Freiberg, Germany (Winkler and Stilp 1992). The materials are largely reasonably high-purity, near-full-density ceramics, but with sufficient microstructure differences to lend insight into the properties governing the shock strength.

Wave profiles representative of the compressive shock wave structure are shown in figure 6.31 for the Cercom titanium diboride (TiB_2) ceramic. For the two experiments, samples of TiB_2 are subjected to planar impacted by samples of TiB_2 at 1.5 km s^{-1}, as illustrated in figure 6.30. Test sample thicknesses were 5.0 mm and 10.1 mm, respectively. The arrival times of the waves in the left-hand plot

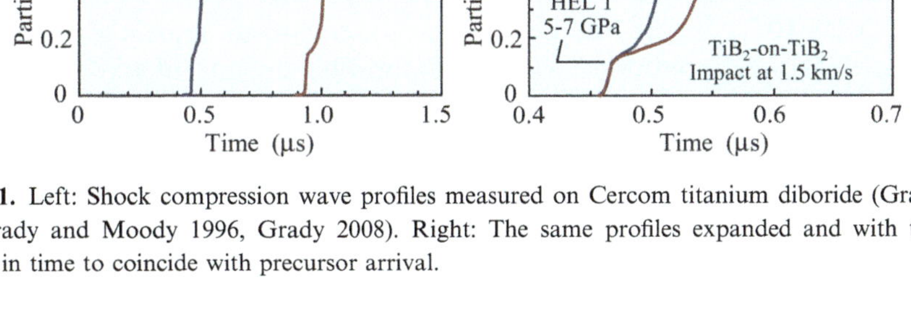

Figure 6.31. Left: Shock compression wave profiles measured on Cercom titanium diboride (Grady 1992a, 1992b, Grady and Moody 1996, Grady 2008). Right: The same profiles expanded and with the second translated in time to coincide with precursor arrival.

correspond to a longitudinal ultrasonic wave speed of 10.79 km s^{-1}. The smaller sample thickness, for which the subsequent decompression wave is also observed in the figure, was impacted by a TiB_2 sample 2.5 mm in thickness. The maximum shock pressure achieved in the two experiments is approximately 30 GPa. The density of the Cercom ceramic was 4.509 g cm^{-3}. The theoretical density of TiB_2 is 4.51 g cm^{-3}.

The Cercom titanium diboride ceramic is unusual in that the uniaxial pressure-assisted preparation of the ceramic led to granular texturing and elastic anisotropy of the material (Manghnani *et al* 1993). Longitudinal ultrasonic velocities along, and orthogonal to, the axis of compression were 10.79 and 11.83 km s^{-1}, respectively. Shock wave profiles measured along and orthogonal to this axis also revealed some dependence of the shock strength on this texture anisotropy.

The same compression profiles are shown on the right-hand side of figure 6.31 expanded in time, and with the second profile translated to the precursor starting time of the first. This display of the profiles shows more clearly the evolution of the wave structure with propagation distance. The shock compression profiles in figure 6.31 show the two-step precursor reported for titanium diboride in earlier work (Grady 1992a, 1992b). Association of this two-wave precursor structure with an upper and lower yield, or a first and second HEL, was suggested in the study. Softening in the first wave (HEL 1) occurs at about 5–7 GPa. Subsequent further softening in the second wave (HEL 2) occurs in the range of 15–17 GPa. Subsequent measurements on TiB_2 ceramics from different suppliers have shown that the stress levels of the two HELs are dependent on the material.

Observations of the precursor structure with propagation distance indicate that the waves for the Cercom ceramic in figure 6.31 leading to both HEL 1 and HEL 2 qualify as structural steady shock waves. The first wave is assigned the ultrasonic longitudinal velocity of 10.79 km s^{-1}. The shock velocity of the second wave is in the range of 9.9–10.0 km s^{-1} while that of the second shock wave to the final Hugoniot state is 9.0–9.2 km s^{-1}. The precursor wave structures observed in the titanium diboride ceramics from other suppliers are qualitatively the same, although variations in amplitudes and velocities are observed.

6.6.3 Microstructure features of the shock strength

There are some unifying similarities in the multiple (two) yield processes exhibited by the compression shock precursor structure of silicon carbide and titanium diboride ceramics, and perhaps other ceramics as well. There is onset of inelastic deformation at a lower yield (HEL 1) that softens the longitudinal compressibility of the material. This lower yield accommodates a limited amount of inelastic shear strain (of order 10^{-3}) before the deformation mechanism saturates. The material stiffens (hardens) and continued loading is through an intermediate compression shock wave supported by increasing elastic shear strain until a second, upper yield (HEL 2) is achieved. Inelastic deformation subsequent to this upper yield is markedly more pervasive, accommodating inelastic strains one to two orders of magnitude higher, dependent on the amplitude of the final shock state.

This sequence of elastic and inelastic deformations in the shock compression process is a sensible interpretation of the response of titanium diboride where a clear structured shock wave between the lower yield and the upper yield can be inferred from the data. The same interpretation for silicon carbide is less certain as the character of the wave (shock wave versus dispersive wave) between the lower and upper yield is not constrained by the data. Still, if the wave in this region is dispersive, only a modest variation on the same interpretation is required. Inelastic deformation subsequent to the lower yield, rather than saturating, entails a sensibly strong rate of pressure (or deformation) hardening that is superseded by a more pervasive deformation mechanism at the upper yield. See also the interpretation of precursor wave structure in U6Nb metal in chapter 11.

The present interpretation of shock strength and yield in certain ceramic materials is provided as a qualitative picture by the sketch on the right in figure 6.32. The picture shows the shear (deviator) stress and strain through the shock compression process. This description assumes no anomalies in the volumetric compression through the shock. More is said on this issue later. Elastic compression up to the lower yield is then supplemented by inelastic deformation identified by either of the dashed curves. The lower curve illustrates a saturation and arrest of the inelastic deformation, providing a region of positive curvature that will sustain a steady shock wave. The upper dashed curve, in contrast, exemplifies a hardening of the deformation process, leading to a region of negative curvature that will not support a shock wave.

Additional insight into the shock strength of silicon carbide ceramics is provided by an examination of the stress versus volume compressibility through the shock compression process. Shock compression profiles for Sic-B silicon carbide ceramic (Grady 1996, 2000a), comparable to those displayed for Sic-N ceramic in figure 6.29, were analyzed to provide the stress versus volume compression states shown on the left in figure 6.32 (Grady 2008). Overlaid on this same plot are direct measurements of longitudinal and transverse stress of Feng *et al* (1998) on SiC-B ceramic subjected to shock compression. A mean stress (or pressure) versus volume

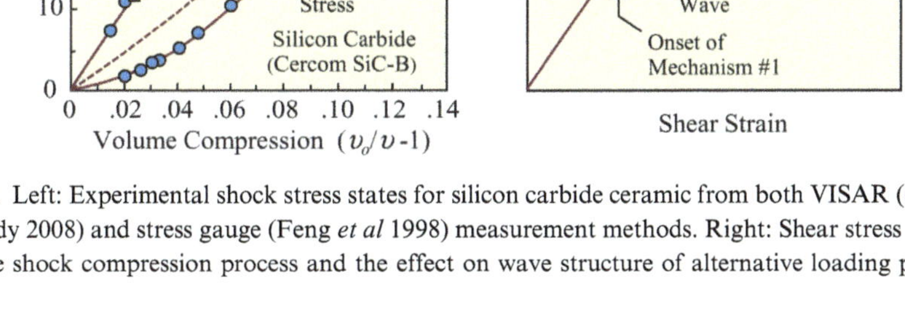

Figure 6.32. Left: Experimental shock stress states for silicon carbide ceramic from both VISAR (Grady 1996, 2000a, Grady 2008) and stress gauge (Feng *et al* 1998) measurement methods. Right: Shear stress versus shear strain in the shock compression process and the effect on wave structure of alternative loading paths.

compression curve estimated from ultrasonic properties is shown with the data. Of particular interest is the reasonable agreement between shock states resulting from analysis of the VISAR data at the upper- and lower-yield states, and those from direct longitudinal stress measurement. Also, the stress difference measured by the orthogonal stress gage continues to increase between HEL 1 and HEL 2, consistent with interpretation of the VISAR wave structure in this range.

Analysis of selected wave profile data for titanium diboride materials also lends insight into the nature of the complex shock wave strength and failure of ceramics. In figure 6.33, the stress versus specific volume states shown in the left-hand plot have been extracted from VISAR wave profile measurements comparable to those displayed in figure 6.31 (Grady and Wise 1993).

Discussion starts with the left-most compression curve and proceeds to the right. The dashed curve provides the hydrodynamic compression of titanium diboride based on a theoretical density of 4.51 g cm^{-3}, and properties for the bulk modulus K_o = 236.3 GPa and its pressure derivative of K_o' = 2.57 (Amulele *et al* 2006). The left-most Hugoniot compression curve is for TiB_2 ceramic provided by Cercom. The wave profiles shown in figure 6.31 were obtained on this ceramic. Metallography revealed a grain size of 10–25 μm and, although close to the theoretical density (4.509 g cm^{-3}), showed some porosity predominantly at grain boundaries and junctions, probably suggesting some higher density impurities. The Hugoniot curve for the Cercom material, shown in the figure, results from the analysis of three experiments. Both a lower yield at 5–7 GPa and an upper yield at approximately 15–17 GPa are identified (arrows) in the plot.

The next Hugoniot compression curve was obtained on an Eagle Picher TiB_2 ceramic with a somewhat lower starting density of 4.452 g cm^{-3}. Metallography shows a somewhat larger grain size of 20–50 μm and a similar grain and pore structure, with a modestly higher level of intercrystalline porosity, consistent with the lower density. This ceramic exhibited a lower yield of 6–7 GPa and an upper yield of about 14 GPa. The two Hugoniot curves show continued offset in specific volume between the Cercom and Eagle Picher ceramics to the upper-yield states.

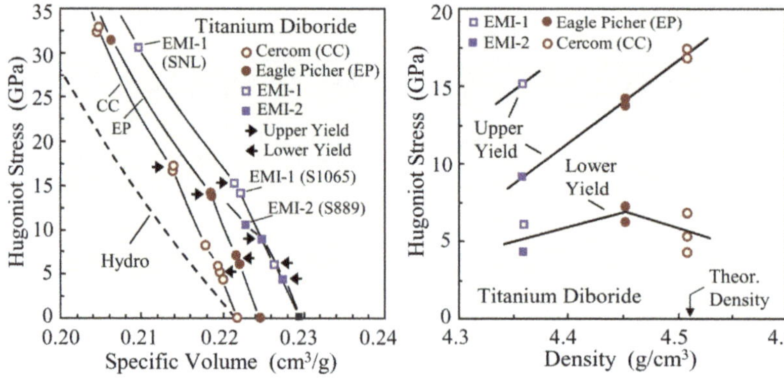

Figure 6.33. Experimental shock stress states for selected titanium diboride ceramics (left), and corresponding upper and lower Hugoniot shock strengths (right) (Grady 2008).

The two curves are nearly coincident at the final Hugoniot states. This observation suggests that the crush of the porosity on shock compression does not initiate until the upper yield is achieved.

The right-most compression curve is populated by Hugoniot compression states from two TiB_2 ceramic materials. Both materials were prepared for and tested under shock compression by the Ernst-Mach Institute (Winkler and Stilp 1992). Both ceramics had a starting density of 4.36 g cm^{-3} and a porosity of approximately 4%–5%. The first material, identified as EMI-1, was a relatively coarse grain ceramic (30–50 μm) with the porosity consisting of near-spherical voids with the pores residing predominantly at the interior of grains (intracrystalline porosity). The second material, identified as EMI-2, was a finer grain ceramic (5–10 μm) with a comparable level of porosity distributed primarily at triple grain junctions (intercrystalline porosity).

Shock compression tests conducted by Winkler and Stilp (1992) on the EMI-2 material reveal both a lower yield (4–5 GPa) and an upper yield of about 8 GPa. A slow third wave carries the material to a final stress of about 12 GPa. This wave indicates marked softening and probably signifies onset of pore compaction. A test performed in the same study on the EMI-1 material to a final stress of 14 GPa reveals only one yield (the reported HEL) at about 6 GPa. The observations that differences in microstructure markedly affected the shock strength, and that one material showed both an upper and lower yield whereas the other material revealed only a single HEL, were significant findings of the study.

A sample of the EMI-1 material was provided to the present author for the purpose of conducting a time-resolve VISAR shock compression experiment to a substantially higher Hugoniot stress. The results of this test reveal that the EMI-1 material did display a two-wave precursor structure, but with a markedly higher upper yield of about 15 GPa (Grady and Wise 1993).

On the right in figure 6.33 are systematically shown the upper- and lower-yield stresses, as measured from structured wave profiles, for the several titanium diboride ceramics that were studied. The yield stress values are plotted against the respective initial densities of the ceramics. For the lower yield there is no clear trend with initial density. The stress versus specific volume paths between the lower and upper yield in the plot on the left clearly suggest that little or no porosity is removed subsequent to the lower yield. The upper yield for the three ceramics with intercrystalline porosity shows a systematic decrease in amplitude with decreasing density (increasing porosity), consistent with onset of crush under shock in other ceramics (e.g. Gust *et al* 1973, Bar-on *et al* 2002). Upper yield for the EMI-1 material with an intracrystalline porous microstructure, however, revealed unexpectedly higher onset of crush. The final shock pressure states for all of the ceramics tested show a significant offset from the hydrodynamic compression curve, suggesting substantial retained strength at the shock state as concluded by Dandekar and Benfanti (1993).

6.6.4 Failure deformation mechanisms in shock compression

The deformation mechanisms responsible for the unusual shock strength and two-step yield in silicon carbide and titanium diboride are not known. Some discussions of the related observations can be offered, and sensible mechanisms proposed, but at this point such efforts are speculative.

Both silicon carbide and titanium diboride are hexagonal crystal structures with limited dislocation slip systems. Modest levels of intracrystalline plasticity have been observed in shock and high-confinement deformation of high-strength ceramics (Cagnoux and Tranchet 1997). Shih (1998) describes the preparation methods and the unique microstructure features of the SiC-B series of ceramics, and has observed intracrystalline plasticity in these silicon carbide ceramics through dislocation and stacking fault processes. Vanderwalker and Croft (1988) have examined shock recovered titanium diboride ceramic and observed dislocation slip on crystal grains favorably oriented with respect to the shock wave direction. From these observations it is reasonable to suggest that the onset and activity of crystal plasticity accounts for the lower-yield HEL 1, and the accompanying modest level of inelastic deformation observed in the shock compression of silicon carbide and titanium diboride ceramic.

The apparent saturation of plastic deformation and subsequent stiffening before the onset of the upper-yield HEL 2 in these same ceramics is a curious effect. Dislocation pile-up unable to penetrate energy barriers at grain boundaries has been suggested. An interesting alternative explanation is provided by the idea of disparate microstructure compliance in the material, a concept that has been exploited to model the volumetric compressibility of solids with microstructure in earlier sections of this chapter. At the onset of lower yield only a limit number of grains have slip systems that are favorably oriented with respect to maximum shear, and in which plastic deformation initiates. Neighboring grains remain elastic and are, consequently, markedly less compliant in shear than grains undergoing plastic deformation. Thus, as deformation proceeds, the stress load driving plastic deformation is gradually transferred to the stiffer elastic grain structure. The concomitant reduction in stress load on active grains arrests plastic deformation and stiffens the body as a whole. Stiffening of the material accounts for the intermediate shock formation following lower yield, but preceding onset of more pervasive deformation at upper yield. A similar description of elastic-plastic deformation on the microscale by Lipkin and Asay (1977) was found to account for fine structure in the shock–reshock and shock–release wave profiles in aluminum.

Mechanisms responsible for the onset of inelastic deformation at the upper yield are equally speculative. Following the previous argument, subsequent to the lower yield further shear stress load is increasingly carried by an elastic granular structure without adequately oriented slip systems to accommodate plastic deformation. It is reasonable to suggest that eventually higher energy secondary slip systems may come into play, accounting for the upper yield through a subsequent plastic deformation process. If a secondary slip mechanism is discounted, it is sensible to suspect that upper yield and subsequent inelastic deformation proceeds through a process of shear fracture. Shear fracture would refer to deformation on the continuum scale. On the microscale all fracture modes would be accommodated. Although confining pressures at the onset of upper yield are high (of order 10 GPa), the granular strengths of the ceramic compounds are markedly higher and are able to accommodate the modest level of dilatant deformations necessary to a fracture mechanism. Grain cleavage, grain separation, grain rotations and relative granular motion will all accompany the fracture process. Fracture near pores will accommodate the crush process. Fracture and subsequent deformation will entail a process of comminution necessarily encompassing a broad span of fragment sizes.

Although speculative, the above description of the complexities of shock strength and deformation is sensible. The description entails inelastic deformation through both plasticity and fracture processes. The two deformation processes are apparently separate in silicon carbide and titanium diboride, leading to the distinct two-wave precursor structure characteristic of these ceramics. Note, however, that this separation is absent, or less pronounced, in the Eagle Picher silicon carbide ceramic. Such a structure is not observed in other high-strength ceramics (e.g. aluminum oxide, boron carbide), suggesting perhaps that one or the other of the deformation mechanisms is absent or, more likely, both participating but inextricably mixed.

Unveiling, and ultimately adequately modeling, the shock strength and deformation mechanisms of high-strength ceramics is challenging. Further studies are needed, including post-shock examination of plasticity mechanisms, such as the work of Vanderwalker and Croft (1988) and Shih (1998). Understanding more intense shock deformation through fracture and comminution will require detailed investigation of the size distributions and morphology of the post-shock material. Very little has so far been undertaken. Of interest in this aspect is the very early paper of Dremin and Breusov (1971) extolling shock comminution of high-strength materials as a means of obtaining granular refinement unachievable by other means.

6.7 Strength and failure models

The section pursues physical theories and models in pursuit of a better understanding of the strength and failure behavior of brittle materials in the shock wave and ballistic environment. In particular the physics underlying failure wave phenomena are investigated and a framework for a physics-based and statistics-based model of failure is explored. The effects of material microstructure again play a key role.

6.7.1 Observations of failure wave phenomena

It has become common in recent years to identify the observed delay in fracture failure, after passage of the shock wave, as a failure wave (Kanel *et al* 1992). Earlier manifestations of the fracture kinetics phenomena were observed in a number of brittle materials. In certain hard rock materials, with compressive and fracture strengths comparable to current ceramics, the existence of upper and lower failure surfaces have been observed due to the delaying kinetics of fracture. These strength features lead to an effective rate-dependent, or time-dependent, failure under shock and high-rate loading (Grady *et al* 1977, Grady 1995b, Grady 1998). Brannon *et al* (1983) observed a process of coarse shear failure features following initially elastic shock compression in single-crystal quartz and lithium niobate crystals with high-resolution photography. Earlier Soviet work inferred the existence of a fracture wave phenomenon in the high-velocity penetration of brittle solids (e.g. Galanov *et al* 1989).

Kanel *et al* (1992) provided the first clear evidence on the nature of the stress state in brittle materials accompanying the failure wave fracture process behind the shock wave. Kanel *et al* used normal- and transverse-oriented piezo-resistive stress gages to monitor the corresponding stresses in glass, and recorded the marked loss in stress difference as the delayed fracture process proceeded within the failure wave. Since this seminal study, further experimental activity has focused on this topic in attempting to better understand its unique dynamic failure features. In spite of this activity, the underlying physics governing the failure wave phenomena are still not well understood.

High-speed photography (Brannon *et al* 1983, Bourne *et al* 1994, Strassburger *et al* 2005, Chocron *et al* 2016) has provided compelling evidence for the delayed failure wave and fracture kinetics phenomena. Other diagnostics are needed, however, to assess stress and deformation features within the failure event.

Alternative and complementary data are provided by the orthogonal stress gage measurements of Kanel *et al* (1992) on K-2 glass, and velocity interferometer records obtained on soda-lime glass (Grady and Chhabildas 1997). The stress profiles of Kanel *et al* in figure 6.34 show an initial abrupt rise of the longitudinal and transverse stress to markedly different amplitudes suggestive of initial elastic response of the glass under the uniaxial-strain shock wave loading. The recorded stress difference behind the initial shock provides a measure for calculating the strain energy stored in elastic distortion. Failure through shear fracture appears to occur several microseconds after passage of the shock wave, as suggested by a rapid drop in stress difference and is accommodated principally by a marked rise in the lateral stress amplitude.

Free surface velocity histories recorded with VISAR diagnostic methods induced through impact shock wave loading of soda-lime glass are also shown in figure 6.34, and bear little resemblance to the profiles of Kanel *et al*. Velocities at a free surface

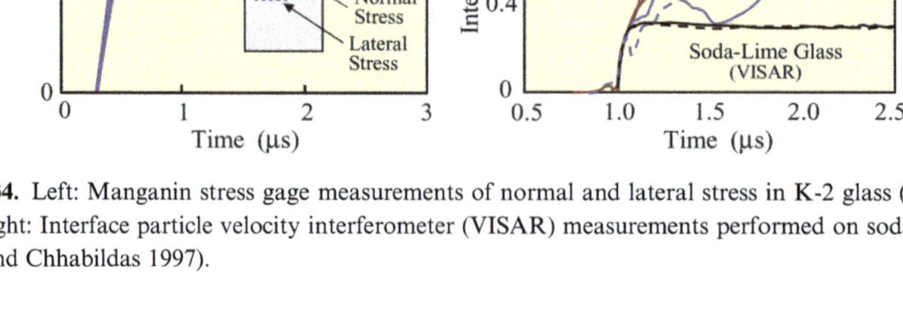

Figure 6.34. Left: Manganin stress gage measurements of normal and lateral stress in K-2 glass (Kanel *et al* 1992). Right: Interface particle velocity interferometer (VISAR) measurements performed on soda-lime glass (Grady and Chhabildas 1997).

are, however, far removed from the in-material measurement of stress histories. It is also important to remember that piezo-resistive gages span a sensible area within the test sample. Such gages encompass several millimeters or more and presumably measure a reasonable average stress over the area subtended. In contrast, the VISAR diagnostic is closer to a point measurement with a laser spot diameter usually considerably less than a tenth of a millimeter.

Attention is focused on the two lowest amplitude profiles on the right in figure 6.34 (distinguished by dashed and solid lines), and the two rather chaotic looking VISAR records next higher in amplitude (also dashed and solid lines). These two sets of records correspond to shock stress amplitudes of about 4.5 GPa and 6.5 GPa, respectively, well within the 4–7 GPa range that other workers have suggested for failure wave behavior in soda-lime glass. The successively higher stress amplitude tests presumably exceed the shock amplitude at which the glass transitions to a more homogenous HEL and post-HEL yield behavior. The two profiles at about 6.5 GPa suggest failure through a coarse fracture process in regions reasonably close to the front of the shock near the point of the VISAR measurement. The lower amplitude profiles suggest nominal elastic response of the glass to the 4.5 GPa shock stress level. The gradual drop in velocity of about 25–35 m s^{-1}, behind the shock front, hints at more complex processes, however. This relaxation in velocity prior to any expected release wave suggests a modest drop in longitude stress and a corresponding increase in the specific volume strain ($\sim$10%). It is reasonable to speculate that the observed behavior is a consequence of delayed shear-fracture-induced dilation brought about by the failure process sometime after passage of the shock wave.

6.7.2 A meso-kinetic model of delayed failure

A physically sensible theory of the dynamic failure of brittle solids should include two principal features. First, the theory should provide an explanation of the underlying physical character of the material that gives rise to the unique dynamic failure behavior. Second, this character of the material should be imbedded within a physics-based model that leads to a satisfactory description of the observed behavior of dynamic failure in brittle solids. Such failure behavior should include the shock wave stress and velocity data for glass discussed previously, and as displayed in figure 6.34, as well as a host of other strength and failure properties unique to brittle solids. A theory of the strength of brittle solids is outlined in the following sections that attempt to address the two requirements (Grady 2006a, 2007).

The theory to be discussed lends insights into the underlying physics and statistics of the delayed failure phenomena in brittle solids subject initially to an elastic shock compaction; namely the failure wave. Before proceeding, it is worth noting that this theory is just one window allowing a view of the governing physics. Other, possibly more profitable, avenues are available. In chapter 8 the physics of Langevin dynamics is applied to the shock-induced high-rate failure of solids. Within the context of Langevin dynamics the delayed failure of a shock compressed brittle solid is associated with the time-dependent post-shock correlation dynamics of an acoustic phonon field impressed on the material by the initial shock. The dynamics of Langevin offers added perspective to the delayed shock failure event.

6.7.2.1 Material features governing failure in brittle solids

The theory that is described here is called a meso-kinetic theory of the strength of brittle solids. As is shown, this name is descriptive of the material aspects of the underlying physics. The single key attribute of brittle solids, which must be accepted before pursuing the theory, is that when a brittle solid is subjected to a mechanical load, there is a long-range (meso-scale) correlation of the induced stress and elastic strain field at any point in the body. That is, the stress at a material point depends not only on properties of the material at that point, but also on properties in some mesoscopic neighborhood of that point. The spatial extent of this correlation length is uncertain, but for a full density granular solid it is envisioned to be tens of grains, and perhaps as much as hundreds of grains or more. This premise is assumed to apply to amorphous, as well as crystalline, brittle materials.

With regard to the former, the issues span the still poorly understood nature of failure producing defects in glassy materials. See, for example, the review of Berthier and Ediger (2016) and references therein. Sites of eminent failure identified in deformed samples have been coined shear transformation zones. Such sites are proposed as soft, poorly organized zones in the glass structure and correlated with low-frequency modes.

To further illustrate the stress state and meso-scale correlation within a brittle body, consider a cylindrical specimen of such a solid (ceramic, glass, concrete, competent rock). Subject the sample to a uniform static axial load within the elastic limits of the material, and further confine the specimen to zero lateral motion. In response to this load, a macroscopic axial stress σ_x and lateral stress σ_y will be developed in the body to support the elastic deformation. This uniaxial loading is appropriate since later in this section the same sample is subjected to shock wave compression under which the same global stress conditions are achieved.

This discussion will shortly examine details of the model where it becomes difficult to view the forest due to too intently focusing on the trees. The essence of the model is depicted in figure 6.35. A heterogeneous material is subjected to uniaxial compression. On a trace through the sample the state of elastic shear stress is illustrated as a function of position. As is shown, the elastic shear stress (and strain), as a consequence of the heterogeneous nature of the material, varies with position. There are two length scales of interest. There are microscale defects of characteristic dimension d. These defects, akin to Griffith flaws, are sites of intense stress concentration and will, at a critical stress intensity, activate failure through brittle fracture (brittle fracture is the only mode of failure considered in the present analysis). In a granular material the defect scale d is envision to be of order granular or sub-granular in dimension. Intensification of stress at the defect is relative to and dependent on a state of stress τ (identified in figure 6.35) in the local neighborhood of the defect. Only the deviator stress τ is of interest, since this will be considered as the stress driving fracture in the present model. The shear stress τ characterizes the

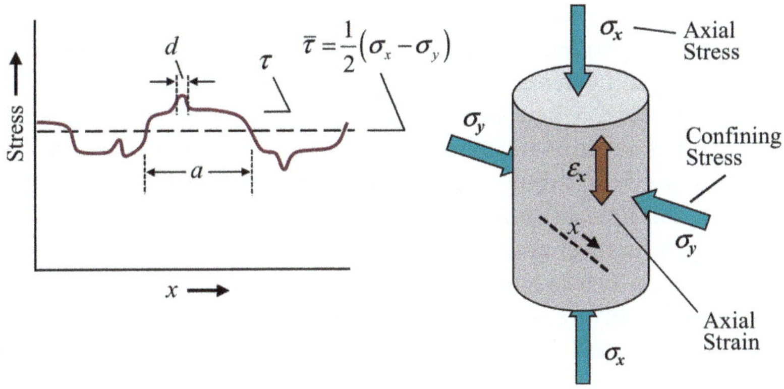

Figure 6.35. Variations of the local shear stress state due to meso-scale heterogeneities in material properties.

nominal shear stress state within a mesoscale region of characteristic dimension a that envelopes the microscale defect. The stress in this mesoscale region differs from the macroscopic mean stress state $\bar{\tau}$ (also identified in figure 6.35) and is a consequence of local statistical correlation characteristics of the heterogeneous material. In a granular material such mesoscale stress variations from the mean would be multi-granular in scale. Such mesoscale stress variations are sensibly expected to be Poisson distributed (an assumption) over the mesoscopic size scale a. When subjected to a shock (Heaviside) compressive load, the change in stress within the region of microscale defects of dimension d and mesoscale regions of dimension a will be time-dependent. Because of the one to two orders of magnitude difference in size scale, equilibration times on the microscale dimension d and on the mesoscale dimension a will correspondingly differ. Time dependence on these several length scales in a heterogeneous solids resides at the heart of dynamic failure in brittle solids.

The material sources leading to the long-range correlations of the stress state and fluctuations of the stress from the mean are many. Qualitatively these sources reside within all of the imperfections inherent to heterogeneous materials. Such imperfections include features such as impurities, chemical inhomogeneity, porosity, second-phase, internal strain and granular correlation, along with hosts of others.

The present theory will sensibly show that the dynamic mesoscale response of the material can explain, at least in part, the unique time-dependent failure behavior observed for brittle solids. It follows that this mesoscale character must be a property of the heterogeneous material and, as such, measurable by some independent means. Experimental measurements such as ultrasonic scattering or line VISAR in shock wave testing are possible methods for assessing the mesoscopic dynamics of heterogeneous materials.

The existence of a mesoscopic stress field with long-range (mesoscale) stress correlations has potential implications for several strength features of brittle solids. Brittle solids are noted for exhibiting a size-dependent strength when self-similar structures are subjected to comparable loads to the point of failure. Size-dependent strength can be expected when the size of the zone in which failure initiates becomes comparable in size to correlation length scales in the mesoscopic stress field. Further, statistical features of mesoscale stress fluctuations will play a role in size-dependent strength. This scale-dependent strength of brittle solids is pursued within the context of the theory. It is not, however, central to the purpose of the present analysis. It is mentioned here because, when size dependence is noted in strength tests such as indentation hardness or triaxial compression, arguments can be made for relating such static size-dependent strength data to time-dependent dynamic strength of the material through the present model.

Another strength feature, and the one that is central to the present development, is that of the failure of brittle solids subjected to strong shock waves. When a shock wave passes through a brittle solid and impresses a continuum axial and lateral stress σ_x and σ_y on the material, comparable to that of the statically loaded specimen described earlier and illustrated in figure 6.35, the same mesoscopic stress field is not immediately achieved. There is instead a time dependence to the stress field as equilibration over the mesoscale correlation length scales is achieved over time scales that are determined by the speed of elastic waves. At the shock state the average values of σ_x and σ_y are constant. Time dependence of the variance occurs as equilibrium on the mesoscale is sought. This process of time-dependent stress equilibration is an essential feature of the present meso-kinetic theory, and is pursued in the following subsections.

6.7.2.2 Fracture at the mesoscale

Fracture after passage of the shock wave will initiate at fracture, producing flaws. A Griffith flaw is commonly considered in linear elastic fracture mechanics, and stresses in the neighborhood of the Griffith crack are written as (e.g. Freund 1990)

$$\sigma_{ij}(r,\theta) \simeq K\frac{f_{ij}(\theta)}{\sqrt{r}}. \tag{6.43}$$

The parameter K is the stress intensity factor and is, in the present application, written as $K \simeq \tau\sqrt{d}$ where, d is the crack dimension and τ is the local mesoscale shear stress field. Depending on the crack orientation, K is some mix of mode I, mode II and mode III stress intensity (Lawn 1993). The contribution from mode I stress intensification in heterogeneous brittle materials is expected to lessen as the confining pressure increases.

When the same Griffith crack is subjected to an instantaneous application of the stress τ, the stress intensity factor is time-dependent with the form shown on the left in figure 6.36. At early times the stress intensity increases with time according to $K \sim \tau\sqrt{ct}$. At late times K equilibrates to the static value of $K \sim \tau\sqrt{d}$. A critical stress intensity factor K_c, at which inelastic growth of the fracture initiates, is identified as a property of the material. Fracture at the flaw initiates when K achieves K_c, either during quasi-static elevation of τ, or during the time-dependent amplification of K.

The response of the same flaw within the far-field mesoscopic stress field $\bar{\tau}$ of the brittle solids is more complex. Stress concentration in the neighborhood of the flaw is still determined by the local or neighboring mesoscopic shear stress τ. However, τ

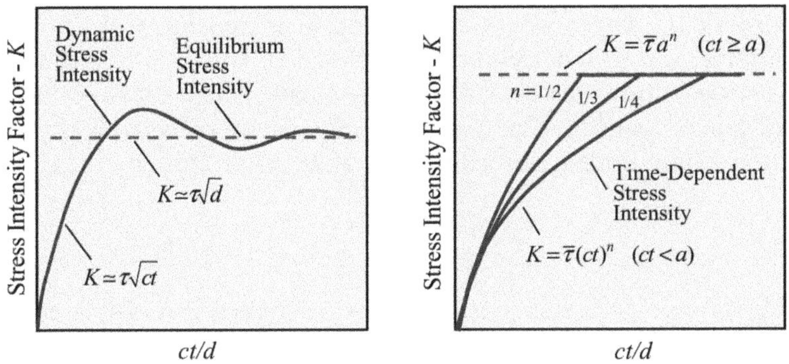

Figure 6.36. Left: The square-root singularity dynamic-stress intensity factor at fracture-initiating defects. Right: Generalization of stress intensification for the model with time-dependent and equilibrium stress intensification in mesoscale regions.

depends in turn on the far field stress $\bar{\tau}$, and on details of the material structure within some mesoscale correlation length $a \gg d$. When subjected to a near-instantaneous stress load, such as passage of a shock wave, time-dependent equilibration of the stress intensity factor over both the flaw and the near-field mesoscale structure within the correlation length a must be accounted for. A more general time dependence of the mesoscale stress intensification is assumed of the form

$$K = \begin{cases} \bar{\tau}(ct)^n & ct < a \\ \bar{\tau}a^n & ct > a \end{cases}. \tag{6.44}$$

The power n is expected to be somewhat less than one-half, reflecting the slower late-time equilibration over the mesoscale structure of dimension a. Stress intensification equilibrates to a constant $K \simeq \bar{\tau}a^n$. Fracture initiates if a critical stress intensification K_c is exceeded. Note that the dimensions of K_c are MPa m^n (pressure times length to the power n).

The dimensions of the stress intensification factor in the present development are not the same as those that result from the square root singularity of a Griffith flaw. This more general stress intensification and failure criterion has similarities to the development of Mott (1947), and is related to the Weibull characterization of the strength of solids (Weibull 1951). Plots of the stress intensification applied within the model, which parallel the Griffith square root singularity, are shown on the right in figure 6.36. The early time amplification of the stress intensification and late time constancy is consistent with the expected behavior over mesoscopic length scales and is reflected by the model.

6.7.2.3 Statistical description of mesoscopic fracture

Fracture-producing defects, and the stress activation of those defects, depend not only on the strength of the impressed stress field $\bar{\tau}$, but also on the character of an overlying microstructure-dependent mesoscopic stress field. In the neighborhood of a fracture-producing defect, stress intensification is dependent on a region of the material characterized by a correlation length scale a and an intensification power n that are properties of the mesoscale field. These parameters are expected to differ at each fracture site of concern. The mesoscale regions are reasonably distributed over the correlation length scale a according to a Poisson probability density

$$p(a) = \frac{1}{a_o} e^{-a/a_o}. \tag{6.45}$$

The intensification power n in equation (6.41) is assumed constant over the range of mesoscopic length scales.

It remains to characterize the onset of fracture at a site. In constructing a statistical theory, the chance of fracture will be specified by a power-law hazard function in the stress intensification of the form

$$k(K) = \frac{m}{K_c}\left(\frac{K}{K_c}\right)^{m-1}. \tag{6.46}$$

Namely, the probability of activation of fracture at a site is determined by a statistical function of the Weibull extreme value type. Onset of fracture is statistically distributed about some critical stress intensification K_c.

Dynamic strength and the meso-kinetic features of the model come into play when the stress loading rates exceed the rate at which the mesoscopic stress field can equilibrate over the correlation length scales. It can be shown that a transition to dynamic-strength behavior occurs when stress wave strain rates achieve levels of order $\dot{\varepsilon} \sim K_c/\rho_o c a_o^{n+1}$. In this relation ρ_o and c are the material density and elastic wave speed, while a_o is the characteristic correlation length from equation (6.45) and K_c is the critical stress intensification (dimensions of MPa m^n).

Shock wave, or Heaviside, loading to an elastic stress state would stimulate a non-equilibrium response of the mesoscale stress field. Consider a shock wave crossing a material plane at time $t = 0$. A fracture activation hazard function as a function of time after passage of the shock wave is calculated from the present meso-kinetic theory. For much of the following analysis, time is considered in units of c^{-1}, where c is the elastic wave velocity and $s = ct$ is the time measure. The temporal hazard function is calculated from the stress intensification hazard function through

$$g(s)ds = k(K)dK, \tag{6.47}$$

or

$$g(s) = k(K(s))K'(s). \tag{6.48}$$

During the time-dependent period of stress intensification about a fracture-producing defect, from equation (6.44),

$$K(s) = \bar{\tau} s^n, \tag{6.49}$$

and

$$K'(s) = n\bar{\tau}s^{n-1}, \tag{6.50}$$

with $\bar{\tau}$ the shear stress amplitude of the Heaviside shock wave. Equations (6.46) through (6.50) lead to

$$g(s) = mn\left(\frac{\bar{\tau}}{K_c}\right)^m s^{nm-1}. \tag{6.51}$$

Letting $r = mn$ and the length scale $a_o = (K_c/\bar{\tau})^{1/n}$ yields for the temporal hazard function for fracture after passage of the shock wave,

$$g(s) = \frac{r}{a_o}\left(\frac{s}{a_o}\right)^{r-1}. \tag{6.52}$$

The distribution of correlation lengths from equation (6.45) implies that fracture flaws associated with correlation lengths of size $a < s$ will have achieved maximum stress intensification. If not already activated, such flaws will no longer provide sites for fracture activation, and are removed from the population. Consequently, the hazard function in equation (6.52) is augmented by the exponential expression to provide

$$h(s) = \frac{r}{a_o}\left(\frac{s}{a_o}\right)^{r-1} e^{-s/a_o}. \tag{6.53}$$

If the number of fracture sites in a volume element dV is $N_o dV = dV/\lambda_o^3$, where λ_o is the mean spacing of such sites, then the statistical rate of fracture activation per unit time, per unit volume, is

$$n(s) = \frac{1}{\lambda_o^3}\frac{r}{a_o}\left(\frac{s}{a_o}\right)^{r-1} e^{-s/a_o}. \tag{6.54}$$

If the body of interest is an area or a line then the pre-factor in equation (6.54) is reduced to $1/\lambda_o^2$, or to $1/\lambda_o$, to reflect the reduced dimensionality. Although not necessary, it seems reasonable that the nominal correlation length a_o and average fracture site spacing λ_o would be comparable, and in the subsequent analysis $\lambda_o = a_o$ is assumed.

6.7.2.4 Fracture growth and interaction

When fractures activate at favored sites the cracks rapidly accelerate to some characteristic velocity and propagate until arrested by some impediment to further crack propagation. In concert, the stress field driving the fracture is relieved as the crack propagates. This facet of the fracture process behind a shock wave has its own set of physical factors. The modeling of fracture growth in the present model is sensibly simple. Once activated, crack acceleration to terminal velocity is governed by the same inertial properties that determine dynamic fracture activation (Freund 1990). Thus, it is reasonable to fold the acceleration growth into the initial activation model and simply assume instantaneous acceleration to a terminal velocity. Both theoretical efforts and experimental observation show that terminal velocity for brittle cracks is some appreciable fraction of the elastic wave speed (proportional to the Rayleigh or shear elastic velocity for example). Here a constant crack growth velocity equal to c_g is assumed. Finally, if a fracture activates at an earlier normalized time $\sigma = ct'$ and propagates to a later time $s = ct$ than a volume surrounding the crack of magnitude,

$$v(s - \sigma) = \frac{4\pi}{3}\eta^3(s - \sigma)^3 \tag{6.55}$$

is assumed stress relieved, where $\eta = c_g/c$ is the ratio of the crack growth velocity and the sound speed. Again, if the interest is fracture on an area, or a line, the power and coefficient in equation (6.55) is modified accordingly.

6.7.2.5 Analytic framework for the meso-kinetic failure model

The essential physics underlying the failure wave phenomenon in brittle solids is now outlined. The physics includes a mesoscopic stress field that overlays the continuum stress field impressed on the body in the shock compression process. Under rapid dynamic loading, such as in shock compression, this mesoscopic stress field does not instantaneously equilibrate, but does so over time as correlation length scales in the field equilibrated at elastic wave speeds. Fractures activate at sites of weakness as stress intensities increase to critical levels over time. Activated brittle fractures propagate at crack growth velocities and relieve the driving deviator stress field.

An analytic solution based on this physics emerges to describe response of the brittle material during fracture failure following a compressive shock wave. First, identifying a parameter $D(s)$, which is the volume fraction of the solid over which shear stress has been relieved by the time-dependent fracture process at time s after passage of the shock wave. It is appropriate to identify $D(s)$ as a fracture damage parameter. The integral

$$D_x(s) = \int_0^s v(s - \sigma)n(\sigma)d\sigma \tag{6.56}$$

provides the sum of the stress relieved volumes from equation (6.55) of all fractures activated at earlier time σ from equation (6.54) integrated from the initial time zero up to the present time s. $D_x(s)$ is identified as the extended damage, and is not the desired damage parameter since the integral is in error by not accounting for further activation of fractures in previously stress relieved regions (exclusion), nor does it account for the overlap of growing stress release regions (impingement) (Grady 1981).

These features are accounted for through the statistical relation (Grady 1981, 2006b)

$$D = 1 - e^{-D_x}, \tag{6.57}$$

relating the extended damage parameter D_x to the damage parameter D. The damage D and the extended damage D_x are nearly equal at early times as they should be when fracture damage is still dilute. At late times D approaches unity whereas D_x is unbounded.

Replacing equation (6.56) with the physical models for fracture activation and growth of stress release from equations (6.54) and (6.55) yields

$$D_x(s) = \int_0^s \frac{4\pi}{3}\eta^3(s-\sigma)^3 \frac{r}{\alpha_o a_o^3}\left(\frac{\sigma}{\alpha_o}\right)^{r-1} e^{-\sigma/a_o} d\sigma. \tag{6.58}$$

Introducing the change of variables $x = \sigma/\alpha_o$ and $y = s/\alpha_o$ results in

$$D_x = \frac{4\pi}{3} r\eta^3\gamma^3 \int_0^y (y-x)^3 x^{r-1} e^{-\gamma x} dx, \tag{6.59}$$

where $\gamma = \alpha_o/a_o$ is the dimensionless ratio of the two characteristic length parameters. The further transformation $z = x/y$ reduces the equation to the convenient form

$$D_x = \frac{4\pi}{3}\frac{r\eta^3}{\gamma^r}(\gamma y)^{r+3} \int_0^1 (1-z)^3 z^{r-1} e^{-\gamma y z} dz. \tag{6.60}$$

Lastly, returning to the time dependence through $\gamma y = s/a_o$,

$$D_x(s) = \frac{4\pi}{3}\frac{r\eta^3}{\gamma^r}(s/a_o)^{r+3} \int_0^1 (1-z)^3 z^{r-1} e^{-(s/a_o)z} dz, \tag{6.61}$$

with $D(s)$provided by equation (6.57).

The continuum shear stress $\bar{\tau}(s)$ relaxes from its initial value $\bar{\tau}$ in response to the time-dependent growth of the damage D, either to zero, or to some reduced residual stress $\bar{\tau}_r$. A reasonable expression of this relaxation of the shear stress is

$$\bar{\tau}(s) = \bar{\tau}(1-D) + \bar{\tau}_r D. \tag{6.62}$$

The statistical number of activated fractures that participate in the failure process can be calculated. The fracture number $N(s)$ at time s is provided by the activation rate from equation (6.54), but reduced by the current volume fraction of stress relieved material

$$N(s) = \int_0^s (1 - D(\sigma))n(\sigma)d\sigma = \int_0^s e^{-D_x(\sigma)} n(\sigma) d\sigma. \tag{6.63}$$

Equation (6.59) is not readily solvable because of the complexity of equation (6.61) for D_x. The fracture number is easily estimated in most cases.

6.7.2.6 Application of the meso-kinetic model

To explore the implications of the present meso-kinetic model, properties for soda-lime glass necessary for the model are developed. These properties are appropriate for soda-lime glass or are acquired through experience with the model. They are sufficient, however, in reproducing some of the features observed in experimental

Table 6.8. Meso-kinetic model properties for soda-lime glass.

Density	$\rho_o = 2.5$ g cm^{-3}
Sound speed	$c = 5$ km s^{-1}
Fracture speed	$c_g = 2$ km s^{-1}
Critical stress intensification	$K_c = 100$ MPa m$^{1/3}$
Intensification modulus	$n = 1/3$
Mesoscale correlation length	$a_o = 0.25$ mm
Mesoscale distribution modulus	$m = 15$

studies of failure waves in glass, and they provide a means for uncovering the behavior of the model under the dynamic mechanical stimuli of interest.

Material and model parameters for soda-lime glass are provided in table 6.8. An elastic sound speed of 5 km s^{-1} at the shock amplitude is reasonable for soda-lime glass. A fracture speed of 2 km s^{-1} is a sensible guess at some value less than the sound speed. Note that the critical stress intensification does not have dimensions of fracture toughness. It is consistent with the present model generalization, where stress intensification is a consequence of both the defect structure and the neighboring mesoscopic correlation field. An intensification modulus of $n = 1/3$ is a reasonable guess and determines the dimensions of K_c. A mesoscale correlation length of 0.25 mm was determined from several iterations of the model, as was the distribution modulus of $m = 15$.

The equations have been specialized to apply to a cross-section 10 mm in width and 1 mm in thickness of an infinite medium subjected to a planar shock wave. The plots are shown in figure 6.37. Time corresponds to duration after passage of the shock wave. Shock amplitude increases from left to right and is identified by the resolved shear stress behind the shock wave in each plot. Time dependence of the fracture number activated at each shock amplitude is shown in the three upper plots. The corresponding position and time of activation for one statistical complexion at each shock amplitude is illustrated in the intermediate plots. Damage growth occurs later in time as fracture growth and interaction proceeds. Reduction of the driving shear stress would correlate with the growth of fracture damage. Time-dependent growth and interaction from the sites of fracture activation are illustrated in the lower plots.

Experiments have been performed focused on the investigation of one-dimensional shock and failure wave effects in glass (Chocron *et al* 2016). Experiments on borosilicate glass with ultra-high-speed camera imaging observed both shock wave passage and features of failure onset behind the shock wave. Images show the onset of failure activation sites following passage of the shock wave. Failure intensity occurs closer behind the shock wave as impact amplitude and shock pressure increase. The experimental observations of failure in glass are in sensible accord with the present model of shock wave failure.

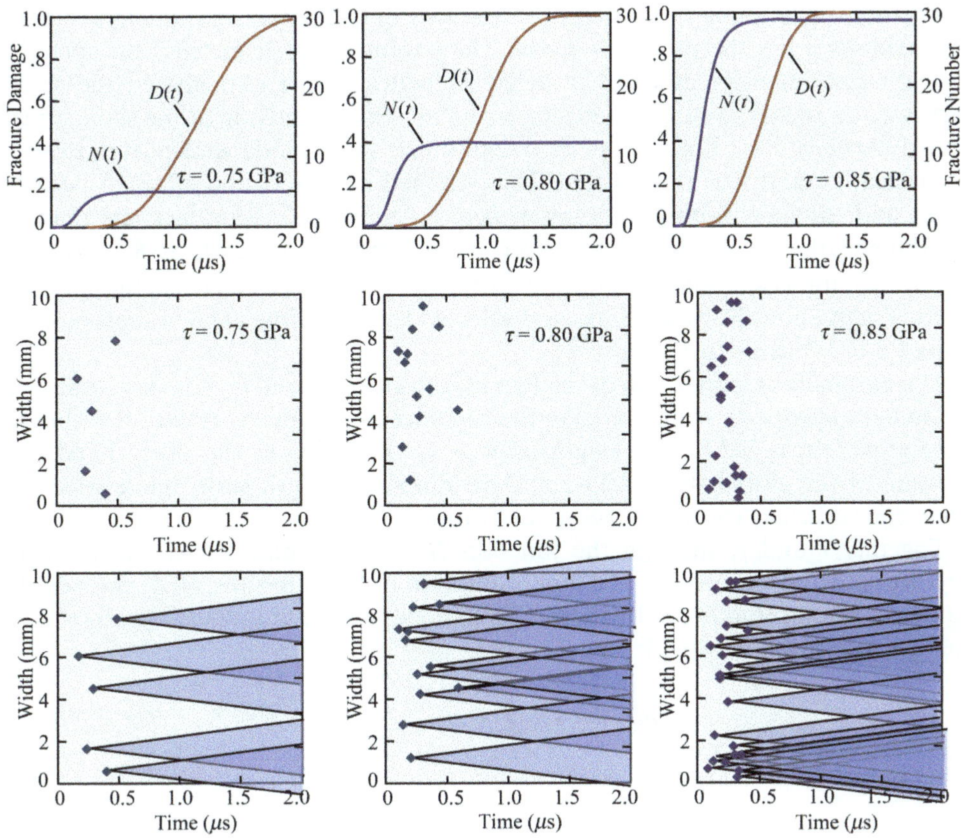

Figure 6.37. Representative application of the fracture kinetics model for soda-lime glass (Grady 2006a, 2007). The panels illustrate the statistical time-dependent fracture activation and damage growth following passage of a shock wave. Increasing shock intensity is noted by the shear stress amplitude. The lower plots illustrate the time-dependent stress release from the site of fracture.

6.7.2.7 Features of the statistical hazard function

Equation (6.53) provides the statistical probability (hazard) function governing the likelihood that a potential fracture site will transition to an active fracture under the conditions imposed. This hazard function has unique features that can be explored in further detail. Letting $S = s/a_o$ and $\gamma = \alpha_o/a_o$, equation (6.53) transforms to a dimensionless hazard function

$$H(S) = \frac{r}{\gamma^r} S^{r-1} e^{-S}. \tag{6.64}$$

The hazard function is the product of two terms. The leading power function determines the increasing time-dependent chance that the defect will achieve microscopic critical stress intensification and activate fracture. The trailing exponential term determines the chance that the site will achieve mesoscopic stress equilibrium and drop out of the population of potential fracture flaws.

The hazard function is governed by the two dimensionless parameters r and γ. The parameter r is the product $r = mn$. The parameter m determines the spread in critical stress intensity at which a defect activates. As m is increased the spread narrows and trends toward conditions where all defects activate at the same critical stress intensification. The parameter n determines the time-dependent rate of stress intensification and it is argued that the combined intensification effect of both the defect and the neighboring correlation field is to reduce n to values less than the Griffith singularity of $n = 1/2$. Since $K(s) \sim s^n$ the rate of growth is $\dot{K} \sim ns^{n-1}$, or $\dot{K} \sim n$ as s approaches unity. The effect of decreasing n is to increase the spread in fracture activation. For the parameters selected for soda-lime glass in table 6.8, $m = 15$ and $n = 1/3$ leads to $r = 5$.

The parameter γ is the ratio of the two length scales α_o and a_o. The length scale a_o is a material property characterizing the nominal long-range correlation within the mesoscopic stress field. The length scale $\alpha_o = (K_c/\bar{\tau})^{1/n}$, on the other hand, is a function of the global stress field and determines the length scale above which the stress $\bar{\tau}$ is sufficient to cause fracture activation.

Together r and γ, through the function $H(S)$, determine the character of the fracture activation process. For a large collection of fracture sites, $H(S)$ provides the statistical rate of fracture activation. For a single site the cumulative probability of fracture activation is provided by

$$P(S) = 1 - e^{-\frac{r}{\gamma^r}\int_0^S H(\sigma)d\sigma}. \tag{6.65}$$

The exponent

$$\frac{r}{\gamma^r}\int_0^S \sigma^{r-1}e^{-\sigma}d\sigma \tag{6.66}$$

includes the complementary incomplete Euler gamma function $\Gamma_c(r,S)$ and equation (6.65) can be written

$$P(S) = 1 - e^{-\frac{r}{\gamma^r}\Gamma_c(r,S)}. \tag{6.67}$$

In the limit of large S equation (6.65) reduces to

$$P(\infty) = P_\infty(\gamma) = 1 - e^{-\frac{r}{\gamma^r}\Gamma(r)} = 1 - e^{-r!/\gamma^r}, \tag{6.68}$$

where $\Gamma(r)$ is the Euler gamma function and the final expression with the factorial term applies when r is an integer. $P_\infty(\gamma)$ provides the probability that a single site will activate under the specified conditions.

The character of the fracture activation hazard function in equation (6.64) is illustrated in figure 6.38. Properties from table 6.8 are used, and a stress of $\bar{\tau} = 0.9\,\text{GPa}$ is assumed, providing a value of approximately $\gamma = 5.5$. The plot shows both the power-law activation and exponential depletion terms separately, in addition to the complete hazard function and the corresponding cumulative probability fracture function. For the selected stress amplitude, a fracture site has

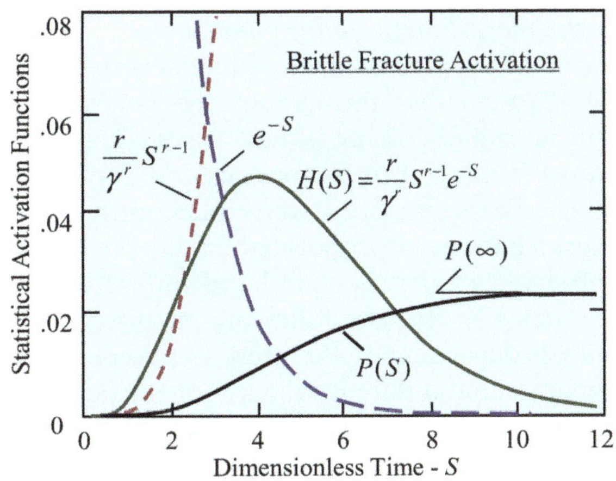

Figure 6.38. Construction of the fracture activation hazard function including power-law activation and exponential depletion.

an approximately 2% probability of fracture and will occur within a time duration of about $1 < S < 10$.

As the driving stress amplitude is raised (decreasing γ) the likelihood that a potential fracture site will activate increases and rapidly approaches unity. This transition to unit cumulative probability is quantified by equating the relevant argument in the exponent of equation (6.68) to one, or

$$\frac{r!}{\gamma^r} = 1. \tag{6.69}$$

The transition curve in the hazard function parameter space identifying the domain in which fracture activation is sparse is shown in the left plot of figure 6.39. The right plot shows an alternative representation of the fractional activation probability for selected values of r. Again, increasing shear stress $\bar{\tau}$ equates to a decreasing $\gamma(r)$.

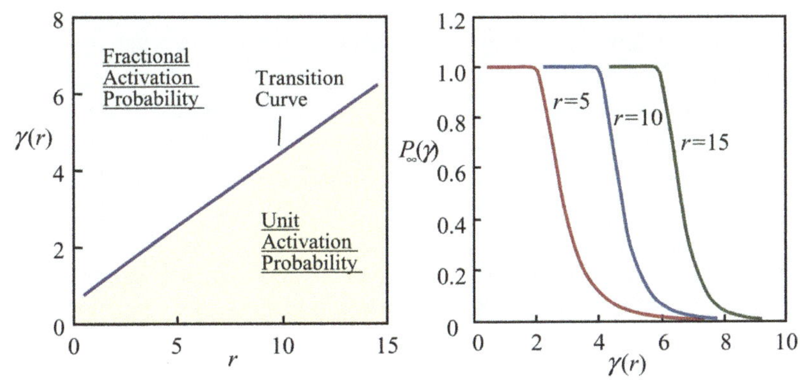

Figure 6.39. Parameter plots of the fracture activation hazard function identifying domains of sparse or fractional fracture activation.

6.7.2.8 Features of the meso-kinetic failure model

The physics emphasized in the present fracture model is the time delay necessary to the equilibration of elastic stress throughout the heterogeneous mesostructure following passage of a rapidly rising shock wave. This correlation dynamics following a shock wave is integral to the physics and plays a fundamental role in a number of shock compression critical state phenomena. The statistical model within which this physics has been incorporated readily predicts the delayed fracture (the failure wave) observed in the shock-induced failure of brittle materials. The model accounts for a delay in fracture following passage of the shock wave. This delay in fracture failure is dependent on the stress amplitude of the shock. The model does not predict a separate and different velocity for the failure wave as reported in several of experimental studies, although this observation is still debated. In any case, this lack of the model may be a consequence of the overly simplistic damage growth model. A growth model that is dependent not only on the driving stress but also on the current damage in the neighborhood of failure may introduce a distinct failure wave velocity.

6.8 Assessment of experimental data and modeling of ceramics with microstructure

Analysis and theoretical modeling provides an improved understanding of the underlying physics governing the compressibility and strength of ceramic materials subjected to shock wave compression. Composition and microstructure such as modest amounts of porosity, additives, or second phases, common in the production of most ceramics, have a marked influence on shock response of ceramic and ceramic-like materials. A span of shock wave data are examined in the chapter, some of which are not readily available elsewhere. Theoretical modeling of the material response is explored in accounting for composition and microstructure effects in the shock compression process. Theoretical models are also used in assessment and physical interpretation of the experimental shock wave data. Some of the key results of the chapter are highlighted in the following summary:

The classic theory of mixtures as applied to the shock compression of solids is shown to have reasonable first-order applicability to the constitutive description of ceramic mixtures. The theory is used here to describe the shock response of a ceramic mixture of silicon carbide and free silicon such as occurs, for example, in RB silicon carbide ceramics. Both EOS (including silicon phase transformation) and strength are incorporated in the theoretical model. Comparison with experimental Hugoniot data shows satisfactory agreement and demonstrates applicability of such models in related uses.

The classical theory of mixtures is further tested against experimental shock Hugoniot and release wave data on metal–ceramic mixtures of copper with boron carbide, silicon carbide and titanium diboride. Again, the theory provides a satisfactory description of the shock-velocity and shock impedance features of the data. Less satisfactory agreement is found with higher order details of the shock decompression data.

The several studies provide support for the use of classical mixture theory for the shock response of microscale and mesoscale ceramic mixtures in selected applications. Early work by others, however, suggests that the additive mixture theory is less appropriate when applied to mixing on the molecular scale (e.g. Kalashnikov *et al* 1973). This and other issues of classical mixture theory are addressed further in chapter 10.

The high crystal strength of ceramics introduces a dimension to the material response that cannot be accounted for in the classic theory of mixtures as applied to shock compression. This feature of ceramic materials affects both volumetric compression and deviator strength. One emphasis in this chapter has focused on the consequences of high crystal strength to the volumetric response. The effect of high crystal strength is most apparent in the shock compaction of porous ceramics, where it is demonstrated that remarkably high shock pressures are needed to achieve full compaction. Compaction of high-strength solids in the shock wave is explored further in chapter 7.

Less apparent is that the same crystal strength and volumetric compaction phenomena can underlie shock response of full-density mixtures of disparate-compliance ceramics, or in mixtures of ceramic with non-ceramics. This behavior is again exemplified by the mixing of silicon carbide and free silicon where the relatively higher compliance of silicon exacerbated by the pressure-induced phase change in silicon leads to volumetric compaction effects and hysteretic dissipation under shock compression.

A theoretical model for mixtures of ceramics on the microscopic and mesoscopic scale is formulated that accounts for the added dimension of crystal strength in the shock compression process. The model determines two bounding nonlinear compression curves for the mixture ceramic. First, the mixture elastic response is calculated through an average of the component partial pressures on the microscale. Second, the mixture equilibrium response is calculated through an average of the component partial volumes. A generalized compaction relation characterizing the material response through the shock compression process is determined through additive mixing of the two bounding compression curves. The model uses functional forms and material constants that are natural to the shock compression properties of solids.

The same theoretical framework can be extended to a description of polymorphic phase transformation under shock compression. The theory recognizes that in ceramics with composition and microstructure complexity, the ceramic mixture features of high crystal strength and microstructure compliance disparities also play a role in the shock pressure-induced phase transformation. The model accounts for such features and describes the broadening of the phase transition pressure–volume response and transition hysteresis observed in high-pressure phase transformation.

Analysis of available shock wave data is undertaken in this chapter in an effort to provide a better understanding of some of the physics that govern the compressibility, strength, and phase transformation features of ceramics in the impact and high-rate regime. Early shock wave and Hugoniot data generated at the LANL and LLNL, as well as more recent shock wave data are included in this investigation.

The theoretical mixture model, emphasizing ceramic composition complexity and microstructure strength, is applied in interpreting the experimental shock wave data. Shock wave data for ceramics of aluminum oxide, boron carbide, silicon carbide, tungsten carbide and titanium diboride, among others, are included in this study.

The dynamic strength of ceramics with the varying levels of porosity common to sintered materials is not commonly available. Not well known is the availability of a theory due to Steinberg (1975) that provides for a prediction of the dynamic strength of a porous ceramic from the known strength of the fully dense ceramic. The Steinberg theory is examined here on data for porous ceramics of boron carbide and aluminum oxide, and is found to provide a reasonable prediction of the measured dynamic strengths. Elastic compressibility of ceramic mixtures is intrinsic to the Steinberg model. A clear relationship between this early study of Steinberg and the more recent mixture model provides insight into the underlying physics governing dynamic strength.

Shock Hugoniot data have been made available by LLNL on aluminum oxide ceramics containing both porosity and a glass second phase. These shock data are examined in the chapter and the microstructure mixture model based on compressibility properties of crystalline Al_2O_3 and shock compression properties of soda-lime glass is applied to the ceramic–glass–porosity composite. The model provides sensible correlation of this seminal shock Hugoniot data.

Early shock Hugoniot data obtained on porous boron carbide ceramic at LANL are joined with more recent shock wave data and pressure-dependent ultrasonic data to interpret the Hugoniot compression behavior of B_4C ceramic. Analysis of these data suggests two shock-pressure driven phase transitions—the first in the pressure range of 25–35 GPa and the second in the range of 45–50 GPa. The microstructure mixture model accounting for phase transformation is applied and provides a natural and satisfactory description of the experimental shock Hugoniot data. There are serious inconsistencies among the boron carbide shock wave data analyzed here and other available boron carbide shock data. Interpretations of the shock wave data offered here, including judgments of shock-induced phase transformation, are certainly arguable (see the text for more detail and related references, e.g. Grady 2015).

Silicon carbide ceramic subjected to shock wave testing in the early investigations at LLNL has come to light as a ceramic mixture of silicon carbide with an appreciable content of free silicon. Phase transformation in silicon carbide ceramic reported in the early work can now be attributed to shock-pressure-induced phase transformation in the free silicon content of the ceramic. The present theoretical model, including both microstructure strength and phase transformation, sensibly describes the available shock wave data. The model is further demonstrated to readily extrapolate to any measure of free silicon, such as ceramic RB production methods.

Shock wave compression and strength data are available for high-purity tungsten carbide ceramic as well as for liquid-phase-sintered tungsten carbide ceramics with varying metal content. A comparison of these Hugoniot data within the framework of the microstructure mixture model reveals that strength at the Hugoniot state is maintained over the range of available data. (The strength on the Hugoniot for tungsten carbide ceramic is explored further in chapter 10.) Increasing metal content in the ceramic is shown to reduce dynamic compressive strength but increase the dynamic tensile strength.

Silicon carbide and titanium diboride ceramics exhibit similar complexities in experimentally observed compression precursor wave structure. These precursor wave features relate to the shock strength and shear deformation characteristics of these two ceramics. Both deformation plasticity and shear fracture play a role in the multiple yield processes observed. It is argued that similar disparities in elastic compliance on the microscale also contribute to the processes of yield and wave structuring. Contrasting multimode failure in the shock event for metallic uranium is explored in chapter 11.

Time-dependent failure through fracture appears to be a fundamental characteristic of brittle solids. Failure waves are one manifestation of this fracture kinetics. A model based on both heterogeneous microstructure and time-dependent stress intensification following shock wave or high-rate loading provides a sensible description of observed fracture kinetics and the delayed failure wave phenomena in brittle solids. The significance of post-shock correlation dynamics that underlies the onset of delayed failure cannot be overemphasized. Evidence is emerging that similar correlation dynamics underlies a number of critical state phenomena that accompany the shock compression event.

Ceramics provide a rich display of features under shock compression that are unique among materials. The extreme crystal strength of the metal–nonmetal compounds lies at the heart of the distinctive behavior of ceramics. Unlike other materials, such as the metals or polymers, the stress state over microscopic and mesoscopic length scales in ceramics subject to shock compression varies markedly about the mean. An understanding of the ceramic microstructure and mesostructure is a prerequisite to an understanding of the shock EOS and dynamic strength of ceramics. Critical state phenomena such as polymorphic phase transformation, plastic yield and fracture are dependent on the local state of stress. The local stress state is in turn uniquely dependent on the ceramic microstructure and mesostructure.

The analysis and interpretation of experimental shock wave data, along with physics-based modeling of the shock compression process, place special emphasis on the role of microstructure and mesostructure in determining the dynamic strength and EOS properties of ceramics. Some achievements are described but many questions remain. The behavior of ceramics in the shock wave environment will continue to provide a productive area of research for some time into the future.

References

Aleksandrov I V, Goncharov A F, Yakovenko E V and Stishov S M 1992 *High-Pressure Research: Applications in Earth and Planetary Sciences* ed Y Syono and M Manghnani (Tokyo: TERRAPUB/American Geophysical Union) 409 p

Alexander C S, Vogler T J, Reinhart W D, Grady D E, Kipp M E and Chhabildas L C 2006 Influence of shock wave measurement technique on the determination of Hugoniot states *AIP Conf Proc.* **845** 1229–32

Altshuler L V and Sharipdzhanov I I 1971 Additive equations of state of silicates at high pressure *Phys. Solid Earth (Izv. Earth Physics)* **3** 11–28

Amulele G M and Manghnani M H 2005 Compression studies of TiB_2 using synchrotron x-ray diffraction and ultrasonic techniques *J. Appl. Phys.* **97** 023506

Amulele G M and Manghnani M H 2006 Application of radial x-ray diffraction to determine the hydrostatic equation of state and strength of TiB2 up to 60 GPa *J. Appl. Phys.* **99** 023522

Barker L M and Hollenbach R E 1972 A laser interferometer for measuring high velocities of any reflecting surface *J. Appl. Phys.* **43** 4669–75

Bar-on E, Partom Y, Rubin M B and Yankelevsky D Z 2002 Porous compaction as the mechanism causing the Hugoniot elastic limit *J. Impact Eng.* **27** 509–20

Bartukowski P and Spletzer S 2001 Porosity effects on the elastic constants of five varieties of silicon carbide ceramic *Army Research Laboratory Report* ARL-TR-2606, November

Bassett W A, Weathers M S and Wu T C 1993 Compressibility of SiC up to 68.4 GPa *J. Appl. Phys.* **74** 3824

Bell R L, Baer M R, Brannon R M, Elrick M G, Hertel E S, Silling S A and Taylor P A 2000 CTH user's manual and input instructions, version 4.00 *Sandia National Laboratories Technical Report* April

Berthier L and Ediger M D 2016 Facets of glass physics *Physics Today* **69** 40–7

Birch F 1978 Finite strain isotherm and velocities in single-crystal and polycrystalline NaCl at high pressures and 300 K *J. Geophys. Res.* **83** 1257–66

Brannon O J, Conrad C, Morris R W, Jones E D and Asay J R 1983 Studies of the spectral and spatial characteristics of shock induced luminescence form x-cut quartz *J. Appl. Phys.* **54** 6374–81

Brar N S, Rosenberg Z and Bless S J 1992 Applying Steinberg's model to the Hugoniot elastic limit of porous boron carbide *Shock Waves in Condensed Matter* (New York: Elsevier) pp 451–4

Bourne N K, Rosenberg Z, Mebar Y, Obara T and Field J E 1994 A high speed photographic study of fracture wave propagation in glass *J. Physique* **04** 635–40

Bourne N K and Rosenberg Z 1995 The dynamic response of soda-lime glass *AIP Conf. Proc.* 370 567.

Bourne N K, Gray G T and Millet J C F 2000 On the failure of shocked titanium diboride *AIP Conf. Proc.* 505 589–92

Budiansky B 1965 *Mech. Phys. Solids* **13** 223

Cagnoux J and Tranchet J Y 1997 Response of high strength ceramics to plane and spherical shock waves *High Pressure Shock Compression* vol 3 ed L Davison and M Shahinpoor (Springer) pp 147–70

Cline C F 2003 private communication

Chocron S, Barnette D D, Holmquist T J, Anderson C E, Bigger R P and Moore T Z 2016 Damage threshold of borosilicate glass under plate impact *J. Dynamic Behavior Mater.* **2** 167–80

Dandekar D P 2001 Shock response of boron carbide *Army Research Laboratory Report* ARL-TR-2456, April

Dandekar D P 2002 A survey of compression studies on silicon carbide *Army Research Laboratory Report* ARL-TR-2695, March
Dandekar D P and Bartkowski P T 2001 Tensile strengths of silicon carbide under shock loading *Army Research Laboratory Report* ARL-TR-2430, March
Dandekar D P and Benfanti D C 1993 Strength of titanium diboride under shock wave loading *J. Appl. Phys.* **73** 673–9
Dandekar D P and Grady D E 2002 Shock equation of state and dynamic strength of tungsten carbide *AIP Conf. Proc.* **620** 783–6
Dremin A N and Breusov O N 1971 The chemistry of shock compression *Priroda* **12** 10–7 (Sandia National Laboratories translation, SAND80-6003, March)
Dremin A N and Karpukhin I A 1960 Method of determining the shock adiabats of dispersed substances *Zhurn. Prikl. Mekh. Tekh. Fiz* **3** 184–8
Duvall G E and Fowles G R 1963 *Shock Waves in High Pressure Physics and Chemistry* ed R S Bradley (New York: Academic) pp 209–92
Duvall G E and Taylor S M 1971 Shock parameters in a two component mixture *J. Composite Mater.* **5** 130–9
Feng R, Raiser G F and Gupta Y M 1996 Shock response of polycrystalline silicon carbide undergoing inelastic deformation *J. Appl. Phys.* **79** 1378–87
Feng R, Raiser G F and Gupta Y M 1998 Material strength and inelastic deformation of silicon carbide under shock compression *J. Appl. Phys.* **83** 79–86
Freund L B 1990 *Dynamic Fracture Mechanics* (New York: Cambridge University Press)
Galanov B A, Grigor'yev O N, Kartuzov V V, Ostapenko V N and Trefilov V I 1989 Impact failure of ceramic plates *Sov. Powder Metall. Ceram.* **4** 299
Gilman J J 1970 Hardness of carbides and other refractory hard metals *J. Appl. Phys.* **41** 1664–6
Goto T, Sato T and Syono Y 1982 Reduction of shear strength and phase-transition in shock loaded silicon, *Japan. J. Appl. Phys.* **21** L369–71
Grady D E 1981 Fragmentation of solids under impulsive stress loading *J. Geophys. Res.* **86** 1047–54
Grady D E 1992a Shock wave properties of high strength ceramics *Conf. Proc. Shock Compression of Condensed Matter* 1991 ed S Schmidt *et al* (Washington, DC: American Institute of Physics) 455 p
Grady D E 1992b Shock-compression properties of ceramics Recent Trends in High-Pressure Research *Proc. Int. Conf. on High Pressure Science and Technology AIRAPT-XIII (Bangalore, India,* 7–11 *October* 1991) ed AK Singh (Oxford and IBH) pp 641–50
Grady D E 1993 Shock compression properties of silicon carbide *Sandia National Laboratories Report* SAND92-1832, July
Grady D E 1994a Shock wave strength properties of boron carbide and silicon carbide *J. Physique* **04** 385–91
Grady D E 1994b Shock compression properties of metal–ceramic mixtures *DYMAT* J **1** 1–13
Grady D E 1994c Hydrodynamic compressibility of silicon carbide through shock compression of metal–ceramic mixtures *J. Appl. Phys.* **75** 197–202
Grady D E 1994d Impact strength and indentation hardness of ceramics *Proc. Joint AIRAPT/APS Topical Conf. on High Pressure Science and Technology (Colorado Springs,* 28 *June–*2 *July,* 1993) pp 741–44
Grady D E 1994e Dynamic failure in brittle solids *Fracture and Damage of Quasibrittle Structures ed Z. Bazant et al* (E&;FN Spon) pp 259–73

Grady D E 1995a Dynamic properties of ceramic materials *Sandia National Laboratories Technical Report* SAND94-3266, February

Grady D E 1995b Shock properties of high-strength ceramics *Constitutive Laws: Theory, Experiment and Numerical Implementation* ed A M Rajendran and R C Batra (Barcelona: CIMNE) pp 35–45

Grady D E 1998 Shock wave compression of brittle solids, *Mech. Materials* **29** 181–203

Grady D E 1999 Impact failure and fragmentation properties of tungsten carbide *Int. J. Impact Eng.* **23** 307–17

Grady D E 2000 Shock profile studies on selected silicon carbide ceramics with application to dynamic yield mechanism in *Shock Compression of Condensed Matter*-1999 American Institute of Physics 629–32

Grady D E 2000a Shock profile studies on selected silicon carbide ceramics with application to dynamic yield mechanisms *AIP Conf. Proc.* **505** 629–32

Grady D E 2000b Shock and impact properties of tungsten carbide *Applied Research Associates Technical Report* August

Grady D E 2001 p-lambda model for material mixtures: application to compaction of porous zirconium dioxide *Applied Research Associates Technical Report for Sandia National Laboratories* October

Grady D E 2002 Analysis of shock and high-rate-date for ceramics: application to boron carbide and silicon carbide *Applied Research Associates Report for U.S. Army TACOM-TARDEC* August

Grady D E 2003 Analysis of shock and high-rate data for ceramics: equation of state properties *Applied Research Associates Report for US Army TACOM-TARDEC* December

Grady D E 2005 Analysis of shock and high-rate data for ceramics: physics based models for armor ceramics *Applied Research Associates Report for US Army TACOM-TARDEC* May

Grady D E 2006a Effects of meso-scale stress equilibration on the delayed failure of brittle solids *Proc. Twelfth Int. Symp. on Plasticity* (*Halifax, Canada,* 17–22 *July*); Khan A S and Kazmi R pp 130–132

Grady D E 2006b *Fragmentation of Rings and Shells* (Springer)

Grady D E 2007 Analysis of shock and high-rate data for ceramics: strength and failure of brittle solids *Applied Research Associates Report for US Army TACOM-TARDEC* July

Grady D E 2008 Analysis of shock and high-rate data for ceramics: failure and fragmentation in the shock and ballistic environment *Applied Research Associates Report for US Army TACOM-TARDEC* April

Grady D E 2009 Analysis of shock and high-rate data for ceramics: equation of state properties and fragmentation in the ballistic environment *Applied Research Associates Report for US Army TACOM-TARDEC* May

Grady D E 2010 Dynamic equation of state and strength of boron carbide *34th Int. Conf. on Advanced Ceramics and Composites (Daytona Beach, FL, 25–29 January 2010)*

Grady D E 2012 Equation of state for solids *AIP Conf. Proc.* 1426 800-803

Grady D E 2015 Hugoniot equation of state and dynamic strength of boron carbide *J. Appl. Phys.* **117** 165904

Grady D E, Hollenbach R E, Schuler K W and Callender J F 1977 Strain rate dependence of dolomite inferred from impact and static compression studies *J. Geophys. Res.* **82** 1325–30

Grady D E and Wise J L 1993 Dynamic properties of ceramic materials *Sandia National Laboratories Report* SAND93-0610, September

Grady D E and Crawford D A 1994 Dynamic properties of armor and warhead materials *DoD-DOE TCG Interim Program Report* 22 November 1994

Grady D E and Moody R L 1996 Shock compression profiles in ceramics *Sandia National Laboratories Report* SAND96-0551, March

Grady D E and Chhabildas L C 1997 Shock-wave properties of soda-lime glass, *Proc. 4th US Symposium on Solid Mechanics (Myrtle Beach, SC, 16–18 October 1996)* (Batelle Press) pp 29–38

Grady D E and Kipp M E 1998 Discrepancies in the interpretation of Hugoniot states from inclined-mirror and VISAR data *Applied Research Associates Report* March

Grady D E, Winfree N A, Kerley G I, Wilson L T and Kuhns L D 2000 Computational modeling and wave propagation in media with inelastic deforming microstructure *DYMAT* 2000, *6th Int. Conf. on Mechanical and Physical Behaviour of Materials under Dynamic Loading,* (*Krakow, Poland,* 25–29 *September* 2000) (EDP Sciences) pp 15–20

Grady D E and Doolittle C 2001 Preliminary compressibility analysis of boron carbide from high-pressure shock and ultrasonic data *Applied Research Associates Report for US Army TACOM-TARDEC* February

Grady D E and Winfree N A 2001 A computational model for polyurethane foam EXPLOMET 2000, *Int. Conf. on Fundamental Issues and Applications of Shock-Wave and High-Strain-Rate Phenomena (Albuquerque, NM,* 18-22 *June 2001*) ed K Staudhammer, L Murr and M Meyers (Elsevier) pp 485–92

Gust W H and Royce E B 1971 Dynamic yield strength of B_4C, BeO and Al_2O_3 ceramics *J. Appl. Phys.* **42** 276

Gust W H and Royce E B 1971 Axial yield strengths and two successive phase transition stresses for crystalline silicon *J. Appl. Phys.* **42** 1897

Gust W H, Holt A C and Royce E B 1973 Dynamic yield, compression, and elastic parameters for several lightweight intermetallic compounds *J. Appl. Phys.* **44** 550–60

Hahn G J and Shapiro S S 1967 *Statistical Models in Engineering* (New York: Wiley)

Herrmann W 1969 Constitutive equation for the dynamic compression of ductile porous materials *J. Appl. Phys.* **40** 2490–9

Hayun S, Paris V, Dariel M P, Frage N and Zaretzky 2009 Static and dynamic mechanical properties of boron carbide processed by spark plasma sintering *J. Eur. Ceram. Soc.* **29** 3395–400

Hill R 1965 *Mech. Phys. Solids* **13** 213

Holmquist T J, Templeton D W and Bishnoi K D 2001 Constitutive modeling of aluminum nitride for large strain, high-strain rate, and high-pressure applications *Int. J. Impact Eng.* **25** 211–31

Holmquist T J and Johnson G R 2002 Response of silicon carbide to high velocity impact *J. Appl. Phys.* **91** 5858–66

Holmquist T J and Johnson G R 2005 Characterization and evaluation of silicon carbide for high-velocity impact *J. Appl. Phys.* **97** 093502

Jeanloz R 1989 Shock wave equation of state and finite strain theory *J. Geophys. Res.* **81** 5873

Johnson J N 1975 Analytic failure surfaces for oil shales of varying kerogen content *High Pressure Science and Technology*vol 2 ed K Timmerhaus and M Barber (New York: Plenum) pp 444–54

Kalashnikov N G, Pavlovsky M N, Simakov G V and Trunin R F 1973 Dynamic compressibility of calcite group minerals *Izv. Earth Phys.* **2** 23–9

Kanel G I, Rasorenov S V and Fortov V E 1992 The failure waves and spallation in homogeneous brittle materials *Conf. Proc. Shock Waves in Condensed Matter*, 1991 (Elsevier) pp 451–454

Kanel G I, Bogach A A, Razorenov S V, Savinykh A S, Chen Z and Rajendran A 2001 A study of the failure wave phenomenon in brittle solids *AIP Conf. Proc.* 1760 020033

Kipp M E and Grady D E 1994 Shock phase transformation and release properties of aluminum nitride *J. Physique* **04** 249–56

Lawn B 1993 *Fracture of Brittle Solids* 2nd edn (Cambridge: Cambridge University Press)

Leger H M, Tomaszewski P E, Atouf A and Pereira A S 1993 Pressure-induced structural phase transitions in zirconia under high pressure *Phys. Rev.* B **47** 14076

Lipkin J and Asay J R 1977 Reshock and release of shock-compressed aluminum *J. Appl. Phys.* **48** 182–9

Manghnani M H 1993 Elastic properties and compressibility of advance ceramic composites at high pressure *Proc. Joint AIRAPT/APS Topical Conf. on High Pressure Science and Technology (Colorado Springs, June 28 June–2 July)*

Manghnani M H, Fisher E S, Li F Y and Grady D E 1993 The elastic moduli of TiB_2 *Ceramic Transactions—Advances in Ceramic–Matrix Composites* ed N P Bansal (American Ceramic Society) pp 771–85

Manghnani M H, Wang Y, Zinin F and Rafaniello W 2000 Elastic and vibrational properties of B4C to 21 GPa Science and Technology of High Pressure vol 2, *Proc. Int. Conf. on High Pressure Science and Technology (AIRAPT-17) (25–30 July)* (University Press) pp 945–48

Manghnani M H 2007 private communication

Marsh S P 1980 *LASL Shock Hugoniot Data* ed S P Marsh (Berkeley, CA: University of California Press)

Maitra A K and Phani K K 1994 Ultrasonic evaluation of elastic parameters of sintered powder compacts *J. Mater. Sci.* **29** 4415

McQueen R G, Marsh S P, Tayler J W, Fritz J N and Carter W J 1970 The equation of state of solids from shock wave studies *High Velocity Impact Phenomena* ed R Kinslow (New York: Academic)

Mott N F 1947 Fragmentation of shell cases *Proc. R. Soc.* A **189** 300–8

Olsson W A 1999 Theoretical and experimental investigation of compaction bands in porous rock *J. Geophys. Res.* **104** 7219–28

Paris V, Frage N, Dariel M P and Zaretsky 2011 Divergent impact study of the compressive failure threshold in SiC and B4C *Int. J. Impact Eng.* **38** 228–37

Roque V, Cross B, Baron D and Dehaudt P 2000 Effects of the porosity in uranium dioxide on microacoustic and elastic properties *J. Nuclear Mat.* **277** 211–6

Rosenberg Z, Yeshurun Y and Brandon D G 1985 Dynamic response and microstructure of commercial alumina *J. Physique* **46** 331–41

Sato Y and Akimoto S 1979 Hydrostatic compression of four corundum-type compounds *J. Appl. Phys.* **50** 5285–91

Sekine T and Kobayashi T 1997 *Phys. Rev.* B **55** 8034

Shih J C 1998 Dynamic deformation of silicon carbide *PhD Thesis* University of California

Sikka S K 1992 Pressure induced glasses, recent trends in high pressure research *Proc. XIII AIRAPT Int. Conf. on High Pressure Science and Technology* ed A K Singh (Oxford and IBH) 254 p

Steinberg D J 1975 Equation of state for B4C and BeO *Lawrence Livermore National Laboratory Report* LLL-UCID-16946

Steinberg D J 1996 Equation of state and strength properties of selected materials *Lawrence Livermore National Laboratory Report* UCRL-MA-106439, February

Strassburger E, Patel P, McCauley J W and Templeton D W 2005 High-speed photographic study of wave and fracture propagation in fused silica *Proc. 22nd Int. Symp. Ballistics (Vancouver, BC, 14–18 November 2005)*

Vanderwalker D M and Croft W J 1988 Dislocations in shock-loaded titanium diboride *J. Mater. Res.* **3** 761–3

Vogler T J, Reinhart W D and Chhabildas L C 2004 Dynamic behavior of boron carbide *J. Appl. Phys.* **95** 4173

Vogler T J, Reinhart W D, Chhabildas L C and Dandekar D P 2006 Hugoniot and strength behavior of silicon carbide *J. Appl. Phys.* **99** 023512

Vogler T J, Lee M Y and Grady D E 2007 Static and dynamic compaction of ceramic powders, *Int. J. Solids Structures* **44** 636–58

Weibull W 1951 A statistical distribution function of wide applicability *J. Appl. Mech.* **18** 293

Wilkins M L 1968 Third progress report of light armor program *Lawrence Radiation Laboratory, Livermore, CA, Report* No. UCRL-50460

Winkler W D and Stilp A L 1992 Pressure induced macro- and micromechanical phenomena in planar impacted TiB2 *Shock Waves in Condensed Matter* (New York: Elsevier) pp 555–8

Yan X Q, Tang Z, Zhang L, Guo J J, Jin C Q, Zhang Y, Goto T, McCauley J W and Chen M W 2009 Depressurization amorphization of single-crystal boron carbide *Phys. Rev. Lett.* **102** 075505

Zhang Y, Fukuoka K, Kikuchi M, Kodama M, Shibata K and Mashimo T 2005 Shock compression behavior of titanium diboride *Int. J. Impact Eng.* **32** 643–9

Zhang Y, Mashimo T, Uemura Y, Uchino M, Kodama M, Shibata K, Fukuoka K, Kikuchi M, Kobayashi T and Sekine T 2006 Shock compression behaviors of boron carbide *J. Appl. Phys.* **100** 113536

IOP Publishing

Physics of Shock and Impact, Volume 2
Materials and shock response
Dennis Grady

Chapter 7

Dynamic compaction of crushable solids

The irreversible dynamic compaction of distended materials has a wide range of science and engineering applications. The present chapter finds motivation primarily from applications concerned with the impact and explosive compaction of ceramic granular materials. Most of the fundamental principles, models and analysis, however, have broader applications to shock compaction of a more diverse range of porous solids. Several applications to porous metals are included. A solution of the shock compaction of crushable media attributed to the analysis of Kompaneets (1956), and briefly described in the second volume of Zeldovich and Raiser (1967), is expanded upon. This solution, intended for purposes of assessing ground shock extent and attenuation in geologic materials, is the solid media companion to the seminal solution for blast waves in air of Sedov (1946) and Taylor (1950), and similarly provides an elegant assessment of the self-similar nature of the resulting shock wave propagation. This solution is used as motivation for exploration of a broader range of crush models and corresponding assessment of compaction effects on shock wave propagation. One analytic solution provides a particularly descriptive example of constrained similarity identified with intermediate asymptotics (Barrenblatt 2006) common to other areas of physics.

Some trends in the experimental shock compaction of distended ceramics are brought to light. These observations and the earlier calculations transition to the long-standing p-alpha model of Herrmann (1969) with contributions by Carroll and Holt (1972). The more recent p-lambda model is described with potential application to a wider range of crushable media. Informative analytic solutions are undertaken based on irreversible crush models that ignore recoverable elasticity. Solutions including recoverable elasticity are undertaken that assess the limits of the applicability of incompressible-crush models. Strong shock dissipation and concomitant heating are examined. Necessary thermodynamic relations are developed relevant to the shock dissipation event. Explicit dissipation mechanisms in the strong shock compaction process are examined, including turbulence in the compaction deformation.

doi:10.1088/978-0-7503-1257-8ch7

7.1 Introduction

The emphasis of the present chapter is principally on the compaction of granular and powdered brittle materials such as high-strength ceramics. Most of the models and analytic solutions, however, have broader applications to the shock and high-rate compaction of any irreversibly crushable media. Applications to irreversible compaction and shock heating of porous solids is addressed in a later section. The chapter opens with an extended analysis of the similarity solution of spherical shock waves in crushable media introduced by Kompaneets (1956) that is most readily accessible in the second volume of the two book set of Zeldovich and Raiser (1967). The methods of Kompaneets are then constrained to planar shock waves where considerably less restriction is imposed on the compaction constitutive model.

The model restriction of an incompressible solid component of a compacting medium—analogous to that of a rigid-plastic material in metal plasticity—considerably simplifies analysis and leads to analytic solutions that are both physically instructive and useful in engineering applications. Several crush constitutive models are investigated that are suggested by systematics observed in experimental compaction data on brittle granular materials. The models are coupled with the conservation relations of planar shock wave propagation to assess shock dissipation and wave attenuation in compacting media.

Analytic solutions are sought for the planar shock compaction of a porous material due to plate impact and explosive loading. Solutions reveal the physics underlying shock propagation and dissipation in distended materials and provide analytic results for benchmarking numerical analysis of the same problems. Length scales are determined that control transitions in compaction as shock propagation and attenuation proceeds. In particular, the field parameters of pressure and mass velocity are shown to transition to self-similar power-law behavior at wave propagation distances well removed from the governing length scales.

Common to the computational modeling of the dynamics of porous materials is the seminal p-alpha model of Herrmann (1969), and the extensions of this model by Carroll and Holt (1972). Concepts of the p-alpha model appropriate to the present purposes concerning shock compaction of brittle granular media are pursed in the chapter and related to comparable features of the p-lambda model introduced in chapter 6, but specialized to porous media.

The simplifications of the incompressible-crush material response allows for the analytic solution of a number of informative problems in shock and dynamic compaction of distended media. In a subsequent section analytic solutions are pursued that include the elastic response of the matrix material in the dynamic compacting solid. The solutions provide an assessment of the approximations inherent in the applications of rigid-compaction models.

Much of the study of the dynamics of compacting media in this chapter is mechanical in nature. Although shock dissipation under dynamic compaction and its thermodynamic consequences are implicit, these issues are not emphasized. In a later section shock dissipation and concomitant temperature increase in the more extreme shock environment are explicitly considered. Thermodynamic relations relevant to shock compaction heating are constructed. An example of explosive shock compaction and dissipative heating is provided for a well-characterized metal. Thermodynamic equation-of-state (EOS) issues in the shock compaction of distended solids are pursued further in chapter 10.

In the closing sections of the chapter miscellaneous topics concerning compaction of distended solids in the intense dynamic environment are briefly explored. The micromechanical mechanisms of dissipation under intense shock compaction are considered, including turbulence in the extreme dynamic deformation environment. Dynamic compaction of disparate compliance mixtures is revisited. The principles of compaction strength attributed to Terzaghi (1925) are joined with the dynamic compaction models.

7.2 Spherical symmetry shock compaction of porous media

The seminal analytic solution of Kompaneets (1956) focuses on the spherically symmetric propagation of shock waves from the deposition of an intense source of energy at a point in a porous medium. Analysis will parallel this early study of Kompaneets while amplifying some of the solution issues and assessing the dependence of attenuation of the field variables on properties of the crushable media. The analysis in this section is restricted to the spherically symmetric shock wave propagation as pursued by Kompaneets.

7.2.1 Shock propagation solution

The present derivation of the outward propagation of a spherical shock wave in a porous medium will approximately replicate the development of Kompaneets (1956) as outlined in the text of Zeldovich and Raizer (1967). The analysis considers the deposition of an intense source of energy at the origin leading to the outward radial propagation of a compaction shock wave, as illustrated in figure 7.1.

A porous medium with porosity $p = 1 - \rho_o/\rho_s$, and with initial density ρ_o, is fully compacted to a final solid density ρ_s through the shock wave at the current radial position X. Incompressibility of the fully compacted material behind the shock wave is assumed and requires the opening of a spherical cavity at current radial position X_o such that

$$\rho_o X^3 = \rho_s\left(X^3 - X_o^3\right), \tag{7.1}$$

and related to the porosity through

$$X_o^3/X^3 = 1 - \rho_o/\rho_s = p. \tag{7.2}$$

The kinetic energy of the system is

$$T = \frac{1}{2}\int_{X_o}^{X} u^2\, dm, \tag{7.3}$$

where integration is over the mass within the region encompassed by the cavity radius X_o and the shock radius X. A spherical mass element at an intermediate

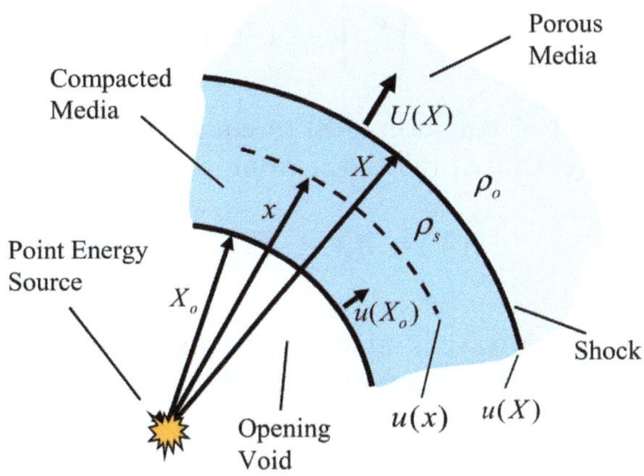

Figure 7.1. Centered spherical shock wave propagation into a porous compacting medium of initial density ρ_o and compacted density ρ_s.

radius x is $dm = 4\pi\rho_s x^2\, dx$, while incompressibility requires that the mass velocity satisfies

$$u(x) = u(X)\left(\frac{X}{x}\right)^2. \tag{7.4}$$

The kinetic energy becomes

$$T = 2\pi\rho_s \int_{X_o}^{X} u^2(X)\left(\frac{X}{x}\right)^4 x^2\, dx. \tag{7.5}$$

Introducing the dimensionless variable $\xi = x/X$ in the integral, equation (7.5) is written as

$$T = \frac{1}{2}\left(3\frac{\rho_s}{\rho_o}\int_{X_o/X}^{1} \xi^{-2}\, d\xi\right)\left(\frac{4\pi}{3}\rho_o X^3\right)u^2(X). \tag{7.6}$$

The second term in parentheses is identified as the mass $M(X)$ encompassed by the shock wave at current radial position X, while the first term in parentheses is the factor β derived by Kompaneets,

$$\beta = 3\frac{\rho_s}{\rho_o}\int_{X_0/X}^{1} \xi^{-2}\, d\xi = \frac{3}{p^{1/3} + p^{2/3} + p}. \tag{7.7}$$

Writing

$$T = \frac{\beta}{2}\frac{4\pi}{3}\rho_o X^3 u^2, \tag{7.8}$$

the reduction in kinetic energy due to shock dissipation as the shock propagates a distance dX is provided by

$$dT = -\left(\frac{u^2}{2}\right)4\pi\rho_o X^2\, dX. \tag{7.9}$$

Equating the differential of equation (7.8) to equation (7.9) leads to a power-law solution for the mass velocity at the shock front,

$$u(X) = AX^{-\frac{3}{2}\frac{(1+\beta)}{\beta}}, \tag{7.10}$$

where A is a constant.

The velocity of the shock wave is related to the mass velocity and the compaction strain $\varepsilon = 1 - \rho_o/\rho_s$ (identical to the initial porosity p) through the shock conservation relation

$$U(X) = u(X)/\varepsilon. \tag{7.11}$$

Similarly, the shock pressure is provided by shock conservation relation

$$P(X) = \rho_o u^2(X)/\varepsilon. \tag{7.12}$$

7.2.2 Flow and pressure field solution

The velocity field behind the shock wave within the region $X_o \leqslant x \leqslant X$ is determined from the incompressibility of the fully compacted material through equation (7.4). The corresponding pressure field under the continued assumption of incompressible inviscid flow is solved through the appropriate governing equations. The general solution for spherically symmetric flow for the pressure field behind the shock wave is

$$\frac{P(x)}{\rho_s} = \frac{\left(\eta - \frac{\eta}{\zeta}\right)}{\eta - 1}\frac{P(X)}{\rho_s} + \frac{(\eta^3 - 1)}{2(\eta^2 + \eta + 1)}\left(\frac{\eta}{\zeta} - 1\right) \times\left[(\eta^2 + 2\eta + 3) - \frac{\left(\frac{\eta}{\zeta} - 1\right)}{\eta - 1}\left(\frac{\eta^2}{\zeta^2} + 2\frac{\eta}{\zeta} + 3\right)\right]u^2(X), \tag{7.13}$$

where $\eta = X/X_o$ and $\zeta = x/X_o$ (Grady and Passman 1990). The first term on the right is the pressure profile component due to the external (shock) pressure load, while the second term is a boundary stress-free inertial component brought about by the incompressible motion of the material behind the shock wave. Representative mass velocity and pressure profiles for an initial 30% porous medium are illustrated in figure 7.2.

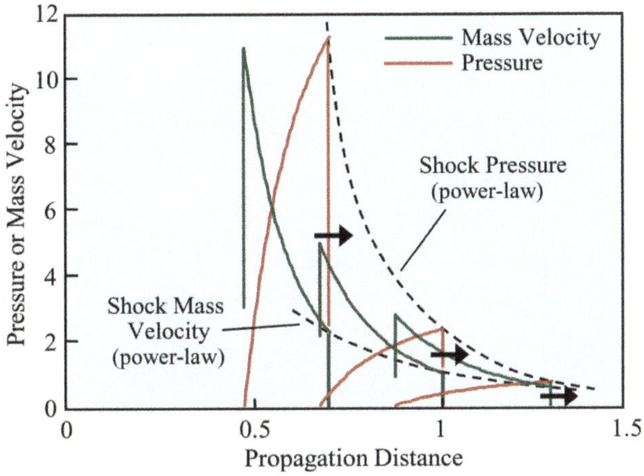

Figure 7.2. Solution for pressure and mass velocity subsequent to outward spherical shock propagation in a 30% initial porosity medium.

7.2.3 Attenuation properties in spherical shocks

From equation (7.10) the mass velocity at the shock front attenuates with range according to

$$u(X) \sim X^{-\frac{3}{2}\frac{(1+\beta)}{\beta}}. \tag{7.14}$$

With this solution and the shock conservation relations the spherical shock attenuation characteristics of other properties of interest are readily calculated. Consider any property $\Pi(X)$ with solution

$$\Pi(X) \sim X^{-n(p)}, \tag{7.15}$$

where $n(p)$ is the attenuation coefficient which depends on the initial porosity p. The attenuation coefficient for mass velocity is from equation (7.14) $n = 3(1+\beta)/2\beta$ where the dependence of β on porosity for spherically divergent flow is provided by (7.7). From (7.12) for the shock pressure

$$P(X) \sim u^2 \sim X^{-3\frac{(1+\beta)}{\beta}} \tag{7.16}$$

and $n = 3(1+\beta)/\beta$.

The kinetic energy behind the shock wave is determined through

$$T \sim Mu^2 \sim X^3 X^{-3\frac{(1+\beta)}{\beta}} \sim X^{-\frac{3}{\beta}}, \tag{7.17}$$

while momentum is provided by

$$S \sim Mu \sim X^3 X^{-\frac{3}{2}\frac{(1+\beta)}{\beta}} \sim X^{-\frac{3}{2}\frac{(1-\beta)}{\beta}}, \tag{7.18}$$

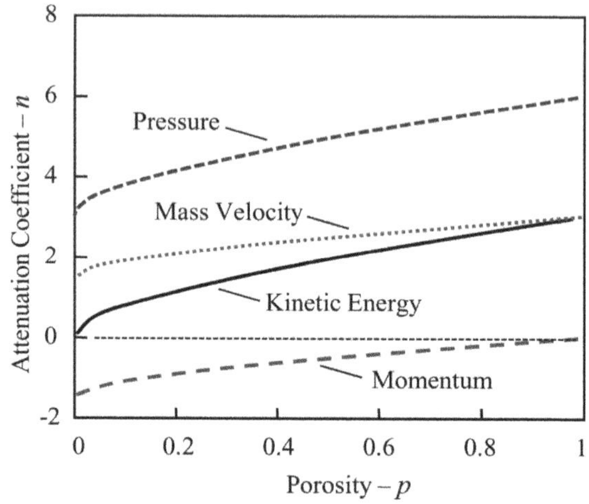

Figure 7.3. Power-law attenuation coefficient dependence on initial porosity for the Kompaneets solution of spherical shock propagation in porous compacting media.

with the corresponding attenuation coefficients of $n = 3/\beta$ for the kinetic energy and $n = 3(1 - \beta)/2\beta$ for the momentum. The range $1 < \beta < \infty$ constrains the momentum attenuation coefficient to a negative number.

Attenuation coefficients for mass velocity, pressure, kinetic energy and momentum as a function of initial porosity are shown in figure 7.3. Kinetic energy is conserved only in the limit of zero porosity ($\beta = \infty$). Momentum is not conserved except in the limit of porosity equal to unity ($\beta = 1$) and the momentum of the system increases with increasing range of the shock wave for any finite porosity. The increasing momentum with propagation distance is a consequence of the curvilinear nature of the flow.

Representative attenuation is illustrated in figure 7.4, where mass velocity at the shock front as a function of radial range from the source is plotted for selected initial porosities. Attenuation coefficients range from $n = 3/2$ at zero porosity where energy is conserved, down to $n = 3$ as porosity of one is approached and momentum is conserved.

7.3 Kompaneets shocks in porous media with other symmetries

The analysis of the spherical shock compression of porous media pursued by Kompaneets is readily extended to shocks in cylindrical and planar symmetries. These extensions have practical applications. The spherical solution offers insight into shock compression due to explosions at depth in porous media and, perhaps, shocks at depth due to the high-velocity impact of compact projectiles onto an infinite half space. Similarly, the cylindrically symmetric solution is a sensible approximation to the stress wave loading in porous media due to the high-velocity penetration of a jet or a rod, or the explosion of a line charge. Planar symmetry approximates the normal impact of a flat plate onto a half space of porous material.

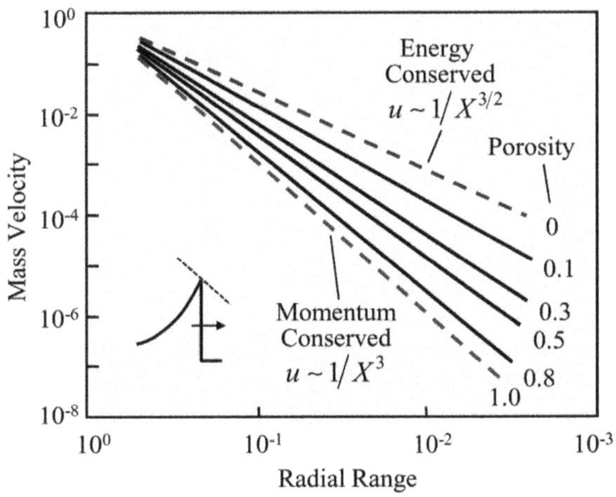

Figure 7.4. Mass velocity at the shock front versus radial range for selected porosities of a Kompaneets crushable medium.

7.3.1 Attenuation properties in other geometries

Let the index $d = 1$, 2 or 3 correspond to planar, cylindrical or spherical symmetry shock compression, respectively. Let $M(X)$ correspond to the mass encompassed by the shock wave at position X as illustrated for spherical symmetry in figure 7.1. Then the kinetic energy that applies to all three geometries is given by

$$T = \frac{1}{2}\beta M(X)u^2(X), \tag{7.19}$$

where

$$\beta(d) = d\frac{\rho_s}{\rho_o}\int_{X_o/X}^{1} \xi^{-(d-1)}\, d\xi, \tag{7.20}$$

and with

$$\begin{aligned} \beta(3) &= \frac{3}{p^{1/3} + p^{2/3} + p} \quad \text{(spherical)}, \\ \beta(2) &= \frac{\ln(1/p)}{1 - p} \quad \text{(cylindrical)}, \\ \beta(1) &= 1 \quad \text{(planar)}. \end{aligned} \tag{7.21}$$

The decay in kinetic energy is equated to the dissipation in the shock wave through

$$dT = \beta Mu\, du + \frac{1}{2}\beta u^2\, dM = -\frac{1}{2}u^2\, dM. \tag{7.22}$$

Integration yields

$$u(M) \sim M^{-\frac{(1+\beta)}{2\beta}}. \tag{7.23}$$

Corresponding expressions for the pressure, kinetic energy and momentum are

$$P(M) \sim M^{-\frac{(1+\beta)}{\beta}}, \tag{7.24}$$

$$T(M) \sim M^{-\frac{1}{\beta}}, \tag{7.25}$$

$$S(M) \sim M^{-\frac{(1-\beta)}{2\beta}}. \tag{7.26}$$

Dependence of the field variables on the shock propagation distance is provided through the correspondence between mass and distance $M \sim X^d$, and where the relations are further specialized to the different geometries through the exponent parameter $\beta(d)$ provided in equation (7.21).

7.3.2 Thermo-mechanics of shock compaction

The Kompaneets solution makes use of a simple but very effective model for the compaction of a distended material. The model is commonly referred to as the snow-plow model. It is described, and quite graphically illustrated, in the article of Duvall and Fowles (1963). A more recent description of the snow-plow model and the underlying physics is provided by Boslough and Asay (1992). In the model, a distended, or porous, material responds to any compressive pressure by fully crushing to the solid density, and staying at the solid density through the remainder of the analysis. The subsequent compacted solid material is commonly considered to be incompressible although, this latter assumption can be relaxed if warranted by the application.

Kompaneets makes effective use of this most elemental compaction model. Dissipation is effected in the solution by the requirement that the pressure traverse the Rayleigh line as the shock wave takes the material from the initial state to the shock state. In the spherical or cylindrical curvilinear geometries of the Kompaneets solution, the compacted material experiences post-shock distortional deformation and dissipation. Subsequent distortion dissipation after passage of the shock wave is an awkward issue in the Kompaneets solution that is not addressed.

To place the Kompaneets, or snow-plow, model in context, along with other approximate models of the dynamic crush process, some general observations of the shock compaction event in distended media is warranted. Thermo-mechanical features of the compressibility of a reasonably well-behaved distended solid material are illustrated in figure 7.5. Large-volume-change polymorphic phase transformations, a complicating feature in many materials under strong shock loading, are not displayed in the plot. Melting under strong shock compression is expected. Compression curves are shown for both the solid material and a distended, or porous, specimen of the same material. Compressibility of the solid includes the isothermal compression curve, the isentropic compression curve, and the shock Hugoniot centered at the initial specific volume of the solid material. The respective compression curves are identified by T, S and H, in the figure. They are offset to

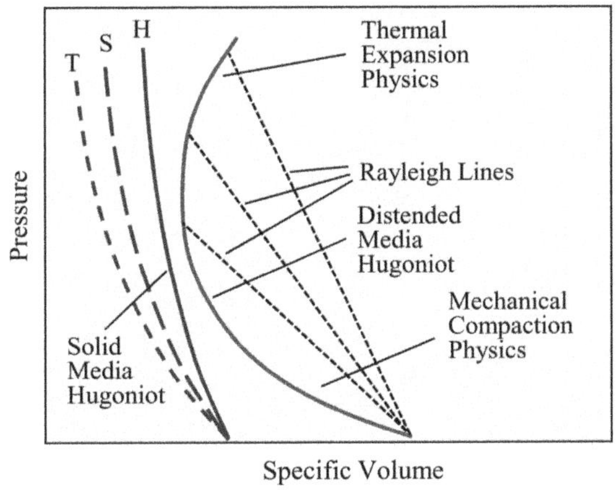

Figure 7.5. Mechanical and thermal EOS features of shock compaction of a distended medium.

lower specific volumes with increasing pressure due to the thermal expansion properties of the material. When the initial state is at absolute zero temperature, the isotherm and the isentrope are one and the same (the cold curve).

The Hugoniot of the distended material is centered at an increased specific volume that is determined by the initial porosity of the material prior to shock compaction. The modulus (slope) of the distended-medium Hugoniot is initially quite low, reflecting the mechanical crush strength properties of the material. Applications in this pressure range are commonly addressed through consideration of details of the crush mechanics, while ignoring thermal aspects of the problem.

As pressure on the Hugoniot increases, the modulus (slope) becomes increasingly larger and eventually changes sign, with specific volume increasing with continued increase in the Hugoniot pressure. In this pressure range thermal expansion on the Hugoniot exceeds lattice compressibility as pressure increases. In the strong shock pressure regime the mechanics of the compaction process can be ignored relative to the thermo-physical properties of the material. Rayleigh lines identify the pressure versus specific volume paths through the steady shock wave, and determine the energy increase at the Hugoniot state. Sensible estimates of energy partitioning between lattice strain energy and thermal energy can be made if the thermodynamic properties of the material are known.

A plot of the thermodynamic surface in pressure, specific internal specific energy and specific volume space in conjunction with the shock wave Hugoniot of the distended material is displayed in figure 7.6. The Hugoniot of the distended material does not achieve the equilibrium thermodynamic surface until compaction of the void volume reaches completion. The Hugoniot of the distended material is identified with a dashed line in the figure until crush is completed. A trace of this portion of the Hugoniot onto the thermodynamic surface at the same pressure and energy displays the enhanced dissipation and thermal energy component in the solid phase due to the compaction process. On completion of compaction the Hugoniot of the distended material resides on the thermodynamic surface at enhanced energy

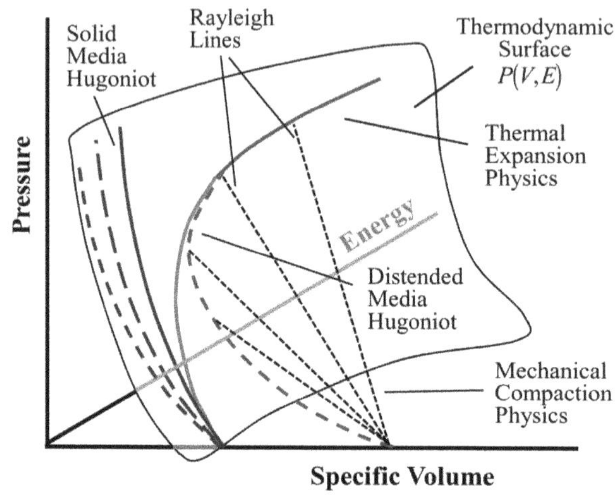

Figure 7.6. Shock compaction of a distended medium and the thermodynamic surface. The distended material Hugoniot is identified by the dashed line over the range of partial crush and as a solid line on the thermodynamic surface when compaction through the shock wave is complete.

states consistent with dissipation in the shock compaction event. Possible deviator strength complications at the Hugoniot state are ignored.

7.3.3 The crush model in distended media

Models of several levels of complexity that have application to both analytic and computational assessment of the dynamic compaction of distended media are illustrated in figure 7.7. On the left is the snow-plow model, and the model assumed in the solutions of Kompaneets. This simplest of models assumes that the material achieves full compaction under any applied pressure. Resistance to compaction occurs only under the extreme rates of shock compression where the stress versus specific volume path traverses the Rayleigh line. The snow-plow model is most appropriate over the range of pressures where full compaction is achieved and the mechanical resistance to crush can be ignored.

A second level of complexity is introduced in the incompressible-crush model. In this model any level of complexity of the crush curve that adequately describes the compaction resistance of the distended material is appropriated. Compressibility of the solid material is ignored, however. Upon decompression from pressure states of maximum compression, whether or not full compaction is achieved, the decompression pressure versus volume path is vertical. Note also that shocks will form over pressure segments where the crush curve is concave upward. Over this range the pressure–volume path again traverses a straight Rayleigh line. The model is appropriate for materials in the mechanical crush region where temperature effects are sensibly small. The model is also appropriate for materials for which the initial distention is reasonably large. It is not appropriate for materials with small amounts of initial porosity. The incompressible-crush model has the useful attribute of providing closed form analytic solutions to realistic shock propagation problems in

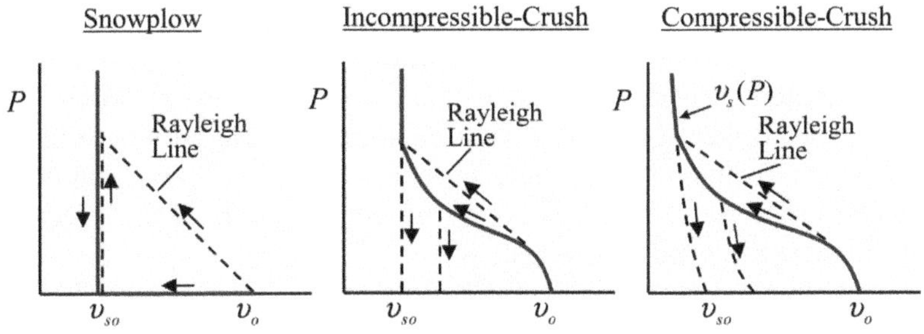

Figure 7.7. Elementary models of shock compaction of distended media.

porous media. Such solutions provide physical insights not readily achieved through computational simulations.

When initial porosity is modest the compressible-crush model, illustrated on the right in figure 7.7, is a more sensible approximation to the dynamic compaction process. Compressibility of both the compacting void and the solid is accounted for, providing a more realistic description of compressibility and wave propagation during decompression from partially or fully compacted states. The compressible-crush model is representative of several of the more common computational crush models in use. The p-alpha model (Herrmann 1969), used in a number of engineering shock wave codes, is of this form, as is the more recent p-lambda model (Grady *et al* 2000, Grady and Winfree 2001). The crush model in the EPIC code (Johnson *et al* 2003) is also of the compressible-crush model form. The latter model in figure 7.7 subsumes an aspect of the first. Namely, snow-plow compaction with subsequent compressibility response of the solid.

Additional physics and other forms can be implemented into modeling the compaction of distended solids if warranted by both the application and knowledge of the material response. The deviator or shear strength is an important consideration both in itself and as an enhancement of the compaction process. Elevated temperature is a natural consequence of the compaction process, and the effects of temperature on both EOS and compaction can be included with the mechanical crush model when necessary. Thermal EOSs are joined with compaction in the computational models noted above. Compaction models based on the maximum pressure achieved in the compression process are the most common, but other assumptions are possible. For example crush models base on the maximum strain, rather than pressure, have been profitably pursued (Wunnemann *et al* 2006).

7.4 Compaction in planar geometries

In planar uniaxial strain flow, a wider family of compaction constitutive responses can be explored. In particular, the assumption of an abrupt compaction to the final density assumed in the spherical and cylindrical Kompaneets solutions can be relaxed, and a continuous dependence of the compaction process on pressure can be considered. Material at different positions behind a propagating and attenuating compaction wave will, therefore, not all be at the same density, but will experience varying levels of partial compaction.

In this section several crush models descriptive of the compaction of distended solids are explored. Emphasis is placed on the compaction of granular mixes and powders of selected brittle compounds, but not entirely. Porous concrete is considered in one application. Other materials are considered where relevant. It is observed that the forms of the crush models considered have application to a broader range of materials. It is the task of the user to decide if the selected model is appropriate for the material of interest.

Also in this section, analytic solutions of compaction wave phenomena are pursued to enlighten the physics of material compaction in the shock wave and high-rate-loading regime. The mechanical crush curves in conjunction with an incompressible-crush model (the center sketch in figure 7.7) are particularly amenable to analytic solutions. Analytic relations resulting from practical compaction problems can be parametrically studied, unencumbered by numerical solution issues. Methods pursued in this section are applied in chapter 4, where Gurney solutions are joined with compaction relations to assess explosive launched plate motions in crushable media.

7.4.1 Variable compaction mass

The kinetic energy is again written as

$$T = \frac{1}{2}\int_{X_o}^{X} u^2\, dm. \tag{7.27}$$

As before, X and X_o span the accelerated and compacted, or partially compacted, material with X identifying the current position of the compaction shock wave. Incompressible flow in planar geometry, dictated by the incompressible-crush assumption, requires that the mass, or particle velocity, behind an attenuating shock wave propagating into a porous medium be independent of position and equal to the Hugoniot mass velocity at the shock front. Consequently,

$$T = \frac{1}{2}u^2(X)\int_{X_o}^{X} dm, \tag{7.28}$$

where X is the current position of the shock wave and X_o is the position where the load imparted to the material is introduced. A single compaction shock wave dictates that the compaction model be everywhere concave upward, or sufficiently overdriven, such that shock compaction traverses a single Rayleigh line. With more complex compaction model structure, multiple wave fronts emerge with shock segments and regions of dispersive wave evolution. Wave analyses towards this end are addressed in chapter 1.

It is now assumed that the compacted density ρ at a position x is dependent on the peak amplitude of the shock pressure P as the shock wave passes over that position. Thus $\rho = \rho(P)$ and the compacted mass element will depend in turn on the position $dm = \rho(x)dx$. The subscripted density ρ_s is here reserved to denote the fully compacted medium or the density of the matrix material.

In equation (7.28), for planar geometry the integral, as before, provides $T = Mu^2/2$, where M is the total mass behind the shock wave. The differential conservation relation in equation (7.22) is also still appropriate with solutions for the mass velocity, kinetic energy and momentum provided by equations (7.23), (7.25) and (7.26). The solution for pressure will now depend on the relation between compaction density and the shock pressure.

7.4.2 Plate impact solutions

The solution for momentum from equation (7.26) reveals the intuitively reasonable observation that momentum is conserved ($\beta = 1$) in the non-curvilinear (planar) Kompaneets problem. In planar geometry momentum conservation can, as an alternative to energy conservation, be invoked to establish the equations governing shock propagation in a porous rigid-compacting medium. An appropriate application of the problem is illustrated in figure 7.8. A rigid solid plate of mass M_p per unit area undergoes planar impact onto an infinite half space $x \geqslant 0$, causing a right facing compaction shock wave at position $x = X(t)$. The material ahead of the shock has the initial density ρ_o, while material behind the shock has a shock-pressure-dependent, and hence a position-dependent, compacted density $\rho_s > \rho(x) > \rho_o$ due to attenuation of the peak shock pressure amplitude.

Decompression incompressibility of the compacted medium requires that the velocity at all positions behind the shock wave be uniform and equal to the mass velocity at the shock state $u(x) = u(X)$. Momentum $S(t) = (M_p + \rho_o X)u$ of the impacting plate and the compacted material behind the shock wave at time t yields the following conservation relation,

$$\frac{dS}{dt} = \left(M_p + \rho_o X\right)\frac{du}{dt} + \rho_o u\frac{dX}{dt} = 0, \qquad (7.29)$$

which in turn provides

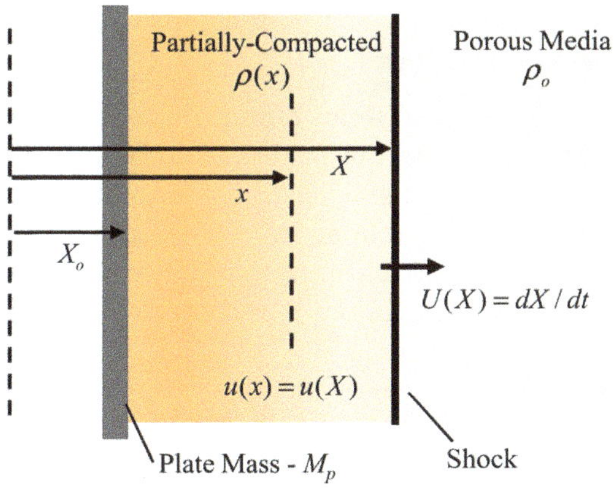

Figure 7.8. Plate impact and subsequent shock propagation within a semi-infinite half space of a porous incompressible-compacting medium.

$$\left(M_p + \rho_o X\right)\frac{du}{dX} + \rho_o u = 0. \tag{7.30}$$

Integration yields for the Hugoniot mass velocity and uniform flow velocity behind the shock

$$u(x) = u(X) = \frac{u_o}{1 + X/b} \tag{7.31}$$

over the range $X_o \leqslant x \leqslant X$, where u_o is the initial impact velocity of the plate and $b = M_p/\rho_o$ provides the problem length scale. For $X/b \gg 1$ the relation becomes asymptotically power-law and the Kompaneets solution for planar geometry in equation (7.23) is recovered. The solution for the particle-velocity field is independent of the crush relation. However, further properties of the shock propagation and planar flow field require knowledge of the compaction properties of the porous material.

The pressure, the compaction strain and the shock velocity are calculated through the solution for the mass velocity provided by equation (7.31), and the Hugoniot relations at the shock front. This step assumes that the compaction curve is concave upward over the compaction range. Otherwise a more complex compaction wave structure ensues. Solution for the pressure and compaction strain at the shock state are determined from the constitutive compaction relation $P(\varepsilon)$ between the pressure P and the strain, $\varepsilon = 1 - \rho_o/\rho$. Pressure and strain are linked to the particle-velocity solution through the Hugoniot relation

$$P(X)\varepsilon(X) = \rho_o u(X)^2. \tag{7.32}$$

Shock velocity is then provided through the further Hugoniot relation

$$U(X)\varepsilon(X) = u(X). \tag{7.33}$$

It remains to develop appropriate constitutive compaction relations $P(\varepsilon)$ descriptive of the pressure versus compaction properties of the materials.

7.4.3 Constitutive compaction relations

There is little theoretical guidance in determining the compaction pressure versus volume response (crush curve) of a porous or distended brittle solid base on strength properties of the constituent materials. There is any number of sensible empirical relations for describing brittle compaction. In this subsection one analytic form is applied for the pressure versus compaction response that is familiar to those in the field of shock compression physics, namely

$$P = \frac{K_o\varepsilon}{(1 - \varepsilon/p)^n}, \tag{7.34}$$

where again $\varepsilon = 1 - \rho_o/\rho$. When $n = 2$ this relation for the pressure versus volume strain is recognized as a consequence of the commonly used linear shock-velocity versus mass-velocity Hugoniot response of a solid where the U–u slope parameter S

is here replaced with $1/p$, with p the initial porosity. In the present context equation (7.34) is regarded as a convenient three-parameter analytic representation for the crush curve of a porous solid. The crush model in equation (7.34) is identified as the *U–u* crush curve, because of its relationship to the corresponding shock wave Hugoniot relation. The initial compaction compressibility and pressure derivative of the compressibility are K_o and $K_o' = 2n/p - 1$, respectively, while the limiting compression $\varepsilon_s = p$ is the initial porosity of the material. Other compaction relations for both brittle and ductile porous solids are explored later in the chapter.

Figure 7.9 shows shock compaction data for a brittle porous solid that is reasonably described by the crush curve in equation (7.34) with $n = 2$. The shock versus mass-velocity data are for SAC-5 concrete, a well-characterized material of earlier research interest (Grady 1993, 1996). These data emphasize the lower shock pressure void-compaction range of this concrete. The data are sensibly described by a linear relation between shock velocity and mass velocity, $U = C_o + Su$, with $C_o =$ 551 m s^{-1} and $S = 4.52$. The constants in equation (7.34) are provided by $K_o = \rho_o C_o^2$, where $\rho_o = 2260$ kg m^{-3}, $n = 2$ and $p = 1/S$. A calculated porosity from $p = 1/S$ of about 22% is in accord with the measured porosity of this material.

The relationship of equation (7.34) to another empirical form used to describe porous compaction of brittle solids warrants mention. There are experimental data that show the compaction of some brittle granular medium is reasonably described by a power-law relation between the compaction pressure and the distention α

$$P = \sigma\alpha^{-\gamma}, \tag{7.35}$$

where σ and γ are constants. The distention is the ratio of the current porous specific volume to the solid specific volume, or conversely, the ratio of the solid to porous density, $\alpha = v/v_s = \rho_s/\rho$. The compaction relation in equation (7.35) is identified as power-law crush. Illustrative are static compaction data for several diverse granular materials. Static compaction pressure versus distention is shown in figure 7.10 for

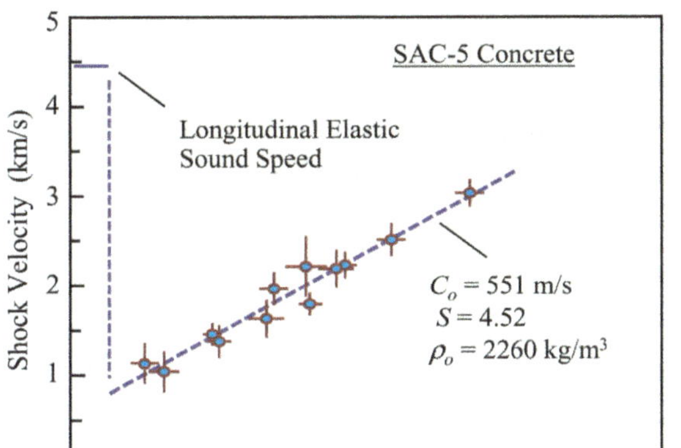

Figure 7.9. Shock wave data for concrete illustrating linear shock-velocity versus mass-velocity behavior in the pressure range of void compaction (Grady 1993, 1996).

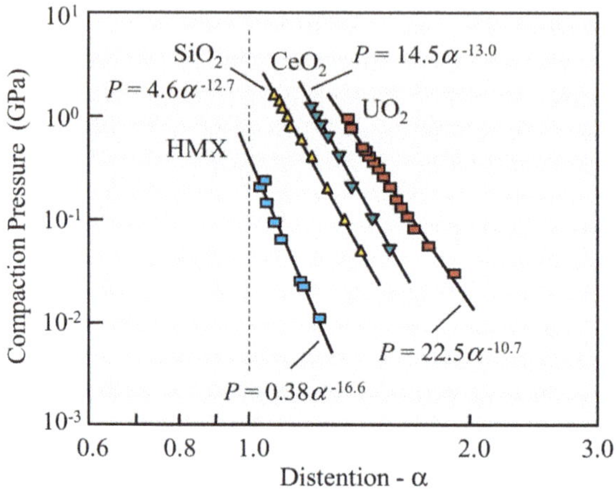

Figure 7.10. Static compaction data for granular media of explosive HMX (Elban and Chiarito 1986) and several ceramic oxide powders (Lee *et al* 2004).

HMX powder and several ceramic oxide powders. Constants for equation (7.35) for several of the materials are shown in the figure.

The compaction modulus for power-law crush is

$$K(\alpha) = -v\frac{dP}{dv} = \gamma P(\alpha), \tag{7.36}$$

with an initial compaction modulus of

$$K_o = K(\alpha_o) = \gamma\sigma\alpha_o^{-\gamma}. \tag{7.37}$$

The pressure derivative of the power-law compaction modulus is accordingly

$$K_o' = \gamma. \tag{7.38}$$

Representative pressure versus compaction curves based on the analytic representation for the *U–u* crush curve from equation (7.34) are shown in figure 7.11. An initial porosity of $p = 0.5$ is used, representative of moderately tamped granular media. An initial modulus of $K_o = 10$ MPa is assumed, representative of a ceramic powder in figure 7.10; compression curves for $n = 1$, 2, 3 and 4 are illustrated on the left in a logarithmic plot.

The corresponding shock-velocity versus mass-velocity plot based on an initial density of 4000 kg m^{-3} and initial shock velocity of $C_o = 50$ m s^{-1} such that $K_o = \rho_o C_o^2 = 10$ MPa are shown in on the right in figure 7.11. Intermediate compaction paths sensibly describe compaction of granular aluminum oxide (Lee *et al* 2004).

A further comparison of *U–u* crush and power-law crush is shown in figure 7.12. The *U–u* curve is based on the properties for SAC-5 concrete identified previously in the text, and in figure 7.9, with $K_o = 686$ GPa, $n = 2$ and $p = 0.22$. The *U–u* crush curve exhibits linear power-law behavior over approximately one and a half decades of compaction pressure. The range of the data from figure 7.9 resides towards the

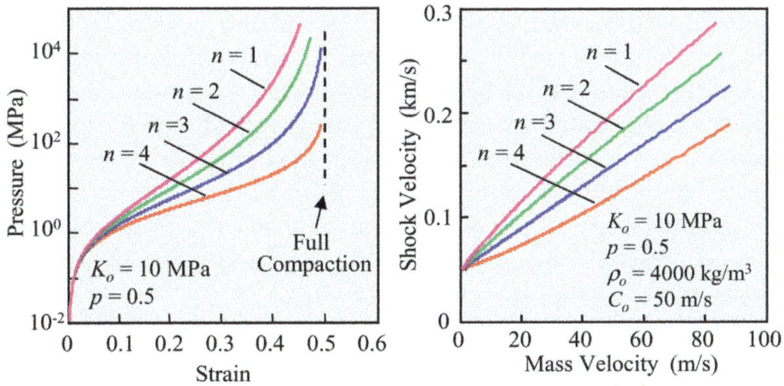

Figure 7.11. Compaction paths (left) and shock velocity versus particle velocity (right) for a porous medium *U–u* crush curve.

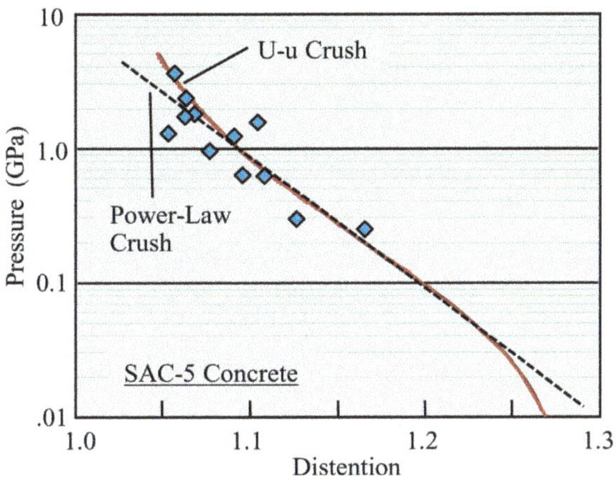

Figure 7.12. The porous medium *U–u* crush curve for SAC-5 concrete is observed to exhibit linear power-law behaviour over approximately one and a half decades in shock pressure.

upper half of the linear behavior. Nonetheless, the power-law behavior observed for other materials as illustrated in figure 7.10 suggest that the data may be equally well fit to a *U–u* crush curve. The *U–u* crush curve has attractive features, including analytic continuity over the full range of compaction. Both compaction relations have found application in dynamic compaction of granular media. Further compaction relations are explored in later sections.

7.4.4 Application of the incompressible-crush model to plate impact on porous media

In an earlier subsection an incompressible-crush approximation to shock compaction was applied to solve the one-dimensional plate-impact-induced compaction wave problem illustrated in figure 7.8. The solution, through momentum balance and the shock conservation relations, is provided by equations (7.29)–(7.33). Here this solution is applied to a specific application for purposes of illustrating issues of compaction wave propagation. In this example a tungsten plate (density 19 000 kg m^{-3}, sonic wave speed 4000 m s^{-1}) 5 mm in thickness is caused to undergo planar (one-dimensional) impact onto the surface of a body of SAC-5 concrete at a velocity of 1000 m s^{-1}. The impact geometry is illustrated in figure 7.8. As described, specific properties of the tungsten plate are not included in the solution, and only the mass of the plate per unit area M = 96 kg m^{-3} is required. The U–u crush model is used to describe compression (compaction) of the SAC-5 concrete with parameters for the model provided by $K_o = 686$ MPa, $n = 2$, $\rho_o = 2260$ kg m^{-3} and $p = 0.22$, which adequately represents the shock data for concrete in figure 7.9.

Particle-velocity and pressure wave profiles that are derived from the solution are shown in the plots in figure 7.13. Shock velocity and the compaction strain as a function of propagation distance are provided in figure 7.14. The constant particle velocity and linear pressure slope of the profiles behind the compaction shock front are consistent with the model assumption of incompressible flow during decompression from the shock state.

The motion (deceleration) of the impact plate from the initial u_o = 1000 m s^{-1} impact velocity is also shown in figure 7.13. The motion of the plate is calculated from

$$X_o(t) = \int_0^t u(t)\,dt = \int_0^X \varepsilon(X)\,dX = X_o(X), \tag{7.39}$$

which provides the displacement of the plate when the shock is at distance X. Deceleration pressure at the plate is calculated from

$$M\frac{du}{dt} = MU(X)\frac{du}{dX} = -P_o(X), \tag{7.40}$$

where $P_o(X)$ is the pressure at the plate–concrete interface when the shock wave is at distance X.

The rapid attenuation of the shock pressure and decrease of the shock velocity with relatively modest drop in the compaction strain (full compaction would be a

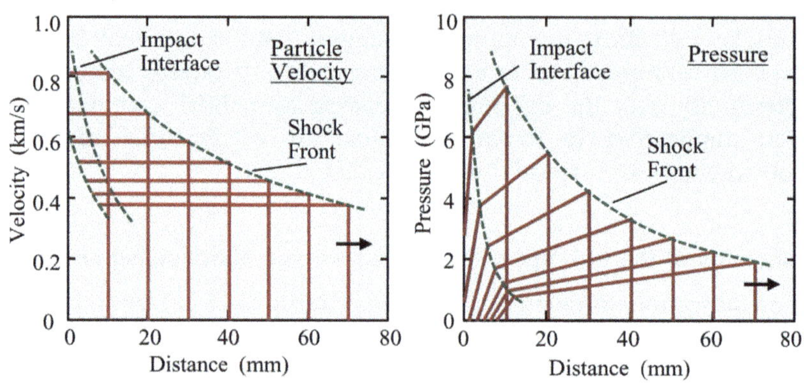

Figure 7.13. Solution for tungsten plate impact on SAC-5 concrete based on an incompressible-crush model of compaction and a U–u description of SAC-5 concrete compaction.

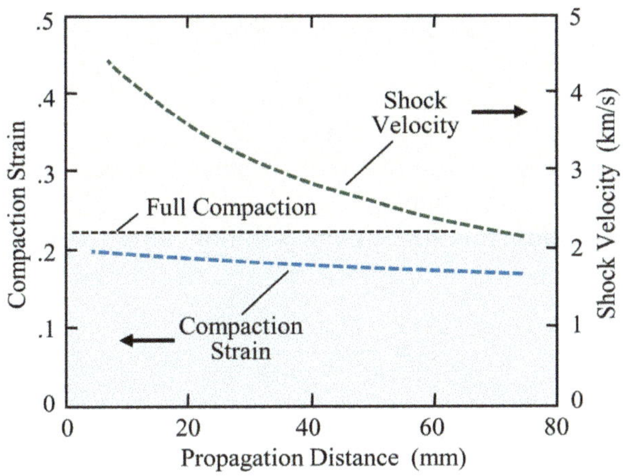

Figure 7.14. Dependence of the compaction strain and shock velocity on propagation distance in the shock compaction of SAC-5 concrete.

strain of 0.22) indicates that compaction dissipation is overdriven within the range of propagation distance. Continued propagation until compaction pressures of 1 GPa and below show a marked drop in the compaction strain. Further features regarding attenuation of the compaction process are revealed, but it is first instructive to develop a non-dimensional representation of the solution.

7.4.5 Non-dimensional solution

Insights into the physics of the dynamic compaction benefit from reducing the analytic solution to dimensionless parameters. Consider further the solution for the planar impact of a metal plate onto a porous medium as depicted in figure 7.8. Let over-bar symbols identify scaled non-dimensional parameters. The solution for the dependence of the shock wave mass velocity on propagation distance corresponding to equation (7.31) is

$$\bar{u} = \frac{1}{1 + \bar{X}}, \tag{7.41}$$

with $\bar{u} = u/u_o$, where u_o is the impact velocity of the plate and the initial particle velocity at the Hugoniot state. The normalized distance $\bar{X} = X/b$ is scaled by the problem length scale $b = M/\rho_o$. The areal mass of the plate is M, while ρ_o is the initial density of the porous material. The distance b translates into a thickness of the porous medium of areal density equal to that of the impacting plate. For example, the distance b for the previous problem of a 5 mm tungsten plate onto SAC-5 concrete is approximately 40 mm.

Scale the strain by the initial porosity (and final compaction strain) such that $\varepsilon = p\bar{\varepsilon}$ with $0 < \bar{\varepsilon} < 1$. Hugoniot conservation of mass and momentum relate pressure, compaction strain and particle velocity through

$$P\varepsilon = \rho_o u^2. \tag{7.42}$$

Normalizing pressure through $pP = \rho_o u_o^2 \bar{P}$ conveniently relates non-dimensional Hugoniot properties through

$$\bar{P}\bar{\varepsilon} = \bar{u}^2. \tag{7.43}$$

The normalizing pressure $\rho_o u_o^2/p$ is that achieved under snow-plow Hugoniot compression to velocity amplitude u_o.

The U–u crush relation applied in the solution of a tungsten plate impacting SAC-5 concrete

$$P = \frac{\rho_o C_o^2 \varepsilon}{(1 - \varepsilon/p)^n} \tag{7.44}$$

becomes, in dimensionless form,

$$\bar{P} = \frac{R^2 \bar{\varepsilon}}{(1 - \bar{\varepsilon})^n}. \tag{7.45}$$

Joining equations (7.43) and (7.45) yields a relationship between the compaction strain and the particle velocity at the Hugoniot state,

$$\bar{u} = \frac{R\bar{\varepsilon}}{(1 - \bar{\varepsilon})^{n/2}}. \tag{7.46}$$

The parameter R is a dimensionless number,

$$R = p\sqrt{\frac{K_o}{\rho_o u_o^2}} = \frac{C_o}{u_o/p}, \tag{7.47}$$

that characterizes the compaction resistance of the material relative to the initial dynamic compaction pressure. Inspection reveals that the smaller the compaction resistance R, the more closely the response replicates snow-plow compaction. The second expression for R on the right in equation (7.47), where $K_o = \rho_o C_o^2$, is recognized as the ratio of the initial compaction wave speed to the shock speed at amplitude u_o under snow-plow compaction.

For the special case of $n = 2$ in equation (7.45), the strain and pressure are directly solved for as a function of mass velocity

$$\bar{\varepsilon} = \frac{\bar{u}}{R + \bar{u}}, \qquad \bar{P} = R\bar{u} + \bar{u}^2. \tag{7.48}$$

The entry normalized compaction strain and pressure are $\bar{\varepsilon}_o = 1/(R + 1)$ and $\bar{P}_o = R + 1$, respectively. As shown in figure 7.9, a value of $n = 2$ is appropriate to the compaction of SAC-5 concrete.

In figure 7.15 solutions for the scaled compaction strain and the shock Hugoniot pressure as a function of scaled propagation distance from the impact plane are calculated from equations (7.41) and (7.48) for increasing compaction resistance (increasing R). Solutions are compared with snow-plow compaction.

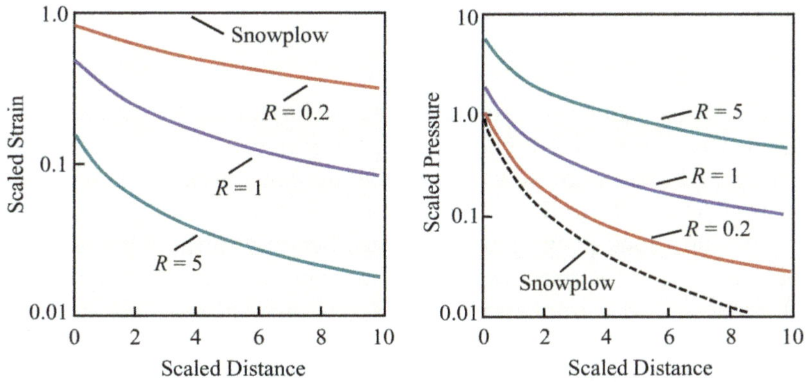

Figure 7.15. Scaled compaction strain (left) and pressure (right) versus distance for select values of the dimensionless compaction resistance R for linear U–u porous compaction.

Increased compaction resistance decreases the magnitude and magnifies the rate of attenuation of the compaction strain. Increased compaction resistance correspondingly increases the scaled compaction pressure. For large compaction resistance, namely $R \gg 1$, the initial scaled compaction strain is $\bar{\varepsilon}_o \simeq 1/R$ and the pressure is $\bar{P}_o \simeq R$, while both decay with range approximately as $\bar{X}^{-1}$. For small compaction resistance ($R \ll 1$) the initial compaction strain and pressure are near unity and reasonably replicated with a snow-plow compaction model. The compaction strain attenuates to one-half at a distance of approximately $\bar{X} = 1/R$. In the more general relation of equation (7.45) the compaction is modestly altered by the additional dimensionless parameter n.

7.4.6 Power-law properties of the limiting solution

In the previous example of a plate impacting an incompressible-compacting material, the distance $b = M/\rho_o$ provides a length scale for the solution. As distances from the impact surface become large compared to the length scale b, the dependence of properties such as the particle velocity, peak pressure, or the compaction strain should transition to a power-law dependence on distance. If sufficiently distant from the impact surface, there is no governing length scale. It is a fundamental law of physics that the character of the field variables must be scale invariant and thus power-law. Similar arguments were made by Sedov (1946) and Tayler (1950) in their analysis of blast waves in air. The same arguments apply here, but the issue is more subtle. It is instructive to examine this transition to power-law behavior using the previous non-dimensional solution of the impact and shock propagation problem. The shock-velocity versus particle-velocity U–u compaction relation from equation (7.34), or in non-dimensional form from equation (7.45), is assumed with $n = 2$ applicable to the SAC-5 concrete since it is readily inverted analytically. The particle-velocity solution in non-dimensional form from equation (7.41) is

$$\bar{u} = \frac{1}{1 + \bar{X}}. \tag{7.49}$$

The particle-velocity transitions to power-law behavior as $\bar{X} = X/b \gg 1$, and this behavior is clearly observed in the left-hand plot of figure 7.16 where the shock front particle-velocity (the same for all values of the resistance parameter R) transitions to an inverse first-power dependence on distance as $\bar{X}$ becomes increasingly greater than unity. As the shock propagation distance becomes large relative to the length scale b, the solution for the particle velocity becomes increasingly less sensitive to the individual values of the impact plate mass M and impact velocity u_o, and is dependent only on the product of the two, or the initial momentum Mu_o.

The corresponding compaction strain is graphed in the center plot and, although transition to a power-law inverse first-power dependence on distance is also observed, the distance at which a transition to power-law behavior occurs depends on the dimensionless parameter R. The transition is seen, at least from small values of R, not to occur until distances many times the characteristic distance (or many times unity in dimensionless units of $\bar{X} = X/b$) are achieved. The strain is solved from equation (7.46) for $n = 2$ yielding

$$\bar{\varepsilon} = \frac{\bar{u}}{R + \bar{u}} \tag{7.50}$$

and, in terms of distance,

$$\bar{\varepsilon} = \frac{1}{1 + R(1 + \bar{X})}. \tag{7.51}$$

When the distance $\bar{X} \gg 1$ equation (7.51) asymptotically approaches

$$\bar{\varepsilon} \approx \frac{1}{1 + R\bar{X}}. \tag{7.52}$$

If the parameter R is large the compaction strain transitions to power-law behavior at about $\bar{X} \sim 1$. If R is small, transition to power-law is not achieved until $R\bar{X} \sim 1$.

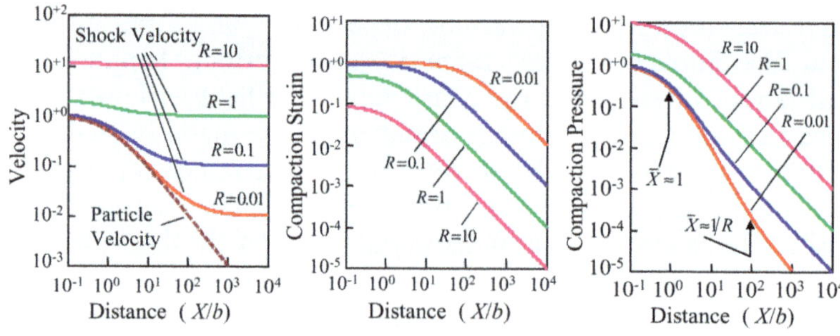

Figure 7.16. Calculated compaction shock wave properties for tungsten plate impact on SAC-5 concrete based on incompressible compaction and the U–u compaction model. Solutions are provided in non-dimensional units for selected values of the non-dimensional compaction resistance R.

Thus, $\bar{X}' = 1/R$, or $b' = b/R$, identifies an additional length scale that scales with the principle length scale through the dimensionless scaling parameter R.

The compaction pressure is similarly written in terms of the particle velocity as

$$\bar{P} = \bar{u}^2(1 + R/\bar{u}), \tag{7.53}$$

and in terms of propagation distance

$$\bar{P} = \frac{1}{(1 + \bar{X})^2}[1 + R(1 + \bar{X})]. \tag{7.54}$$

For $\bar{X} \gg 1$ equation (7.54) becomes

$$\bar{P} \approx \frac{1}{\bar{X}^2}[1 + R\bar{X}] = \frac{1}{\bar{X}^2} + \frac{R}{\bar{X}}. \tag{7.55}$$

When the parameter R is large the pressure transitions to a power law $1/\bar{X}$ at $\bar{X} \sim 1$ as observed in the right-hand plot in figure 7.16. However, when R is sufficiently small there is a range between the two length scales identified by the dimensionless range $1 < \bar{X} < 1/R$ in which $1/\bar{X}^2$ power-law behavior is exhibited by the solution. This complexity of the solution is demonstrated by the curve for $R = 0.01$ in the right-hand plot of figure 7.16. This behavior is similarly demonstrated by the corresponding shock-velocity plot in the left-hand graph.

The regions of different behavior are understood as follows. In the range $0 < \bar{X} < 1$ the intrinsic problem length scale b governs shock attenuation. In the range $1 < \bar{X} < 1/R$ and sufficiently removed from either of the two length scales b and $b' = b/R$, the solution for the pressure is self-similar, and is power-law with a $1/\bar{X}^2$ behavior that is insensitive to the form of the crush model (snow-plow crush). This limiting intermediate trend is common to other areas of physics and is identified with the term intermediate asymptotics (Barrenblatt 2006). (See for example the topic of catastrophic brittle fragmentation in chapter 2.)

In the range $\bar{X} > 1/R$ attenuation again becomes scale invariant and power-law but with a $1/\bar{X}$ dependence on distance that does depend on the crush model. The second length scale $b' = b/R$ is identified as the propagation distance at which attenuation transitions from snow-plow crush to attenuation that depends on details of the crush model.

7.4.7 P-lambda compaction

Discussions have focused on two empirical laws for the crush of distended material. They are the power-law crush curve and the *U–u* crush curve, with much of the analysis dedicated to the latter. Although adequate for describing compaction for many materials, there are arguments for introducing yet a third crush law for distended material.

A compaction model for mixtures of disparate compliance materials, identified as the p-lambda model (Grady *et al* 2000, Grady and Winfree 2001), was developed based on needs similar to that motivating the Herrmann (1969) p-alpha model for porous media. Although developed for describing the compaction of mixtures, the p-lambda compaction model can be specialized to porous solids. Resulting closed-form solutions for porous compaction wave propagation can be used to bench mark computational solutions. (See also discussions of the model in chapter 6.) Further, knowledge of the compaction properties of a porous granular medium can then be extended to mixtures of the same medium with other components through effective pressure methods that parallels the theoretical ideas of Terzaghi (1925) (see sections 7.9.3.2 and 7.9.3.3).

For porous media the p-lambda model is based on the superposition of specific volumes

$$v(P) = \lambda v_s(P) + (1 - \lambda)v_o(P), \tag{7.56}$$

where $v_s(P)$ and $v_o(P)$ are elastic compression relations for the pressure dependence of the specific volume for the fully compacted solid medium, and the initially distended solid, respectively. The parameter λ is a pressure-dependent function ranging from zero for the initial distended material to unity for the fully compacted material. The functional form used for λ based on statistical strength of materials concepts is

$$\lambda = 1 - e^{-(P/\varphi)^r}. \tag{7.57}$$

The parameters φ and r are properties of the material that characterize compaction. Other functional forms can be used where appropriate.

The parameter φ is a pressure scale that characterizes the compaction strength of the material. The dimensionless power r determines the extent of spread of the compaction process over pressure. The parameter r typically ranges over about $0.5 < r < 2$. Unlike the power-law crush curve and the U–u crush curve, for $r > 1$, the p-lambda crush curve leads to structured compaction waves. The model is specialized to incompressible compaction (corresponding to the middle sketch in figure 7.7) by assuming $v_s(P)$ and $v_o(P)$ in equation (7.56) are constant, independent of the pressure P. This restriction results in a crush curve for the compaction strain $\varepsilon = 1 - v/v_o$ of the form

$$\varepsilon(P) = p\lambda = p(1 - \mathrm{e}^{-(P/\varphi)^r}), \tag{7.58}$$

where again p is the initial porosity of the distended medium.

Solutions relating the mass velocity, the compaction pressure and the compaction strain in the shock propagation process are provided by the Hugoniot relations as in the previous sections (assuming $r < 1$ and structured waves are excluded). The pressure is calculated from

$$P\varepsilon = Pp(1 - \mathrm{e}^{-(P/\varphi)^r}) = \rho_o u^2, \tag{7.59}$$

where dependence of the mass velocity on range for the planar impact of a plate onto porous media is provided by equation (7.31).

7.4.8 Non-dimensional solution for p-lambda compaction

The p-lambda incompressible compaction provides a two-parameter functional form for the crush curve and compaction resistance of the material. The same non-dimensional parameters of the previous section for $\bar{X}$, $\bar{u}$, $\bar{\varepsilon}$ and $\bar{P}$ are also appropriate. The non-dimensional compaction strain corresponding to equation (7.58) is

$$\bar{\varepsilon} = 1 - e^{-(\bar{P}/R)^r}. \tag{7.60}$$

Again, R is a dimensionless number characterizing the compaction resistance of the material relative to the driving dynamic pressure. The parameter R for the present incompressible p-lambda model is

$$R = p\frac{\varphi}{\rho_o u_o^2}. \tag{7.61}$$

Combining the compaction resistance function with the shock Hugoniot equation relates the pressure and mass velocity

$$\bar{P}(1 - e^{-(\bar{P}/R)^r}) = \bar{u}^2. \tag{7.62}$$

Solution for the mass velocity at the shock state as a function of propagation distance is provided by equation (7.41) for the plate impact onto porous material. Corresponding expressions for compaction strain, pressure and shock velocity are solved for. Solutions are shown in figure 7.17 where the compaction resistance R increases from left to right. With increasing compaction resistance, both the compaction shock pressure and the shock velocity increase while the compaction strain decreases.

7.4.9 Explosive compaction of porous media

An instructive analysis of the compaction of porous media is performed that combines both the launch of the compacting projectile and the subsequent

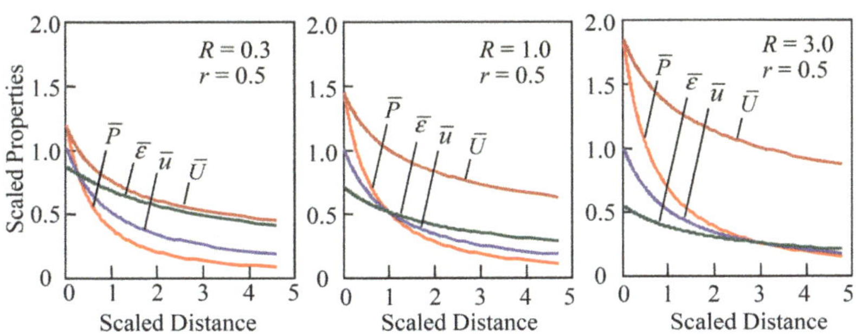

Figure 7.17. Solutions for the compaction strain, pressure, mass-velocity and shock-velocity attenuation versus propagation distance for the planar plate impact problem based on the incompressible p-lambda crush model. Plots from left to right illustrate increasing values of the compaction resistance parameter R. The compaction index of $r = 0.5$ is appropriate for ceramic powders.

compaction resistance of the material. Solutions result that can be compared with complimentary computational analysis. An explosive assembly is illustrated in figure 7.18 where two metal plates with mass per unit area M_A and M_B enclose an explosive of mass C per unit area. Plate M_B interfaces with the porous medium. Either plate M_A or M_B, or both, may be eliminated from the problem.

Detonation and launch of the plates and explosive products occur rapidly relative to the compaction of the porous medium. The two events are nearly decoupled in time. Analysis first focuses on the explosive acceleration of the metal plates and explosive products. The solution methods of Gurney (1943) (see also Jones *et al* (1980) and Kennedy (1997)) are used, and both momentum and energy balance are applied to solve the problem. (The analysis applies the Gurney solutions introduced in chapter 4.)

Consider the explosive charge and plates assembly alone, independent of the distended medium. A stationary plane will exist within the assembly to the right of which plate M_B and explosive products C_B are launched to the right, and to the left of which plate M_A and explosive products C_A are launched to the left. Explosive products in the separate regions satisfy $C_A + C_B = C$. Only momentum to the right of the stationary plane directed toward the porous medium will contribute to the compaction process.

This momentum is written

$$M_B V_B + \int^{C_B} u \, dm. \tag{7.63}$$

The integral is over the thickness of the mass C_B. In the Gurney derivation, velocity is assumed to be linear through the detonating explosive from zero at the stationary plane to velocity V_B at the plate mass M_B interface. In mass coordinates this is expressed as

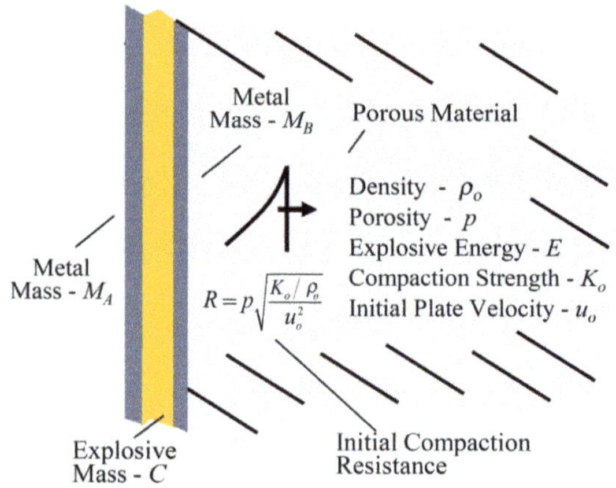

Figure 7.18. Compaction of a porous granular material from planar detonation of an explosive charge and acceleration of metal plates. Relative compaction resistance is determined by the compaction strength of the porous medium, and the energy and mass of the explosive assembly.

$$u(m) = \frac{m}{C_B} V_B \tag{7.64}$$

over $0 < m < C_B$. Completing the integral yields

$$M_B V_B + \frac{1}{2} C_B V_B. \tag{7.65}$$

A corresponding expression for the outward momentum is obtained and momentum balance requires

$$\left(M_B + \frac{1}{2} C_B\right) V_B = \left(M_A + \frac{1}{2} C_A\right) V_A. \tag{7.66}$$

Linearity of the velocity through the explosive thickness yields the proportionality

$$C_A / C_B = V_A / V_B. \tag{7.67}$$

With $C_A + C_B = C$ the above relations are used to solve for the explosive mass C_B,

$$C_B = \frac{C}{2} \frac{2M_A + C}{M_A + M_B + C}. \tag{7.68}$$

The velocity V_B directed towards the porous medium is calculated from energy conservation

$$\frac{1}{2} M_B V_B^2 + \frac{1}{2} \int^{C_B} u^2 \, dm = C_B E, \tag{7.69}$$

where E is a Gurney specific energy characteristic of the explosive. Completing the integral and solving for V_B yields

$$V_B = \sqrt{2E} \bigg/ \left(\frac{M_B}{C_B} + \frac{1}{3}\right)^{1/2}. \tag{7.70}$$

In the earlier solution of compaction caused by an impacting plate, the initial momentum of the plate prior to onset of compaction is identified as $S = Mu_o$. The equivalent momentum for the present explosive compaction event requires the explosive and metal mass contribution

$$M = M_B + \frac{1}{2} C_B, \tag{7.71}$$

and the initial velocity

$$u_o = \sqrt{2E} \bigg/ \left(\frac{M_B}{C_B} + \frac{1}{3}\right)^{1/2}. \tag{7.72}$$

The contributing explosive mass C_B is given by equation (7.68). Representative Gurney properties (Cooper 1997) for selected explosives are provided in table 7.1.

Table 7.1. Gurney properties for representative explosives.

Explosive	Density(kg m^{-3})	Detonation velocity(km s^{-1})	Gurney velocity $\sqrt{2E}$(km s^{-1})
Comp-B	1720	7.9	2.7
HMX	1840	8.8	2.8
LX-14	1890	9.1	3.0
PBX-9404	1840	8.8	2.9
TNT	1630	6.7	2.4
Tritonal	1720	6.7	2.3

An infinite half space of a porous medium subjected to a strong impulsive load and subsequent shock compaction through the detonation of an explosive assembly is illustrated in figure 7.18. For a porous material consider a granular body of silicon dioxide (SiO_2). The compaction response is characterized through static compaction experiments presented earlier in figure 7.10. Some complexities of compaction of distended SiO_2 are noted later in the chapter (see figure 7.47). Compaction of SiO_2 is subjected to analytic description in section 7.6.2.

Alternatively, the extreme shock Hugoniot response of porous SiO_2 is illustrated in figure 7.19 (Trunin *et al* 2001; see also Grady *et al* 2013). Shock data for a span of initial porosities are shown. These Hugoniot compaction data illustrate the substantial thermal expansion characteristics inherent to the extreme shock compaction event. These Hugoniot data exemplify the shock compaction EOS issues addressed in section 7.8. The present analysis, however, addresses the lower pressure mechanical compaction of granular material, and the compaction data in figure 7.10 sufficiently represent the expected compaction behavior

The *U–u* compaction relation provided in equation (7.34) is used to describe the data. A compaction modulus for SiO_2 of K_o = 50 MPa and a strengthening

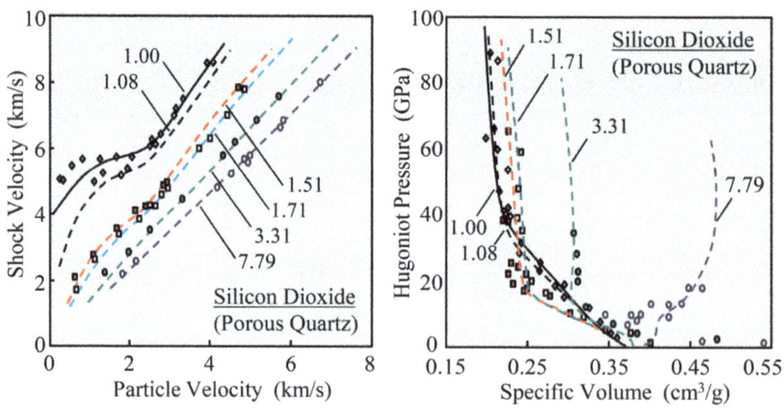

Figure 7.19. Shock Hugoniot response of porous silicon dioxide. Data from Trunin *et al* (2001).

parameter of $n = 2$ adequately represent the data. The initial porosity for the material is $p = 0.47$ (density of 1.4 g cm^{-3}).

The solution for the present compaction problem is provided by equations (7.41) and (7.48). The normalizing length scale is $b = M/\rho_o$ with ρ_o the initial density of the distended SiO_2 ceramic, while the mass M is provided by equation (7.71). The compaction resistance from equation (7.47) is

$$R = p\sqrt{\frac{K_o}{\rho_o u_o^2}}, \tag{7.73}$$

with u_o provided from the (7.72) Gurney equation. Compaction curves for selected explosive system conditions are shown in figure 7.20.

For an assembly with an outer metal plate and no inner plate ($M_B = 0$), an explosive pad of TNT with a Gurney velocity from table 7.1 of $\sqrt{2E} = 2.4$ km s^{-1} and SiO_2 powder with a porosity of $p = 0.47$, the initial compaction resistance parameter from equations (7.72) and (7.73) reduces to

$$R = p\sqrt{\frac{K_o/\rho_o}{6E}}, \tag{7.74}$$

and provides a value of R of 0.021, indicating near-complete initial compaction.

To provide a quantitative assessment of the solution a one centimeter thick plate of TNT is assumed with metal plates M_A and M_B opposite and adjacent to the SiO_2 porous medium, respectively, as illustrated in figure 7.18. Identify $M_{AB} = M_A + M_B$ with the total metal mass in the metal–charge assembly. Four calculations are illustrated in figure 7.20. For one pair of calculations the metal mass is comparable to the charge mass ($M_{AB}/C = 1$). For the second pair the metal mass is a factor of ten larger than the charge mass ($M_{AB}/C = 10$). For each set of calculations the metal mass adjacent to the porous medium is alternatively modest ($M_B/M_{AB} = 0.1$) or dominant ($M_B/M_{AB} = 0.9$). When most of the metal mass is opposite to the porous

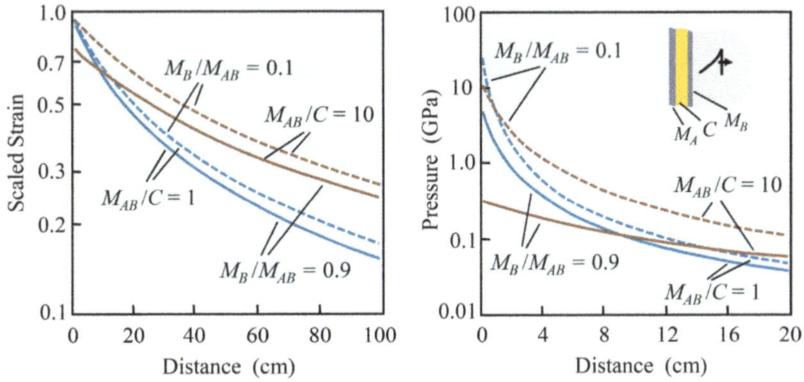

Figure 7.20. Solutions for planar metal and explosive assembly shock compaction of porous silicon dioxide. One centimeter thickness of TNT explosive charge with various metal mass configurations. Total metal mass is $M_{AB} = M_A + M_B$.

medium ($M_B/M_{AB} = 0.1$) the assembly configuration more closely replicates a tamped system where effort is made to direct explosive energy for the purpose of maximum effect—explosive compaction in the present case.

Plots of calculated compaction of the porous SiO_2 medium in figure 7.20 display the compaction strain as a function of distance on the left and compaction pressure on the right. The pressure plot is more closely focused on the first twenty centimeters of propagation distance where initial velocity and imparted mass separately influence the compaction. The results are interesting. Irrespective of the disposition of the metal mass, more effective compaction is achieved when the metal mass is significantly greater than the charge mass ($M_{AB}/C = 10$). Effective tamping with most metal mass opposite to the porous medium ($M_B/M_{AB} = 0.1$) modestly improves compaction (dashed curves), but the metal–charge mass ratio is the dominant influence. The calculated results are of course subject to limitations of the Gurney solution and the separation in time of the Gurney explosive acceleration and the subsequent compaction calculation.

7.5 Compaction resistance in non-planar geometries

Analytic solutions to problems involving the shock compaction of porous media exhibiting compaction resistance in planar one-dimensional geometries are tractable, as illustrate by the several examples pursued in the previous section. Analytic solutions would have decidedly wider application if they could be extended to non-planar geometries, such as the outwardly directed (explosive) or inwardly directed (implosive) shock compaction of porous media in spherical or cylindrical geometries. In this section several approximate analytic solutions in non-planar geometries are explored. This brief effort points to possible solution methods, and identifies the complications of analysis in non-planar geometries.

In the first effort the point source spherical explosion geometry first solved by Kompaneets (1956) for the non-resisting compaction of a porous body is extended to a compaction resistant material. The second problem considers the cylindrical compaction of a metal tube that is fully loaded with a porous medium. The introduction of a crush curve in problems with curvilinear geometries adds considerably to the analytic complexity. Computational simulations become increasingly attractive.

7.5.1 Kompaneets spherical geometry with compaction resistance

The objective is to extend the solution of Kompaneets (1956) to porous media that exhibit compaction resistance through a representative crush curve. The analysis focuses on outwardly directed one-dimensional spherical compaction waves generated by a near-point deposition of energy. Extension to cylindrical compaction is readily apparent.

Starting with the expression for kinetic energy for the mass within the region enclosed by the outwardly moving cavity radius X_o and the shock wave radius X as illustrated in figure 7.1,

$$T = \frac{1}{2}\int_{X_o}^{X} u^2\, dm. \tag{7.75}$$

Parameters identified in figure 7.1 continue to apply. The mass element is $dm = 4\pi\rho(x)x^2dx$ and differs from the earlier derivation in that the compacted density $\rho(x)$ is now a function of range through dependence of the extent of compaction on the amplitude of the shock wave. After passage of the shock the material is again assumed to be incompressible and the expression for the kinetic energy is

$$T = \frac{1}{2}\left(\frac{3}{\rho_o}\int_{X_o/X}^{1} \rho(\xi)\xi^{-2}\, d\xi\right)\left(\frac{4\pi}{3}\rho_o X^3\right)u^2(X). \tag{7.76}$$

Equation (7.76) differs from the earlier equation (7.6) only in the dependence of the density $\rho(\xi)$ on propagation distance.

Density is removed from the integral through the mean value theorem

$$\beta = 3\frac{\bar{\rho}}{\rho_o}\int_{X_o/X}^{1} \xi^{-2}\, d\xi, \tag{7.77}$$

where $\bar{\rho}$ is an average value of the density over the interval $X_o/X \leqslant \xi \leqslant 1$. A mean compaction within the interval is identified as

$$\bar{\varepsilon}(M) = \left(\frac{X_o}{X}\right)^3 = 1 - \frac{\rho_o}{\bar{\rho}}, \tag{7.78}$$

where the mean compaction depends on the position of the shock X, or equivalently the mass $M = 4\pi\rho_o X^3/3$ encompassed by the shock wave.

The approximation of incompressible flow behind the shock wave which implied no further compaction, or dilation, with subsequent deformation is questionable. It is worth noting that this approximation is comparable to computational models such as the p-alpha model (Herrmann 1969) where the extent of compaction is determined strictly by the maximum pressure achieved in the compression wave. Subsequent deformation after passage of the peak compaction pressure is assumed to not further affect the level of compaction.

Completing the integral yields

$$\beta(M) = \frac{3}{\bar{\varepsilon}^{1/3} + \bar{\varepsilon}^{2/3} + \bar{\varepsilon}}, \tag{7.79}$$

which has the same functional form as in the previous development of the Kompaneets solution but contains a dependence on the range of the shock wave through the average compaction $\bar{\varepsilon}(M)$. The kinetic energy becomes, as before,

$$T = \frac{1}{2}\beta M u^2, \tag{7.80}$$

where mass velocity u and the parameter β are dependent on the position M of the shock front.

Also, as in the Kompaneets analysis, kinetic energy is reduced as the shock propagates according to the relation $dT = -u^2 dm/2$, or

$$\frac{1}{2}\beta u^2\, dM + \beta M u\, du + \frac{1}{2}M u^2\, d\beta = -\frac{1}{2}u^2\, dM, \tag{7.81}$$

and differs in the dependence on the variation of β. Equation (7.81) is solved for the mass velocity in the form

$$\frac{du}{u} = -\frac{(1+\beta)}{2\beta}\frac{dM}{M} - \frac{d\beta}{2\beta}. \tag{7.82}$$

The parameter β depends on M through the average compaction $\bar{\varepsilon}(M)$.

Identify V as the volume enclosed by the shock at position X (proportional to X^3) and V_o as the corresponding volume enclosed by the cavity radius X_o (proportional to X_o^3). The volume occupied by the mass M enclosed by the shock wave is then $V_M = V - V_o$. The average compaction is given by

$$\bar{\varepsilon} = V_o/V. \tag{7.83}$$

The change in the average compaction as the shock propagates through a mass M is

$$d\bar{\varepsilon} = \frac{dV_o}{V} - \frac{V_o dV}{V^2} = \frac{dV_o}{V} - \bar{\varepsilon}\frac{dV}{V}. \tag{7.84}$$

Note the relations $M = \rho_o V$, $dM = \rho_o\, dV$ and $dM = \rho(M) dV_M$. The last expression is key in relating the mass increment encompassed by the shock to the Hugoniot density immediately behind the shock. With these expressions equation (7.84) is written

$$d\bar{\varepsilon} = \left(1 - \frac{\rho_o}{\rho(M)}\right)\frac{dM}{M} - \bar{\varepsilon}\frac{dV}{V}. \tag{7.85}$$

The quantity within the parentheses is the Hugoniot strain $\varepsilon(M)$. The rate of change of the average compaction is written

$$\frac{d\bar{\varepsilon}}{\bar{\varepsilon}} = -\left(1 - \frac{\varepsilon(M)}{\bar{\varepsilon}}\right)\frac{dM}{M}. \tag{7.86}$$

Equations (7.82) and (7.86) provide two ordinary differential equations for the solutions for the mass velocity u at the shock front and the average compaction strain $\bar{\varepsilon}$. Numerical solution of the two ordinary differential equations is required.

7.5.2 Compaction of a metal tube with porous media

The problem is that of a thin-walled metal tube filled with a porous material. Compaction of the material occurs when the metal tube is given an initial inward velocity u_o such as might be induced by detonation of a thin explosive sheet enveloping the tube. Particle motion and shock propagation is assumed one-dimensional cylindrical. The problem is illustrated in figure 7.21.

The initial radius of the tube is X_o, while X_s is the current radius of the inward-directed compaction shock wave, and X is a Lagrangian coordinate within the interval $X_s \leqslant X \leqslant X_o$. The mass

$$m(X) = \rho_o \pi \left(X_o^2 - X^2\right) = M_o\left(1 - X^2/X_o^2\right) \tag{7.87}$$

per unit length of the tube is an alternative mass coordinate corresponding to Lagrangian coordinate X. $M_o = \rho_o \pi X_o^2$ is the mass of porous material per unit length of the tube. The initial density of the porous medium is ρ_o, while $M = m(X_s)$ is the current mass encompassed by the shock wave. After initiation of the problem, when the shock is at a radial position X_s, the kinetic energy of the system is

$$T = \frac{1}{2}\int_0^M u^2(m)\, dm + \frac{1}{2}u^2(0)M_T. \tag{7.88}$$

The mass per unit length of the tube is M_T. The inward-directed particle velocity at coordinate m is $u(m)$.

As previously, assumption of a porous material that compacts within the shock wave, and remains incompressible during pressure decompression (neither further compaction nor dilation) is assumed. As noted earlier there are concerns with such a model in curvilinear geometries. Incompressible flow yields

$$\frac{u(0)}{u(m)} = \frac{x}{x_o} \cong \frac{X}{X_o}. \tag{7.89}$$

Eulerian coordinates x and x_o determine the current positions of particles identified by the Lagrangian coordinates X and X_o. The approximation in equation (7.89) leads to a modest under-prediction of the kinetic energy in equation (7.88). It

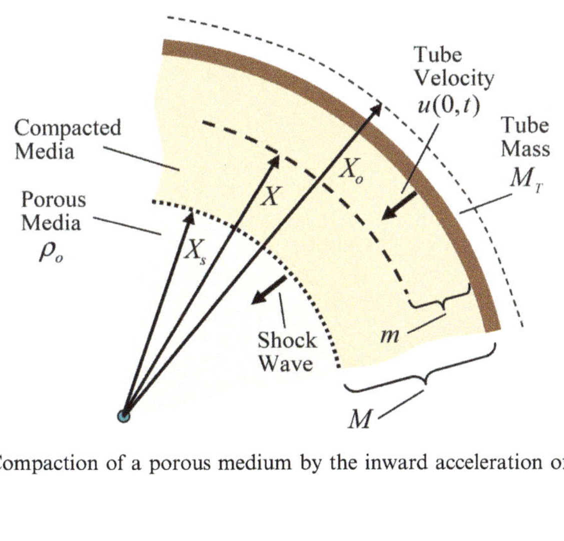

Figure 7.21. Compaction of a porous medium by the inward acceleration of a metal tube.

compensates by leading to analytically tractable equations and provides in most cases reasonable estimates of the compaction profile. Combining equations (7.87) and (7.89) yields

$$u(m) = \frac{u(0)}{(1 - m/M_o)^{1/2}}, \tag{7.90}$$

and the expression for the kinetic energy behind the shock wave encompassing mass M becomes

$$T(M) = \frac{1}{2}u^2(0)\int_0^M \frac{dm}{1 - m/M_o} + \frac{1}{2}u^2(0)M_T. \tag{7.91}$$

Identify the dimensionless ratio of the porous filler and metal tube mass $R = M_o/M_T$. Then

$$T(M) = \frac{1}{2}u^2(0)M_T\left(1 + R\int_0^{M/M_o} \frac{d\xi}{1 - \xi}\right), \tag{7.92}$$

with integration variable $\xi = m/M_o$. The solution yields

$$T(M) = \frac{1}{2}u^2(0)M_T(1 - R\ln(1 - M/M_o)). \tag{7.93}$$

Over the range $0 \leqslant m \leqslant M$, the particle velocity is a function of both the mass coordinate and the time $u(m, t)$. At $m = 0$ the velocity $u(0, t)$ is the $u(0)$ identified in equation (7.93), and is the time-dependent inward velocity of the tube wall as compaction and deceleration proceeds. In equation (7.93) and henceforth, $u(0, t)$ will be identified as u with initial value $u(0, 0) = u(0) = u_o$.

The governing compaction equations are obtained from the energy relation depleting kinetic energy through dissipation in the shock front,

$$dT = -\frac{1}{2}u_s^2\, dM, \tag{7.94}$$

where dM is the increment of mass encompassed by the propagating shock over a corresponding time increment, and u_s is the particle velocity at the shock front. From equation (7.90)

$$u_s(M) = \frac{u}{(1 - M/M_o)^{1/2}}. \tag{7.95}$$

Equations (7.93), (7.94) and (7.95) yield

$$M_T[1 - R\ln(1 - M/M_o)]u\, du + \frac{u^2}{(1 - M/M_o)}dM = 0. \tag{7.96}$$

Integration provides for the velocity of the tube wall when the shock encompasses mass M,

$$u(M) = \frac{u_o}{1 - R\ln(1 - M/M_o)}, \tag{7.97}$$

and for the particle velocity at the shock front,

$$u_s(M) = \frac{u_o}{(1 - R\ln(1 - M/M_o))(1 - M/M_o)^{1/2}}. \tag{7.98}$$

Let $d_o = X_o$ be the initial radius of the metal tube. The propagation depth of the shock wave is given by

$$d/d_o = 1 - (1 - M/M_o)^{1/2}. \tag{7.99}$$

Tube wall velocity and particle velocity at the shock front are shown as a function of depth of the propagating shock in the plot on the left in figure 7.22 for several values of the mass ratio $R = M_o/M_T$. For the calculation depicted in figure 7.22, the initial inward velocity of the metal tube wall is $u_o = 700$ m s^{-1}. Tungsten carbide powder at a porosity of $p = 0.5$ is the filler. Values of $R = 3$ and 10 correspond to a 50 mm diameter tantalum tube containing the tungsten carbide powder with wall thicknesses of 3.9 mm and 1.2 mm, respectively. The velocity relations in equations (7.97) and (7.98) are independent of the compaction constitutive relation. This independence of the compaction constitutive relation also occurs in analytic solutions of planar compaction waves. This result is a consequence of the assumption of incompressible flow subsequent to shock compaction.

Increasing values of the parameter R leads to increasing rates of attenuation of the shock particle velocity with depth of penetration of the shock into the porous medium. In all cases, however, cylindrical convergence of the shock wave leads to a transition to shock amplification as the shock wave converges on the centerline.

As in earlier sections, the propagation shock velocity and shock pressure along with the compaction strain are determined from the velocity solution, a compaction constitutive relation $P(\varepsilon)$ and the shock conservation relations

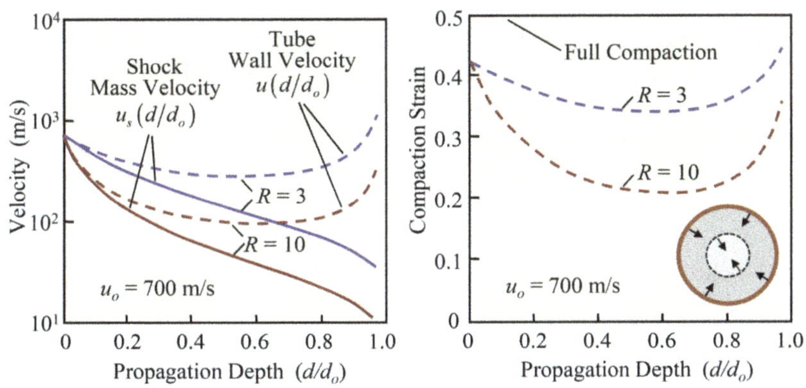

Figure 7.22. Cylindrical compaction of a tube of tungsten carbide powder for several initial tube masses to powder mass ratios. Left: Cylinder wall-velocity and shock particle-velocity dependence on shock propagation depth. Right: Compaction strain as a function of depth.

$$P(M)\varepsilon(M) = \rho_o u_s(M)^2 \tag{7.100}$$

and

$$U(M) = u_s(M)/\varepsilon(M). \tag{7.101}$$

Assuming a compaction constitutive relation of the form

$$P(\varepsilon) = \frac{K\varepsilon}{(1 - \varepsilon/p)^2}, \tag{7.102}$$

a value of $K = 500$ MPa with a porosity of $p = 0.5$ provides a reasonable fit to experimental shock and static compaction data for tungsten carbide powder (Vogler *et al* 2007). (See also section 9.2.) Equations (7.98), (7.100) and (7.102) are combined to provide a prediction of the compaction strain profile in tungsten carbide for the initial inward wall velocity of $u_o = 700$ m s^{-1} and for the previous values of the parameter R. Compaction strain profiles are shown on the right in figure 7.22.

7.6 Physics-based compaction models

A compaction EOS refers to a thermo-mechanical, or strictly mechanical, constitutive relation for describing the compaction resistance of a porous solid to an applied dynamic or static stress load. Selected compaction EOSs have been introduced in earlier sections. The relations considered have been empirical or, at best, semi-empirical. These constitutive relations are revisited in the present section. Consideration is also given to more physically based compaction EOSs. In addressing the full thermo-mechanical response depicted in figure 7.6, the present discussions pertain to the path by which compaction and the thermodynamic surface are joined.

7.6.1 Carroll and Holt porous compaction model

A theoretical development for the dynamic compaction of porous materials was put forth by Carroll and Holt (1972). Their model was posed within the framework of the, at that time, recently developed p-alpha model of Herrmann (1969). The development of the model proceeds through analysis of the symmetric plastic compaction of a spherical ductile shell.

Carroll and Holt restricted thinking to a competent porous elastic perfectly plastic metal with spherical hole-like voids. It is intuitive, however, that after the initial compaction of a ductile metal powder, subsequent void compaction would approach the material behavior considered by Carroll and Holt. Although initial details of their model (such as the initial elastic response) may not apply to compaction of a porous metal powder, the later ductile compaction should be readily captured. Discussions here follow the development of Carroll and Holt. The same model was developed earlier by Chadwick (1963) for application to static compaction of porous metals based on symmetric compression of the spherical shell assuming an incompressible, work-hardening material. Applicability of the model to more brittle materials, in which fracture is expected to dominate the compressive failure and compaction process, is less clear.

Intrinsic to the Carroll–Holt model is the assumption that the symmetric response of a hollow spherical element of inner radius a and outer radius b when subjected to a pressure P on the outer boundary adequately approximates the average response of a corresponding element of the porous medium. Whether the corresponding material element contains one void, or many voids, is open to discussion. This issue relates to whether or not the material response, and the model, is dependent on the microstructure length scale.

Subject to a uniform outer pressure P and zero pressure on the inner surface of the spherical shell, the stress components (positive in compression) over the radius r are

$$\sigma_r = P\frac{b^3}{r^3}\frac{(r^3 - a^3)}{(b^3 - a^3)} \quad (7.103)$$

and

$$\sigma_\theta = P\frac{b^3}{2r^3}\frac{(2r^3 + a^3)}{(b^3 - a^3)}. \quad (7.104)$$

Identify the porosity of the hollow sphere as $p = a^3/b^3$. It is convenient to write the stresses in the form

$$\sigma_r = \frac{P}{\xi}\frac{(\xi - p)}{(1 - p)} \quad (7.105)$$

and

$$\sigma_\theta = \frac{P}{2\xi}\frac{(2\xi + p)}{(1 - p)}, \quad (7.106)$$

where $\xi = r^3/b^3$ and $p \leqslant \xi \leqslant 1$. Defining the mean stress $\bar{\sigma} = (\sigma_r + 2\sigma_\theta)/3$ and the shear stress $\tau = (\sigma_\theta - \sigma_r)/2$ yields

$$\bar{\sigma} = \frac{P}{1 - p} \quad (7.107)$$

and

$$\tau = \frac{3}{4}\frac{P}{\xi}\frac{p}{1 - p}. \quad (7.108)$$

The distention α of the porous medium and the porosity p are related through $1/\alpha = 1 - p$, providing alternative expressions for the stress terms,

$$\bar{\sigma} = \alpha P \quad (7.109)$$

and

$$\tau = \frac{3}{4}\frac{P}{\xi}(\alpha - 1). \quad (7.110)$$

The mean stress is constant through the thickness of the shell. The shear stress is a maximum at the inner radius. For increasingly larger porosity, both mean stress and shear stress increase relative to the magnitude of the applied pressure.

If ductile flow is characterized by a constant von Mises stress Y, then yield first initiates at the inner boundary when the shear stress achieves $\tau = Y/2$. As the boundary pressure P is further increased yield onset propagates toward the outer boundary until plastic flow subsumes the shell as compaction proceeds. The stress field within this transition regime is provided in the paper by Carroll and Holt (1972). They also note that little deformation occurs within the elastic and the elastic-plastic regime relative to the full compaction process. This observation simplifies the governing equations emerging from the model through assumption of incompressibility of the matrix material.

Static equilibrium requires

$$\frac{\partial \sigma_r}{\partial r} - \frac{4\tau}{r} = 0. \tag{7.111}$$

In the fully plastic regime $\tau = Y/2$ and, in terms of the independent variable $\xi = r^3/b^3$, equation (7.111) becomes

$$\frac{\partial \sigma_r}{\partial \xi} - \frac{2Y}{3\xi} = 0. \tag{7.112}$$

Integration yields

$$\sigma_r(\xi) = \frac{2Y}{3} \ln(\xi/p) \tag{7.113}$$

and, when evaluated at the outer boundary, this provides the relation between applied pressure and porosity as compaction of the ductile shell proceeds (e.g. Tone 1948, Tong and Ravichandran 1997),

$$P = -\frac{2Y}{3} \ln p. \tag{7.114}$$

The elastic strain energy density at a point within the hollow sphere in response to the applied pressure P is written

$$u(r) = \frac{1}{2E}\left(\sigma_r^2 + 2\sigma_\theta^2\right) - \frac{\nu}{E}\left(\sigma_\theta^2 + 2\sigma_r\sigma_\theta\right), \tag{7.115}$$

where E and ν are the Young's modulus and Poisson's ratio of the solid, respectively, while σ_r and σ_θ are provided by equations (7.105) and (7.106). Introducing

$$\sigma_r = \bar{\sigma} + 4\tau/3 \tag{7.116}$$

and

$$\sigma_\theta = \bar{\sigma} - 2\tau/3, \tag{7.117}$$

equation (7.115) becomes

$$u(r) = \frac{1}{2K}\bar{\sigma}^2 + \frac{2}{3G}\tau^2, \tag{7.118}$$

with K and G the elastic bulk and shear modulus, respectively. The total strain energy within the compressed sphere is the integral

$$U = \int_a^b 4\pi r^2 u(r)\, dr \tag{7.119}$$

or, after changing to integration variable ξ,

$$U = \frac{4\pi}{3}b^3 \int_p^1 u(\xi)d\xi. \tag{7.120}$$

Completing the integral provides the total elastic strain energy

$$U = \frac{1}{2}\alpha\left(\frac{4\pi}{3}b^3\right)\left(\frac{1}{K} + \frac{p}{4G/3}\right)P^2. \tag{7.121}$$

From equation (7.121) the elastic bulk modulus $\bar{K}$ of the spherical shell element, and by inference that of the porous medium, is provided by

$$\frac{1}{\bar{K}} = \alpha\left(\frac{1}{K} + \frac{p}{4G/3}\right) \tag{7.122}$$

and depends on both the bulk and shear moduli of the matrix material, as well as on the porosity of the shell. If incompressibility of the matrix material is assumed ($K \to \infty$), then

$$\bar{K} = \frac{4}{3}\frac{G}{\alpha p} = \frac{4}{3}\frac{G}{\alpha - 1}. \tag{7.123}$$

Whether or not incompressibility is assumed, it is the elastic shear strain energy that initiates failure and compaction of the spherical element, either through plastic flow or through fracture in the case of a brittle material. The shear elastic strain energy from equation (7.121) is written

$$U_d = \frac{1}{2}\alpha\left(\frac{4\pi}{3}b^3\right)\left(\frac{p}{4G/3}\right)P^2 = \frac{1}{2}(\alpha - 1)\left(\frac{4\pi}{3}b^3\right)\frac{3}{4G}P^2. \tag{7.124}$$

The total volume compression strain ε of the shell element is calculated from P and $\bar{K}$. Assuming compressibility of the matrix material, the volume strain in the solid ε_1 is calculated from $\bar{\sigma}$ and K. The volume strain of the void ε_2 is then calculated from the volume fraction averaged strain $\varepsilon = \eta_1\varepsilon_1 + \eta_2\varepsilon_2$. Compaction is the diminution of the volume strain ε_2 as outer pressure on the shell increases.

Either equation (7.114) or equation (7.124) provides a model-based crush curve that can be tested with experimental compaction data. Equation (7.114) is a material strength-based crush curve that provides a dependence of porosity on pressure as compaction proceeds. The compaction model has been used by others and has been found to describe reasonably well the compaction of porous ductile metals. Equation (7.124), when constrained to a constant shear strain energy provides a relation between porosity and compaction pressure, and an alternative energy-based crush curve.

Such crush models based on the Carroll and Holt principles demonstrate reasonable agreement with compaction of porous metals (e.g. Tong and Ravichandran 1997). Less certain is the applicability of the model to the compaction of granular brittle materials. As compaction of brittle media dominates the discussions in the present chapter, comparisons of the Carroll and Holt model with compaction data for a brittle material is offered. Such comparisons illuminate the marked differences between the crush of ductile solids and that of brittle granular materials.

Shock compaction studies on dry quartz sand (Brown *et al* 2007) provide a set of dynamic crush data to compare with the Carroll and Holt compaction models. In this investigation a quartz sand with a distended density of 1.57 kg m^{-3} and a porosity of 41% based on a crystal density of 2.66 kg m^{-3} was subjected to shock compaction testing. The measured shock wave states for the dry quartz sand are plotted on the left in figure 7.23 in the shock-velocity versus particle-velocity plane common to the display of shock Hugoniot data. These same data through the appropriate shock wave analysis are mapped into a shock pressure versus compacted distention α plot on the right in the figure. The initial distention before compaction is approximately $\alpha_o = 1.7$. Also shown with the data is a power-law crush curve of the form $P = \sigma\alpha^{-\gamma}$, described previously in equation (7.35), that has been found to sensibly describe the compaction behavior of a number of brittle granular solids. It is also notable that the slope of $S = 2.35$ with a U–u crush relation from equation (7.34) with $n = 2$ equates to an initial porosity of p = $1/S = 0.43$, remarkably close to the initial porosity of 0.41 for the SiO_2 test samples.

The same data are again plotted in figure 7.24 and compared to several models descriptive of the compaction event. Both the stress-based compaction from equation (7.114) and the energy-based crush curve from equation (7.124) based on the Carroll–Holt spherical shell element analysis are compared with the dynamic compaction data. In each description incompressibility of the matrix material and a von Mises stress onset of compaction is assumed. Neither of these single-parameter physics-based crush curves provide an adequate representation of the experimental dynamic compaction data for quartz sand. Although sensible for the compaction of ductile metals, the Carroll and Holt model does not capture the compaction properties of most granular brittle materials.

Reasonably simple physical arguments suggest an explanation for the marked difference between ductile compaction and the crush of brittle materials. Strength of brittle solids is known to depend on the size scale of the failing element and is commonly characterized by a Weibull power-law relation of the form $\sigma_f a^m = \beta$. The

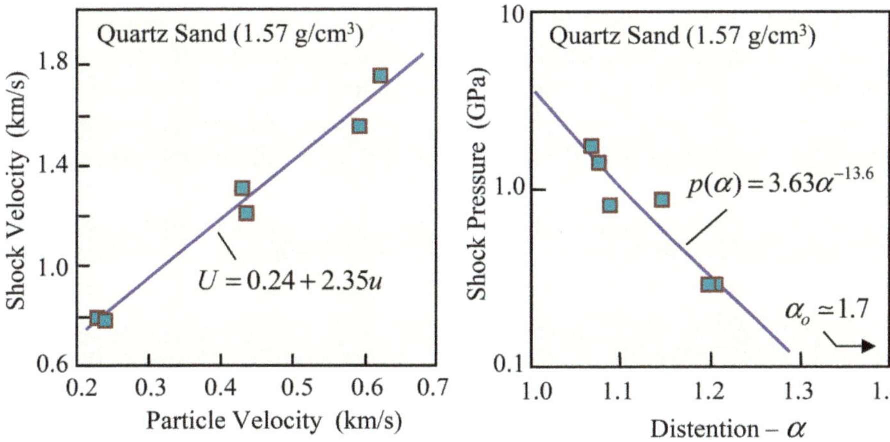

Figure 7.23. Left: Shock Hugoniot data for dry quartz sand (Brown *et al* 2007). Right: The shock compaction states described by a power-law crush curve.

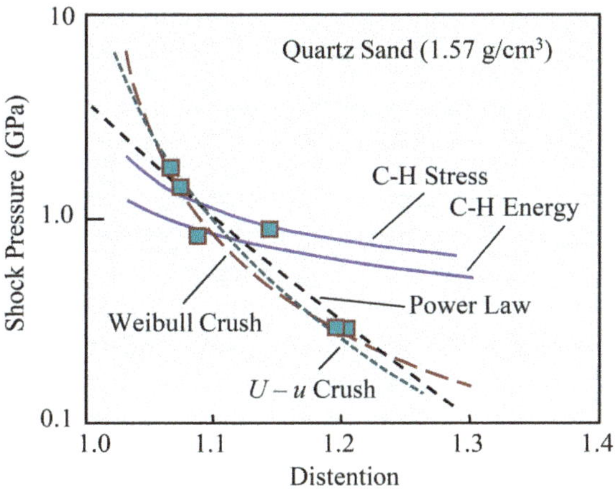

Figure 7.24. Comparison of selected compaction models with shock compaction data for quartz sand. Data from Brown *et al* (2007).

stress property σ_f is the fracture strength, while a is the characteristic size of the failing element and β is a constant property of the material. Consider again the Carroll–Holt spherical shell model with a the size of the cavity and b the size of the shell. Excepting the Weibull strength concept, it is reasonable to propose that the compaction pressure and cavity size are related through $Pa^m = \beta$. Porosity is related to cavity size and shell size through $p \sim (a/b)^3$. The analysis leads to a relation between compaction pressure and porosity,

$$P = \frac{\beta}{a^m} = \frac{\beta}{b^m p^{m/3}} = \frac{\beta}{b^m}\left(\frac{\alpha}{1-\alpha}\right)^{m/3}. \tag{7.125}$$

The size scale b is determined by the nominal grain size of the porous body, for example, the characteristic particle size of a powder. A compaction curve of the form of equation (7.125) is shown with the quartz sand data in figure 7.24 with values of $m = 2$ and $\beta/b^m = 7$ MPa. This elementary model provides a better representation of the data than do those based on the ductile strength properties of a material.

Also shown with the data is the linear *U–u* crush curve. Good agreement with the data is expected from the linearity displayed in the shock-velocity versus particle-velocity plot shown on the left in figure 7.23. Further compaction models for brittle solids are examined in the subsequent sections.

The application of Weibull concepts to the compaction resistance of brittle solids suggests that dependence on the size scale b such as the grain size of a granular powder would be observed. Studies on the compaction resistance of cerium oxide powders with differing powder morphologies have been performed (Fredenburg *et al* 2014). Differing compaction resistance with different morphologies is observed. Dependence of compaction resistance on particle size is reported in static compaction studies on granular SiO_2 (Lee *et al* 2004). Further applications of Weibull concepts to brittle compaction are explored in section 7.9.2.

7.6.2 Continuum compaction models with microstructure

The previous subsection examines the elastic and inelastic compaction of a hollow shell of solid material for the purpose of quantifying the compaction resistance of a porous, or distended, medium. In this, and the next few subsections, attention is focused on physical models that provide a description of the continuum resistance to compaction of a distended solid to an applied pressure. The emphasis focuses principally on models of p-alpha (Herrmann 1969) and the p-lambda (Grady *et al* 2000) type. The latter model has appeared previously in this chapter and in chapter 6.

7.6.2.1 The p-alpha porous compaction model

The seminal model description of the continuum compaction of porous solids (Herrmann 1969) has found wide application in the numerical simulation of the response of such materials subjected to impact or explosive loading. The reasons for the popularity of the p-alpha model are reasonably clear. Although the model does not provide the highest fidelity representation of the compaction and deformation of crushable media in the shock and high-rate environment, the model is intuitive and readily implemented. Further, except in a few isolated cases, available material property data characterizing compaction do not justify a higher fidelity model. Parameters needed for the p-alpha model can be extracted from other available data or sensibly estimated. For a phenomenon where the necessary material property data are not commonly available, the latter advantage provides strong incentive for use of the model.

P-alpha model framework. The p-alpha model builds on the compressibility relation for the solid material. This can be as simple as a monotonic relation between the pressure and the specific volume of the solid material identified here as $\upsilon_s(P)$. The relation may also include a thermal energy dependence where warranted. When a tensor description of the stress and strain is implemented, this compression relation refers to the volume compression and the appropriate mean stress.

A second volumetric expression $\upsilon_d(P)$ describes the content of void in the distended material at any pressure P. The specific volume of the porous material is then the sum of the solid and distended components $\upsilon(P) = \upsilon_s(P) + \upsilon_d(P)$. The pressure versus volume for both the distended and the fully dense solid is illustrated on the left in figure 7.25.

The functional property central to construction of the p-alpha compaction model is the distention parameter $\alpha(P)$ and its dependence on the pressure. Again, dependence on temperature or other kinematic parameters can be included if warranted. The distention provides a relative measure of the void content of the porous medium through the ratio

$$\alpha(P) = \frac{\upsilon(P)}{\upsilon_s(P)} = 1 + \frac{\upsilon_d(P)}{\upsilon_s(P)}. \tag{7.126}$$

In principal, any thermodynamically and mechanically acceptable functional form that captures the observed compaction behavior of the material can be implemented into the p-alpha model. That shown on the left in figure 7.25 is representative of the quadratic form commonly used in application of the model.

In the p-alpha formulation the distention $\alpha(P)$ has both an elastic and an inelastic (compaction) dependence on the pressure. The extent to which attention need be put on the initial elastic relation is dependent on the material of concern. For a competent ceramic with modest porosity where reasonable fidelity in propagation of the precursor wave is required, a degree of attention to the elastic dependence of the distention $\alpha_e(P)$ is warranted. Less attention to this aspect of the model is needed for a powder, or for any distended material that initiates compaction early in the pressure loading process. If dependence on pressure is linear in the elastic range then

$$\alpha_e(P) = \alpha_o(1 - P/K(\alpha_o)), \tag{7.127}$$

where α_o is the initial distention. The elastic modulus of the fully dense and the corresponding distended solid are provided by K_s and $K(\alpha_o)$, respectively. Several

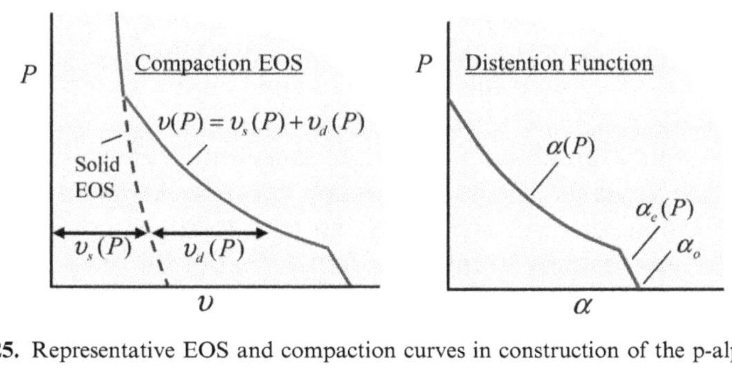

Figure 7.25. Representative EOS and compaction curves in construction of the p-alpha model.

useful models for the elastic modulus of a competent distended solid are captured by the functional dependence on the initial distention

$$K(\alpha_o) = K_s \alpha_o^{-b}. \tag{7.128}$$

The power of $b = 1$ provides the familiar Voigt upper-bound iso-strain compressibility of a distended solid. The compliance of modestly porous ceramics is considered further in chapter 6.

Following the onset of compaction, the p-alpha model allows any physically sensible functional form for the compaction resistance that is descriptive of the dynamic compaction data. The logarithmic form of Carroll and Holt (1972) based on the analytic solution of the crush of a spherical plastic shell is described in the previous subsection. A power-law relation of the form $P = \sigma\alpha^{-\gamma}$ (equation (7.35)) sensibly describes the compaction of selected ceramic powders. A compaction expression of the form

$$\alpha(P) = 1 + (\alpha_e - 1)\left(\frac{P_s - P}{P_s - P_e}\right)^n \tag{7.129}$$

is provided in the CTH shock physics code (Kerley 1992, Bell *et al* 2000), where P_e and P_s define the pressure at onset and completion of compaction, respectively.

In figure 7.26 two further features required in the simulation of compaction with the p-alpha model are illustrated. First, the extent of compaction of the p-alpha model is determined by the maximum pressure in the dynamic loading process. When pressure subsequently decreases from this maximum pressure a decompression path must be specified. The functional form for the decompression modulus is commonly determined by a suitable average of the solid material modulus and that of the initial elastic modulus of the distended solid. The average depends on the extent of compaction preceding decompression. Equation (7.129) for example, determines the minimum distention achieved on compaction.

Second, the deviator or shear strength of the compacting solid must be addressed in the computational simulation of a dynamic compaction event. In application of the p-alpha model this aspect of the mechanics is usually dealt with in an approximate way. Compaction involves the average stress (pressure) and the volumetric strain. Shear strength addresses the deviator stress and the distortional (shear) deformation of the material. Commonly an elementary von Mises strength Y is used to bound the shear strength of the compacting solid. When warranted, and suggested by the data, a pressure-dependent shear strength $Y(P)$ is more appropriate. Some issues concerning the strength of solids in the compaction regime are pursued in section 7.9.

Lastly, when shock amplitudes are high, substantial entropy production and shock heating occurs in the shock compaction process. At intense levels of shock heating the associated thermal expansion can lead to shock compacted Hugoniot states that are at substantially lower density than that of the Hugoniot state for the initially full-density solid at a corresponding shock pressure level. This is demonstrated, for example, by the schematic compaction curve depicted in figure 7.5. This

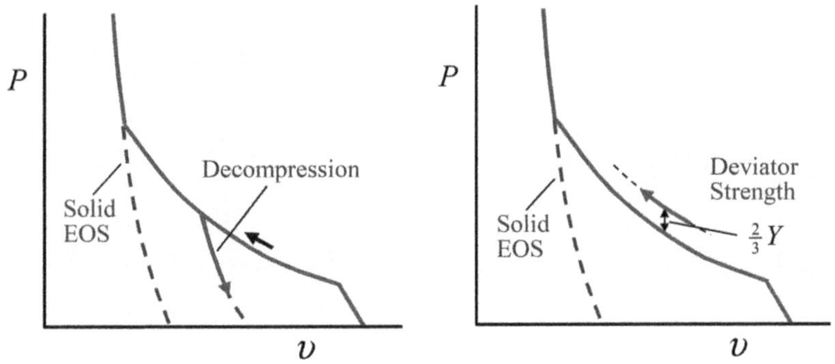

Figure 7.26. Application of the p-alpha model requires characterization of decompression paths from states of partial compaction (left) and deviator strength of the compacting distended solid (right).

aspect of the compaction is addressed by ascribing a thermo-elastic EOS to the solid material. The thermodynamics of shock compaction heating is addressed further in section 7.8 and chapter 10.

P-alpha model for shock compaction of granular and powdered media. The p-alpha compaction model is a commonly used and well-vetted description of the crush of distended media under shock and high-strain-rate deformation. The model quite effectively replicates the compaction response of distended metals, porous foams and geological materials under a wide range of explosive and impact conditions. The common form of the model in equation (7.129) is not, however, functionally appropriate for describing the compaction of powdered and granular brittle solids. Rather, an alternative power-law model provides a more satisfactory description of the compaction.

An analytic representation of both static and dynamic compaction of powdered and granular media discussed earlier in the chapter that adequately describes such data is of the form

$$P(\alpha) = P_s\alpha^{-n}. \tag{7.130}$$

In this expression α is the distention at pressure P valid over the range $1 \leqslant \alpha \leqslant \alpha_o$ where $\alpha_o = \rho_o/\rho_{oo}$ is the initial distention of the material. Here ρ_o and ρ_{oo} refer to the ambient pressure density of the solid and distended material, respectively. Parameters in the compaction expression are the intercept compaction strength P_s and the compaction power n. The latter is a reasonably large number commonly in the range of five to fifteen. The compaction relation is not referenced to an initial distention. Available compaction data, albeit not extensive, indicate no or at best only modest dependence on the initial distention late in the compaction event. A p-alpha functional form that does address early compaction differences is described later in the section. The compaction relation in equation (7.130) is compared with selected static and dynamic compaction data in figure 7.27, replicating some of the data previously plotted in figure 7.10.

The compaction relation in equation (7.130) as represented is not appropriate for implementation into the p-alpha compaction model. Rather the inverted relation

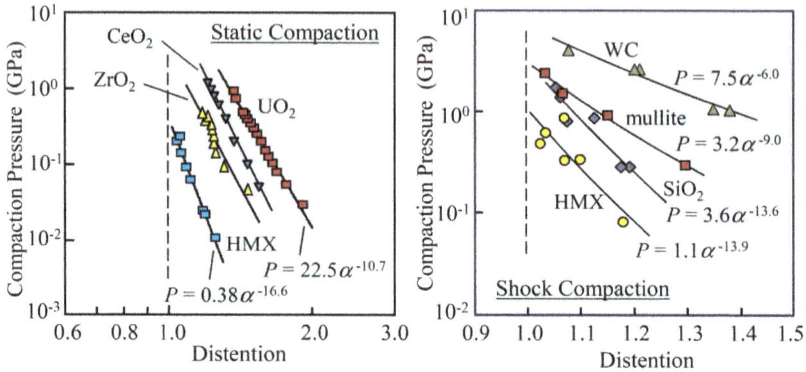

Figure 7.27. Static (left) and shock (right) compaction data for powdered and granular media (Elban and Chiarito 1986, Sheffield *et al* 1995, Lee *et al* 2004, Brown *et al* 1987, Vogler *et al* 2007).

$$\alpha(P) = \left(\frac{P_s + \eta}{P + \eta}\right)^{1/n} \tag{7.131}$$

is employed. The intercept compaction strength P_s and the compaction power n are the same as extracted from the (7.130) representation of the data. (Actually, almost the same. More is said on this later.) The parameter η is the configuration pressure. It is a fictitious property that is adjusted after the strength and power parameters in the (7.131) compaction relation have been determined. The parameter η assures that the distention is equal to the initial media distention $\alpha = \alpha_o$ at the pressure-equals-zero initiation of compaction. The (7.131) representation effectively models the measured compaction behavior of powders and granular materials.

The compaction compliance, or modulus, of the power-law model governs the details of wave propagation in materials subjected to impact or explosive shock loading. A closer examination of the compaction compliance properties of the model is warranted to assure behavior that is sensibly consistent with observation. From equation (7.126) the specific volume of the distended medium is

$$v(P) = v_s(P)\alpha(P). \tag{7.132}$$

The compaction compliance (inverse of the bulk modulus) is then

$$\frac{1}{K} = -\frac{d \ln v}{dP} = -\frac{d \ln v_s}{dP} - \frac{d \ln \alpha}{dP}. \tag{7.133}$$

The differential of the second term on the right follows from equation (7.131) and yields

$$\frac{1}{K} = \frac{1}{K_s} + \frac{1}{n(P + \eta)}. \tag{7.134}$$

The first term on the right is the compliance of the crystal lattice, while the second provides the compliance of the compacting void. The compaction modulus (that associated with void compaction) is then $K_c(P) = n(P + \eta)$. Moduli are plotted as a

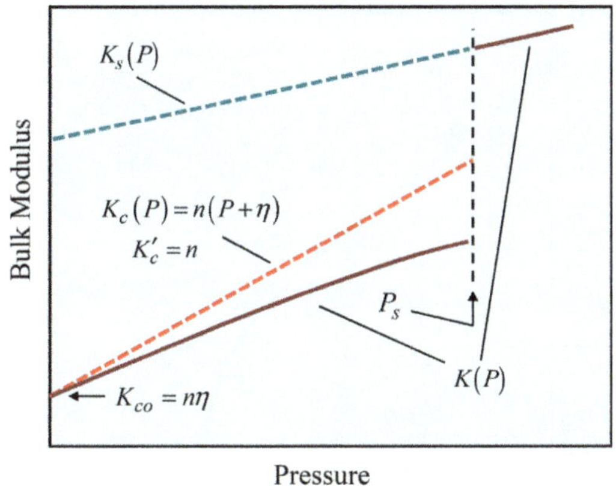

Figure 7.28. Decomposition of the power-law compression modulus into lattice and compaction components.

function of pressure to illustrate compliance characteristics of the power-law compaction model in figure 7.28.

Shock compaction of silica sand. The plot in figure 7.28 is representative of the shock compaction of silica sand for which the Hugoniot state data within the compaction regime are shown on the right in figure 7.27 and previously in figure 7.23. Both the individual component moduli (crystal lattice and void), and the composite modulus are monotonic, leading to formation of a single shock wave. The component compaction modulus is a linear function of the compaction pressure with intercept $K_{co} = n\eta$ and pressure derivative $K_c' = n$. The intercept modulus determines the speed of the foot of the wave $c_o = \sqrt{n\eta/\rho_{oo}}$ under non-shock loading conditions. The density ρ_{oo} is that of the initially distended solid. For the SiO_2 fit in the figure the wave speed is approximately $c_o = 150$ m s^{-1}. Further, the pressure derivative $K_c' = n$ provides a measure of the rapidity with which an initially ramped input will shock up. This number is typically about 3–6 for full-density solids. Values range from about 6–14 for the shock data shown on the right in figure 7.27. As is characteristic of the p-alpha model, the compression modulus jumps discontinuously to the crystal lattice modulus at the intercept pressure P_s.

The discussion following the power-law compaction model introduced in equation (7.131) identifies two parameters characterizing the model. Namely, the intercept compaction strength P_s and the compaction power n. The configuration pressure η is a dependent parameter adjusted to initiate the $\alpha(P)$ compaction curve at $\alpha_o = \rho_o/\rho_{oo}$, at pressure equal to zero. The configuration pressure is solved from

$$\eta = \frac{P_s}{\alpha_o^n - 1}. \tag{7.135}$$

The p-alpha power-law compaction model is illustrated for granular silicon dioxide in figure 7.29. Crystal density of SiO_2 is $\rho_o = 2.65$ g cm^{-3}. The starting density for the compaction data shown in the figure is $\rho_o = 1.57$ g cm^{-3}. Compaction curves for two further initial densities are illustrated in the figure. In each case the configuration pressure is calculated from equation (7.135).

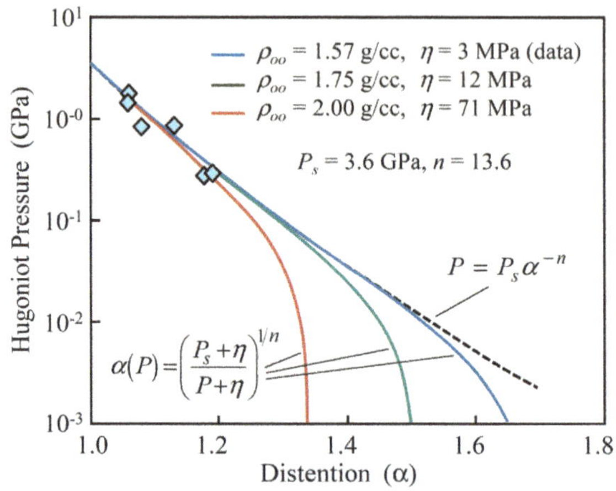

Figure 7.29. Power-law compaction of granular silicon dioxide for selected initial distention densities. The initial density of shock compaction data (Brown *et al* 2007) is provided in the plot.

If, however, data are available that clearly demonstrate that the compaction curve is dependent on the initial state of distention, then the power-law model has the latitude of treating the configuration pressure η as a third parameter.

The experimental compaction as plotted in figure 7.27 does not fully utilize the available test data. The plots clearly demonstrate the power-law character of the compaction data and provide reasonable first-order values of the intercept compaction strength and compaction power. This representation of the experimental data, however, is a plot of $\alpha(P) = \upsilon(P)/\upsilon_{so}$ where $\upsilon(P)$ is the measured Hugoniot specific volume and $\upsilon_{so} = 1/\rho_o$ is the ambient specific volume of the solid. This representation is independent of the lattice compliance of the solid, but is not a correct representation of the true distention as defined in equation (7.132). The distention data could be replotted as in figure 7.27 using an appropriate crystal compressibility for the solid component. More convenient is to effect a fit directly to the experimental Hugoniot data.

The application of this method for Hugoniot compaction data on granular silicon dioxide is illustrated in figure 7.30. Dashed curves in the figure result from the application of intercept strength and compaction power parameters directly from the curve fit in figure 7.27. Solid curves are a fit to the experimental Hugoniot data. Model parameters are shown in the figure. A constant bulk modulus of 36 GPa is used to account for the lattice compliance of the quartz. Compression of the solid component is illustrated with the dashed line to the right in the right-hand figure.

Fading memory compaction. The p-alpha model is a robust, physically intuitive engineering description of the compaction of distended solids subject to static or impulsive pressure loads. As such, provided basic mechanical or thermodynamic principles are not violated, the user is free to cast the p-alpha compaction function in forms that conveniently replicate the compaction data or the compaction application of concern.

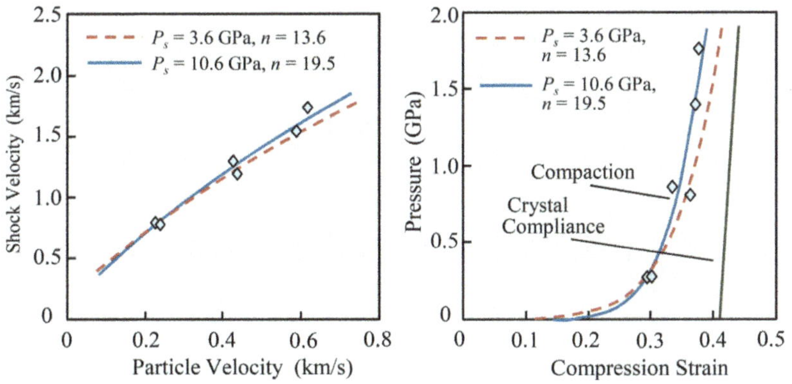

Figure 7.30. Power-law compaction of granular silicon dioxide with an initial density of 1.57 kg m^{-3} and distention of 1.69 (Brown *et al* 2007).

Most forms for the alpha function generally replicate the stiffening of compaction as the extent of distention lessens, and includes features for identifying onset and completion of compaction. The alpha function in the previous subsection was framed with the intent of modelling compaction of a porous material where any arbitrary initial porosity could be specified.

A further form of the p-alpha compaction function provides features useful in representing compaction in certain compaction applications. As in previous forms, alpha is a decreasing function of the compaction pressure within the range $1 < \alpha < \alpha_O$. The fading memory alpha function is

$$\alpha(P) = 1 + \left[\frac{1}{(\alpha_o - 1)^{nm}} + \left(\frac{P}{\sigma}\right)^n\right]^{-1/nm}. \tag{7.136}$$

The compaction pressure scale is σ, whereas n and m are function shape parameters. For pressures significantly less than σ, alpha takes the power-law form

$$\alpha = \alpha_o - f(\alpha_o, m, n)\left(\frac{P}{\sigma}\right)^n, \qquad P \ll \sigma, \tag{7.137}$$

and allows reasonable flexibility in describing details of early compaction. For pressures significantly greater than σ, alpha approaches power-law compaction independent of the compaction details at low pressure,

$$\alpha = 1 + \left(\frac{P}{\sigma}\right)^{-m}, \qquad P \gg \sigma. \tag{7.138}$$

Several sets of compaction data, noteworthy in their own right, exhibit application of the p-alpha model. One is shock compaction data on porous 2024 aluminum (Butcher *et al* 1974). This aluminum was identified as a high-pressure EOS standard and extensive shock Hugoniot data are available for solid and porous 2024 aluminum (McQueen *et al* 1967, Marsh 1980). The compaction data for porous 2024 aluminum with an initial density of 2.16 g cm^{-3} (approximately 78%

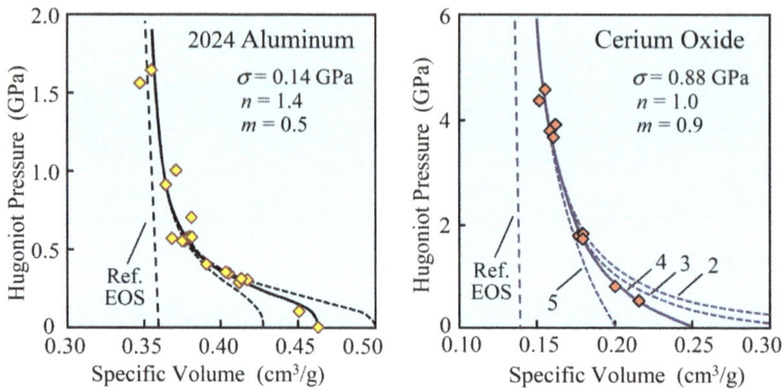

Figure 7.31. Hugoniot compaction data for aluminum (left; Butcher *et al* 1974) and cerium oxide (right; Fredenburg and Dennis-Koller 2013). The solid line is the p-alpha representation of compaction data provided by parameters in the graph. Dashed lines identify compaction from difference initial densities. Initial densities in g cm^{-3} are identified for cerium oxide on the right.

theoretical) is shown on the left in figure 7.31. The p-alpha model representation from equation (7.136) that overlies the data accounts in the low-pressure range for the approximately 0.1 GPa initial crush onset for this aluminum. Additionally, predicted curves are shown for modestly high and lower initial densities. Compaction curves converge to power-law compaction at higher pressures.

On the right are shown compaction data for cerium oxide with an initial density of 4.0 g cm^{-3} (approximately 55% theoretical) (Fredenburg and Dennis-Koller 2013, Fredenburg *et al* 2014). Again, the p-alpha representation in the form of equation (7.136) along with compaction paths for different initial densities are shown. Compaction tests on different morphology CeO_2 powders reveal differences early in the compaction process, but are not apparent as later higher pressure compaction proceeds. Such details in the early compaction can be addressed with the present p-alpha model.

7.6.3 The p-lambda porous compaction model

The p-lambda model was developed to replicate the compaction of mixtures of widely disparate compressibility materials (Grady *et al* 2000). The p-lambda model also addresses the compaction of porous solids with a level of engineering fidelity comparable to the p-alpha model. The physics underlying the p-lambda compaction model differs, however, from that motivating the p-alpha model.

The p-lambda porous compaction model framework. Referring to figure 7.32, the p-lambda compaction model can be likened to a generalized phase transformation between two distinct states of the same material. In application of the model to porous compaction, the two states refer, respectively, to the initial distended solid and the final fully compacted non-porous solid. Each of the two states has its own thermo-elastic, or strictly elastic, EOS identified by the two compressibility relations in the pressure versus specific volume plot shown on the left in figure 7.32.

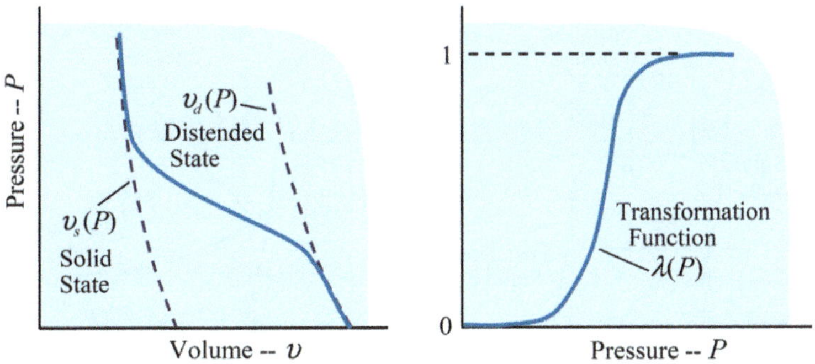

Figure 7.32. The p-lambda compaction model illustrating the two-state character of the compacting solid (left) and the compaction transformation function (right).

The compaction transition is accomplished in the p-lambda model through specification of a transformation function $\lambda(P)$ shown on the right in the figure. The specific volume of the compacting solid through states of partial compaction at any pressure is provided through the superposition of the two material states through

$$v(P) = \lambda(P)v_s(P) + (1 - \lambda(P))v_d(P). \tag{7.139}$$

In this relation $v_d(P)$ and $v_s(P)$ are the thermo-elastic compression of the initial distended solid and fully compacted solid, respectively. $\lambda(P)$ and $1 - \lambda(P)$, respectively, provide the mass fraction of the compacted (transformed) and the distended (untransformed) material.

Any lambda transformation function that adequately describes the compaction data can, in principle, be incorporated into the p-lambda compaction model. Several examples are illustrated in figure 7.33. On the left a transformation function of the Heaviside form,

$$\lambda(P) = H(P - P_t), \tag{7.140}$$

and the corresponding compaction function, exhibit a homogeneous compaction process experiencing compaction onset and completion at a unique transformation pressure P_t. The transformation function and compaction curve illustrated on the right in figure 7.33 sensibly replicates the p-alpha compaction model implemented in the CTH shock physics code (Kerley 1992, Bell *et al* 2000). This compaction curve follows from the transformation function of the form

$$\lambda(P) = \left(\frac{P - P_e}{P_s - P_e}\right)^n. \tag{7.141}$$

As in the p-alpha model, compressibility of the distended solid is elastic until compaction onset at pressure P_e. Compaction completes at pressure P_s.

The transformation function of choice in the application of the p-lambda model has become the exponential lambda function (ELF),

$$\lambda(P) = 1 - e^{-(P/\sigma)^n}. \tag{7.142}$$

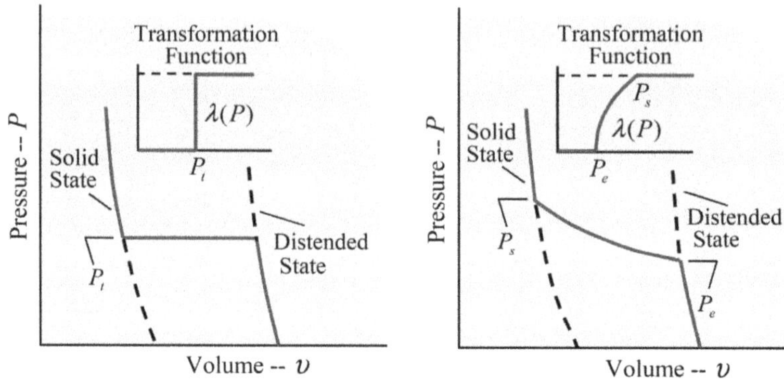

Figure 7.33. Representative transformation functions for the p-lambda compaction model and corresponding compaction curves.

The ELF is found to quite effectively replicate most of the experimental compaction data that have been generated. The functional form of the ELF is illustrated in figure 7.34 over a range of values for the power n. On the left is shown the transformation function. In the middle the pressure derivative of the transformation function illustrates the rate of compaction with increasing pressure. Corresponding compaction curves base on the ELF are shown on the right. The ELF uniformly converges to a Heaviside function as the power n becomes arbitrarily large, approaching an increasingly flat homogeneous compaction process. In contrast, as the power n becomes small, compaction is spread over an increasingly larger pressure range. For $n < 1$ the compaction curve is concave upward over the entire pressure range.

The ELF also benefits from a familiar strength-of-materials interpretation. At any point in the porous body, compaction can be identified as a failure process determined by a statistical power-law hazard function,

$$h(P)dP = \frac{n}{\sigma}\left(\frac{P}{\sigma}\right)^{n-1} dP. \tag{7.143}$$

The term $h(P)dP$ provides the statistical chance of failure (compaction) at a specified point in the body as pressure is elevated from P to $P + dP$. A power-law relation for the failure hazard function follows from the seminal strength-of-materials study of Weibull (1951). The ELF,

$$\lambda(P) = 1 - \mathrm{e}^{-\int h(P)dP} = 1 - \mathrm{e}^{-(P/\sigma)^n}, \tag{7.144}$$

in this interpretation is recognized as the cumulative probability of failure with the constants σ and n determining the distribution scale (average) and shape (dispersion), respectively.

Higher levels of physics could, in principle, be incorporated into the transformation function. The transformation function could, for example, be expressed as a rate law,

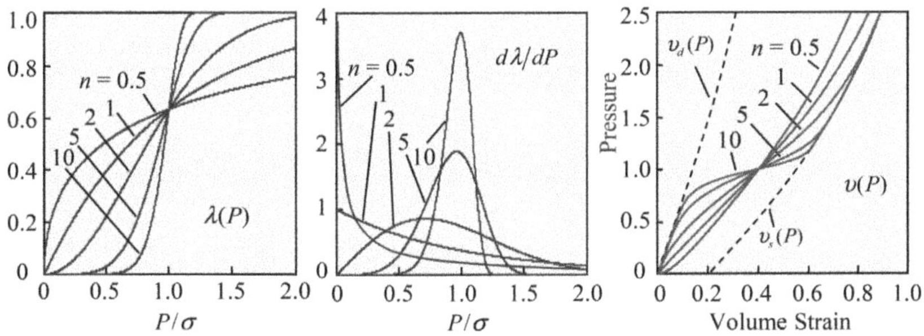

Figure 7.34. Functional forms for the exponential lambda transformation function.

$$\dot{\lambda} = f(\lambda, P, \tau, T, \dot{\varepsilon}), \tag{7.145}$$

in analogy with other kinetics-governed physical transformation processes. The additional constitutive variables included in the functional dependence of equation (7.145) are the shear stress τ, the temperature T, and the deformation strain rate $\dot{\varepsilon}$. Pursuit of equation (7.145) as a transformation law in the description of dynamic compaction would only be warranted if justified by available experimental compaction data.

Implicit in equation (7.142) is that the ELF is the natural exponential function with e the Euler number. This is not necessary to the model and other exponential functions offer convenience in certain applications. For example, an exponential function with $e = 2$ in equation (7.142) yields the median compaction when $P = \sigma$ with half of the material achieving the compaction state.

As with the p-alpha model, elasticity must be accounted for on decompression from the compaction state. Bulk elasticity is naturally achieved in the p-lambda model on decompression from a compaction state through freezing the value of lambda at the maximum compaction state achieved $\lambda(P_{max}) = \lambda_{max}$. Bulk decompression elasticity is then provided by

$$\upsilon(P) = \lambda_{max}\,\upsilon_s(P) + (1 - \lambda_{max})\upsilon_d(P). \tag{7.146}$$

When a strength model is incorporated with the p-lambda model, specification of a Poisson's ratio appropriate to the material completes the elastic description.

P-lambda compaction of porous media. The p-lambda model with the ELF is used in figure 7.35 to describe compaction of the same porous metal and granular oxide represented with the p-alpha description in the previous figure 7.31. Parameters for the lambda function appropriate to the curve fit are provided in the plots. Description of the compaction data with either the p-alpha model or the p-lambda are comparable. Application of either can be expected to exhibit both the strengths and the weaknesses of this level of engineering model. The compaction function of the p-alpha and the p-lambda form can be related through the equivalence $\lambda(P, \alpha_o) = (\alpha_o - \alpha(P, \alpha_o))/(\alpha_o - 1)$.

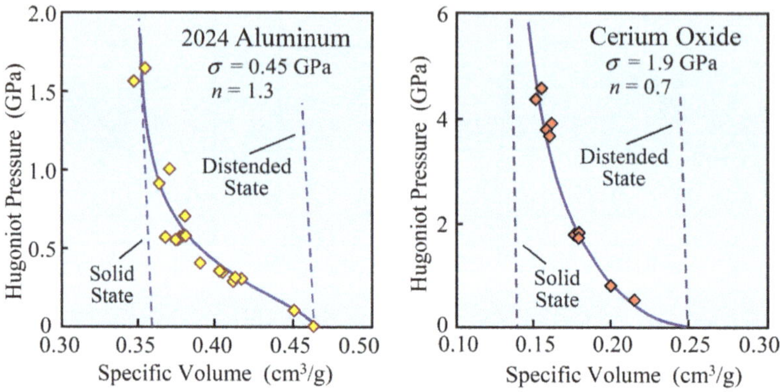

Figure 7.35. P-lambda description of Hugoniot compaction data for aluminum (left; Butcher *et al* 1974) and cerium oxide (right; Fredenburg and Dennis-Koller 2013). P-lambda Weibull parameters for the compaction relations are provided in the plots.

7.7 Compaction with elastic recovery

In much of the previous analysis of the dynamic compaction of crushable media a more basic model assuming rigid-compaction behavior of the material is applied. Specifically, the material is considered to have some functional dependence of the compaction on the applied pressure that carries the material from an initial density ρ_0 to the current density ρ. On subsequent depressurization from the compacted state the material is assumed to be rigid, maintaining the same density achieved at the final compacted state as the pressure is reduced. A rigid decompression path is illustrated in figure 7.36.

Through the assumption of a rigid-compaction material model, the governing partial differential equations of mass, momentum and energy reduce to ordinary differential equations and yield analytic solutions for a number of interesting problems. Analysis with the rigid-compaction model is informative and provides useful first-order solutions.

In the compaction of crushable media the response is more complex. Upon compaction to a more densified state, decompression will result in some amount of elastic recovery. The contribution of elasticity increases as the extent of compaction in the problem lessens. Elastic decompression is also illustrated in figure 7.36. Analysis of the elastic-compaction problem would provide for assessment of the extent of error in assuming a rigid-compaction model. Treatment of an elastic-compaction response leads to markedly more complex analysis. Computational analysis can be pursued. Analytic methods are nonetheless instructive and sensible analytic solutions can be obtained for certain compaction problems. One such problem is treated here to illustrate the method, and is compared with the corresponding rigid-compaction model solution.

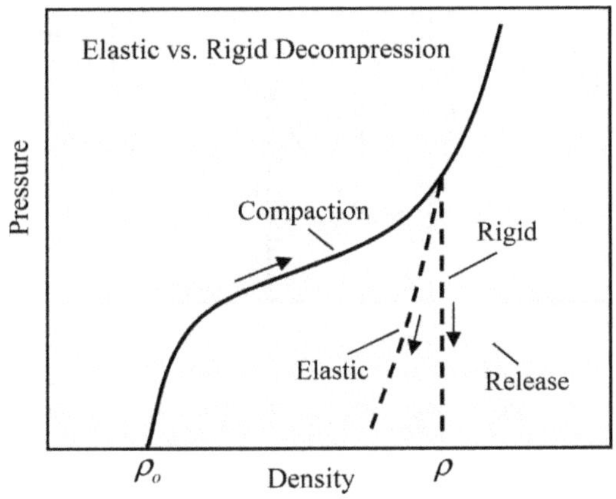

Figure 7.36. Contrasting the rigid and elastic release subsequent to compaction in porous media.

7.7.1 Governing equations

The problem treated is that of an attenuating planar one-dimensional stress pulse that compacts the crushable medium through a propagating shock wave and subsequently decompresses the compacted material through the following stress unloading wave. The differential equations governing the shock compression and subsequent decompression in this one-dimensional problem arise from conservation of mass and momentum, and are

$$\frac{\partial \varepsilon}{\partial t} + \frac{\partial u}{\partial x} = 0, \tag{7.147}$$

$$\frac{\partial u}{\partial t} + \frac{\partial \sigma}{\partial x} = 0. \tag{7.148}$$

The finite uniaxial strain is provided by $\varepsilon = 1 - \rho_o/\rho$, with ρ_o and ρ the initial distended density and current compacted density, respectively, while u is the particle velocity and σ is the normal stress or pressure. The Lagrangian distance and time coordinates are x and t, respectively. The above conservation equations, with an appropriate constitutive description of the material deformation relating the stress and strain, provide a complete set of equations for determine uniaxial wave propagation given a set of initial and boundary conditions.

The constitutive equation that is used to describe compaction of a crushable medium will be

$$\frac{d\sigma}{d\varepsilon} = \begin{cases} \rho_o U^2 \text{ (loading)}, \\ \rho_o C^2 \text{ (unloading)}. \end{cases} \tag{7.149}$$

This constitutive relation is illustrated in figure 7.37. The expressions in equation (7.149) provide both the modulus for shock compaction $\rho_o U^2$ and the elevated elastic modulus on decompression $\rho_o C^2$, where the shock velocity U and elastic wave

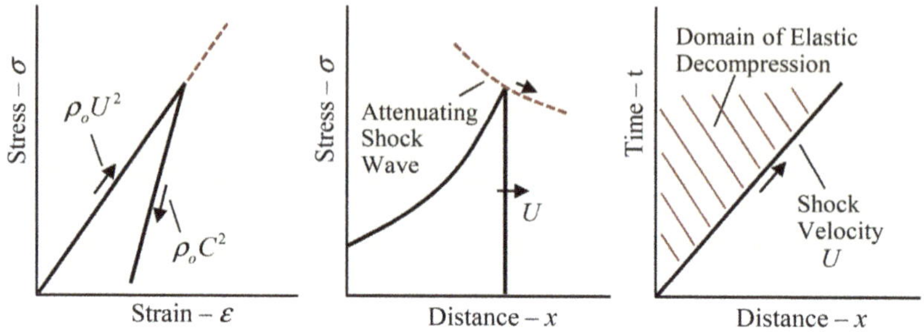

Figure 7.37. Stress versus strain compaction and decompression of a simple compacting material (left). Corresponding compaction shock wave attenuation (middle) and elastic release (right).

velocity C are restricted to the range $C > U$. Both U and C are assumed constant in the analysis to follow. A more general constitutive response with a dependence of the compression modulus, or shock velocity, on amplitude $U(\sigma)$ can be treated through the same formalism, however, the relations become analytically more complex. It is also noted that as the elastic velocity C becomes arbitrarily large the solution will converge uniformly to that of the rigid-compaction model. At the other bound, as C approaches the shock velocity U from above, the material approaches a reversible linear elastic material in both compression and decompression.

7.7.2 Boundary and initial conditions

The compaction problem occupies the domain $x \geqslant 0$ and $t \geqslant 0$. The material in the region $x > 0$ at the time $t = 0$ is in a uniform state with initial conditions $\rho = \rho_o$, $\varepsilon = 0$, $u = 0$ and $\sigma = 0$. The boundary condition at $x = 0$ is some arbitrary pressure load of the form

$$\sigma(0, t) = \psi(t). \tag{7.150}$$

It is equally appropriate to place a boundary condition on the particle velocity, or specify some relation between the stress and particle velocity. The pressure boundary condition on the $x = 0$ surface of the crushable material initiates a pressure pulse that propagates an attenuating compaction wave into the $x > 0$ domain of the material. Solution of the above governing equations provides predictions of the resulting compaction wave propagation.

The boundary pressure function $\psi(t)$ is specified such that the initial pressure leads to shock compaction of the crushable medium followed by time-dependent decompression. An x versus t plot shown in figure 7.37 identifies the trajectory of the shock wave, and the domain $x < Ut$ over which the solution is sought.

7.7.3 Elastic solution

In the domain of decompression $x < Ut$ elastic wave propagation is governed by the unloading modulus provided in equation (7.149) and the corresponding elastic wave speed C. The governing relations from equations (7.147) and (7.148) reduce to the wave equation for a linear elastic solid and have a general solution of the D'Alembert form

$$\sigma(x, t) = f(t - x/C) + g(t + x/C), \tag{7.151}$$

$$\rho_o Cu(x, t) = f(t - x/C) - g(t + x/C). \tag{7.152}$$

If the solution for the elastic strain is desired, it is noted that hysteresis in the compression and decompression process requires the solution,

$$\rho_o C^2 \varepsilon(x, t) = f(t - x/C) + g(t + x/C) + h(x), \tag{7.153}$$

including an additional function of the distance as well as the wave propagation terms.

7.7.4 Application of the boundary conditions

The solution domain to which the elastic solutions provided in equations (7.151) and (7.152) apply is bounded by the crushable material Lagrangian interface at $x = 0$ and the path of the compaction shock wave $x = Ut$. The boundary condition on the interface path $x = 0$ is provided by equation (7.150). The solution on the shock path $x = Ut$ is constrained by the shock jump relation between stress and particle velocity,

$$\sigma(Ut, t) = \rho_o Cu(Ut, t). \tag{7.154}$$

Application of the boundary conditions result in

$$f(t) + g(t) = \psi(t), \tag{7.155}$$

$$\alpha f(\alpha t) + \beta g(\beta t) = 0, \tag{7.156}$$

where the latter relation follows from the equation (7.154) shock relation. The new variables are defined by $\alpha = 1 - U/C$ and $\beta = 1 + U/C$. Equations (7.155) and (7.156) are rewritten as

$$f(t) = -\frac{\beta}{\alpha} g\left(\frac{\beta}{\alpha} t\right), \tag{7.157}$$

$$g(t) - \frac{\beta}{\alpha} g\left(\frac{\beta}{\alpha} t\right) = \psi(t). \tag{7.158}$$

The boundary conditions have provided two relations to solve for the two unknown functions f and g. There is not a general technique to solve for f and g in closed form for an arbitrary specification of $\psi(t)$. One satisfactory technique is to expand $f(t)$, $g(t)$ and the boundary pressure $\psi(t)$ in a Taylor series about $t = 0$:

$$f(t) = \sum_{j=0}^{\infty} \frac{1}{j!} f^j(0) t^j, \tag{7.159}$$

$$g(t) = \sum_{j=0}^{\infty} \frac{1}{j!} g^j(0) t^j, \qquad (7.160)$$

$$\psi(t) = \sum_{j=0}^{\infty} \frac{1}{j!} \psi^j(0) t^j. \qquad (7.161)$$

The solution will of course be closed form if $\psi(t)$ is specified as a polynomial of finite degree. The relations in equations (7.157) and (7.158) can then be differentiated j times and evaluated at $t = 0$ to provide the expansion coefficients $f^j(0)$ and $g^j(0)$. This exercise provides

$$f^j(0) = [1 - (\alpha/\beta)^{j+1}]^{-1} \psi^j(0), \qquad (7.162)$$

$$g^j(0) = [1 - (\beta/\alpha)^{j+1}]^{-1} \psi^j(0). \qquad (7.163)$$

Using the elastic solution for the pressure provided in equation (7.151), the general solution is obtained,

$$\sigma(x, t) = \sum_{j=0}^{\infty} \frac{1}{j!} \psi^j(0) \left\{ \frac{1}{1 - (\alpha/\beta)^{j+1}} \left(t - \frac{x}{C} \right)^j + \frac{1}{1 - (\beta/\alpha)^{j+1}} \left(t + \frac{x}{C} \right)^j \right\}. \qquad (7.164)$$

The attenuation of the pressure of the compaction shock is calculated by specifying $x = Ut$ in equation (7.164) and results in

$$\sigma_p(x) = \sigma(x, x/U) = \sum_{j=0}^{\infty} \frac{1}{j!} \psi^j(0) \left\{ \frac{\beta^j \alpha^j (\beta - \alpha)}{\beta^{j+1} - \alpha^{j+1}} \right\} (x/U)^j. \qquad (7.165)$$

7.7.5 Rigid and elastic comparisons

For a boundary pressure history of polynomial form, the solutions provided by equations (7.164) and (7.165) continue to apply, and simplify considerably. A quadratic boundary pressure history,

$$\psi(t) = \sigma_o(1 - 2t/\tau + (t/\tau)^2), \qquad (7.166)$$

valid over $0 \leqslant t \leqslant \tau$, provides for an instantaneous pressure jump to σ_o, reducing to zero at time $t = \tau$.

The momentum of a shock compacted material encompassed by the shock wave is provided by $\rho_o x u$ when a rigid-compaction model of the shock wave compaction event is applied. Momentum conservation yields

$$\frac{d}{dt}(\rho_o x u) = \sigma(0, t) = \psi(t). \qquad (7.167)$$

Equations (7.166) and (7.167) are readily solved for a solution to the rigid-compacting problem, yielding for attenuation of the shock pressure with propagation distance

$$\sigma_p(x) = \sigma_o\left(1 - x/U\tau + \frac{1}{3}(x/U\tau)^2\right). \tag{7.168}$$

The corresponding solution to the elastic-compaction model is provided by the first three terms of equation (7.164),

$$\sigma(x, t) = \sigma_o\left\{\begin{matrix} 1 - \dfrac{2}{\tau}\left[\dfrac{\beta^2}{\beta^2 - \alpha^2}\left(t - \dfrac{x}{C}\right) + \dfrac{\beta^2}{\alpha^2 - \beta^2}\left(t + \dfrac{x}{C}\right)\right] \\ + \dfrac{1}{\tau^2}\left[\dfrac{\beta^3}{\beta^3 - \alpha^3}\left(t - \dfrac{x}{C}\right)^2 + \dfrac{\beta^3}{\alpha^3 - \beta^3}\left(t + \dfrac{x}{C}\right)^2\right] \end{matrix}\right\}. \tag{7.169}$$

The rigid and elastic solutions are compared in figure 7.38. On the left is a snapshot of the compaction wave profile at a specified propagation distance for values of the elastic-release-velocity to the shock-velocity ratio of $C/U = 1.5$, 2, 3 and 5, as well as the corresponding wave profile from the rigid solution. Also shown on the left are peak shock stress attenuation curves for both the rigid solution and elastic solution for $C/U = 1.5$. In the right-hand plot, profiles at successive propagation distances are compared for both the rigid solution and the elastic solution for a velocity ratio of $C/U = 1.5$.

As intuition suggests, the rigid-compaction model becomes an increasingly adequate representation of the compaction behavior when the ratio C/U is large. This would be the case, for example, for a porous material with initial large distention early in the compaction process. On the other hand, when C/U is small, as would be the case for only a modestly distended porous solid, or when the shock amplitude becomes large, the rigid-compaction model becomes increasingly

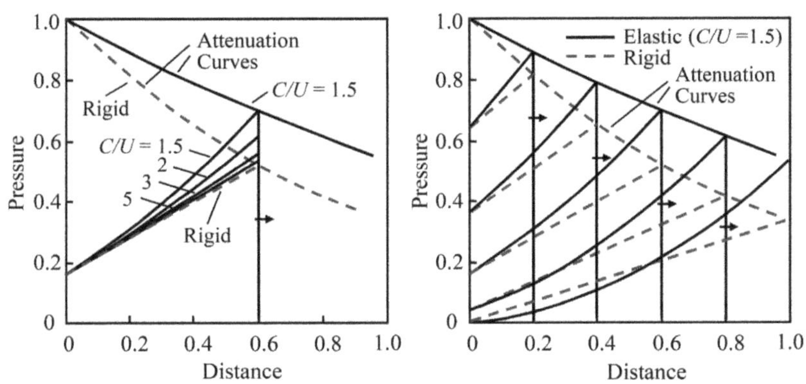

Figure 7.38. A comparison of shock compaction and release wave profiles base on assumptions of both rigid-compaction and elastic-compaction material responses.

inadequate and elastic behavior of the compacting material must be reasonably accounted.

7.7.6 Further elastic-compaction solutions

Lastly, it is worth noting that the formalism used to provide an analytic solution to the present compaction problem with elastic decompression can be extended to a broader range of one-dimensional compaction problems. For example, the compaction curve defined by the loading modulus and corresponding shock velocity can be extended to a nonlinear dependence on compaction amplitude. Other boundary conditions can be explored, such as the impact and subsequent deceleration of a plate, which places a constraining relation between the boundary pressure and particle velocity. This compaction problem was explored with a rigid-compaction model in earlier sections.

The elastic model and analytic solution method remain constrained to the linear elastic response of the material in the domain of decompression wave propagation. The different compression models and boundary conditions then lead to different restricting relations on the D'Alembert elastic wave equation solution. These relations are again solved through polynomial expansion of the D'Alembert functions and the boundary conditions, as illustrated in the earlier solution. The solutions become analytically more complex but fully tractable with today's computer technology.

7.8 Dissipation and heating in shock compaction

Distended solids when subjected to strong shock compression dissipate an inordinate amount of energy and experience a corresponding increase in temperature. Shock-induced melt, vaporization and dissociation are possible consequences of the extreme shock compaction of a porous solid. The physical processes through which such dissipation is achieved are, however, not fully appreciated and warrant discussion.

Early in this chapter (see figures 7.5 and 7.6) the full thermo-mechanical consequences of the shock compaction of distended solids is briefly surveyed. Intense shock dissipation and accompanying temperature increase result from strong shock compression of porous materials. Highly heated Hugoniot states are achieved on the EOS surface that are difficult to achieve by other experimental methods.

Much of the present chapter is focused on the athermal mechanical consequences of shock compaction. In the present section, issues of shock heating brought about by strong shock compaction are the topics of interest. Thermodynamic relations pertinent to shock dissipation, temperature increase and thermal expansion are developed. Thermodynamic relations, where applied, assume a linear shock-velocity versus particle-velocity response of the material. The assumed shock EOS form is for the calculation example and does not constrain the generality of this section. (EOS issues in the shock event are expanded on in chapter 10.) The isentropic bulk modulus on the Hugoniot is calculated. Relations are derived connecting the thermodynamic Gruneisen parameter with several less commonly used properties.

Methods are developed for calculating the shock Hugoniot for distended materials and corresponding levels of shock heating. These methods are applied to the explosive shock compaction of a distended solid.

7.8.1 Shock equation of state for distended solids

The effect of initial distention (porosity) on the shock compressibility is illustrated in figure 7.39 (compare with figures 7.5 and 7.6). The Hugoniot compressibility of the full-density material is provided by the principal Hugoniot curve (the cold Hugoniot) emanating from the initial specific volume $v_o = 1/\rho_o$. A representative Hugoniot for the same material initially distended to a specific volume $v_{oo} = 1/\rho_{oo}$ is shown by the dashed curve also emanating from the initial specific volume v_o. Initial crush strength of the porous solid is ignored (snow-plow crush) and is not critical to the discussions in this section. The straight Rayleigh line from the initial distended specific volume v_{oo}, to the Hugoniot pressure, provides the pressure P versus specific volume v compression path through a steady structure shock wave.

In figure 7.39 the principal Hugoniot of the initially fully dense material (the cold Hugoniot) and the Hugoniot of the same material in an initially distended state (the hot Hugoniot) are contrasted at any Hugoniot pressure state in common through the specific volume difference brought about by differing thermal states of the matter. This volume difference $\Delta v_h(P)$ relates to the initial volume distention Δv_o through thermodynamic differential expressions in the neighborhood of the principal Hugoniot. Complications of phase transitions within the shock can be treated but are not explicitly discussing in the following.

7.8.1.1 Differential properties of the shock Hugoniot

The Hugoniot for the initially distended material lies above, or to the right of, the Hugoniot of the full-density solid because of the markedly larger irreversible $P\,dv$ work expended through the compaction shock wave and the additional thermal

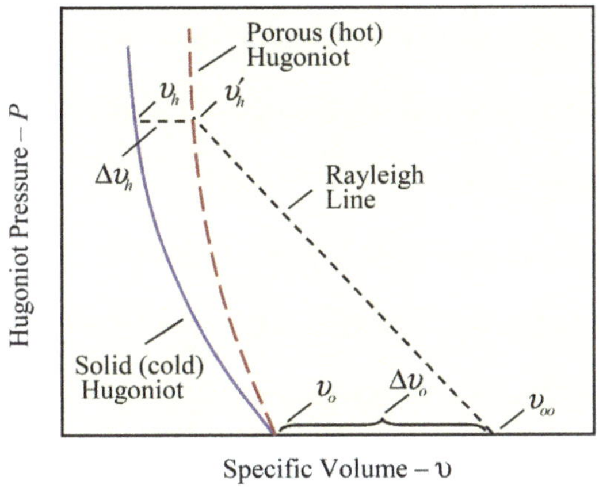

Figure 7.39. Illustrates relative Hugoniot compressibility of the fully solid and an initially distended body of the same material. Compaction resistance is ignored (snow-plow model).

pressure generated. The thermal pressure at any specific volume is calculated with the differential relation

$$\Gamma = \upsilon \frac{\partial P}{\partial E}\bigg|_{\upsilon}, \qquad (7.170)$$

where Γ is the thermodynamic Gruneisen parameter. Both theoretical arguments and experience suggest that Γ is not sensitive to reasonable excursions of internal energy E for many materials, and is usually assume to be a volume-dependent-only function $\Gamma(\upsilon)$. It is also common that the functional volume dependence is not known and constant assumptions of either $\Gamma = \Gamma_o$ or $\rho\Gamma = \rho_o\Gamma_o$ are made.

For increasing levels of initial distension of the porous body, Hugoniot compression curves become increasingly vertical in P versus υ space. Extreme conditions are achieved that can lead to a change in the sign of the slope (or backward leaning) of the shock Hugoniot, as was previously illustrated in figure 7.5. As a consequence, the use of equation (7.170) as a difference expression for relating Hugoniot curves can be inconvenient.

The thermal beta parameter. Experimentally more expedient is the dimensionless parameter,

$$\beta = P \frac{\partial \upsilon}{\partial E}\bigg|_{P}, \qquad (7.171)$$

which allows convenient estimates of the volumetric displacement of adjacent Hugoniot curves. Values for this thermal expansion property are not readily available but can be calculated from other more common thermodynamic properties. Starting with a $P(\upsilon, E)$ representation of the EOS, the differential is written

$$dP = \frac{\partial P}{\partial \upsilon}\bigg|_{E} d\upsilon + \frac{\partial P}{\partial E}\bigg|_{\upsilon} dE. \qquad (7.172)$$

An isentrope that intersects the Hugoniot state is then

$$\frac{\partial P}{\partial \upsilon}\bigg|_{S} = \frac{\partial P}{\partial \upsilon}\bigg|_{E} + \frac{\partial P}{\partial E}\bigg|_{\upsilon} \frac{\partial E}{\partial \upsilon}\bigg|_{S}. \qquad (7.173)$$

Equation (7.173) can be expressed in terms of the isentropic bulk modulus K_s, and the pressure P, yielding

$$-\frac{K_s}{\upsilon} = \frac{\partial P}{\partial \upsilon}\bigg|_{E} - P \frac{\partial P}{\partial E}\bigg|_{\upsilon}. \qquad (7.174)$$

The first derivative on the right is reduced through the thermodynamic identity

$$\frac{\partial P}{\partial \upsilon}\bigg|_{E} = -\frac{\partial P/\partial E\,|_{\upsilon}}{\partial \upsilon/\partial E\,|_{P}} = -\frac{\Gamma/\upsilon}{\beta/P}. \qquad (7.175)$$

Equation (7.174) becomes

$$\frac{K_s}{v} = \frac{\Gamma}{v}\frac{P}{\beta} + \frac{\Gamma}{v}P \tag{7.176}$$

or

$$\beta = \frac{\Gamma P}{K_s - \Gamma P}. \tag{7.177}$$

The beta parameter goes to zero with the pressure and is of order unity at high pressure. As an example, for a Gruneisen gamma based on Murnaghan EOS nonlinearity and a linear U versus u Hugoniot relation, a limiting $\beta = 2S/(2S - 1)$ at high pressure is calculated.

The Rice–Walsh parameter. Although the β parameter serves the present purpose, an alternative form is

$$\zeta = P\frac{\partial v}{\partial H}\bigg|_P, \tag{7.178}$$

where $H = E + Pv$ is the EOS enthalpy. The parameter ζ leads to a direct measure of the thermal energy differences between Hugoniot states at the same pressure. It is readily shown that

$$\zeta = \frac{\beta}{1 + \beta} = P\frac{\Gamma}{K_s}. \tag{7.179}$$

Rice and Walsh (1957) developed a shock wave EOS for water based on the assumption that $\zeta = \zeta(P)$ is solely a function of pressure. The Rice–Walsh EOS is pursued further in chapter 10 on shock wave EOS issues.

The isentropic bulk modulus on the Hugoniot. The isentropic bulk modulus K_s at the Hugoniot state, and the initial isentropic decompression path, can be estimated from the slope of the Hugoniot at that state through appropriate thermodynamic relations (Duvall and Fowles 1963). Consider again the differential of a shock EOS of the form $P(v, E)$,

$$dP = \frac{\partial P}{\partial v}\bigg|_E dv + \frac{\partial P}{\partial E}\bigg|_v dE. \tag{7.180}$$

From this relation the isentropic derivative is

$$\frac{\partial P}{\partial v}\bigg|_{\bar{S}} = \frac{\partial P}{\partial v}\bigg|_E + \frac{\partial P}{\partial E}\bigg|_v \frac{\partial E}{\partial v}\bigg|_{\bar{S}} = \frac{\partial P}{\partial v}\bigg|_E - P\frac{\partial P}{\partial E}\bigg|_v, \tag{7.181}$$

where the second equality makes use of the thermodynamic law $dE = Td\bar{S} - Pdv$ and the isentropic relation $\partial E/\partial v|_{\bar{S}} = -P$. Where necessary the overbar symbol $\bar{S}$ distinguishes entropy from the linear shock velocity versus particle velocity S in the relation $U = C_o + Su$ used in later analysis.

Combining equations (7.180) and (7.181) yields

$$dP = \frac{\partial P}{\partial v}\bigg|_{\bar{S}} dv + \frac{\partial P}{\partial E}\bigg|_{v} (Pdv + dE) = \frac{\partial P}{\partial v}\bigg|_{\bar{S}} dv + \frac{\partial P}{\partial E}\bigg|_{v} Td\bar{S}. \tag{7.182}$$

With the thermodynamic identities

$$\frac{\partial P}{\partial E}\bigg|_{v} = \frac{\Gamma}{v} \quad \text{and} \quad \frac{\partial P}{\partial v}\bigg|_{\bar{S}} = -\frac{K_s}{v}, \tag{7.183}$$

there follows a differential expression for the entropy,

$$\frac{\Gamma}{v} T\, d\bar{S} = dP + \frac{K_s}{v} dv. \tag{7.184}$$

Consider the Hugoniot state referenced from the initial state $P = 0$ and $v = v_o$. The Hugoniot energy is

$$E = E_o + \frac{1}{2} P(v_o - v). \tag{7.185}$$

The energy differential in the neighborhood of the Hugoniot state is

$$dE = \frac{1}{2}(v_o - v)dP - \frac{1}{2} P\, dv, \tag{7.186}$$

while from the first law

$$T\, d\bar{S} = dE + P\, dv = \frac{1}{2}(v_o - v)dP + \frac{1}{2} P\, dv. \tag{7.187}$$

Eliminate the entropy differential between equations (7.184) and (7.187),

$$\left(\frac{K_s}{v} - \frac{\Gamma}{2v} P\right) dv = \left(1 - \frac{\Gamma}{2v}(v_o - v)\right) dP, \tag{7.188}$$

or,

$$\frac{K_s}{v} = \frac{K_h}{v}\left[1 - \frac{\Gamma}{2v}(v_o - v)\left(1 - \frac{Pv}{K_h(v_o - v)}\right)\right], \tag{7.189}$$

where the volume derivative on the Hugoniot (the local slope of the Hugoniot) is defined as $K_h = -v\, dP/dv|_h$.

The Hugoniot momentum conservation equation yields

$$P = \rho_o U^2 \frac{v_o - v}{v_o}, \tag{7.190}$$

where U is the shock velocity. Equation (7.189) becomes

$$K_s = K_h\left[1 - \frac{\rho\Gamma}{2}(v_o - v)\left(1 - \frac{\rho_o^2 U^2}{\rho K_h}\right)\right]. \tag{7.191}$$

Equation (7.191), as written, reproduces the relation developed by Duvall and Fowles (1963) where the difference in isentropic and Hugoniot compliance at the Hugoniot state is determined by a correction dependent on the ratio of the slope of the Rayleigh line and the local slope of the Hugoniot.

The uranium alloy U6Nb is a reasonably well characterized metal for present discussions. Hugoniot, ultrasonic and thermal properties are provided in table 7.2. Shock properties for U6Nb and other uranium alloys are addressed further in chapter 11. The isentropic bulk modulus on the Hugoniot for U6Nb is shown on the left in figure 7.40 based on the relations provided in equations (7.189) or (7.191) to a compression strain of 0.25, or a Hugoniot pressure of about 75 GPa. Calculations assume both Γ and $\rho\Gamma$ equal to a constant. The isentropic modulus is contrasted with local slope of the Hugoniot at the same compression strain.

The distended solid. Analysis leading to equations (7.189), or (7.191), applies to a material that is shocked to states on the principal Hugoniot from an initial crystal density state. The same analysis is applicable to a distended material. Distention is introduced in equation (7.185) for the Hugoniot energy through replacement of the crystal specific volume v_o with the specific volume v_{oo} for the distended solid. Equation (7.189), through the same replacement, applies directly for calculating the Hugoniot state isentropic compliance on the Hugoniot of the distended material. Although a relation paralleling equation (7.191) can be developed, equation (7.189) is more readily framed in a form more useful for calculation. Energy on the distended material Hugoniot comparable to equation (7.185) is written

$$E - E_o = \frac{1}{2}P(v_{oo} - v) = \frac{1}{2}P(v_{oo} - v_o + v_o - v) = \frac{1}{2}\frac{P}{\rho_o}(k + \varepsilon), \tag{7.192}$$

with $\varepsilon = 1 - v/v_o$ a measure of the lattice strain in the solid material and $k = (v_{oo} - v_o)/v_o$ a measure of the initial distention of the material. In the p-alpha model notation $k = \alpha_o - 1$ with α_o the initial distention parameter. (Implicit is the

Table 7.2. Shock EOS properties for U6Nb.

ρ_o	17 440 kg m^{-3}
C_l	2950 m s^{-1}
C_s	1310 m s^{-1}
C_o	2530 m s^{-1}
ν	0.377
G_o	29.9 GPa
K_{ho}	111.6 GPa
S	1.57
Γ_o	2.18
c_p	117 J kg^{-1} K
T_m	1500 K

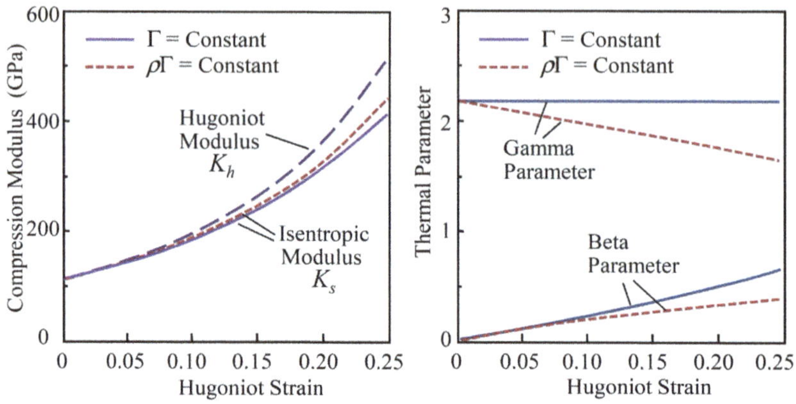

Figure 7.40. Left: Hugoniot and isentropic compression moduli for uranium U6Nb metal based on linear shock-velocity versus particle-velocity Hugoniot, and assumption of either a constant Gruneisen gamma or a constant product of the density and Gruneisen gamma. Right: A comparison of the thermal Gruneisen gamma parameter and corresponding beta parameter for U6Nb.

assumption that $E_{oo} = E_o$ or that any difference is inconsequential.) Comparable replacement in equation (7.189) yields

$$K_s = K_h\left[1 - \frac{1}{2}\frac{\upsilon_o}{\upsilon}\Gamma(k+\varepsilon)\left(1 - \frac{\upsilon}{\upsilon_o}\frac{P}{K_h(k+\varepsilon)}\right)\right], \tag{7.193}$$

and with modest analytic effort

$$K_s = K_h\left[1 - \frac{1}{2}\frac{\rho}{\rho_o}\Gamma(k+\varepsilon)\left(1 - \left[\frac{d\ln P}{d\ln(k+\varepsilon)}\right]^{-1}\right)\right]. \tag{7.194}$$

For the special case of a linear shock-velocity versus particle-velocity Hugoniot,

$$P = \frac{\rho_o C_o^2 \varepsilon}{(1 - S\varepsilon)^2}, \tag{7.195}$$

Equation (7.191), or equation (7.194), with $k = 0$ becomes

$$K_s = K_h\left[1 - \Gamma\frac{S\varepsilon^2}{(1-\varepsilon)(1+S\varepsilon)}\right]. \tag{7.196}$$

A Taylor expansion in the lattice strain yields

$$K_s = K_h[1 - \Gamma S\varepsilon^2(1 + (1-S)\varepsilon + (1 - S + S^2)\varepsilon^2) + \cdots]. \tag{7.197}$$

Entropy increase on the Hugoniot from equation (7.184) leads to

$$T\,d\bar{S} = \frac{1}{\Gamma}(K_s - K_h)d\upsilon. \tag{7.198}$$

From equation (7.197) to first-order in the neighborhood of the origin,

$$T_o \, d\bar{S} = -K_{ho} S \varepsilon^2 \, dv = \frac{1}{\rho_o} K_{ho} S \varepsilon^2 \, d\varepsilon, \tag{7.199}$$

with $K_{ho} = \rho_o C_o^2$ corresponding to K_h at $P = 0$. Entropy on the Hugoniot increases as

$$\bar{S}(\varepsilon) = \bar{S}_o + \frac{K_{ho} S}{3\rho_o T_o} \varepsilon^3. \tag{7.200}$$

Equation (7.200) illustrates the general relation that the isentrope and Hugoniot are second-order tangent at the origin (Duvall and Fowles 1963).

For a distended material with k other than zero similar analysis applies. If $P_{ho}(\varepsilon)$ provides the Hugoniot functional form for the crystal density solid, and the reference relation in a Mie–Gruneisen EOS, then the Hugoniot expression for any level of distention is

$$P_h(\varepsilon) = P_{ho}(\varepsilon) \frac{1 - \dfrac{1}{2}(\rho/\rho_o)\Gamma\varepsilon}{1 - \dfrac{1}{2}(\rho/\rho_o)\Gamma(\varepsilon + k)}. \tag{7.201}$$

Applying the linear shock-velocity versus particle-velocity Hugoniot from equation (196) to the crystal density response, the limiting relation for the entropy on the Hugoniot corresponding to equation (7.200) becomes

$$\bar{S}(\varepsilon) = \bar{S}_o + \frac{K_{ho} S}{3\rho_o T_o} \varepsilon^3 + \frac{K_{ho} k}{2\rho_o T_o} \varepsilon. \tag{7.202}$$

Entropy production due to pore compaction increases as the first power of lattice compression on the Hugoniot in contrast to third-power dependence on Hugoniot nonlinearity. Entropy production terms are comparable at a lattice strain of order $\varepsilon = \sqrt{3k/2S}$.

As an example, commercial sintered ceramics commonly retain a few percent porosity. For ceramics the Hugoniot S parameter is typically about one. One to two percent porosity would suggest a transition strain of about $\varepsilon = 0.12$ to 0.17. Ignoring shear strength, transition from compaction dominated entropy production to Hugoniot nonlinearity dominated entropy production in a ceramic such as silicon carbide would occur at Hugoniot pressures of about 35–50 GPa.

7.8.1.2 Distended Hugoniot

Analysis within the preceding section relates thermo-elastic derivatives (the Gruneisen Γ and β) to the shock wave EOS of solids. Both the Gruneisen Γ and the β thermal parameters are dimensionless and of order unity. These thermal parameters are plotted and compared on the right in figure 7.40, again using properties for the metal U6Nb provided in table 7.2 (Grady 2005) assuming either Γ or $\rho\Gamma$ constant over the selected range of Hugoniot compression strain.

For any pressure P the specific volume difference Δv_h between states on the full-density and distended metal Hugoniot curves, as illustrated in figure 7.39, can be calculated from the energy difference from the relation for β provided in equation (7.177),

$$\Delta v_h = \frac{\beta}{P}\Delta E. \tag{7.203}$$

The energy difference ΔE at the respective Hugoniot state can be calculated. Assuming a negligible difference in the initial state energy, the Hugoniot energies at pressure P are

$$E = \frac{1}{2}P(v_o - v_h) \tag{7.204}$$

and

$$E' = \frac{1}{2}P\left(v_{oo} - v_h'\right) \tag{7.205}$$

for the solid (cold) and distended (hot) Hugoniot states, respectively. The energy difference is

$$\Delta E = E' - E = \frac{1}{2}P\left(\left(v_{oo} - v_h'\right) - (v_o - v_h)\right) = \frac{1}{2}P(\Delta v_o - \Delta v_h), \tag{7.206}$$

where $\Delta v_o = v_{oo} - v_o = kv_o$ and $\Delta v_h = v_h' - v_h$, respectively. k is the distention parameter defined previously. Combining equations (6.203) and (7.206) yields

$$\Delta v_h = \frac{\beta}{2+\beta}\Delta v_o, \tag{7.207}$$

with β provided from equation (7.177).

The specific volume on the distended metal Hugoniot at pressure P is then

$$v_h'(P) = v_h(P) + \Delta v_h = v_h(P) + \frac{\beta}{2+\beta}kv_o. \tag{7.208}$$

Equation (7.208) provides the Hugoniot relation for the distended solid based on the Hugoniot of the full-density solid and thermal properties of the material characterized by the parameter $\beta(P)$. Equation (7.208) is an alternative calculation of the distended Hugoniot that can be contrasted with equation (7.201).

The distended material particle-velocity and shock-velocity relations follow from

$$u' = \sqrt{\frac{P}{\rho_{oo}}\left(1 - v_h'/v_{oo}\right)} \tag{7.209}$$

and

$$U' = \sqrt{\frac{P}{\rho_{oo}}\left(1 - v_h'/v_{oo}\right)^{-1}}. \tag{7.210}$$

Both the Gruneisen parameter Γ and the isentropic bulk modulus K_s are necessary to calculate the β parameter at the Hugoniot pressure P. The isentropic modulus K_s on the solid Hugoniot is provided by either equations (7.189) or (7.192). For a linear shock versus particle-velocity representation $U = C_o + Su$ of the cold Hugoniot, K_s is provided by equation (7.196). Based on equation (7.196), the $\beta(P)$ parameter on the cold Hugoniot for uranium U6Nb is shown on the right in figure 7.40. Shock Hugoniot curves for U6Nb assuming constant $\rho\Gamma$ for selected initial distensions based on equation (7.208) are shown in figure 7.41. EOS properties are provided in table 7.2.

7.8.2 Temperature and shock compaction

Shock compression to an elevated pressure Hugoniot state of any material will lead to a temperature rise dictated by the EOS of the material. An elevated temperature will persist upon decompression and return to ambient pressure. Residual temperature subsequent to a shock event is a consequence of irreversible dissipation impressed on the material through the shock. The following several subsections pertain to residual temperature in solid material after a shock event. Hugoniot EOS and temperature issues are addressed further in chapter 10.

7.8.2.1 Post-shock temperature

Estimates can be made of the post-shock temperature resulting from shock compaction of a distended solid. It is assumed that the shock is strong enough that full compaction is achieved. This is not the temperature at the shock Hugoniot state. Rather, it is the temperature of the material following full decompression from the shock state.

Although a portion of the energy dissipated in the shock compression process resides in an athermal form associated with configurational energy of the assorted defects created in the shock, this portion is usually small and it is sensible to assume

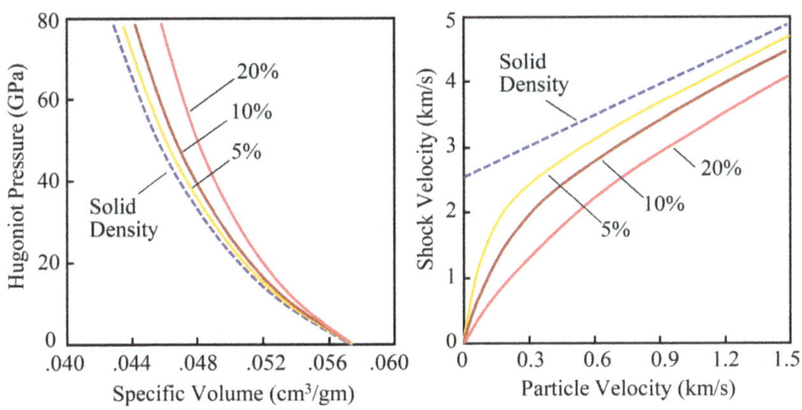

Figure 7.41. Hugoniot pressure versus volume (left) and shock-velocity versus particle-velocity (right) curves for full-density and distended U6Nb uranium alloy.

that all the energy dissipated goes into thermal energy and corresponding temperature rise. (There are exceptions, discussed further in chapter 8, that pose concerns with excepting this oft used assumption.) When a change of phase occurs the latent heat of transformation is accounted for.

Energy dissipated in the shock compression of the distended solid is the integral over the specific volume υ,

$$\mathcal{E} = -\int P(\upsilon)d\upsilon, \tag{7.211}$$

with integration over the full compression and decompression cycle. The pressure $P(\upsilon)$ is the dynamic pressure including both the lattice and viscous resistance to volume change during the shock compaction and decompression event. Pressure in the steady wave shock traverses the Rayleigh line path on compression and the isentrope centered at the Hugoniot state on decompression. The dissipation energy in equation (7.211) is the area enclosed by the Rayleigh line to the Hugoniot state from above, and the decompression isentrope from the Hugoniot state from below as illustrated in figure 7.42. It is the shock Hugoniot energy minus the recovered lattice strain energy on isentropic decompression. Equation (7.211) takes the form

$$\mathcal{E} = \frac{1}{2}P_h\left(\upsilon_{oo} - \upsilon_h'\right) - \int_{\upsilon_o'}^{\upsilon_h'} P_s(\upsilon)d\upsilon. \tag{7.212}$$

The Hugoniot state of shock compaction of the distended medium is first established. (Note the inversion of integration limits to emphasize the recovery of lattice strain energy on decompression.) The Hugoniot state specific volume υ_h' at Hugoniot pressure P_h identified in figure 7.42 is calculated from equation (7.208). The first term on the right in equation (7.212) provides the Hugoniot state energy.

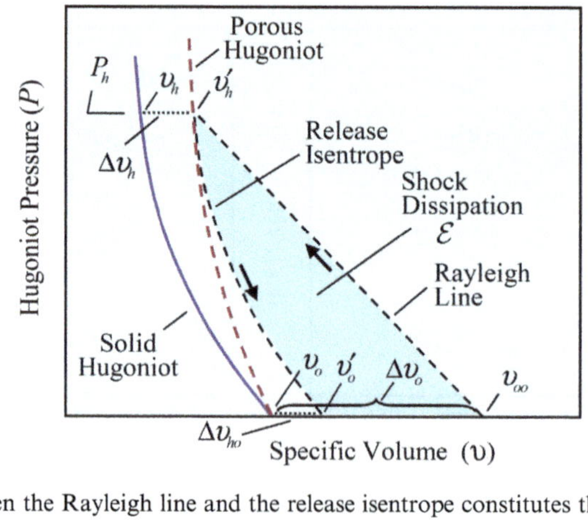

Figure 7.42. Area between the Rayleigh line and the release isentrope constitutes the energy dissipated in the shock compaction and decompression of a distended material.

The isentropic release path from the Hugoniot state requires analysis of the thermodynamic EOS. More extensive discussion is found in chapter 10. A sensible approximation to the release isentrope follows from the assumption that thermal expansion due to the shock compaction heating is sensibly the same at the Hugoniot pressure and at ambient pressure. This assumption modestly over-estimates the dissipation energy and temperature, as is shown for example through application of the Anderson–Gruneisen relation for thermal expansion (Anderson 1967). For present purposes the full-density (cold) principal Hugoniot reasonably approximates the principal isentrope. The release isentrope is then approximated by an appropriate translation of the principle solid material cold Hugoniot (and isentrope) as illustrated in the figure with $\Delta v_{ho} = \Delta v_h$. Note also that to this approximation K_s is replaced by K_h, markedly simplifying the estimate of β used in equation (7.208). The second term in equation (7.212) is then the area under the principal Hugoniot.

For a linear shock-velocity versus particle-velocity Hugoniot the isentropic energy recovered on decompression (the second term on the right in equation (7.212)) becomes

$$\mathcal{E}_s = \int_0^{\varepsilon_h} \frac{C_o^2 \varepsilon}{(1 - S\varepsilon)^2} d\varepsilon, \tag{7.213}$$

with $\varepsilon = 1 - v/v_o$ the compression strain. Integration yields

$$\mathcal{E}_s = \frac{C_o^2}{S^2}\left[\frac{S\varepsilon_h}{1 - S\varepsilon_h} + \ln(1 - S\varepsilon_h)\right], \tag{7.214}$$

with ε_h the compression strain at the principal Hugoniot state.

Equation (7.212) with equation (7.214) provide a sensible assessment of the dissipation energy incurred in the shock compression and decompression cycle of an initially distended solid material,

$$\mathcal{E} = \frac{1}{2} P_h\left(v_{oo} - v_h'\right) - \mathcal{E}_s. \tag{7.215}$$

Temperature of the shock compacted material at ambient conditions is then

$$T = T_r + \mathcal{E}/c_p, \tag{7.216}$$

where T_r is the initial starting temperature and c_p is a measure of the ambient specific heat.

7.8.2.2 Explosive shock compaction

An appropriate example of shock wave heating of a distended solid is provided by analysis of the coupling of an explosive detonation wave into a distended material. Calculations of the planar coupling of a detonation shock wave into the distended material are performed in the pressure versus particle-velocity plane. The required EOS properties are the pressure versus particle-velocity Hugoniot for the distended metal, and the shock adiabat for the explosive defined here as the joining of the

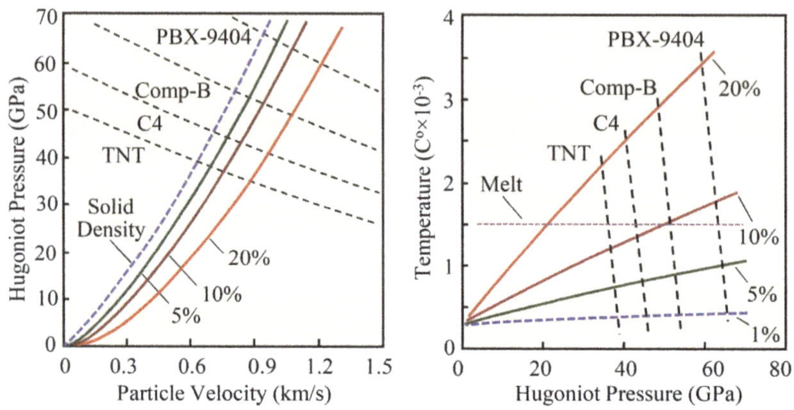

Figure 7.43. (a) Pressure versus particle-velocity plot providing shock amplitude explosively coupled into distended U6Nb metal. (b) Calculated post-shock temperatures in distended U6Nb and temperature states achieved through explosive shock loading. Percentages identify the initial porosity of the metal (Grady 2005).

reshock response and the isentropic release path about the explosive Chapman–Jouget (CJ) state for the explosive products.

The pressure versus particle-velocity Hugoniot for a selection of distensions of U6Nb metal are shown on the left in figure 7.43 and correspond to the Hugoniot curves plotted in figure 7.41. The reduced pressure versus particle-velocity relation of Cooper (1997) provides the explosive shock adiabat

$$P/P_{CJ} = 2.412 - 1.732(u/u_{CJ}) + 0.3195(u/u_{CJ})^2. \tag{7.217}$$

In this relation the CJ pressure and CJ particle velocity are P_{CJ} and u_{CJ}, respectively, with $u_{CJ} = P_{CJ}/\rho_o D$ in terms of the initial density ρ_o of the explosive and the detonation velocity D.

The respective curves in the pressure versus particle-velocity plane are shown in figure 7.43 for selected explosives along with the porous Hugoniot curves for distended U6Nb metal. Pressure amplitudes for any combination of the selected explosives and levels of distention are readily extracted from the plot.

Post-shock temperature versus pressure following shock compaction to respective Hugoniot states is calculated from equation (7.216), while shock dissipation $\mathcal{E}$ is obtained from the preceding relations. Representative calculations of temperature on the distended material Hugoniot and the corresponding Hugoniot pressures achieved by selected explosives are shown on the right in figure 7.43. The calculations are compared with the ambient pressure melt temperature of U6Nb metal.

7.9 Dynamic compaction and dissipation in distended solids

Analysis leading to predictions of post-shock temperatures in metals reveals that modest levels of porosity give rise to a marked increase in temperature. The increased temperature is a consequence of amplified material dissipation resulting from shock compression of the distended medium. The nature of shock-induced dissipation in distended ductile and brittle solids, and the underlying physics, is fundamental to the shock compaction event. The dissipation mechanisms in the two classes of materials differ markedly. Dissipation is notably modified in the dynamic compaction event when an entrained substance, such as water, replaces the void. Deviator strength, in conjunction with dynamic compaction, is recognized as necessary to the analysis of applications including the dynamic crush of distended media. Selected topics concerning these and related issues of dynamic compaction constitute the content of this closing section. Dissipation in the shock event resides at the heart of considerations of structured shock waves in chapter 8.

7.9.1 Compaction of the distended ductile solid

Consider a porous metal that is adequately characterized by the initial porous density ρ_o, a solid density of the matrix metal ρ_s, a characteristic length scale a associated with the porous microstructure, and a plastic flow stress Y that captures the deformation dissipation of the metal. Other thermo-mechanical properties such as elastic modulus, deformation hardening, melt temperature and so forth, although relevant, are ignored. The test metal is subjected to a reasonably strong compression wave sufficient to achieve levels of compaction near completion in the porous material. The particle-velocity amplitude of the wave is u and the wave risetime is τ. The compaction wave carries the material to a pressure P. The velocity of the wave is U and consequently the wavelength, and span of the metal encompassed by the wave, is $\lambda = U\tau$. The compaction event is illustrated in figure 7.44.

7.9.1.1 Ductile compaction dissipation

First consider the situation where the wavelength λ is much greater than the microstructure length scale a. In the limit of large λ the rate of compaction does not differ markedly from that of quasi-static compaction of the porous solid. The material will exhibit some pressure versus compaction relation. That of Carroll and Holt (1972) relating compaction pressure to current porosity is representative,

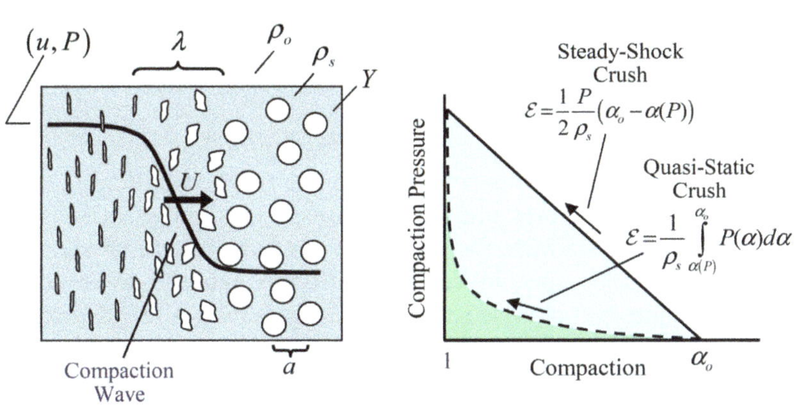

Figure 7.44. Dissipation due to near-quasi-static compaction and steady wave shock compaction is dependent on the wavelength of the compaction wave relative to the porous medium micro-length scale.

$$P = \frac{2}{3}Y\ \ln(1/p) = \frac{2}{3}Y\ \ln\left(\frac{\alpha}{\alpha - 1}\right), \tag{7.218}$$

and has demonstrated some success in characterizing compaction of porous metals. See for example the experimental study of Butcher *et al* (1972) on 2024 aluminum. The second step in equation (7.218) relates the compaction pressure to the current value of the p-alpha distention parameter $\alpha(P) = v(P)/v_s = \rho_s/\rho(P)$. This lower-bound near-quasi-static compaction curve is illustrated in figure 7.44 on the right.

Dissipation in the compaction event, with α_o the initial distention of the porous metal, is then

$$\mathcal{E}(P) = \frac{1}{\rho_s}\int_{\alpha(P)}^{\alpha_o} P(\alpha)d\alpha. \tag{7.219}$$

As the wavelength λ diminishes and the rate of compaction increases, stiffening of the compaction curve occurs. Initial stiffening would probably reflect a logarithmic dependence of the compaction resistance on the rate of compaction, in line with other strength-of-materials measurements, where thermal activation governs deformations accommodating the compaction. As the width λ of the compaction wave approaches that of the microscale, a marked stiffening of the compaction curve is necessitated by the physics. The upper-bound compaction curve is that of the Rayleigh line path where the width of the wave is commensurate with a steady structured compaction shock wave. Dissipation through the shock is then

$$\mathcal{E}(P) = \frac{1}{2}\frac{P}{\rho_s}(\alpha_o - \alpha(P)). \tag{7.220}$$

with the P versus α path necessarily tracing the linear Rayleigh line. (The modest lattice elasticity is ignored in the present discussion.) It is unlikely that the measured quasi-static compaction curve and the Hugoniot through the compaction domain will overlie. Although certainly a constitutive modeling issue, this observation is not pertinent to the present considerations. Dissipation under steady shock compaction will significantly exceed the quasi-static compaction dissipation as illustrated in the figure.

7.9.1.2 Onset of compaction turbulence

A compaction wave such as the profile shown in figure 7.44 rapidly steepens with propagation distance in a porous solid due to the marked stiffening of the crush curve. As steepening of the pressure wave proceeds, dissipative stresses must come into play that eventually stabilize the wave into a steady structured shock wave and cause the pressure versus specific volume path through the wave to traverse a straight Rayleigh line. Shock viscosity in some general sense, as discussed more extensively in both chapters 1 and 8, is inferred. Dissipation caused by compaction of the distended solid through a steady shock wave is overwhelmingly larger than the corresponding quasi-static compaction dissipation to the same pressure.

The bounding compaction events are illustrated in figure 7.44. Early in the wave progression the wavelength of the unsteady wave is large relative to the characteristic microstructure compaction length scale, $\lambda \gg a$. Compaction is nominally quasi-static and compaction dissipation is provided by equation (7.219) and the area under the compaction curve in figure 7.44. The wave steepens as propagation progresses until the pressure versus strain path transits the shock Rayleigh line, and the shock dissipation is that expressed in equation (7.220). The relative magnitudes of the dissipation of quasi-static compaction and steady wave shock compaction are qualitatively illustrated on the right in figure 7.44.

Differences in the microstructure dissipation mechanisms emerge from consideration of a dimensionless property that has similarities to the Reynolds' number in hydrodynamic turbulence. Namely, the ratio

$$N = \frac{\rho_o \dot{\varepsilon}^2 a^2}{Y}. \tag{7.221}$$

In this expression $\dot{\varepsilon} = u/U\tau = u/\lambda$ is a measure of the nominal rate of compaction or the strain rate in the structured shock wave, a is the characteristic length scale of the material porous microstructure, and Y is a measure of the plastic strength of the solid material determining dissipation during compaction. As the shock wave steepens and the strain rate increases, the value of N will approach some critical value N_c. For all compaction events for which N is markedly less than N_c, crush on the micro-scale is comparable to quasi-static compaction. As N approaches and exceeds the critical N_c, turbulent compaction ensues that is necessary for the amplification of dissipation commensurate with the steady compaction shock wave.

The nature of turbulent compaction will involve various mechanisms not occurring in quasi-static compaction and will be dependent on the extent to which N exceeds N_c. Rayleigh–Taylor and/or Helmholtz instability in the plastic deformation can occur. Micro-jetting and far-wall splash within the pore microstructure will follow. The differences between the quasi-static compaction of a void in a ductile material and the shock compaction of the same void are given via the illustrations in figure 7.45. All such compaction turbulence mechanisms will magnify the dissipative plastic deformation during compaction and lead to dynamic dissipation far in excess of quasi-static compaction. Again, the material model assumed considers only a constant Y. Clearly, however, the excessive dissipation will lead to far more complication constitutive response including extreme rate plasticity and melting.

Turbulence in the dynamic compaction event emerges because larger scale compaction processes are not sufficient to dissipate the requisite energy. Turbulence in the shock compaction event is then the activation, growth and motion of chaotic plastic deformation on successively smaller length scales necessitated by transition to steady wave shock compression. Turbulent deformation will cascade to successively smaller length scales until dissipation is adequate to support the steady wave shock. For $N \gg N_c$ a lower-bound length scale λ is determined from $N = \rho_o \dot{\varepsilon}^2 \lambda^2 / Y$ of order unity, or $\lambda \simeq \sqrt{Y/\rho_o \dot{\varepsilon}^2}$. Relating this turbulence length scale

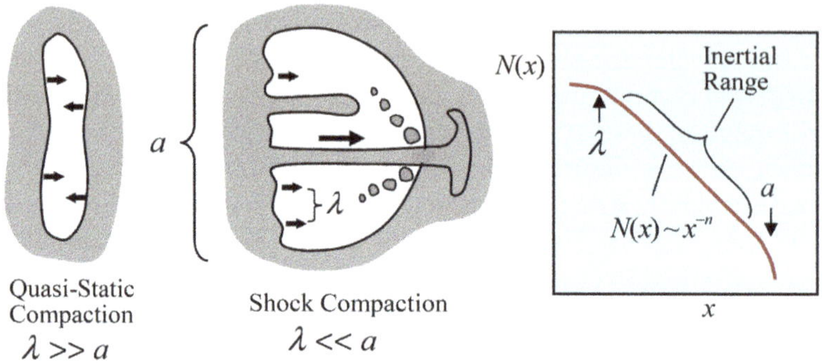

Figure 7.45. Processes are illustrated for the quasi-static compaction (left) and shock compaction (middle) of a ductile material void. Different compaction mechanisms account for the marked disparity in compaction dissipation. Steady wave shock compaction dictates plasticity turbulence over a range of length scales.

to the width of the structured shock is problematic. The compaction shock wave on the microscale is chaotic and the shock width, and an associated strain rate, is meaningful only in a statistical sense.

Turbulent dissipation will be active over the inertial range $\lambda < x < a$ (Falkovich and Sreenivasan 2006). The bounding length scales are illustrated in figure 7.45. Physics dictates that microstructure features of the dynamic compaction process exhibit power-law dependence on size within the inertial range. The number of transient hot spots would, for example, show a power-law dependence on size.

7.9.2 Compaction of the distended brittle solid

In brittle solids the processes of compaction and dissipation in both the quasi-static and the shock environment differ markedly from that of the ductile solid. Consider an accumulation of cemented or unbonded particles, or a powder, of a brittle material such as glass or ceramic. A characteristic microstructure length scale a also describes the initial particle size in the distended aggregate of brittle particles. The extent of distention is again defined by the initial density ρ_o and solid density ρ_s. Crush under pressure loading proceeds through the collective failure and fragmentation of individual particles when the stress load on elemental particles achieves a critical failure level. In characterizing the strength and failure properties of a porous brittle material, such as was done for a ductile material with a single yield stress Y, it is necessary to characterize both the failure strength σ of individual particles (a statistical measure), and the energy γ expended in production of new fracture surface area in the compaction event.

7.9.2.1 Brittle compaction strength and dissipation

As an external load is applied to a granular body of a brittle solid, stress loading of individual particles to failure proceeds through contact forces, as perhaps captured in figure 7.46. Failure proceeds when a critical stress, or possibly a critical elastic strain energy, is exceeded. Failure of any particular particle is inherently statistical. In highly brittle solids the strain energy achieved at failure of a particle can

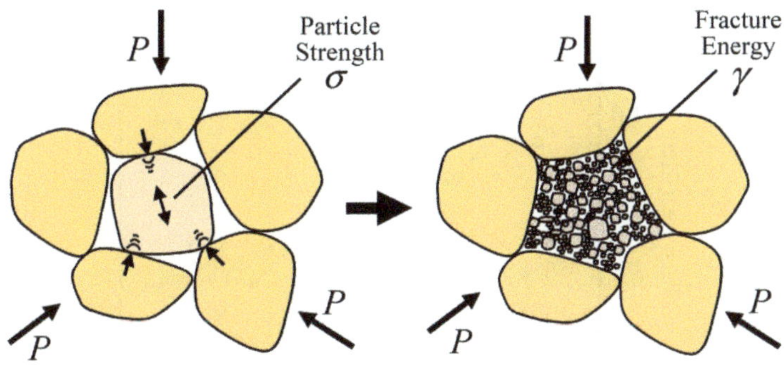

Figure 7.46. Catastrophic failure and explosive fragmentation of a particulate element in the compaction of a highly brittle material as a consequence of both large elastic strain energy characterized by the strength property σ and weak fracture resistance captured by the fracture energy γ.

significantly exceed the ability of the material to dissipate energy through creation of new fracture surface. Local failure can be intense. Strain energy is converted into both fracture surface energy of the distribution of fragments resulting from failure of the particle, and into kinetic energy of the fragments. The latter leads to local secondary impact and further particulation or into acoustic elastic energy. The fragments created in the failure of any single particle exhibit a power-law distribution in size. See chapter 2 and also Grady (2010). The number of fragments and extent of the distribution are dependent on the intensity of the strain energy (characterized by the strength property σ) achieved before failure, and the magnitude of the fracture surface energy γ as illustrated in figure 7.46.

The power-law nature of the compaction comminution extends to the description of the evolving particle size distribution through compaction of brittle media. Both the static compaction and dynamic (Hopkinson bar) compaction studies of Huang *et al* (2017) on SiO_2 quartz sand demonstrate power-law particle size distribution over an extended particle size range. The power-law distribution is bounded from above by the initial particle size and below by a minimum particle size that evolves downward as compaction proceeds. It is reasonable that the lower power-law cutoff size is determined by the extent of compaction and the limiting fracture surface energy γ.

Additional compaction experiments on granular SiO_2 are shown in figure 7.47, further illustrating the complexities of brittle compaction. Shock compression data were previously displayed in figure 7.23 and subjected to analytic compaction description in section 7.6.2. Dynamic compaction is contrasted with comparable static compaction measurements. In one static compaction test initial granular particle size ranges down from an upper particle size of 44 μm. In the second the granular SiO_2 is sieved to include only particles in the range of 1–5 μm. The latter exhibits significantly higher compaction resistance. Evolving power-law particulation in the former, as noted by Huang *et al* (2017), more effectively fills the void as compaction proceeds. Size-dependent strength and delayed formation of power-law particulation in the 1–5 μm test leads to markedly stiffer compaction resistance.

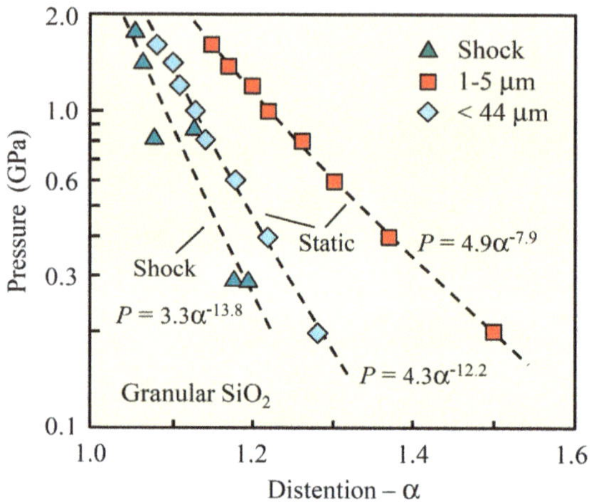

Figure 7.47. Impact shock (Brown *et al* 2007) and static (Lee *et al* 2004) compaction of granular silicon dioxide. Initial particle size ranges down from 44 μm in static test 1. Particle size is sieved to between 1–5 μm in static test 2. The initial distended density is 1.57, 1.40 and 1.07 g cm^{-3} for the shock test, static test 1 and static test 2, respectively.

Dimensional consideration of the brittle compaction event is less definitive than for compaction of a ductile porous solid. Considerations of dimensional issues are nonetheless worthwhile. A possible non-dimensional expression comparable to equation (7.221) is $\rho_o\dot{\varepsilon}^2a^2/\sigma$, which identifies a ratio of a dynamic pressure at the granular level to a measure of the statistical strength of the grain. Here $\dot{\varepsilon} = u/U\tau = u/\lambda$, as before, is a measure of the continuum loading rate as the compaction shock wave envelopes the material, a is the characteristic particle size and σ is a stress measure of the brittle material strength. Perhaps more appropriate is the dimensionless ratio is $\rho_o\dot{\varepsilon}^2a^2/(\sigma^2/E)$ with E an elastic modulus of the material. As such, the ratio relates a measure of the intensity of local kinetic energy to elastic strain energy of the grain commensurate with failure. Elasticity is judged as less important in the mechanics of compaction of ductile solids. Elasticity, as a measure of stored strain energy, is central to the failure of high-strength solids.

Also relevant to the brittle compaction event is the dimensionless ratio, independent of the loading rate, $\sigma/\gamma a$, or $(\sigma^2/E)/\gamma a$, relating failure strength or energy to brittle fracture resistance of the material. Large numbers for either ratio relate to the explosive fragmentation intensity of the brittle failure event. Higher values lead to more intense fragmentation. The apportion of strain energy into fracture surface energy and kinetic energy at failure of an elemental particle is not known, although there are arguments that suggest the partitioning in such abrupt transitions is approximately equal (e.g. Archambeau *et al* 1976). Kinetic energy imparted to resulting fragments leads to subsequent impact and secondary breakage within the pore space. Most kinetic energy transitions to incoherent acoustic energy and ultimately thermalizes.

Key to the compaction strength of a brittle granular medium is the strength parameter σ. This strength measure is inherently statistical. Strength testing is commonly accomplished through measurements on elemental units and quantified through statistical descriptions. Two-parameter Weibull relations are representative and conveniently capture such strength test data. The Weibull distribution

$$N(p) = N_o(1 - e^{-(P/\sigma)^m}), \tag{7.222}$$

with a strength scale parameter σ and Weibull modulus m, reasonably describes the statistical cumulative failure number N_o of similar spherical (or comparably equidimensional) particles of a brittle material subjected to compression to a load P. If such particles are of a size a, the strength of brittle materials commonly depends on particle size. Particle size dependence can be accommodated in the Weibull relation when warranted (e.g. Strack *et al* 2015).

Consider a porous bed of such particles. Equation (7.222) can be applied to provide a functional description of the compaction strength and compaction history of such a material. If failure and crush of a single element yields an average contribution ε_c/N_o to the total compaction strain, then as successive probabilistic failure of elements within the porous bed ensues, equation (7.222) yields a compaction strain versus compaction pressure relation,

$$d\varepsilon = \varepsilon_c \frac{dN}{N_o} = \varepsilon_c \frac{m}{\sigma}\left(\frac{P}{\sigma}\right)^{m-1} e^{-(P/\sigma)^m} dp. \tag{7.223}$$

A representative compaction relation from equation (7.223) is illustrated in figure 7.48. The shape and amplitude of the crush curve are determined by the Weibull parameters σ and m. In this rigid-compaction model, ε is a measure of the volume compaction strain with ε_c corresponding to full closure of the void space.

Dissipation in the static compaction event is provided by the integral of the compaction curve to the requisite compaction strain. In terms of the elemental compaction number relation dependence on the load pressure $N(P)$, the differential dissipation expression is

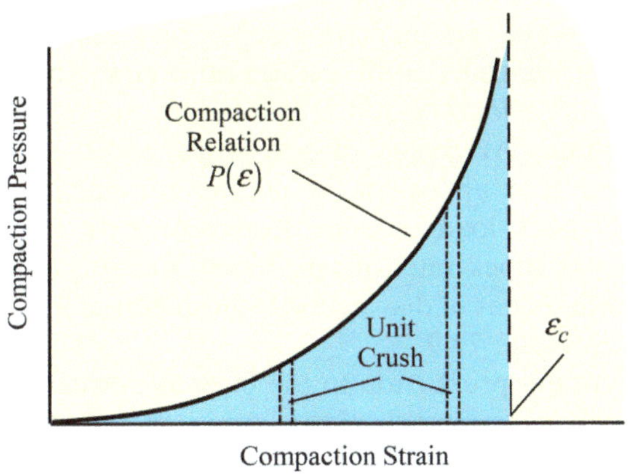

Figure 7.48. Pressure versus compaction strain of brittle media based on a Weibull description of continuum compaction. Illustrated are the compaction contribution of successive unit elements.

$$d\mathcal{E} = \frac{P}{\rho_s} d\varepsilon = \frac{P}{\rho_s} \varepsilon_c \frac{dN}{N_o}. \tag{7.224}$$

Energy dissipation associated with the crush of one element is then $\Delta\mathcal{E} = P\varepsilon_c/N_o$. Dissipation energy of two such elements is illustrated in figure 7.48; one earlier in the statistical failure event and one substantially later. Dissipation associated with failure and crush of the latter element is markedly larger than that of the earlier, which is indicative of higher strain energy release and more extensive fragmentation.

Several weaknesses of the present model are apparent. The failure and more refined fragmentation of the particle element at high compaction pressure should accommodate a larger contribution to the compaction strain than would the same element that fails at a lower pressure. Second, to achieve full compaction, secondary, tertiary and perhaps further particle breakage is necessary to achieve full elimination of void volume. Local stresses approaching theoretic crystal strength would be required. Nonetheless, equation (7.223) sensible reproduces compaction measurements of brittle granular media (see figure 7.50), and the underlying physics yields some intuitive understanding of the brittle compaction event.

The same porous medium when subjected to a dynamic load that transitions to an evolving compaction wave will exhibit a compaction path and dissipation comparable to that illustrated in figure 7.48 when rise of the wave to the compaction pressure is sufficiently slow. Namely, the compaction wavelength λ is substantially larger than the microscale length a. The compaction wave will steepen with propagation distance, and dissipation will accordingly increase as transition to a steady compaction shock and Rayleigh-line dissipation ensues. At a specific pressure the extent of compaction achieved through a steady shock wave and that achieve at more modest compaction rates need not be the same. In general they are not.

Mechanisms accounting for the excess dynamic dissipation are several and depend on the extent of steepening of the compaction wave. Early in the steepening event only modest increases in dissipation occur. Gradients in the compaction load on elemental particles lead to statistically higher strain energy conditions before particle failure and yield corresponding increases in energy dissipation.

As the compaction wave continues to steepen and approaches a strong shock wave, more extreme mechanisms of dissipation must come into play. On the microscale, complexities of stress loading and brittle failure can occur under the influence of strong shock loading. At points of low-angle grain contact, intense jetting fragmentation akin to that appearing in ductile compaction can occur. Within void regions far-wall shock loading from upstream ejecta fragment debris will be increasingly common as shock intensity increases. Both shock-induced jetting and far-wall impact as mechanisms of dissipation intensification in the shock compaction event are illustrated in figure 7.49.

Dissipation intensifies in proportion to the increased intensity of the compaction shock wave. The shock wave imparts elastic strain energy to the brittle particulate constituting the distended material. Particulate failure is followed by explosive fragmentation with strain energy partitioned between fracture surface energy and fragment kinetic energy. The kinetic energy becomes random acoustic energy that

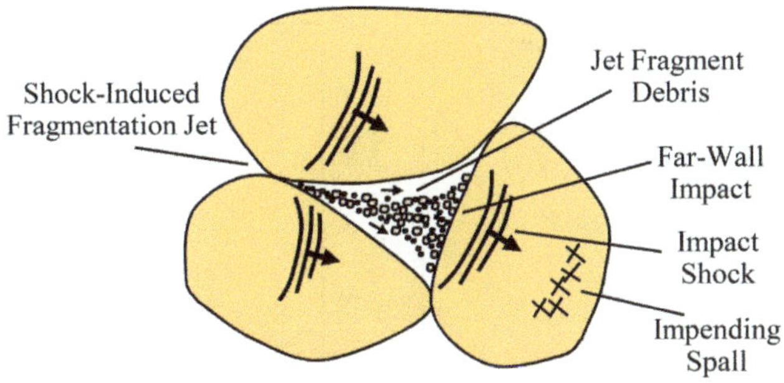

Figure 7.49. An illustration of microscale intense shock compaction mechanisms in the extreme shock compaction event, including shock-induced jetting and secondary fragment debris impact.

percolates up thorough higher frequency normal modes as thermodynamic equilibrium ensues subsequent to shock passage. This energy is thermalized and accounts for the temperature rise experienced in the shock compaction of brittle solids. Depending on conductivity properties of the material, local thermalization can precede continuum thermalization. Localized hot spots and melting are potential intermediate stages in the equilibration process subsequent to extreme shock compaction. The relationship of compaction shock structure to the microscale in brittle shock compaction is illustrated in the next subsection.

7.9.2.2 Compaction of tungsten carbide powder

Shock and static compaction testing on granular tungsten carbide (WC) are illustrative of the issues governing high-pressure compaction of brittle solids. Both dynamic and static experimental compaction data are shown in figure 7.50 (Vogler *et al* 2007). The test powder was prepared with initial near-equidimensional granularity (see the inset photograph in the figure) and was sieved to a particle size range of 20–32 μm. Powder morphology is illustrated in the inset on the left in figure 7.50. The initial powder density of the material prior to shock compaction testing was 8.65–8.85 g cm^{-3}, corresponding to a porosity of about 45%. A comparable initial density was used for static compaction testing of similarly prepared material. The crystal density of tungsten carbide is 15.7 g cm^{-3}.

Five impact shock experiments defined a Hugoniot compaction of WC powder to in excess of 4 GPa and a maximum compaction density of 14.7 g cm^{-3} (approximately 10% residual porosity). A compaction curve of the form of equation (7.223) is shown with the data. A strength factor of $\sigma = 1.6$ GPa and a Weibull modulus of $m = 0.8$ yield an optimal fit to the first-shock Hugoniot compaction data. The test configuration also achieved reshock states that can be assessed from the data through appropriate analysis. Reshock states for the five tests are similarly shown in the plot on the left.

Static compaction tests to approximately 1.5 GPa reveal a markedly softer static compaction resistance. This observation is in accord with dynamic and static compaction testing on other brittle granular compounds. (See earlier sections in the present chapter.) Differences may reflect the dissipation intensification

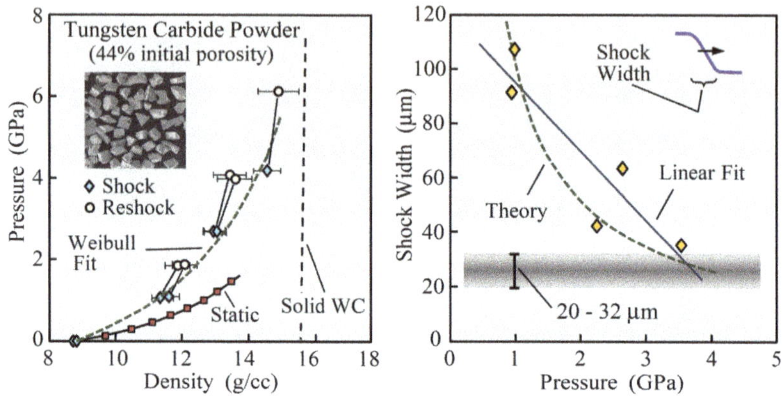

Figure 7.50. Left: Shock compaction of tungsten carbide powder (Vogler *et al* 2007) illustrating steady shock compaction and reshock states. Right: Steady shock wave width versus shock pressure in comparison with particle size of the test material.

necessitated by the Rayleigh line compaction path discussed in the previous subsection. Alternatively, complexities inherent to the static compaction event discussed in the next subsection may underlie differences observed in static and dynamic compaction.

The steady structured compaction shock waves were measured with velocity interferometry (VISAR) technology in the dynamic testing. This wave structure is captured through a measure of the width of the shock wave and plotted against the Hugoniot pressure on the right in figure 7.50. Measured shock width is contrasted with the initial granular particle size of the test WC powder. Decreasing shock width with increasing shock pressure is necessary to accommodate the energy dissipation commensurate with Rayleigh-line transit of the shock path. Shock widths are a few times the initial particle size, and approach the grain size of the powder at the highest Hugoniot pressures. Compaction phenomena such as those contemplated in figure 7.49 should be expected at the intense compaction loading rates achieved.

These same data are examined further in chapter 8 in accompaniment with shock data on other granular powders. Compaction wave structure is captured through shock pressure versus strain-rate dependence. Underlying energy principles are found consistent with the observed compaction wave structure trends. The curve identified as theory relates to steady compaction wave systematics identified in chapter 8.

The reshock states shown on the left in figure 7.50 do not reside on the principal compaction Hugoniot. Deviator strength effecting the second wave structure is a possible source of the reshock state offset. This character can be viewed as the shock production of a competent ceramic of the initial powder in the first wave. The second wave testing the elastic-plastic strength of the newly created ceramic.

7.9.2.3 Structure features in brittle compaction

Static and near-static compaction of brittle solid materials exhibits structural features not readily captured by the continuum compaction models explored in earlier sections. The nature of such compaction behavior is varied and apparently dependent on both the brittle material properties and the loading conditions leading to compaction.

Triaxial compaction testing of porous sandstone is illustrative of one form that such structural compaction takes (Wang *et al* 1992, Olsson 1999, 2001). At the onset of compaction failure, compaction proceeds through the propagation of compaction bands. Compaction bands may or may not include a shear deformation component. Load stress remains constant or nearly constant through the compaction event. Tests on sandstone reveal extensive compaction through the compaction band, reducing the initial porosity of about 25% to a few percent. One sketch on the left in figure 7.51 is illustrative of testing on Castlegate sandstone (Olsson 1999) arrested part way through the compaction event and examined post-test. Compaction bands indicative of normal compaction, and with both normal and shear components of compaction, are observed in the tests. Such compaction banding under tectonic forces has been observed in geologic formations of porous quartz sandstone (Mollema and Antonellini 1996).

In less competent brittle material markedly different compaction behavior is observed. Compaction activation occurs at or near structural boundaries. Compaction fronts propagate through the material. Only modest levels of distention reduction are accommodated, and numerous transits of the compaction fronts are necessary to approach full compaction. Succinct experiments demonstrating this repetitive multiple compaction front process have been perform and diagnosed through optical imaging of experiments on initially highly distended breakfast cereal (Valdes *et al* 2011, Guillard *et al* 2015). The structure and interference patterns of the compaction fronts are observed and shown to be dependent on the rate of

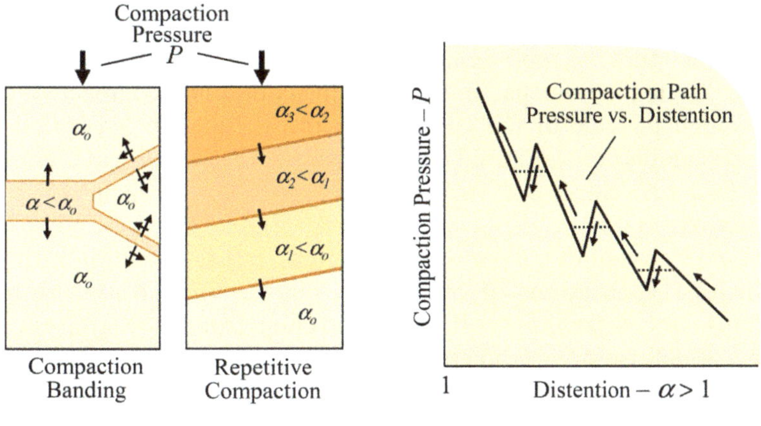

Figure 7.51. The sketches on the left illustrate two modes of structural compaction exhibited by brittle solids. The compaction path on the right exhibits compaction instability leading to the formation and propagation of compaction bands.

compaction loading. The second sketch on the left of figure 7.51 is depictive of this repetitive compaction process in brittle solids as distention α in the porous body is sequentially reduced.

In both of the several types of compaction experiments on brittle materials, the onset and continued compaction is a non-local structure-dependent event. At least a qualitative explanation for the compaction behavior is suggested by the compaction pressure versus distention plots shown on the right in figure 7.51. The plot illustrates regions of compression in which loading is elastic, or near-elastic. In this phase, strain energy is accrued in the brittle particulate constituent of the porous solid. Also illustrated are abrupt states of compaction onset and accompanying loss of load supporting ability. Failure constitutes the fracture and refinement of a particulate with corresponding reduction in porosity. The process can repeat as the pressure load is increased, as is necessary to accommodate the required compaction. Whether few or many such abrupt events occurs during compaction is dependent on a number of material property and pressure-loading conditions that are not fully understood. Some aspects of this dependence have been explored in computer studies of the brittle compaction event (Guillard *et al* 2015).

The tests of Olsson (1999) are on well-cemented quartz sandstone. The stress load is elastic to approximately 150 MPa failure stress. Quartz granules are strong. On the other hand, quartz fracture energy is low. Local fracture is pervasive and hence porosity reduction from an initial 20%–25% to a few percent following compaction is intuitively sensible. Figure 7.46 is perhaps illustrative. Conversely, although the strength and fracture properties of breakfast cereal are not known (Guillard *et al* 2015), it seems unlikely that significant strain energy is stored preceding the repetitive failure event accommodating compaction in this material.

Onset of brittle compaction, as illustrated on the right in figure 7.51, is an unstable critical state event. Onset of compaction at statistically weaker sites is not self-stabilizing. The formation and propagation of compaction from such initiation sites is a natural consequence of unstable critical state phenomena. That compaction onset morphs into planar compaction fronts is consistent with the necessary density difference accommodated in the compaction transition. The dashed line paths illustrated in the figure are representative of the Maxwell construction characteristic of such critical state phenomena, and capture the continuum compaction path of the material through passage of the compaction front. Pressure is continuous across the front and remains constant until compaction within the interval is complete. The tests of Olsson are consistent with constant pressure through the compaction propagation event.

Theoretical treatment of compatibility through such deformation fronts has been pursued through applications of bifurcation theories (e.g. Rudnicki and Olsson 1998). Details of compaction within the traveling fronts are addressed through applications of deformation gradient models. Such unstable compaction has been likened by Guillard *et al* (2015) to a Portevin–Le Chatelier phenomenon; Le Chatelier identified with the instability nature of compaction onset, while Portevin with comparisons of similarities to the jerky character of certain metal plasticity behavior.

The compaction tests of Vogler *et al* (2007) on granular tungsten carbide shown in figure 7.50 reveal that in the strong shock compaction event with shock widths of a few grain dimensions, instability and structural aspects in the compaction process are unlikely. Both theory and experiment show, however, that the shock width increases as the compaction pressure is reduced. Thus, in the weak shock compaction regime with shock widths significantly exceeding the characteristic grain size, compaction instability issues may play a more important role. Or, similarly, if dynamic compaction entails multiple shock or ramp type pressure wave loading.

7.9.3 Dynamic compaction in crushable mixtures

An added dimension to the shock compaction of solid materials is encountered when mixtures of disparate compliance materials are of interest. Examples might include a mixture of silicate sand and water, particulate ceramic and epoxy, or molecular crystals of a reactive material and binder. A number of constitutive modeling methods have been pursued to address the response of such mixtures to strong shock compression loading. See, for example, the work of Bedford and Drumheller (1983).

The present chapter is focused principally on the dynamic compaction of solid material with entrained void (or air). This solid–void structure is inherent in many of the engineering level models developed to describe dynamic compaction; the p-alpha model for example. Compaction is also a consequence of compression loading when compliant substances other than the void are entrained. Models such as the p-alpha models are applied with modest adjustments to account for properties of the entrained media, but are not explicitly developed for such purposes.

The p-lambda model was developed specifically for the purpose of addressing dynamic compaction in disparate compliance mixtures; a solid with entrained water for example. The p-lambda model is discussed in further detail in chapter 6 with application to ceramic materials, and is also applied to porous media earlier in the present chapter. It is nonetheless instructive to re-examine the p-lambda model with attention focused on the dynamic compaction of distended solids with entrained media.

The intent of this subsection is to explore several issues that emerge, both in experiments and in reflection, in the high-rate and shock compaction of disparate compliance mixtures. The first is the recognition that a concept identified as an effective stress in the shearing strength of a granular medium with an entrained fluid, introduced by Terzaghi some decades ago, also has application to compaction of mixtures. As the effective stress principle simplifies the constitutive description of the shearing strength of a granular-fluid medium, so do similar principles simplify construction of the constitutive description of the compaction of mixtures. Second, in the compaction of a mixture comprised of a strong component and a lower strength more compliant component, stress states on the mixture granular scale can differ substantially from the pressure load effecting compaction of the mixture. Critical state physical phenomena, such as phase transformations, can be markedly effected by this disparity. Lastly, applications involving the compaction of mixtures composed of components with strength have demonstrated that system calculations using pressure–volume compaction models alone are inadequate. System level calculations require deviator strength models in addition to pressure compaction models to adequately represent observations.

7.9.3.1 P-lambda mixture compaction model

Like other models in physics, the p-lambda model is characterized by two bounding states of the material along with a transition relation that transforms the material from one state to the second. Relations for the elastic compressibility of the mixture and compression of the fully crushed composite constitute the upper and lower bounding states, respectively. Expressions convenient for the bounding states are the elastic iso-strain compressibility and fully crushed iso-pressure compressibility, although other relations are appropriate for certain applications. Dynamic compaction is modeled through a functional relation that transforms the mixture from the elastic state to the fully crushed state. A functional dependence of the compaction parameter on the continuum compressive pressure most commonly completes the model. Relative motion of the solid and entrained media is not an aspect of the model. Material is considered occluded, either physically, or due to inertia in the extreme loading rates of the shock event.

The p-lambda mixture compaction model is framed by identifying compression relations for the upper-bound (uncompacted) compression and lower-bound (compacted) compression of the mixture. Upper-bound compression assumes equal elastic compression strain $\mu_s = \mu_i = v_i/v_{io}$ among mixture components. The upper-bound compression relation follows from volume fraction weighted arithmetic average of component pressures at equal elastic compression strains,

$$P(\mu_s) = \sum_i \eta_i P_i(\mu_s). \tag{7.225}$$

Inversion of equation (7.225) provides the upper-bound compression relation $\mu_s(P)$. The subscript s identifies equal component compression strain (iso-strain or iso-compression). Lower-bound compression assumes equal pressures $P = P_i$ among mixture components. The lower-bound compression relation follows from volume fraction weighted arithmetic average of component compressions

$$\mu_p(P) = \sum_i \eta_i \mu_i(P). \tag{7.226}$$

The subscript p identifies equal component pressure (iso-pressure).

Compaction is identified as a transition of the mixture material from the upper-bound iso-compression state to the lower-bound iso-pressure state. A transition parameter $\lambda(p)$ is identified over the range $0 \leqslant \lambda \leqslant 1$ such that compression of the mixture at any pressure is provided by additive superposition of the iso-strain and iso-pressure compressions,

$$\mu_c(P) = \lambda \mu_p(P) + (1 - \lambda)\mu_s(P), \tag{7.227}$$

where the subscript c identifies volume compression on the compaction curve.

A power-law dependence of the transition parameter on pressure of the form

$$\frac{d\lambda}{dP} = \frac{1}{\sigma}\left(\frac{P}{\sigma}\right)^{n-1}(1 - \lambda) \tag{7.228}$$

provides the compaction transition parameter constitutive relation. Compaction depletion is addressed through the $(1 - \lambda)$ term. Integration yields

$$\lambda(P) = 1 - e^{-(P/\sigma)^n}. \tag{7.229}$$

Equation (7.229) applies as pressure increases. The transition parameter is frozen at the maximum compaction pressure achieved $\lambda(P_{max})$ on decompression, or recompression if $P < P_{max}$, and equation (7.227) provides the decompression/recompression path

$$\mu_c(P) = \lambda(P_{max})\mu_p(P) + (1 - \lambda(P_{max}))\mu_s(P). \tag{7.230}$$

A representative application of the p-lambda mixture compaction model is illustrated in figure 7.52. Shown are the component and mixture volume compression relations for a three-component mixture of tungsten carbide, paraffin and air. Component compression relations are identified as dashed curves on the left in the figure and are constructed from available EOS data. Iso-strain (compression) and iso-pressure solid curves on the left are calculated through equations (7.225) and (7.226). Compaction relations follow from reasonable crush strength properties of tungsten carbide through equations (7.227) and (7.229) with $\sigma = 3$ GPa and $n = 1.5$. Compaction to 4 GPa and subsequent decompression are illustrated on the right in figure 7.52.

7.9.3.2 Compaction and the Terzaghi principle

The shearing strength of a bed of granular medium is determined by the contact strength of the material across a shearing interface. If the confining pressure load is increased, then the contact strength across the interface increases and

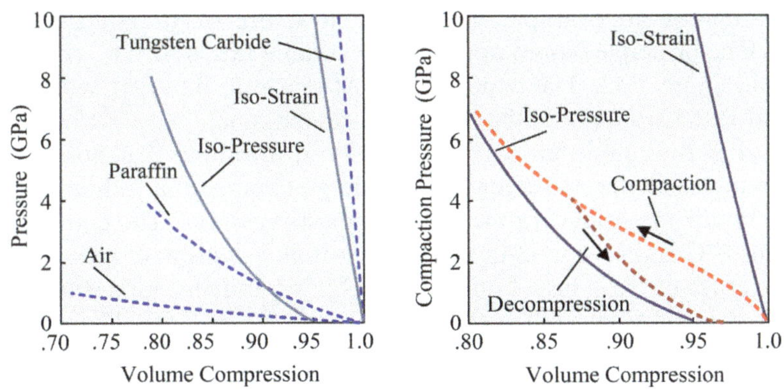

Figure 7.52. Representative p-lambda compaction of a three-component mixture by percentage volume of air (5%), tungsten carbide (45%) and paraffin (50%).

correspondingly the shear strength of the medium increases. If a fluid is introduced in the granular medium and fluid pressure is elevated to a pressure p_w, the contact strength is reduced and similarly the shear strength of the material is reduced. The confining pressure load must be increased by an amount comparable to the fluid pressure p_w to recover the same contact strength and shearing strength of the medium. From this logic Terzaghi (1925) proposed that the effective shearing strength of the granular medium containing fluid at pressure is governed by the difference

$$\tau_e = \tau - p_w . \tag{7.231}$$

This Terzaghi principle has since found application in condensing the constitutive description of fluidized granular media.

The Terzaghi principle similarly has application to the compaction of granular material (Grady 2001, Fenton *et al* 2015). Compaction resistance to an applied pressure load is a consequence of failure characteristics and strength of the granular material as void diminishes and contact strength increases during compaction. If fluid permeates the void space at fluid pressure p_w, then the pressure load must correspondingly increase to achieve a comparable level of compaction. Accepting logic similar to that proposed by Terzaghi, a pressure expression comparable to equation (7.231),

$$P_e = P - p_w , \tag{7.232}$$

governs the effective compaction pressure of a granular medium with fluid or fluid-like material in the void space.

There is a central difference between the two applications. In the shearing strength of granular medium considered by Terzaghi, the fluid pressure is considered independently specified. In the compaction of granular media, void volume is considered occluded, due either to a closed cell porous structure or the extreme dynamics of the shock compaction event. Fluid pressure is thus a unique function of the diminishing void volume as compaction proceeds.

Application of the Terzaghi compaction principle is illustrated in figure 7.53. Again, consider as an example a granular structure of tungsten carbide (WC) particulate. The principle would apply to the volume ratios of WC, paraffin and air considered in figure 7.52. The concept is most relevant, however, when the higher strength and stiffer material is the dominate component.

A body of porous tungsten carbide initially at an initial distended volume V_{oo} subjected to increasing pressure load will undergo compaction trending towards the compressibility of the solid ceramic at the highest pressures. The compaction curve of the porous WC is shown in figure 7.53. An initial distention of $\alpha_o = V_{oo}/V_o = 1.15$ is used. The crush curve is based on a reasonable assessment of the tungsten carbide compaction data of Vogler *et al* (2007) shown in figure 7.50. The abscissa is the relative volume compression V/V_{oo}.

Consider the same porous tungsten carbide with the pore volume filled with paraffin. Compression from initial volume V_{oo} exhibits a compaction curve

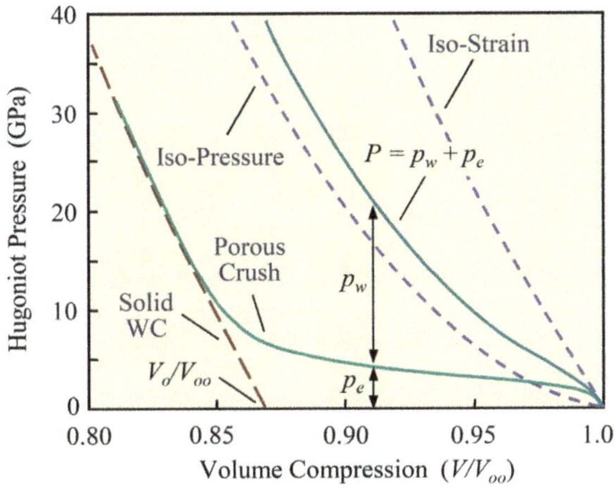

Figure 7.53. Application of the Terzaghi compaction principle to the shock compression of tungsen carbide and paraffin mixture.

illustrated in the figure that resides above that of the porous ceramic. The Terzaghi principle proposes that at any level of compaction the pressure according to equation (7.232) is $P = p_e + p_w$, with p_e that corresponding the effective compaction strength of the porous WC and p_w the back pressure determined by the compression EOS of the paraffin to that level of compaction. Compaction of the mixture will approach the iso-pressure compressibility of the WC–paraffin mixture at high pressure. Or, more precisely, the difference between the compaction curve and the iso-pressure curve becomes increasing less consequential with increasing pressure.

Accepting the Terzaghi principle as applied to compaction allows application of the crush properties of the porous tungsten carbide and EOS properties of the mixture components to estimate the p-lambda parameters σ and n in equation (7.229), characterizing compaction of the mixture (Grady 2001, Fenton *et al* 2015).

7.9.3.3 Terzaghi effective pressure and critical state events

A material mixture in which the dominant component is a stiff, high-strength solid, while the minor component supports minimal strength and is substantially more compliant, undergoes dissipative compaction when subject to intense compression. The response of such mixtures subject to compaction exhibit the Terzaghi principle as applied to mixture compaction. Namely, back pressure from the more compliant component delays compaction to higher pressure levels determined by the compaction strength and relative compliances of the mixture components.

There is a physically significant inverse effect that can occur in such mixtures when subject to compaction level pressures. The pressure in the compliant component resides below that of the load pressure P applied to the mixture by an amount determined by the effective compaction pressure of the dominant component. Namely

$$p_w = P - p_e, \tag{7.233}$$

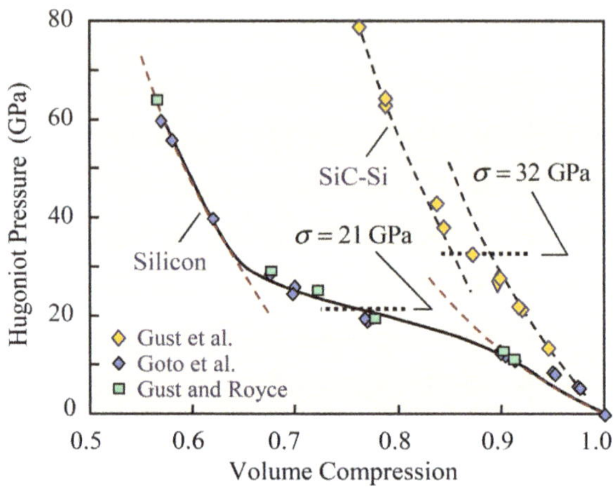

Figure 7.54. Shock Hugoniot data for KT silicon carbide from Gust *et al* (1973) and for silicon (Gust and Royce 1971a, 1971b Goto *et al* 1982). Effects of the silicon phase transformation is apparent for both the silicon and the silicon carbide–silicon mixture.

with p_e a function of the extent of compaction achieved.

If critical state physics of the compliant component is dependent on the local pressure p_w achieved, then the compaction strength of the dominant component will delay onset of the critical state event. A clear case is that of a pressure-dependent first-order phase transformation in the more compliant component of the material mixture. Several examples exhibit the consequences of this mixture compaction effect.

In figure 7.54 are plotted shock Hugoniot data for crystalline silicon (Gust and Royce 1971a, 1971b, Goto *et al* 1982) and silicon carbide (Gust *et al* 1973), pursued previously in earlier chapters. The silicon carbide, as a consequence of production methods at the time, contains approximately 10%–15% free silicon. The ceramic is therefore a microstructure mixture of silicon carbide and silicon. Silicon is more compliant than silicon carbide and markedly so when undergoing phase transformation under pressure.

Silicon carbide must accordingly undergo compaction to accommodate the higher volumetric compression of the mixture silicon component including the volume contraction accompanying the phase transformation. Silicon carbide is a high-strength ceramic. Appreciable strength necessary to achieve compaction is expected. Although phase transformation on the Hugoniot is distributed over an appreciable pressure range, the effects of the silicon phase transformation is apparent on both the silicon Hugoniot, and the silicon carbide and silicon mixture Hugoniot. The pressure difference of approximately 10 GPa is attributed to the reduced pressure in the silicon in the mixture as a consequence of compaction strength of the silicon carbide.

A further, perhaps more controversial, example of the same phenomenon is illustrated in figure 7.55 through comparison of the principle Hugoniot response of alpha-quartz and mullite. Mullite, named after the island of Mull off the west coast of Scotland where it was first discovered, is a solid solution phase of alumina and

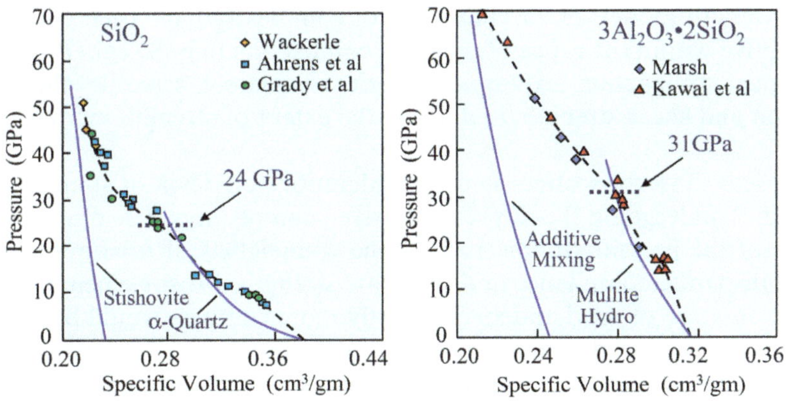

Figure 7.55. Shock Hugoniot data for quartz (left; Wackerley 1962, Ahrens and Rosenberg 1968, Grady *et al* 1974) and for mullite (right; Marsh 1980, Kawai *et al* 2004).

silica that occurs naturally and is also fabricated for ceramic applications (Duval *et al* 2008). Shock Hugoniot measurements have been performed on silica rich mullite ($3Al_2O_3{\cdot}2SiO_2$) with an initial solid density of 3.15 g cm^{-3} (Marsh 1980, Kawai *et al* 2004). Hugoniot data displayed on the right in figure 7.55 reveal the onset of phase transformation on the Hugoniot near 31 GPa shock pressure. Reference curves are provided by the mullite hydro-compression and an additive mixing of stishovite and aluminum oxide for the high-pressure phase (Kawai *et al* 2004).

Based on a density average, the initial volume fraction of Al_2O_3 is 0.37. This relative volume fraction changes with pressure in favor of the aluminum oxide due to the marked difference in compressibility of the two compounds. Corresponding Hugoniot compression of SiO_2 is shown on the left where onset of transition from α-quartz to stishovite is identified at 24 GPa shock pressure. The elevated transformation pressure of the four-to-six fold silica coordination transformation in the mullite structure on the right in the figure may be a consequence of the lattice strength the non-transforming Al_2O_3, comparable to the silicon and silicon carbide composition transformation displayed in figure 7.54. The former is a solid solution, however, while the latter is particulate on the microscale. Thus, other explanations for the pressure disparity in mullite should be entertained.

7.9.3.4 Strength and dynamic compaction

Discussions of compaction in the present chapter are limited to geometries in which compaction pressure can be identified with the component of stress collinear with the direction of shock propagation. Decomposition into mean stress and shear stress components is not addressed. In computational applications of the engineering-level compaction models considered in the chapter, compaction is commonly identified with the mean stress. Shear strength during compaction is modelled separately. Compaction experiments of sufficient complexity demonstrate that shear strength, in addition to the compaction constitutive description, must be addressed for successful computational replication. Shear strength in conjunction with compaction is alluded to briefly regarding application of the p-alpha model (see figure 7.26).

Shear strength models of varying levels of complexity have been developed that are descriptive within the range of dynamic compaction (e.g. Swegle 1980). Lacking are dynamic compaction experiments that adequately separate the effects of compaction and shear strength, and assess the extent of strength model complexity necessary.

The present subsection offers some considerations on deviator strength modeling within bodies undergoing shock-pressure-driven compaction. The discussions retain the theme of the present section, namely the compaction of mixtures of materials with both disparate compliance and disparate strength. Most elemental is a porous material, a mixture of solid and air. Modestly more complex would be a mixture of solid and water, or solid, water and air. Also of interest is the mixture of a solid compound and a metal, such as liquid-phase sintered engineering ceramic. Representative, for example, is the tungsten carbide ceramic. Tungsten carbide is commonly prepared through liquid-phase sintering methods with percentage levels of metal. Cobalt metal is frequently the additive metal. The Hugoniot properties of tungsten carbides and the strength on the Hugoniot are pursued in both chapters 6 and 10. The two-component mixture of tungsten carbide and paraffin considered in section 7.9.3.2 is retained here for the purpose of an example.

A description of the strength of a compacting mixture is based on the assumption that each component of the mixture contributes in some sense strength (or lack of strength; fragility) to the strength of the mixture as a whole. The description then requires specification of the independent strengths of the participating components of the mixture. Then a recipe is specified for relating component strengths to the deviator strength of the mixture. A requirement of the description is that any relative fractions of the participating components can be specified. For example, if shear strength through a dynamic compaction event involves quartz sand, water and air, since the strength of air and water is negligible, only the strength of quartz sand and its proportion of the mixture need to be specified.

An arbitrary interior surface of a mixture shown on the left in figure 7.56 illustrates the relative area contributions of two mixture components. A reasonable

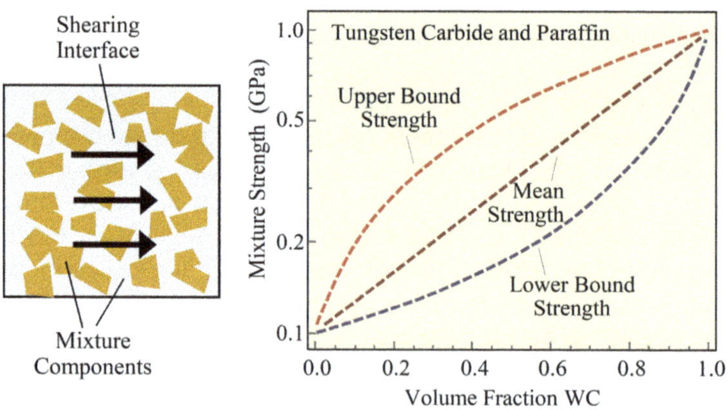

Figure 7.56. Measures of average strength for a two-component mixture.

assumption is that strength of the mixture depends on the strength of each component and on the relative area fraction on the shearing surface of each of the components. Although a sensible assumption, analysis is more amenable through a dependence on volume fraction, as opposed to area fraction.

Several averaging schemes are applied in assessing mixture strength. In one a volume weighted arithmetic average of mixture component strength is assumed. In the second, a volume weighted arithmetic average of the inverse strength (the fragility) is applied. The two averages provide measures of the upper- and lower-bound strength of the mixture. Each mixture average can have application in aspects of the strength model.

Component strength is identified by y_i where i ranges over the number of components in the mixture. The fragility of component i is the inverse $f_i = 1/y_i$. Component strength y_i and fragility f_i are considered independent of the state of confining pressure of the mixture. Upper-bound mixture strength is given by

$$\tau_u(P) = \sum_i \eta_i(P) y_i, \tag{7.234}$$

where η_i is the volume fraction of component i. The corresponding lower-bound strength is given by

$$\tau_l(P) = \left(\sum_i \eta_i(P) f_i \right)^{-1}. \tag{7.235}$$

Mixture strengths τ_u or τ_l are dependent on pressure through pressure dependence of the component volume fractions η_i. The upper- and lower-bound strength measures for a tungsten carbide and paraffin mixture is illustrated on the right in figure 7.56. Component strengths of 1 GPa and 0.1 GPa are arbitrarily assigned to the tungsten carbide and paraffin, respectively.

A further mixture strength measure, also shown in figure 7.56, that renders an equitable weighting of both the weak and the strong mixture components is provided by the geometric volume weighted average

$$\tau_m(P) = \prod_i y_i^{\eta_i(P)} = \left(\prod_i f_i^{\eta_i(P)} \right)^{-1}. \tag{7.236}$$

Under compaction, strength increases as the more compliant component (paraffin) becomes an increasingly smaller volume fraction of the mixture. Compaction pressure dependence of strength is accomplished through a joining with the p-lambda compaction model. Accordingly, strength is provided by

$$\tau(P) = \lambda(P)\tau_p + (1 - \lambda(P))\tau_s, \tag{7.237}$$

where $\lambda(P)$ from equation (7.229) provides the mixture fraction reduced in the compaction event to the pressure-equilibrated state. Under compression, strength is assumed to be dominated by the higher strength component and thus strength

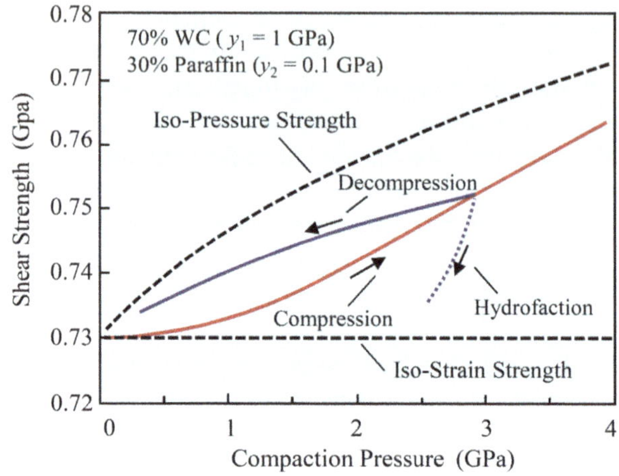

Figure 7.57. Strength of tungsten carbide and paraffin mixture based on p-lambda compaction.

contribution from the iso-pressure component is $\tau_p(P) = \tau_u(P)$ from equation (7.234). Strength contribution from the iso-strain component is $\tau_s = \tau_u(0)$ since volume fractions in the iso-strain component remain the same irrespective of pressure. Explicitly introducing equation (7.229) yields

$$\tau(P) = \tau_p(P) - \left(\tau_p(P) - \tau_s\right)e^{-(P/\sigma)^n}. \tag{7.238}$$

Application of equation (7.238) is illustrated in figure 7.57 for a 70% tungsten carbide and 30% paraffin mixture. Both iso-pressure strength and iso-strain strength provide the bounding curves. Strength as a function of compaction to a maximum pressure of 3 GPa is shown. The compaction parameters provided in section 7.9.3.2 are applied. Not addressed in this assessment is the concept of shear enhance compaction as a consequence of non-hydrostatic compression of the compacting medium (e.g. Johnson and Green 1976).

A further matter is the description of the strength subsequent to compaction as pressure is reduced from the compacted state. One solution is to retain the mixture strength relation provided by equation (7.237), but freeze the lambda parameter at the maximum compaction pressure achieved $\lambda(P_{\max})$ as is done in the p-lambda compaction model. This strength curve on decompression from 3 GPa is illustrated in figure 7.57.

There are physically sensible arguments for rejecting this description of deviator strength on decompression. In the compaction of granular mixtures, deviator strength is dominated by the strength of the stronger and stiffer mixture components. This dominance of the stronger components is not realized on decompression from higher pressure levels in the compaction event. Weaker materials, which are commonly also more compliant, expand and separate the stronger and stiffer component particles. The weaker more compliant components consequently increasingly dominate the deviator strength of the mixture as confining pressure reduces below the peak compaction pressure achieved.

This strength phenomenon in mixtures is identified as decompression hydrofaction, a term that refers to effects that cause a material to behave increasingly hydro-like in subsequent shearing deformations as the confining pressure is reduced from an elevated compaction pressure state. A more commonly recognized and related effect is that of soil liquefaction wherein a wet soil geology subject to a stress disturbance such as an earthquake event is reduced to subsequent liquid-like behavior.

Strength modelling accounting for hydrofaction is developed through appropriate relations describing the strength of the material at decompression pressures less than the maximum compaction pressure achieved. As such, it represents a significant modification of the decompression strength described in equations (7.237) and (7.238). Both the relative differences in compliance and strength will play a role. Strength tending towards the lower-bound strength of equation (7.235) seems sensible. Whether a hydrofaction modelling effort is warranted or not is arguable. Little test data is available to support the occurrence of such strength phenomena. The effect is nonetheless intuitively reasonable. The alternative strength reduction reflecting hydrofaction processes shown in figure 7.57 is illustrative.

References

Ahrens T J and Rosenberg J T 1968 Shock metamorphism: experiments on quartz and plagioclase *Shock Metamorphism of Natural Materials* ed B M French and N M Short (Baltimore, MD: Mono Book Corp) pp 59–81

Anderson L 1967 Equation for thermal expansivity in planetary interiors *J. Geophys. Res.* **72** 3661–8

Archambeau C B, Minster J B, Achenbach J D, Pao Y-H and Tiersten H F 1976 *Theory of Failure as a Transition Process in a Stressed Medium* (Washington, DC: National Science Foundation)

Barenblatt G I 2006 Scaling phenomena in fatigue and fracture *Int. J. Frac.* **138** 19–35

Bedford A and Drumheller D S 1983 Theories of immiscible and structured mixtures *Int. J. Eng. Sci.* **21** 863–960

Bell R L, Baer M, Brannon R M, Elrick M G, Hertel E S, Silling S A and Taylor P A (2000) CTH user's manual and input instructions, version 4.00, *Sandia National Laboratories Technical Report* April.

Boslough M B and Asay J R 1992 Basic principles of shock compression *High-Pressure Shock Compression of Solids* ed J R Asay and M Shahinpoor (Berlin: Springer) pp 7–42

Brown J L, Vogler T J, Grady D E, Reinhart W D, Chhabildas L C and Thornhill T F 2007 Dynamic compaction of sand *AIP Conf. Proc.* **955** 1363–66.

Butcher B M, Carroll M M and Holt A C 1972 Shock wave compaction of porous aluminum *J. Appl. Phys.* **45** 3864–75

Chadwick P 1963 *Int. J. Mech. Sci.* **5** 165

Carroll M M and Holt A C 1972 Static and dynamic pore-collapse relations for ductile porous materials *J Appl. Phys.* **43** 1626–35

Cooper P W 1997 *Explosives Engineering* (New York: Wiley)

Duval D J, Risbud S H and Shackelford J F 2008 Mullite *Ceramic and Glass Materials* ed J F Shackelford and R H Doremus (Berlin: Springer)

Duvall G E and Fowles G R 1963 Shock waves *High Pressure Physics and Chemistry* vol 2, ed Bradley (New York: Academic).

Elban W L and Chiarito M A 1986 Quasi-static compaction study of coarse HMX explosive *Powder Technol.* **46** 181–93

Falkovich G and Sreenivasan K R 2006 Lessons from hydrodynamic turbulence *Physics Today* **59** 43–9

Fenton G, Grady D E and Vogler T J 2015 Shock compression modeling of distended mixtures *J. Dynamic Behav. Mater.* **1** 103–13

Fredenburg D A and Dennis-Koller D D 2013 Consolidation response of brittle particulate materials to shock loading, *Proc. Eng.* 58, 401-408.

Fredenburg D A, Koller D D, Coe J D and Kiyanda C B 2014 The Influence of morphology on the low- and high-strain-rate compaction response of CeO_2 powders *J. Appl. Phys.* **115** 123511

Goto T, Sato T and Syono Y 1982 Reduction of shear strength and phase-transition in shock loaded silicon, *Japan. J. Appl. Phys.* **21** L369–71

Grady D E 1993 Impact compression properties of concrete, *Proc. Sixth Int. Symp. on Interactions of Nonnuclear Munitions with Structures (Panama City Beach, FL, 3–7 May 1993).*

Grady D E 1996 Shock equation of state properties of concrete *Structures under Shock and Impact* vol 4, ed N. Jones, C. A. Brebbia and A. J. Watson (Southampton: Computational Mechanics) pp 405–14.

Grady D E 2001 P-lambda model for material mixtures: application to compaction of porous zirconium dioxide, *Applied Research Associates Technical Report for Sandia National Laboratories*, October.

Grady D E 2004 Analytic solutions and constitutive relations for shock propagation in porous media *AIP Conf. Proc.* **706** 205–08.

Grady D E 2005 Investigation of explosively driven fragmentation of metals—metal breach, enhanced ductility, adiabatic shear, shock structure and strength, prepared for University of California, Lawrence Livermore National Laboratory Contract No. B545288 and Sandia National Laboratories, Contract No. 29523 *Applied Research Associates Technical* Report September.

Grady D E 2008 Fragment size distributions from the dynamic fragmentation of brittle solids *Int. J. Impact Eng.* **35** 1557–62

Grady D E 2010 Length scales and size distributions in dynamic fragmentation *Int. J. Fracture* **163** 85–99

Grady D E, Murri W J and Fowles G R 1974 Quartz to stishovite: wave propagation in the mixed phase region *J. Geophys. Res.* **79** 332–8

Grady D E and Passman S L 1990 Stability and fragmentation of ejecta in hypervelocity impact *Int. J Impact Eng.* **10** 197–212

Grady D E, Winfree N A, Kerley G I, Wilson L T and Kuhns L D 2000 Computational modeling and wave propagation in media with inelastic deforming microstructure *DYMAT 2000 (25–29 September)* (EDP Sciences) pp 15–20

Grady D E and Winfree N A 2001 A computational model for polyurethane foam *EXPLOMET 2000, Int. Conf. on Fundamental Issues and Applications of Shock-Wave and High-Strain-Rate Phenomena (Albuquerque, New Mexico, 18–22 June)* ed K. Staudhammer, L. Murr, and M. Meyers, (Elsevier) pp 485–92.

Grady D E, Fenton G and Vogler T J 2013 Equation of state and evidence of enhanced phase transformation for shock compression of distended compounds *Int. J. Impact Eng.* **56** 19–26

Guillard F, Golshan P, Shen L, Valdes J R and Einav I 2015 Dynamic patterns of compaction in brittle porous media *Nature Phys.* **11** 835–9

Gurney R W 1943 The initial velocity of fragments from bombs, shells and grenades *Army Ballistic Research Laboratory Report* BRL 405.

Gust W H and Royce E B 1971a Dynamic yield strength of B_4C, BeO and Al_2O_3 ceramics *J. Appl. Phys.* **42** 276

Gust W H and Royce E B 1971b Axial yield strengths and two successive phase transition stresses for crystalline silicon *J. Appl. Phys.* **42** 1897

Gust W H, Holt A C and Royce E B 1973 Dynamic yield, compression, and elastic parameters for several lightweight intermetallic compounds *J. Appl. Phys.* **44** 550–60

Herrmann W 1969 Constitutive equation for the dynamic compression of ductile porous materials *J Appl. Phys.* **40** 2490–9

Huang J Y, Hub S S, Xub S L and Luo S N 2017 Fractal crushing of granular materials under confined compression at different strain rates, *Int. J. Impact Eng.* in press

Johnson G R, Holmquist T J and Beissel S R 2003 Response of aluminum nitride (including a phase change) to large strains, high strain rates, and high pressures *J. Appl. Phys.* **94** 1639–46

Johnson J N and Green S J 1976 *The Effects of Voids on Material Deformation* (New York: ASME)

Jones G E, Kennedy J E and Bertholf L D 1980 Ballistic calculations of R W Gurney *Am. J. Phys.* **48** 264–9

Kawai N, Nakamura K G and Kondo K 2004 High-pressure phase transition of mullite under shock compression *J. Appl. Phys.* **96** 4126–30

Kennedy J E 1997 The Gurney model for explosive output for driving metal *Explosive Effects and Applications* (Berlin: Springer) pp 221–58

Kerley G I 1992) CTH EOS package: porosity and reactive burn models *Sandia National Laboratories Report* SAND92-0553, April.

Kompaneets A S 1956 shock waves in a plastic compacting medium *Dokl. Akad. Nauk. SSSR* **109** 49–52

Lee M Y, Pfeifle T W and Hardy R D 2004 Hydrostatic and uniaxial strain compaction of eight metal oxides and tungsten carbide, *Sandia National Laboratories Report* SAND2004-5854.

Marsh S P 1980 *LASL Shock Hugoniot Data* ed S P Marsh (Berkeley, CA: University of California Press)

McQueen R G, Marsh S P and Carter W J 1968 The determination of new standards for shock wave equation-of-state work *Behavior of Dense Media under High Dynamic Pressures, Symposium on the Behavior of Dense Media under High Dynamic Pressures, (Paris, September 1967)* (New York: Gordon and Breach) pp 67–83.

Mollema P N and Antonellini M A 1996 Compaction bands: a structural analog for anti-mode i cracks in aeolian sandstone *Tectonophysics* **267** 209–28

Olsson W A 1999 Theoretical and experimental investigation of compaction bands *J. Geophys. Res.* **104** 7219–28

Olsson W A 2001 Quasistatic propagation of compaction fronts in porous rock *Mech. Mater.* **33** 659–68

Rice M H and Walsh J M 1957 Equation of state of water to 250 kilobars *J. Chem. Phys.* **26** 824–30

Rudnicki J W and Olsson W A 1998 Reexamination of fault angles predicted by shear localization theory *Int. J. Rock Mech. Min. Sci.* **4-5** 88
Sedov L I 1946 Propagation of strong shock waves *J. Appl. Math. Mech* **10** 241–50
Sheffield S A, Gustavsen R I, Alcon R R, Graham R A and Anderson M U 1995 Shock initiation studies of low density HMX using electromagnetic particle velocity and PVDF stress gauges, *Proc. Tenth Int. Detonation Symp., Office of Naval Research Report* ONR 33395-12, 166-174.
Strack O E, Leavy R B and Brannon R M 2015 Aleatory uncertainty and scale effects in computational damage models for failure and fragmentation *Int. J. Numer. Methods Eng.* **102** 468–95
Swegle J W 1980 Constitutive equation for porous materials with strength *J. Appl. Phys.* **51** 2574–80
Taylor G I 1950 The formation of a blast wave by a very intense explosion: theoretical discussion, Proc. R. Soc. London, A, 201, 1065, 159.
Terzaghi K 1925 *Erdbaumechanik auf Bodenphysikalischer Grundlage* (Leipzig: Franz Deuticke)
Tone C 1948 Theorie und Verhalten der zusammengepressten Pulver *Berg-Hüttenmänn. Monatsh. Montan. Hochschule Leoben* **93** 62–7
Tong W and Ravichandran G 1997 Recent developments in modeling shock compression of porous media *High-Pressure Shock Compression of Solids* vol 4 (Springer) pp 177–204
Trunin R F, Gudarenko L F, Zhernokletov M V and Simakov G V 2001 *Experimental Data on Shock Compression and Adiabatic Expansion of Condensed Matter* (Sarov: RFNC-VNIIF)
Valdes J R, Fernandes F L and Einav I 2011 Periodic propagation of localized compaction in a brittle granular material *Granular Matter* **14** 71–6
Vogler T J, Gluth J W, Grady D E and Hall C W 2005 Dynamic compaction of tungsten carbide powder, *Sandia National Laboratories Technical Report* SAND2005-1510, April.
Vogler T J, Lee M Y and Grady D E 2007 Static and dynamic compaction of ceramic powders *Int. J. Solids Structures* 636–58
Wackerle J 1962 Shock wave compression of quartz *J. Appl. Phys.* **33** 922–37
Weibull W 1951 A statistical distribution function of wide applicability *J. Appl. Mech.* **18** 293
Wong T F, Szeto H and Zhang J 1992 Effect of loading path and porosity on the failure mode of porous rocks *Appl. Mech. Rev.* **45** 281–93
Wunnemann K, Collins G S and Melosh H J 2006 A strain-based porosity model for use in hydrocode simulations of impacts and implications for transient crater growth in porous targets *Icarus* **180** 514–27
Zeldovich Y B and Raizer Y P 1967 *Physics of Shock Waves and High-Temperature Hydrodynamic Phenomena* vol 2 (New York: Academic).

Chapter 8

The nature of structured shock waves

Within the wave structure of nonlinear shock waves in solids lies much of the physics of shock-induced material state transformation and failure. Relating shock wave structure to the underlying physics resides at the heart of research into the science of shock waves in solids. One central outcome has been a fourth-power relation linking the stress jump through a steady structured shock wave and the maximum strain rate within the shock transition. This synthesis of structured shock behavior has received recognition as a unifying relation over a sensibly wide range of materials, material conditions and shock compression amplitudes. Less widely recognized is the complementary applicability of the invariance of the product of the energy dissipated and the time over which this energy is dissipated through a structure shock wave—a property with the dimensions of specific action. Whereas the latter invariance, coined the dissipative action, implies the fourth-power behavior for steady shock waves, this invariance also has apparent application to unsteady structured waves, as well as other shock transition phenomena. Applications to fracture spall waves and to failure through dynamic adiabatic shear are observed. Over the past several decades theories have been proposed to explain the fourth-power trend observed in the shock response of solid materials. Most are based on the dynamics of failure on the microscale. Despite credible theoretical efforts, a consensus on the underlying responsible physics has not been achieved.

A brief history of the origin of the fourth-power relation is provided in the present chapter. Commentary is offered on the physical principles of solid viscosity in the shock wave and the underlying invariance of the energy–time product in the shock wave event. Results for steady structured waves in porous and composite solids are examined that do not in general exhibit fourth-power behavior. Observations of trends with component cell size and impedance disparities suggest acoustic scattering may contribute to wave structuring in composite and layered materials. Deformation turbulence in the shock front may also play a wave structuring role. Fundamental statistic principles, such as those embedded within the dynamics of

doi:10.1088/978-0-7503-1257-8ch8

Langevin, are investigated. These and other physical explanations are explored as possible underlying causes of the systematics observed in structured shock waves.

8.1 Introduction

The chapter addresses the structure of shock waves within single-phase solid materials. The phase transformation and strength of solids, and the rich influences on the structure of shock waves of these critical state physical processes, are not considered here; the effects of strength and phase transformation on shock wave structure are considered in chapter 1 and for selected solids in chapters 6 and 9. Instead, this chapter explores observations of the smooth structuring of large-amplitude shock waves in solids usually attributed to an underlying solid viscosity and systematics of these test observations gleaned from half a century of experimental effort.

The hydrodynamic approximation of shock waves in condensed matter and the accompanying discontinuous shock wave have found useful applications in many calculations and in assessing the basics of shock physics. In other applications shock wave structure brought about by strength, viscosity, change of state and heat conduction must be accounted for. A feature unique to the realm of nonlinear wave propagation is the steady wave or the wave of permanent regime. The steady wave profile, as its name implies, propagates without change in form and is a consequence of a stable balance between the competing processes of stress and strain nonlinearity, and the dispersing properties of viscous dissipation and heat conduction.

The existence of steady wave solutions of the Navier–Stokes equations has been known since the independent works of Rayleigh (1910) and Taylor (1910). Recognition that steady wave shocks also play a significant role in solids apparently emerged with the work of Band (1960) who outlined the formal solution for a visco-elastic shear yielding solid. The concepts of steady wave shocks in solids was treated further in the study of Band and Duvall (1961). Issues of the transition to steady wave behavior in solids was undertaken in an investigation by Bland (1965).

The first definitive experimental evidence for steady structured waves in solids appears to be the seminal study on aluminum by Barker (1968) using the then newly developed laser interferometer measurement technique. Since that time a vast experimental base of structured wave profile measurements on solids has evolved using a number of experimental methods.

Over that time a unifying character to structured waves in solids has emerged. This universal character has become most commonly known as the fourth-power law of steady shock waves. The observed power-law behavior relates the driving shock pressure to the maximum strain rate in the steady shock wave. This nature of structured waves in solids was noted in the study of Grady (1981), applying the earlier data of Barker. This commonality was supported with further experimental wave profile data in the investigations of Swegle and Grady (1985, 1986). These intriguing features of steady waves in solids have been investigated by numerous workers since that time. A review of this history is included in later papers (Grady 2010, 2015).

The origin of the fourth-power law, however, has roots in earlier observations (Grady 1980). Shock wave experiments and material response modeling for crystalline quartz performed by Hayes and Graham (1978) provided solutions for the temporal history of energy dissipation as a structured shock wave penetrates the material. Although dissipation decreased as the depth of shock penetration increased, it was observed that the product of the energy dissipated and the time over which the energy dissipated remained constant.

The solution of Hayes and Graham (1978) on which this observation was based is that of an unsteady decaying precursor wave. The question of whether the same invariance of energy dissipated multiplied by the time over which dissipation occurs would similarly hold for a steady structured wave was raised. Assuming some simplifications of the shock compression process, it is shown (Grady 1981) that invariance of this product implied to first-order a fourth-power dependence of the steady wave strain rate on the shock wave pressure. The only steady wave data (Barker 1968) available at the time were consistent with the fourth-power behavior and hence invariance of the energy–time product. Concurrent studies of spall and fragmentation, and the emergence of an action principle in this field of shock failure (Grady and Lipkin 1980, Grady 1982), provided further motivation (see chapter 2).

In the present chapter some of the history of the fourth-power law for structured shock waves is appraised. Further perspective is put forth on invariance of the energy–time product (which has the dimensions of specific action as defined in the field of classical mechanics) in the shock compression event. The role and underlying physical sources of solid viscosity in the structuring of shock waves in solid matter are explored within the context of the fourth-power law. Limitations to the fourth-power behavior of solid viscosity have been sought. Such limitations may occur in the upper or lower limits of the span of the pressure range tested for any given material. Limitations may also occur in the types of materials and material conditions tested.

The structured shock wave introduces a significant amount of energy into the material of concern in a brief period of time. The response of the material in the disposition of this energy is inherently governed by irreversible statistical physics. Measurements of structured shock waves, as well as other modes of shock-induced failure, reveal systematics that bear striking resemblance to relations that emerge from the developments of Langevin statistical physics. Ideas of statistical physics are explored that may have application to the extreme dynamic failure inherent to the shock event.

Porous and composite solids do not in general exhibit fourth-power behavior in the propagation of steady structured waves. Several diverse composite materials, however, demonstrate fourth-power steady wave behavior and offer insights into underlying mechanisms. One-dimensional regular laminate composites do not exhibit fourth-power steady waves, although the observed systematics furthers our understanding of the physics leading to wave structuring. Some systematic behaviors are observed concerning component impedance differences and unit cell sizes that suggest contributions to shock viscosity through momentum diffusion brought about by acoustic scattering. Steady structured shock waves in granular powders do not exhibit fourth-power behavior. Invariance of the dissipative action is observed, however. Such trends may lend insight into the basic physics governing effective viscosity and wave structuring in solids.

8.2 Steady wave profiles and the fourth-power law

A unifying feature of shock compression in solids is a relationship that is observed between the risetime of a shock wave and the amplitude of the pressure change through the shock wave. When diagnostics are used that are capable of measuring the time-resolved pressure or particle velocity through the shock wave with reasonable fidelity, it is observed that the profiles are sensibly sigmoidal in shape, and it has become common to quantify the wave risetime through an assessment of the strain rate of the fastest rising portion near the midpoint of the wave. This strain rate on the abscissa is plotted against the Hugoniot shock pressure jump on the ordinate and a relationship between the two is determined.

8.2.1 Experimental evidence for fourth-power dependence in the steady wave shock

Experimental data of this type for metals obtained with velocity interferometer instrumentation are shown in figure 8.1 on the left. This plot includes most of the data originally used to point out the unique features of this representation of steady structured wave profile data (Grady 1981, Swegle and Grady 1985).

In the early presentation of the data the Hugoniot pressure state, rather than the pressure jump from the Hugoniot elastic limit (HEL) state to the Hugoniot state, was plotted and this systematic is maintained in the present plot for the metals of figure 8.1. For the preponderance of the data for the metals, this correction does not make a noticeable difference in the plot. Similar data for a sparser set of nonmetals are shown in figure 8.1 in the right panel (Swegle and Grady 1985, Grady 1998a). The brittle materials have markedly larger HEL amplitudes and here the pressure jump from the HEL state when related to the peak strain rate in the steady wave retains the fourth-power dependence. It is believed that the pressure jump through

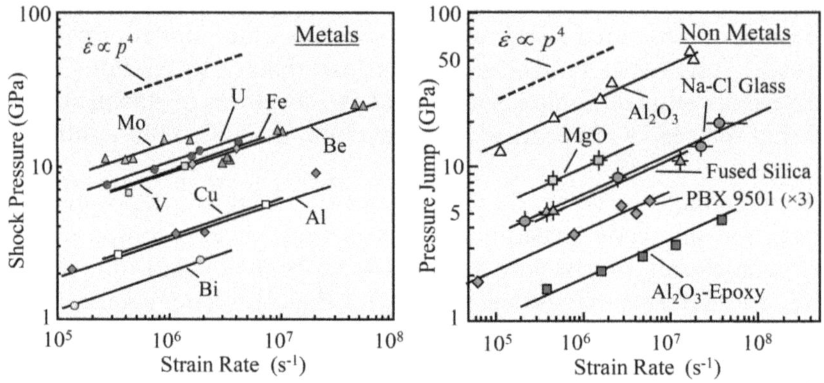

Figure 8.1. Steady wave pressure versus strain rate for selected metals (left) and nonmetals (right) demonstrating the fourth-power behaviour observed in many materials (Grady 2010). The pressure jump for PBX 9501 is multiplied by three.

the steady structured wave is the more appropriate parameter in assessing fourth-power behavior. Remarkable additions to the nonmetals are the steady wave profile data for an unreacting PBX 9501 composite of HMX and binder (Dick *et al* 1998) and an epoxy–43vol.% aluminum oxide composite (Setchell and Anderson 2005).

A further observation is illustrated in figure 8.2. Shock wave experiments were performed by Asay and Chhabildas (1981) on aluminum where the material is shock compressed to Hugoniot pressure levels well in excess of the HEL (approximately 20 GPa for the experiments reported) and then subjected to a second reshock centered on the initial shock state. Structured steady waves following a second precursor wave are again observed. Plotted in figure 8.2 are experimental measures of the steady wave viscous pressure versus strain rate (see section 8.5.1). The first shock data are from the study of Barker (1968). The second shock data are from the Asay and Chhabildas (1981) study. Although sparse, the second shock data are not inconsistent with fourth-power behavior. Steady wave fourth-power requires a second-power dependence between viscous pressure and strain rate (Grady 1998b). The data do not, however, overlap those of the initial shock compression data. The first shock amplitudes for the two tests were reasonably close but were not the same.

Several questions can be raised relating to the universality of the fourth-power relation. First, are there materials, or material conditions, yet untested that exhibit a steady wave structure inconsistent with the fourth-power behavior? Second, are there restrictions in the range of the shock amplitude over which the fourth-power law applies? This can included both the lower amplitude range as the plastic wave merges with the HEL amplitude, as well as the higher amplitude range where risetimes exceed the capability of current instrumentation. The first concern is addressed in part in later sections.

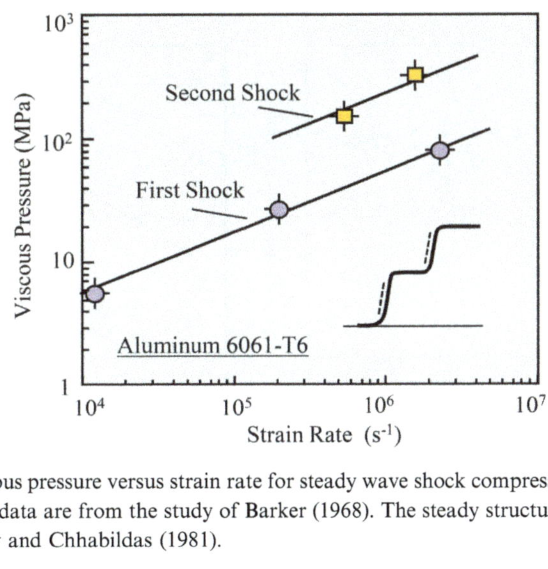

Figure 8.2. Plot of viscous pressure versus strain rate for steady wave shock compression in aluminum (Grady 1998b). The first shock data are from the study of Barker (1968). The steady structured second wave data for aluminum is from Asay and Chhabildas (1981).

The experimental work of Crowhurst *et al* (2011) relates to the second. The preponderance of structured shock wave data on metals, such as those plotted in figure 8.1, for which assessments are made of the steady wave risetime, achieve Hugoniot pressure states well below the pressure level at which the elastic precursor wave is overdriven. Crowhurst *et al* (2011) have performed structured wave measurements on aluminum at Hugoniot pressure states that exceed the transition to overdriven shock waves. Wave profiles are shown on the left in figure 8.3. The authors have resolves steady structured shock waves in pure aluminum in the range of approximately 25–45 GPa and with shock risetimes down to several tens of picoseconds. Strain rates in the shock wave at the upper end of the pressure range studied are in excess of 10^{10} s^{-1}, exceeding the highest strain rates achieved in the work of Barker (1968) on aluminum by four orders of magnitude.

In contrast to gun or explosive launch and projectile impact methods for generating shock waves, the work of Crowhurst *et al* (2011) uses a high-energy laser to launch a projectile or, alternatively, to impart a shock directly into the target, through the rapid expansion of a hot dense plasma. These data are analyzed by the authors for direct comparison with the earlier data of Barker (1968). The data are compared on the right in figure 8.3 where Hugoniot pressure is plotted against nominal strain rate in the structured steady wave. The new data suggest continuation of the fourth-power trend inferred from earlier data to markedly higher strain rates and to Hugoniot pressure levels well in excess of the overdriven shock transition.

8.2.2 Non-steady waves

Chhabildas and Hills (1986) conducted a study of structured shock waves in a well-characterized vanadium metal. They examined, among other shock properties, the relationship between the Hugoniot pressure and the strain rate, as well as the shock viscosity, for this metal. Shock pressure versus strain rate for the three shock wave

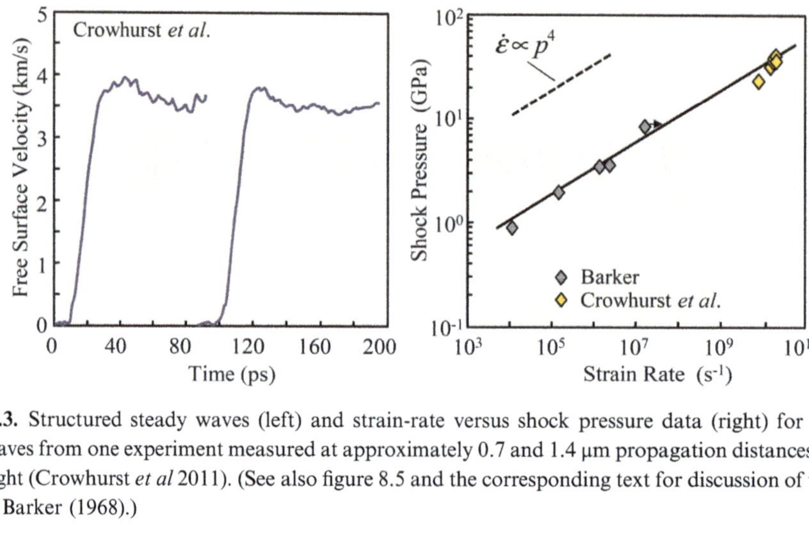

Figure 8.3. Structured steady waves (left) and strain-rate versus shock pressure data (right) for aluminum. Shock waves from one experiment measured at approximately 0.7 and 1.4 μm propagation distances are shown on the right (Crowhurst *et al* 2011). (See also figure 8.5 and the corresponding text for discussion of the seminal study of Barker (1968).)

measurements reported by Chhabildas and Hills (1986) are shown in figure 8.4. These data are not observed to be in accord with the fourth-power behavior. The discrepancy is more pronounced if the pressure jump from the HEL state (now believed to be the more appropriate pressure measure) is plotted on the ordinate. The data suggest that the fourth-power behavior observed in previous structured wave data for other materials, as plotted in figure 8.1, does not extend to metallic vanadium.

That vanadium does in fact exhibit fourth-power behavior is demonstrated in a later study by Swegle (1991). Swegle applies a rate-dependent plasticity model, developed in an earlier investigation of the fourth-power law (Swegle and Grady 1985), of the form

$$\sigma(\varepsilon, \dot{\varepsilon}) = Y(\varepsilon) + A\dot{\varepsilon}^{1/2}. \tag{8.1}$$

In equation (8.1) σ and ε are the appropriate effective shear stress and shear strain. A strain-hardening term of the form

$$Y(\varepsilon) = Y_1 + (Y_0 - Y_1)e^{k\varepsilon/(Y_0 - Y_1)} \tag{8.2}$$

was used in the earlier study of Swegle and Grady (1985). Equation (8.2) is reduced to a constant $Y(\varepsilon) = Y_o$ in Swegle's later analysis of the vanadium wave profile data.

The strain-rate power to the one-half dependence provided by the visco-plastic model in equation (8.1) captures the fourth-power behavior exhibited by steady structured shock waves as presented in figure 8.1. A strain-rate dependence of the form $h(\varepsilon)\dot{\varepsilon}^m$ was shown in the analysis of Dunn and Grady (1986) to encompass the Swegle plasticity model. While allowing for a richer description of shock profile structure, little new information is added by the generalization.

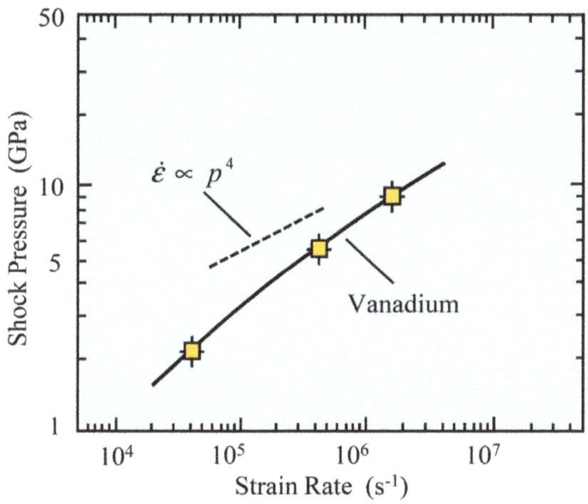

Figure 8.4. Shock pressure versus strain rate for structured waves in vanadium (Chhabildas and Hills 1986). Calculations by Swegle (1991) show that measured profiles have not achieved steady wave structure.

The Swegle visco-plasticity model of the form of equation (8.1) predicts the non-steady evolution of waves towards steady wave behavior as well as the intrinsic steady wave structure. Typically, in such structured shock wave experiments, the input wave structure is a discontinuous step to the final pressure state caused by the planar impact of surfaces of the test materials. Evolution of the wave requires separation of the elastic precursor and spreading of the plastic wave until steady wave conditions are achieved. Swegle demonstrated that the profiles measured on metallic vanadium by Chhabildas and Hills (1986) had not achieved steady wave structure. The Swegle model readily calculates the structure of the measured non-steady waves and provides predictions of the propagation distance necessary to achieve sensibly steady wave behavior.

The universal nature of the fourth-power behavior of steady structured shock waves can probably never be fully demonstrated. The study of Swegle (1991) shows, however, that care must be taken in assessing structured waves for steady wave character. The transition from non-steady to sensibly steady wave behavior can be estimated through analysis (e.g. Bland 1965, Swegle and Grady 1985), but is demonstrated most certainly only through experimental measurements. Aspects of the analysis of Bland relating to shock viscosity and structured shock waves are treated in chapter 1.

8.3 Origin of the fourth-power law

A seminal experimental study focused on the structure of shock waves in solids is that of Barker (1968) on aluminum 6061-T6 using both the Michelson displacement and spectral velocity interferometer techniques of Barker and Hollenbach (1965). Barker provided a plot of the strain rate in the rapidly rising portion of the steady plastic wave versus the Hugoniot pressure and noted that the data exhibited a power-law behavior. The data provided by Barker are plotted in figure 8.5. A steady

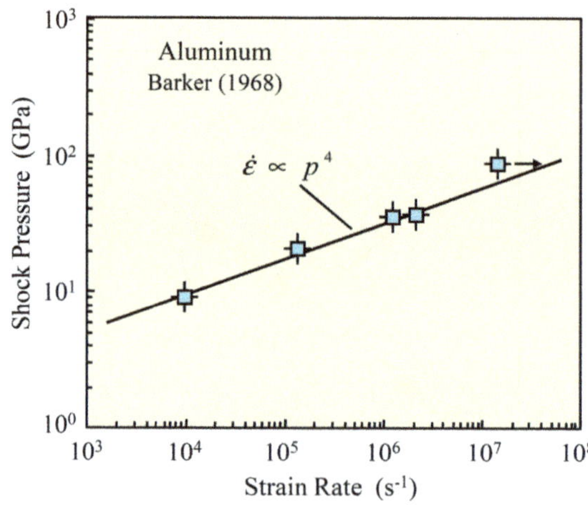

Figure 8.5. Strain rate versus shock pressure of the steady structure shock waves measured by Barker (1968) with interferometer methods on 6061-T6 aluminum. Barker noted both power-law behavior and that the highest point was uncertain due to limitations in instrumentation resolution.

wave analysis of the time-resolved aluminum wave profiles applying a rate-dependent dislocation plasticity model was undertaken by Johnson and Barker (1969). These seminal experimental and modeling studies of the viscous properties of metals, as inferred from structured shock waves in aluminum, were apparently not pursued further over the following ten years or so. Interest re-emerged from a different application in the nature of structured shock waves.

8.3.1 Precursor attenuation in quartz

In the mid to late 1970s efforts were focus on achieving an understanding of deformation in the shock through heterogeneous shear in both crystalline solids of geophysical interest (Grady *et al* 1975, Grady and Murri 1976, Grady 1977, 1980) and in metals (Hayes and Grady 1982, Grady and Asay 1982). Quartz was the prototypical material for the crystalline solid. Hayes and Graham (1978) reported on an experimental investigation exploring elastic precursor attenuation in x-cut quartz. In that same study Hayes applied a general nonlinear visco-plastic Maxwell equation in which functional expressions and constants were selected to provide an optimum description of the experimental data. Based on a Maxwell model, and the fit to experimental data, solutions for the propagation and evolution of structured shock waves resulting from planar impact were obtained. Wave profiles from a 9 GPa peak-shock-pressure impact experiment on quartz, provided by the Hayes and Graham (1978) computational model and experimental data fit, are shown in figure 8.6.

The solution of Hayes and Graham (1978) was used to assess the magnitude and time history of energy dissipation in the quartz shock compression for purposes of

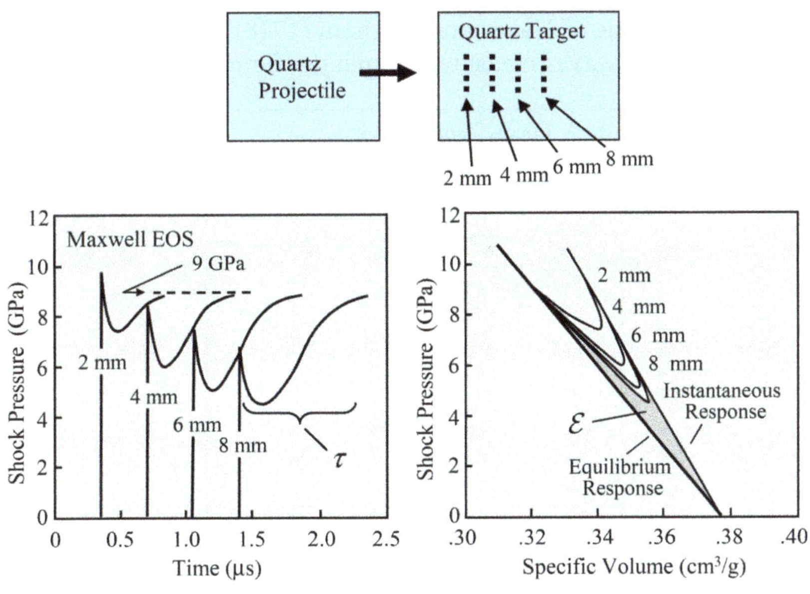

Figure 8.6. Left: Stress wave profiles to 9 GPa shock pressure in crystalline quartz based on a Maxwell visco-plastic model optimized to experimental stress wave data for quartz (Hayes and Graham 1978). Stress versus volume histories corresponding to the same wave profiles are shown on the right.

calculating energy partitioning in the heterogeneous deformation process (Grady 1980). Energy dissipation at successive distances from the impact plane can be calculated from the stress versus specific volume history, as illustrated on the right in figure 8.6.

A plot of the dissipation energy $\mathcal{E}$ as a function of distance from the impact plane is shown in figure 8.7. Also shown in the figure is the temporal thickness τ of the precursor shock wave at which the energy $\mathcal{E}$ was dissipated (see figure 8.6 for the wave thickness at the 8 mm position) at the same Lagrangian position. It was observed that the product of the dissipation energy and the dissipation time, $\mathcal{E}\tau$ (also shown in figure 8.7) was nearly constant over the calculated range of stress wave propagation in the quartz (Grady 1980). Motivations for exploring the nature of this energy–time property emerged from concurrent studies on the dynamic fracture and fragmentation of solids (Grady and Lipkin 1980, Grady 1982).

8.3.2 Steady waves in aluminum

Chhabildas and Asay (1979) reported on time-resolved steady wave profile measurements on several metals including several tests on aluminum, and were responsible for the resurfacing of the earlier conference report by Barker (1968). The Hugoniot pressure versus strain-rate data plotted in figure 8.5 are the same as those plotted in figure 2.5 of Chhabildas and Asay (1979), in which comparison is made to their own higher pressure data.

8.3.3 Equivalence of the fourth-power relation and the shock invariant

An idea was posed that the property $\mathcal{A} = \mathcal{E}\tau$ observed to be nearly invariant in the attenuating wave analysis of Hayes and Graham (1978) for quartz might extend to the character of steady waves measured in metals (Grady 1981). A hydrodynamic

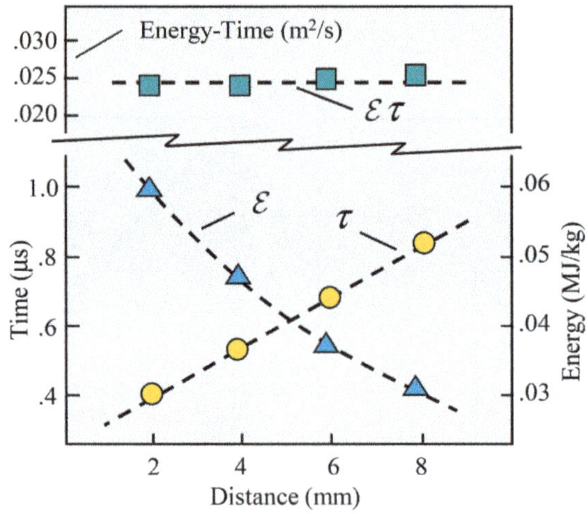

Figure 8.7. Energy dissipation and wave duration at each of the solution stations as provided in the previous figure. The product of energy and duration is observed to be nearly constant for quartz over the range of the experimental data and calculation (2–8 mm).

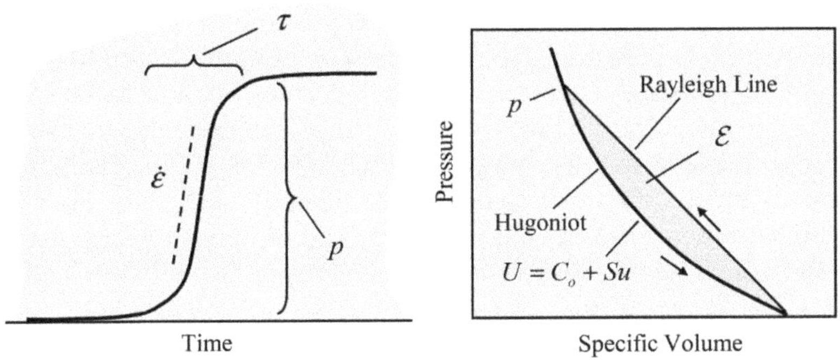

Figure 8.8. A steady structured shock wave with finite risetime (right). Dissipation in the shock process is provided by the region enclosed by the Hugoniot (approximately the decompression isentrope) and the Rayleigh line (left).

model was applied, with a linear shock-velocity versus particle-velocity Hugoniot $U = C_o + Su$, to steady structured shock waves as illustrated in figure 8.8. The influence of the Gruneisen properties of the material on decompression is ignored leaving the shock Hugoniot and the decompression isentrope coincident.

Within constraints imposed by the simplifying assumptions, dissipation energy $\mathcal{E}$ is provided by the area enclosed by the Rayleigh line and the Hugoniot intersecting at a shock pressure p. Taylor expansion in powers of the normalized Hugoniot pressure $p/\rho C_o^2$ yields for the dissipation energy in units of specific energy

$$\mathcal{E} = \left[S/3\left(\rho C_o^2\right)^2\right]p^3 + O(p^4). \tag{8.3}$$

The period τ of the steady wave is related to the Hugoniot pressure and the strain rate, as illustrated on the left in figure 8.8, through

$$\tau \simeq p/\rho C_o^2 \dot{\varepsilon}. \tag{8.4}$$

Forming the product of the dissipation energy and the dissipation period,

$$\mathcal{E}\tau = \mathcal{A}, \tag{8.5}$$

and solving for the functional dependence of the steady wave strain rate on the Hugoniot pressure yields

$$\dot{\varepsilon} = [S/3\mathcal{A}(\rho C^2)^3]p^4 + O(p^5). \tag{8.6}$$

Thus, to lowest order the fourth-power dependence of strain rate on the shock pressure, and the invariance of $\mathcal{E}\tau = \mathcal{A}$, are equivalent. Whether one is more fundamental than the other is arguable. The latter, however, makes no recourse to steady wave behavior, whereas the fourth-power law is restricted to steady waves. Also not known are the range of materials and range of shock amplitudes over which the observed criteria apply. A number of authors have, over the intervening years, undertaken physical and mechanistic explanations of the observed fourth-power universality of structured shock waves. The literature should be consulted in evaluating these subsequent efforts. Most often cited in these later studies is the paper of Swegle and Grady (1985) where further experimental measurements on metals are reported. It is worth noting that if the constancy of the dissipative energy–time product is indeed the more fundamental then, as the steady shock amplitude increases, higher order terms in equations (8.3) and (8.6) must be accounted for and a divergence from fourth-power is predicted. This concern is pursued further in section 8.6.1.3.

8.4 Dissipative action and invariance in the structured shock wave

It is noted in the opening of the present chapter that a fourth-power relation between the driving pressure and the steady wave strain rate is not the original expression proposed to encompass the observed commonality in structured wave risetime over different materials. Rather, early experimental work and analysis on structured waves suggested that the product of the energy dissipation in the shock wave, and a measure of the temporal period of the wave, exhibits a degree of constancy. This product of the energy and the time over which this energy is dissipated was proposed as a possible invariant in the shock process (Grady 1981). It was subsequently shown that for steady structured shock waves, invariance of this product of dissipation energy and wave duration is equivalent, at least to first-order, to the fourth-power law. The shock invariant appears to have application to unsteady as well as steady structured shock waves.

The physics underlying the observed invariance of the parameter $\mathcal{A} = \mathcal{E}\tau$, or the more restrictive fourth-power law, is not fully understood. The fundamental mechanisms of dislocation dynamics have played a central role in a number of efforts put forth to explain the fourth-power behavior (e.g. Johnson and Barker 1969, Rubin 1990, Partom 1990, Johnson 1992, Malygin *et al* 2013, Armstrong *et al* 2007). The dispersive features of wave scattering in heterogeneous media has been proposed (Grady 1998b). A model of failure in the shock wave through distributed adiabatic shear has also demonstrated invariance of the dissipation energy–time product (Grady and Kipp 1987, Grady 2010). In the present and subsequent section further responsible physics is explored.

8.4.1 Action principles and implications for the shock wave

Some preliminary comments on this invariant property of the structured shock wave are offered. The product is written more formally as

$$\mathcal{A} = \int_0^{\varepsilon_h} \delta\varepsilon \, \mathrm{d}t, \tag{8.7}$$

where the integrated increments of energy and time are identified in figure 8.9. On the left a temporal history of a structured shock wave, that in general need not be

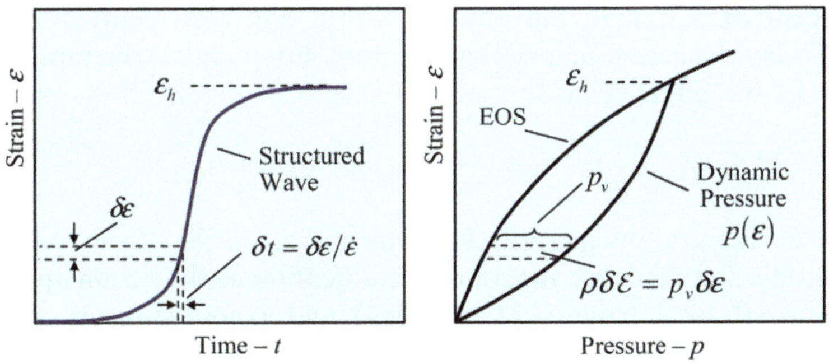

Figure 8.9. The dissipative action incurred within the material subject to structured wave shock compression.

steady, is illustrated. A time element δt and the corresponding increment in compression strain is identified. The resulting increment in dissipation $\delta\mathcal{E}$ is identified in the plot on the right. The pressure path $p(\varepsilon)$ corresponds to the dynamic pressure versus strain response of the material at some Lagrangian position within the wave. The pressure $p_v(\varepsilon)$ is the dissipation or viscous pressure that determines the amplitude of the dynamic pressure in excess of the equilibrium equation-of-state (EOS) pressure. The expression in equation (8.7) corresponds to the simpler product $\mathcal{E}\tau$ considered earlier where $\mathcal{E}$ is the total dissipation through the shock wave and τ is the wave period. This dissipation as depicted constitutes both thermal energy produced as a consequence of irreversible deformation processes in the shock wave and configurational energy resulting from athermal shock-induced defects introduced in the lattice structure of the solid. It does not include the change in the elastic EOS energy relating to compression of the principal lattice. Discussions here, although potentially relevant, do not extend to more extreme changes of state in solids subject to shock compression, including crystal structure phase changes, melting, amorphization, ionization and other consequences occurring under extreme shock compression of solids.

The property $\mathcal{A}$ has dimensions of action in the context of classical mechanics. Actually the dimensions are that of specific action where $\mathcal{E}(t)$ provides energy per unit mass. As specific action the dimensions are that of J-s kg^{-1} (and interestingly that of diffusion, namely m^2 s^{-1}). In classical physics this action, depending on the application, takes on several distinct meanings. When converging to the quantum mechanical limit the action approaches the magnitude of Plank's constant. In the area of classical Hamiltonian mechanics and Hamilton–Jacobi theory the action continues to play a fundamental role. Appropriate to the present application is the action integral defined as (Ter Haar 1964)

$$\mathcal{A} = \int_{t_0}^{t_2} 2K\,dt, \tag{8.8}$$

where K is the kinetic energy.

The concept of action in the physical world has been controversial since its inception. Today the action of a system is commonly posed as the time integral over the period τ of the Lagrangian,

$$\mathcal{A} = \int^{\tau} L\, dt. \tag{8.9}$$

Historically the integral in equation (8.9) was not called the action but was referred to as Hamilton's first principle function. Identification as the action appears to have been popularized by Feymann *et al* (1964) and contemporaries in the 1960s. Governing equations for the motion of the system follow from the conditions under which the action integral in equation (8.9) is stationary.

Action, and the principle of least action, was apparently independently proposed in alternative but similar forms in 1744 by Pierre Maupertuis and Leonhard Euler. Arguments over priority of the discovery emerged within several years. There is some evidence that Gottfried Leibniz posed the principle several decades earlier. The action as posed in the earlier historic works relates more closely to that currently referred to as the abbreviated action,

$$\mathcal{A} = \int 2K\, dt = \int (L + H) dt, \tag{8.10}$$

and best relates to the property explored here.

In equation (8.7) the energy $\delta\mathcal{E}$ results from that portion of the work done by the shock wave that contributes to dissipation and does not add to the EOS lattice energy of the material. It is noted that dissipation can also contribute to defect production, which is a form of potential energy or configurational energy not accounted for in the continuum EOS of the material. In other deformation-of-materials fields, defect production is commonly known to be a small, and frequently ignorable, contributor to dissipation when compared to the generation of heat energy. An assumption of a comparable partitioning of thermal and defect dissipation to the shock compression event is open to question. The effects of shock defect production on EOS uncertainties was considered by Ruoff (1980). Issues of extensive defect production in the shock wave was raised in the study of Graham (1981) where evidence of residual resistivity, shock polarization and shock color changes, combine to support excessive shock-induced defect production. More recently, studies of Nellis *et al* (2009) on crystalline sapphire interpret the onset of shock-induced opacity and shock wave structure through intense defect production approaching amorphization in the shock dissipation process. Despite evidence of substantial shock-induced defect dissipation in certain materials, it is principally ignored in the following, and thermal dissipation is assumed to be the first-order contribution in the shock compression event.

Dissipation in the shock wave, as considered here, is the accumulation of collective non-conservative deformation mechanisms that will ultimately thermalize as thermodynamic equilibrium behind the shock wave is achieved. Dissipative deformation mechanisms would include individual dislocation motions, twinning, slip bands, microfracture and pervasive adiabatic shear, to name a few. The specific

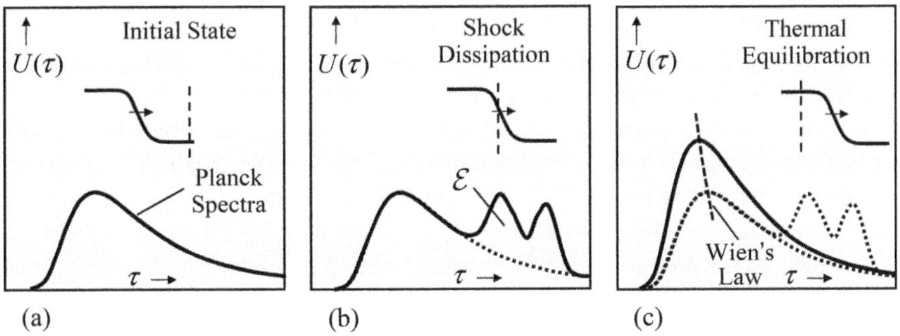

Figure 8.10. From left to right the plots represent, as depicted by the solid curve in each: (a) ambient Planck thermal phonon energy versus phonon period before arrival of the shock wave; (b) transient nonequilibrium energy distribution induced by failure dissipation within the passing structured shock wave and (c) energy distribution following post-shock entropy production and subsequent equilibration over the thermal spectrum.

deformation mechanisms stimulate dissipation within the lattice kinetic energy field at phonon frequencies characteristic of each unique mechanism.

A sensible plot is provided in figure 8.10 for discussion of dissipation within the shock event. The graph is that of the Planck thermal phonon energy spectrum of a solid. On the left is shown the equilibrium thermal Planck distribution at some initial temperature preceding arrival of the shock wave. Subsequent plots are descriptive of the event as passage of the shock wave proceeds.

As dissipation in the shock wave ensues, two distinct, and most probably inseparable, processes take place. First, non-conservative deformation events raise the kinetic energy of normal modes within a spectrum of frequencies commensurate with the unique dissipation mechanisms of the material. For example, the non-conservative movement of a single dislocation would be expected to stimulate higher frequency normal modes than would the cooperative motion of a collection of dislocations in a deformation slip band. The activation of a brittle micro-fracture would excite yet another range of frequencies. Second, kinetic energy will be passed between normal modes as statistical mechanical equilibrium is sought until an equilibrium equipartition of thermal energy is again achieved after passage of the shock wave. The latter process is a consequence of material nonlinearity and a weak coupling between normal modes.

The sequence of such events through a structured shock wave is illustrated in figure 8.10. Before arrival of the shock wave the distribution of kinetic energy over the normal modes is described by the Planck energy distribution, consistent with equilibrium at ambient temperature, as shown on the left. The transient distribution of energy over the normal modes as a consequence of shock dissipation in the solid in the passing shock wave is shown in the middle and reflects the unique failure mechanisms of the material involved. Subsequent entropy production and transition to statistical mechanical equilibrium at an elevated shock temperature is illustrated in the next step transitioning from the central plot to the graph on the right.

The distinct temporal separation of the processes as suggested by the illustrations displayed in figure 8.10 oversimplifies the event. The transition to statistical equilibrium would initiate early in the structured wave when deformation-induced dissipation begins. Nonequilibrium dissipation and the quest towards equilibrium will proceed in tandem as passage of the shock wave proceeds.

The invariant energy–time product that has evolved from experimental observations of shocks in solids has been identified as an action property, or dissipative action, strictly through dimensional considerations. Invariance, or the constancy of an action property as the state of a physical system changes also has history. Namely, in the dynamics of periodic systems. In the Hamiltonian description of a periodic system, transformation to an alternative canonical set of momentum–position coordinates yields benefits in describing the dynamics of the system. Alternative coordinates are the action and angle variables. In particular the alternative momentum coordinate, namely the action variable, is found invariant under adiabatic changes in the periodic system.

The periodic system of focus in the present discussion is that of sound waves or the thermal and acoustic phonons in the lattice motions of a crystal structure. Many of the physical principles developed in the description of electromagnetic radiation carry over directly to the characterization of thermal phonon energy in the solid crystal lattice. Thermal phonon energy exhibits the same equilibrium Planck energy distribution as that of electromagnetic thermal radiation. Correspondingly, thermal phonons individually and collectively exhibit adiabatic invariance of the action variable.

In thermal equilibrium the temperature of the material is related to the spectral period of the phonon field through invariance of the action variable. With a spectral period at the maximum of $\tau_m = 2\pi/\omega_m$, Wien's law expressed as $kT\tau_m$ = constant manifests invariance of the phonon field action. It follows that if shock compression to a Hugoniot state leads to a thermal dissipation energy characterized by an equilibrium temperature T, then shock compression to an incrementally higher Hugoniot state and corresponding Hugoniot dissipation temperature would be governed by Wien's law. Wien's law applies to the peaks of the Planck curves shown on the left (pre-shock) and on the right (post-shock) in figure 8.10. Here it is worth noting that the Wien's law thermal-energy phonon-period constancy exhibits the same dissipative-energy wavelength constancy observed in steady structured shock waves. The dominant phonon period, however, is several orders of magnitude less than the steady wave risetimes measured in common metals. Viscosity of the equilibrium phonon gas is not expected to account for structuring of the steady shock wave within the range of current experimental data.

A further aspect of the experimental fourth-power data is noted. Namely, the observed behavior between the steady wave strain rate and the pressure through a steady structured wave is a power law of the form $\dot{\varepsilon} \propto p^4$. Power-law relations in physics have special implications. Power-law behavior implies some universality of the underlying physics. In the present case a power-law relation suggests that there is no governing pressure scale over the span of the power-law behavior. In many applications in physics power-law behavior appears over a limited range of the scaling parameter. This behavior is observed, for example in hydrodynamic turbulence (Falkovich and Sreenivasan 2006) and fatigue fracture (Barrenblatt 2006), and is commonly referred to as a law of intermediate asymptotics.

The pressure scales bounding the fourth-power behavior observed in the structure of steady shock waves are uncertain. There are several possibilities. As a lower bound a measure of the yield stress Y negates power-law behavior as shock pressures approach from above the neighborhood of the strength of the material governing dynamic failure. Armstrong *et al* (2007) propose a dislocation-dynamics-based model that describes the asymptotic approach from below to higher pressure power-law behavior. The elastic modulus M capturing the lattice binding energy strength of the lattice is a sensible candidate for an upper-bound pressure to power-law behavior. Except for some of the lighter solids the modulus M is usually in excess of 100 GPa. The range in pressure between the two pressure scales, Y and M, in typical metals is commonly one to two decades.

The data of Barker (1968) and Crowhurst *et al* (2011) for aluminum shown in figure 8.3 is representative. These data collectively exhibit nominal fourth-power behavior over Hugoniot pressures of approximately 1–40 GPa. The HEL of aluminum is a few tenths of a GPa whereas the bulk modulus is 76 GPa.

8.4.2 Correlation dynamics and length scales

Earlier chapters and previous sections in the present chapter have revealed that the product of the dissipated energy and the time duration to failure, identified as the dissipative action, is potentially a unifying criterion for the dynamic failure of solid materials (Grady 2015). Failure modes include the steady and unsteady shock compression of condensed matter, failure through tensile spall, adiabatic shear failure, shock compaction and possibly the compressive shock failure waves observed in brittle solids. This energy–time product remains constant, or nearly constant, as some measure of the rate or intensity of loading to and through failure is varied.

Invariance of the dissipative action for structured shock waves is an experimental observation. The fourth-power behavior of steady structured shock waves in solid materials and the near-first-power dependence in granular compounds (section 8.6.1) provide direct evidence for invariance of the dissipative action. Invariance of the dissipative action under failure through tensile spall or failure through adiabatic shear follows from energy-based models of the failure process (see chapters 2 and 5). Experimental spall data and, to a lesser extent, adiabatic shear failure data, support the energy-based models and, by extension, invariance of the dissipative action.

In both the dynamic spall failure and adiabatic shear failure models, a dissipative action emerges that is a composite of two properties of the material. The first is an interfacial energy Γ that characterizes dissipation on sites of failure. The second is an impedance z that determines the speed at which momentum and energy can be transported to the zones of failure. This impedance can depend on either elastic or plastic momentum diffusive properties of the material. Alternatively, a mobility $\mu = 1/z$ quantifies the ease with which the material reconfigures local momentum and energy requisite to the dynamic failure event. The dissipative action then decomposes into the product of two distinct properties of the material that characterize the failure dissipation and energy/momentum transport

$$A = \varepsilon\tau \sim \Gamma\mu. \tag{8.11}$$

The several dynamic failure modes, including dynamic spall (chapter 2), adiabatic shear (chapter 5) and failure in the steady shock wave (this chapter) are each characterized by a dissipative action property. It is reasonable to question whether there is a relationship among the action values associated with the different modes of failure for a specific material. For example, does the dissipative action for dynamic spall in a specific metal relate in any way to that inferred from the steady wave shock measurement on the same metal? This question has been explored for two metals (Grady 2015). They are aluminum (6061-T6) and alpha-uranium (depleted). Action determined for each of the three failure modes are remarkably close for uranium. The same calculation for aluminum yields values of the dissipation action that differ markedly. Corresponding calculations provide the interface dissipation Γ associated with the several modes of failure. For uranium, interface dissipation inferred for failure in the steady shock and that under spall or adiabatic shear failure are comparable, and sensibly close to that determined from the static fracture toughness of the metal. For aluminum, interface dissipation in the steady shock is markedly lower than in the other modes of failure. The calculations show that interface dissipation in the shock in uranium exceeds that in aluminum by better than one order of magnitude. The physics underlying differences in the two metals is not understood.

Predictions of the failure fabric resulting from the tensile spall and adiabatic shear failure models can be applied to the steady structured shock wave. Both models yield a measure of the nominal spacing of deformation features characterizing the deformation fabric of the failure event. Additionally, the model of adiabatic shear failure provides an estimate of the spatial thickness of individual deformation features. A frame of reference for comparison of the calculated length scales is the width of the steady structured shock wave within which the deformation is constrained.

Calculated spacing of the planar deformation features in the steady shock for 6061-T6 aluminum and alpha uranium are compared with the width of the steady shock wave in figure 8.11. The spatial scales plotted on the left are depicted in the sketch on the right. Calculations span the strain-rate range of the experimental steady wave data for the same metals as in figure 8.1. For both metals the models calculate that the dynamic self-correlation process within the failure event leads to a characteristic spacing of deformation features that is of the same order as the width of the shock wave. Logic dictates that the spacing does not significantly exceed the shock width. The models dictate that the spatial scale of the deformation fabric is also not substantially less than the shock width.

The adiabatic shear failure model also provides an estimate of the thickness of the deformation features accommodating failure in the shock wave. This, of course, assumes that deformation within the shock wave continues to be governed by similar

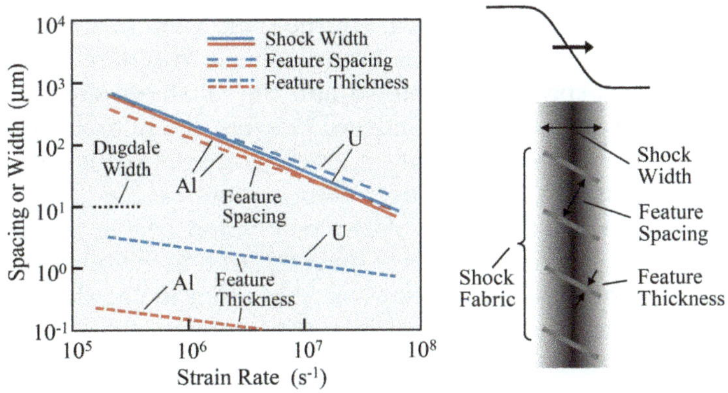

Figure 8.11. Model-based calculations of the spacing and thickness of planar deformation features are compared with the shock width in the steady shock compression of aluminum (red curves) and uranium (blue curves).

continuum heterogeneous two-temperature thermo-plasticity that is intrinsic to the adiabatic shear model. Supporting evidence for heterogeneous thermo-plastic shock deformation in aluminum has been explored (Grady and Asay 1982). The underlying assumption of thermally dependent deformation zones may still be appropriate within the shock. Supporting evidence for thermal disparity on length scales commensurate with the shock wave width also follows from experimental and theoretical work exploring statistical Boltzmann heat transport (Maasilta and Minnich 2014).

Deformation feature thicknesses calculated from the model are of the order of a few micrometers for uranium and a few tenths of a micrometer for aluminum. Curves of feature thickness are also plotted in figure 8.11. Thickness estimates of the plastic deformation features as well as the width of the steady shock can be compared with the width of the plastic zone of a quasi-static fracture in a ductile solid (Dugdale 1960). The Dugdale relation provides a fracture plastic zone width of order 10^{-5} m for both aluminum and uranium.

The concept of deformation features and deformation fabric, and the evolution of such in the dynamic failure event, is intrinsic to all of the physical thinking and failure modeling described in earlier chapters. A deformation feature, depending on the mode of failure, is a fracture or thin planar shear band whose inelastic dissipation is characterized by an energy Γ (units of J m^{-2}) that remains constant, or nearly constant, over a broad range of loading rates. The deformation fabric is the collection of deformation features characterized by a number density and spatial distribution necessary to accommodate the dynamic failure event. The deformation fabric can vary markedly as the range of loading rates is spanned. The deformation fabric addressed in the failure models is inherently a consequence of the failure dynamics. Correlation dynamics and dynamic self-organization subject to time and spatial constraints are the determining causes.

Correlation dynamics, dictated by the competition of the energy and momentum needs of the heterogeneous failure features and the time-constrained redistribution of energy, lies at the heart of the spall fracture and adiabatic shear failure models. The same dynamics is here extended to failure in the structured shock wave. Both the spall failure model and the adiabatic shear failure model assume a simple one-to-one correspondence between the number of deformation features necessary to the failure event and the energy redistribution length scale. More realistic in application, many more deformation features than necessary are seeded, but most quickly drop out of the population as the energy balance is sought. This is very apparent in experimental testing involving the dynamic fragmentation of expanding metal rings or exploding metal shells where many fractures are activated, but few survive to complete the breakup process. Comparable observations of failure activation and survival are noted in the adiabatic shear failure of imploded metal tubes (see chapter 5).

A less apparent, but similar failure dynamics occurs in brittle materials. Here an initially sparse activation of fractures is further seeded through the multiple branching of existing fractures. This failure process cascades until the fracture density is consistent with the load intensity and dissipation requirements of the failure event. Similar multiplication of failure features through branching activation occurs in the adiabatic shear failure of metals subject to high-velocity impact.

8.4.3 Langevin dynamics and the dissipative action

The shock wave inputs a large amount of energy into the material in a brief period of time. The response of the material in disposing of this energy is an irreversible statistical physics event of the highest order. A subset of such statistical physics finding wide application to the micro-dynamics of material systems is identified with the topic of Langevin dynamics. Applications of the principles of Langevin dynamics to the dynamic failure event is not immediately apparent. Exploration of the application of Langevin dynamics is justified by observations on structured shock waves and other dynamic transition measurements that display experimental results suggestive of similar governing statistical physics. See for example the statistical mechanical text of Reif (1965) for a succinct presentation of the physics of Langevin dynamics.

The statistical physics of Langevin dynamics has found analogous application to a range of physical phenomena. The classic example of Langevin physics, from which numerous other applications have been abstracted, is that of the Brownian motion and viscous drag of a meso-particle within a thermal fluid of micro-particles. Perhaps more relevant to applications in the science of solids, that offers parallels conceivably appropriate for a metaphorical transition to dynamic failure, is that of the dynamics of alloy precipitates in a crystal solid (Cottrell and Bilby 1949). At finite temperature an alloy precipitate is subject to statistical diffusion within the crystal lattice governed by physics comparable to that of Brownian motion in a dilute fluid. Further, when subject to a stress gradient such as that imposed by a nearby dislocation, the precipitate will undergo transport under the imposed force. Both transport mobility and precipitate diffusion are captured through the physics of Langevin dynamics.

For a specific precipitate particle of mass M, the equation of motion is written

$$\frac{dMV}{dt} = F_E + F_T, \tag{8.12}$$

with velocities and forces as appropriate vector properties where necessary. External forces imposed on the particle are F_E. Of specific interest is the force imposed by the stress gradient from a neighboring dislocation. Other external forces might include, for example, those imparted by an electric or magnetic field. The force F_T, on the other hand, is the sum of forces acting on the precipitate particle brought about by the internal phonon field of the crystal solid at finite temperature. This internal force term is further partitioned into $F_T = F_V + F$. First is a continuum force characterizing the resistance (or the mobility) of the particle to transport through the lattice in response to the external forces. This is commonly approximated through a linear dependence on the velocity of the particle in the form

$$F_V = -\alpha V, \tag{8.13}$$

with α the governing resistance coefficient. If a size a can be sensibly associated with the particle, then the coefficient α can be related to a measure of the phonon viscosity η acting on the particle through the Stokes relation $\alpha = 6\pi a\eta$.

The continuum influence of the random phonon forces on the precipitate particle is accounted for through the resistance force $F_V(t)$. The remaining random force $F(t)$ imparts no motion on average to the particle and is hypothesized at any time to exhibit a statistical first moment of $\langle F(t)\rangle = 0$ irrespective of the velocity of the particle. The second moment, or variance, is non-zero and accounts for particle diffusion. Writing equation (8.12) as

$$\frac{dMV}{dt} = F_E - \alpha V + F(t) \tag{8.14}$$

yields the Langevin equation. The Langevin equation is a stochastic differential equation with solutions dependent on the statistical treatment of the random force $F(t)$.

The interdependence of transport mobility and diffusivity is captured through the Einstein relation, which can be derived through an appropriate averaging solution of equation (8.14) (e.g. Reif 1965)

$$D = kT/\alpha. \tag{8.15}$$

Further, a measure of the impedance property α, which governs the resistance to transport within the stress field, is related to correlation of the underlying random thermal phonon interactions through a powerful fundamental result identified as the dissipation–fluctuation theorem,

$$\alpha = \frac{1}{kT}\int_0^{\infty} \langle F(0)F(\tau)\rangle d\tau. \tag{8.16}$$

The integrand of equation (8.16) is the statistical correlation function of the random phonon force $F(t)$ while kT captures the finite thermal energy of the crystal lattice. It is worth noting the markedly different time scales that enter the dynamics of Langevin. Continuum level motions of the particle are governed by the time scale $t_1 = M/\alpha$. Chaotic motions about the continuum are determined by the correlation time scale t_2 governed by the random imparting of the phonon forces. In applications adequately described by the Langevin theory $t_1 \gg t_2$.

Material failure in the shock environment, including structured shock waves, dynamic spall and adiabatic shear, exhibit experimental results that are suggestive of the principle outcomes of Langevin dynamics. Experimental results emerge that mirror both the Einstein equation and the dissipation–fluctuation relation of equations (8.15) and (8.16). Consideration of dynamic failure within the framework of the Langevin theory is warranted.

Identification of a variable quantifying the failure transition is necessary. Consider the dynamics of a single representative failure feature such as illustrated in the center of figure 8.12. Here the topic is that of the dynamics of failure of a solid in the presence of a rapidly applied shearing force (such as a shock wave) or a tensile force (such as in dynamic spall). The Langevin particle is associated with the average and chaotic dynamics of a representative single planar failure feature in the presence of both the driving shearing or tensile forces, as well as the random forces brought about by the multitude of neighboring failure sites inducing chaotic forces and motions to the representative failure feature.

Figure 8.12 provides a depiction of such a dynamic failure event. A single representative failure site, among many, is singled out for analysis. The failure site is characterized by a size a and a rate of growth $\dot{a} = da/dt$. An interfacial dissipation Γ, a property unique to the material, is incurred at the site as growth proceeds. Growth at the site is driven by the prevailing stress field imparted, for example, by the shearing stresses of a shock wave immersing the material or dynamic tension

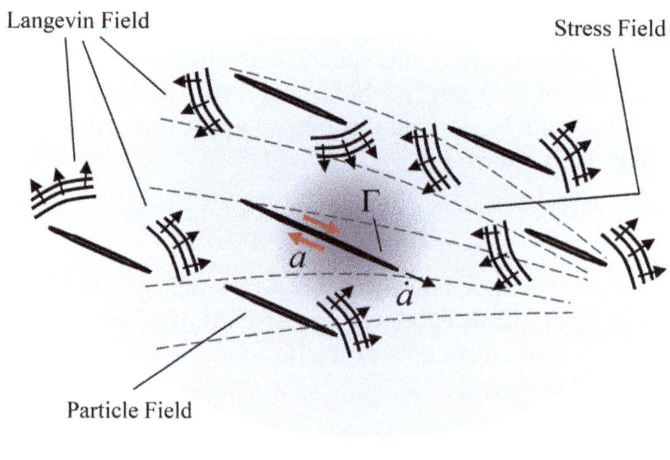

Figure 8.12. Growth of failure at a representative site in the material is a consequence of a prevailing stress field and the statistically chaotic Langevin stress field resulting from concurrent failure at neighboring sites.

imparted by the interaction of opposing shock-induced release waves. Failure growth at the site is also influenced by the random stress field generated by the concurrent and chaotic failure of neighboring sites.

Dynamics of any failure site can be developed by building on fracture failure notions offered by Mott (1948). (Discussion of the Mott analysis is provided in chapter 3.) Mott in turn draws insights from earlier developments of Griffith (1924) to identify the potential energy and kinetic energy of a crack through which the governing equation of motion can be written. The local failure (crack or shear zone) as idealized by Mott (1948) dissipates energy through two mechanisms. (Although motivated by the failure dynamics suggested by Mott, the ideas explored here extend somewhat beyond the analysis undertaken by Mott.) First is interface deformation dissipation, such as dislocation plasticity or the energy of newly fractured local free surface. Dissipation through local deformation is captured through an interface dissipation energy constant Γ, which is an intrinsic property of the material. Second is radiated acoustic dissipation originating from the dynamics of stress release at the local failure site. Dimensional arguments require that the radiated dissipation also scales with the interface energy Γ. Analysis demonstrates that the energies of interface dissipation and radiation dissipation are comparable in magnitude. (Supporting analysis is provided in chapter 3 on Mott crack dynamics.) Dissipation by the two modes of failure is illustrated in figure 8.13.

Markedly different length scales are associated with the two modes of dissipation. In tension and propagation of a ductile tensile crack, the two length scales (x_d and x_r in figure 8.13) are identified with crack opening displacement and with the crack growth process zone such as quantified in the fracture model of Dugdale (1960). Under confined shearing failure two comparable dissipation length scales are identified with the width and process zone length of a propagating adiabatic shear

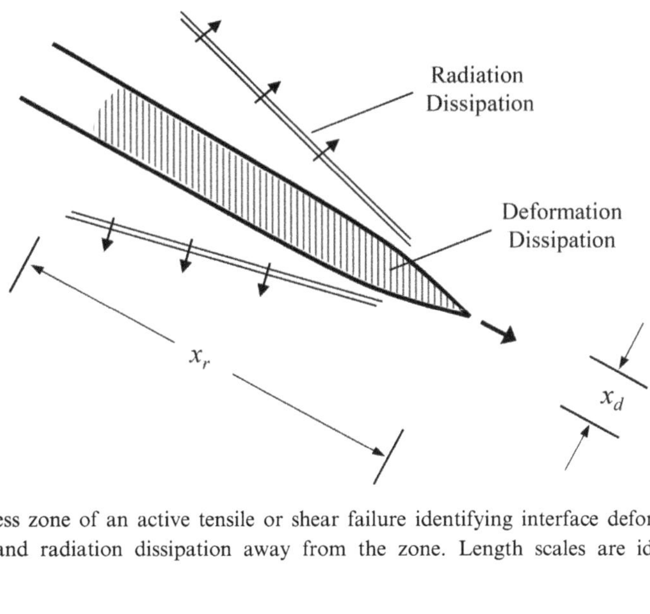

Figure 8.13. Process zone of an active tensile or shear failure identifying interface deformation dissipation within the zone and radiation dissipation away from the zone. Length scales are identified with each dissipation mode.

band (Grady 1992). Radiation dissipation is associated with the coarser characteristic length scale x_r of the propagating process zone.

Both deformation and radiation dissipation contribute to the transient phonon spectrum as dynamic failure of the solid ensues. The dissipation dimensions x_d and x_r provide length scales for the initial transient spectrum of the dissipation-induced phonon field. This bimodal dissipation spectrum is aptly pictured by the transient field depicted in the center plot in figure 8.10. As illustrated in the figure, this transient dissipation spectrum will ultimately equilibrate to the equilibrium thermal Planck distribution. Before equilibration, however, the coarser of these transient spectra has an important function to perform. It represents the chaotic Langevin forces, or pressures, influencing growth of the representative failure feature in figure 8.12.

In chapter 3 the Mott dynamic crack is described whereby the activation and growth of such a shear or tensile failure feature subject to a prevailing stress field is described through specification of the kinetic energy K and potential energy V of the crack, and solving the equation of motion for the time history subsequent to failure activation. In analogy with equation (8.12), the governing equation is extended to account for the random stress field impinging on the crack from the many similarly and concurrently failing fracture sites through the addition of a random force F_T. (Note also an unfortunate change in symbolism for the kinetic energy between the two chapters.) Thus, the Langevin equation descriptive of the failure site dynamics is constructed.

The force F_T acting on the Mott dynamic crack is a consequence of the randomly varying stress field imposed by the failure dynamics and acoustic energies resulting from the neighboring failure field. This field is expected to impose an impeding force on the failure feature as well as a fluctuating force that leads to a spectrum of crack growth histories including many that drop from the population. The random force accounts for the correlation dynamics in establishing the failure fabric in the event.

If dynamic failure is adequately described by Langevin dynamics, then the dissipation–fluctuation relation yields a property impeding and introducing time-dependent resistance to the growth of failure in the dynamic event. Provided by equation (8.16), it is more appropriately written in terms of the energy and pressures of this transient Langevin field,

$$\alpha \sim \frac{1}{\Gamma}\int_0^{\infty} \langle p(0)p(\tau)\rangle d\tau. \tag{8.17}$$

Again, the integrand is the statistical correlation function descriptive of the random phonon pressure field incident on the failure site. This acoustic phonon field, however, is that generated by the transient radiation dissipation of the neighboring sites undergoing comparable failure, and the coarser of the two transient spectra depicted in figure 8.10. The energy of this acoustic phonon field scales with the dissipation parameter Γ, replacing kT in equation (8.16) appropriate to the equilibrium Planck spectrum.

Equation (8.17) relates the correlated transient pressure spectra to an impeding parameter α that determines the material resistance to the redistribution of momentum necessary to the failure event. There is ample experimental evidence that this resistance to momentum redistribution in shock failure events scales with the acoustic impedance, namely $\alpha \sim z = \rho c$. Thus, the dissipation–fluctuation theorem in the present application provides an expression relating acoustic impedance, dissipation energy and the correlation function of the pressure field.

Identifying a characteristic correlation time τ (and correlation distance $\lambda = c\tau$), and approximating the integral in equation (8.17), results in $z \sim p^2\tau/\Gamma$, with $p^2 = \langle p(0)p(0)\rangle$ a measure of the initial dispersion of the random pressure field. In this simplified form correspondence with relations emerging from energy-based theories of spall fracture and fragmentation are clearly evident (see section 2.3.6). It is also of interest to contemplate similarities with other well-known criteria in shock physics, including the Tuler–Butcher spall criteria (Tuler and Butcher 1968) and the delayed shock-to-detonation $p^2\tau$ criteria (Walker and Wasley 1969).

A further fundamental result emerging from solution of the Langevin equation is the Einstein relation (Einstein 1905, Smoluchowski 1906) in which the product of the Langevin field energy and the particle mobility yields a constant characteristic of some measure of diffusivity. In applications where thermal energy determines the Langevin energy field, the Einstein relation from equation (8.15) is written $D = kT\mu$. Molecular thermal energy is identified with kT, while $\mu \sim 1/\alpha$ is a measure of the particulate mobility and D is a diffusion property of the process. Dividing by the molecular mass to provide a specific energy of the Langevin field $\mathcal{E} \sim kT/m$ and identifying a characteristic mobility time $\tau \sim m\mu$ yields an alternative expression for the Einstein relation

$$D \sim \mathcal{E}\tau. \tag{8.18}$$

The Einstein relation emerging from Langevin dynamics as expressed in equation (8.18) and the property identified as the dissipative action in equation (8.11) bear a clear resemblance. The energy $\mathcal{E}$ in equation (8.18) is a kinetic particle energy or phonon energy associated with the equilibrium thermal state of the matter that governs particle transport and momentum diffusion in a Langevin application. Energy dissipation in equation (8.11), as emerges from models of the dynamic failure event, is identified with the interfacial dissipation Γ. The interface energy Γ also provides a measure of the energy intensity of the transient acoustic field radiated from sites of failure and depicted in figure 8.13. Mott (1948), in analysis of the acceleration history of the Mott crack (see chapter 3), recognized the occurrence of radiated strain energy and incorporates this into the kinetic energy term characterizing the Hamiltonian of the Mott crack. Archambeau *et al* (1976) in a more formal analysis of any abrupt failure event shows that radiated kinetic energy scales with, and is comparable in magnitude to, the failure dissipation. A less formal analysis addressing the same issue is pursued in chapter 3 subsequent to discussion of the Mott crack. Dynamic failure as illustrated in figures 8.12 and 8.13 creates a transient acoustic kinetic energy field. The intensity of this field scales with the dissipation energy Γ. This energy field is identified as the Langevin field that accounts for the additional random stress load any individual failure site is subjected to in addition to the prevailing stress field.

Solution of the appropriately averaged form of the Langevin equation yields the average motion history of the Mott crack. The solution, accounting for the non-zero variance, correspondingly yields the dispersion in motion histories of the representative failure feature about the average. Thus, the dissipative action $\mathcal{A}$, with dimensions of diffusivity, and outward similarities to the Einstein relation emerging from Langevin dynamics, reflects a measure of diffusion, or kinematic viscosity, governing the temporal development of the dynamic failure event. Namely, the dissipative action is the property that determines the time-dependent reorganization of momentum and kinetic energy as failure ensues. If so, the process relates to the establishment and coarsening of the deformation fabric as the failure event proceeds to completion. Representative of such dynamic failure is the correlation dynamics exhibited in the adiabatic shear failure of an imploding hollow cylinder of stainless steel illustrated in chapter 5, or the failure and fragmentation of expanding metal cylinders in figures 2.14 and 2.18, where many sites of failure initially participate but few carry through to completion. Similar correlation dynamics can be envisioned governing failure within the structured shock wave.

A further central principle emerging from the statistical consequences of Langevin dynamics is the relationship between dissipation and fluctuation. Namely, the intensity of fluctuations in the Langevin field correlates with the strength of dissipation. This principle is identified as the dissipation–fluctuation theorem (Nyquist 1928, Callen and Welton 1951, Kubo 1957, 1959). The analysis undertaken would indicate that the width of the steady structured shock wave, and the fourth-power law, is a manifestation of the Nyquist–Callen–Welton dissipation–fluctuation theorem. As such, in the present application to failure dynamics, identification as a dissipation–correlation relation is probably more fitting.

The analogy of dynamic failure in solids with more traditional applications of Langevin dynamics reveals limitations. The Langevin field of a dilute gas is the classical random kinetics (the temperature) of the gas particulate. The energy of the Langevin field pre-exists and overwhelms dissipation introduced in the Langevin event, dissipation in the field due to viscous drag, for example. Similarly, in solid state precipitate diffusion the Langevin field is identified with the quasi-particle phonon field. It is also temperature determined, with energy that overwhelms the dissipation incurred in transport of the alloy precipitate. In stark contrast, that proposed as the Langevin field in the dynamic failure event is the acoustic phonon field generated by the failure event itself. The dissipation in the failure event and energy of the Langevin field are comparable. Further, the prevailing stress field responsible for onset and initial growth of failure is itself mitigated by the failure process in later stages of the event. Consequently metaphorical considerations are at best oblique and the applicability of principles such as the Einstein relation and the dissipation–fluctuation theorem to dynamic failure remains tentative.

Nonetheless, the foregoing analysis sensibly supports the premise that the empirical dissipative action relates to concepts of classical action physics. Namely, this property identifies with the canonical momentum action variable of the transient acoustic phonon field that emerges from the dynamic failure event. The dissipative action is also identified with a diffusion, or kinematic viscosity, that bears responsibility for the time-dependent correlation and organization of the failure fabric necessary to the dynamic event.

8.4.4 Further applications in the shock transition

Discussions of the physics of Langevin dynamics in the present section has been biased towards the dissipation action invariance with applications to observations of the fourth-power trend in the structure of steady shock waves. There are other shock transition events in which the underlying dynamics may be profitably explored through the physics of Langevin statistics. Dynamic spall, failure waves in brittle solids and shock-to-detonation waves in energetic materials are illustrative.

The interaction of shock-induced opposing planar release waves within the elastic regime of a competent rock or ceramic leads to a sharp elastic tension wave. If the stress level exceeds the dynamic tensile strength of the solid, spall failure ensues. Spall failure at the tensile state is not immediate. Dynamics on some time scale is necessary to bring about the onset and growth of spall failure. Consequently failure is delayed until some later time after passage of the tensile shock wave. The planar wave that carries the material to the tensile elastic stress state is of a finite width. This width of the wave is determined by nonlinear wave mechanics and the intrinsic dispersive properties of the material. Consequently, the elastic tensile stress achieved subsequent to wave passage is not an equilibrium state. Elastic correlation over length scales greater than the width of the wave is necessary to the ensuing equilibration of the stress state. Elastic equilibration occurs through acoustic wave propagation over the requisite time and length scales. Thus, even in the realm of elastic dynamic tension, dissipation is present. In addition to the elastic state of lattice tension, an acoustic phonon spectrum is produced by the elastic shock. The nature of this spectrum is determined by the width of the tensile shock and the unique micro-structure of the elastic material. If spall failure is not achieved this induced phonon field will eventually thermalize. When spall failure occurs this phonon field is responsible for the redistribution of momentum and kinetic energy on the micro-scale necessary to the accommodation of stress concentration and failure activation at potential site of tensile failure onset. Global spall failure occurs after some substantial time scale. A significantly smaller time scale is associated with the chaotic stress fluctuations of the phonon field responsible for the momentum redistribution necessary to the onset of spall failure. As outlined, this shock event appears amenable to the statistical tools of Langevin physics.

Failure waves in brittle solids and the delayed localized compressive shear failure observed in competent elastic materials including glass and ceramic largely mirrors the discussions of tensile spall in the preceding paragraph. A structured compressive shock wave of some small but finite wave width achieves a compressive elastic stress state within the material. The longitudinal and transverse stress levels achieved reflect the elastic properties of the solid. Stress on the microscale behind the shock is not in equilibrium. Again, an acoustic phonon field undertakes elastic equilibration over longer length scales behind the shock wave. As momentum redistributes on the microscale, stress concentrations at sites of weakness can activate the compressive delayed shear failure characteristic of the failure wave phenomena. As in the application of tensile spall, the global failure event and the random phonon field responsible for momentum rearrangement occur on two markedly different time scales. One statistics-based analysis of the failure wave process is described in chapter 6. In the present context pursuit through the alternative of Langevin dynamics may be comparably profitable.

Application to the shock-to-detonation wave in energetic materials offers further physics to consider within the framework of Langevin statistics. Under appropriate conditions the shock compression of an energetic material does not immediately initiate reaction. Energetic reaction, depending on the initial shock pressure, is delayed for some period after passage of the shock wave. Experimental studies of this reaction phenomena are quantified in measurements of run distance to detonation as a function of initial shock pressure. Representative, for example, is the study of Jackson *et al* (1976) on TATB, where measurements of the run distance to reaction are reported for several initial densities of this explosive. Such data speak to the delay in reaction transformation of the material subsequent to passage of the reaction-inducing shock wave. The rate of reaction at any point in the material subsequent to shock passage is determined by the magnitude of the Arrhenius parameter, A/kT, the ratio of the activation energy to the thermal energy. Thermal hot spots generated within the compression shock wave markedly reduce the magnitude of the Arrhenius parameter and increase the rate of reaction at select sites in the material. Hot spots are well-known as sites for the onset of reaction activation. Creation of thermal hot spots ceases after passage of the shock wave. If reaction does not follow, such hot spots fade in time through thermal diffusion. Not in equilibrium after passage of the shock wave is the stress at the Hugoniot state. As with tensile spall and the compressive failure wave, stress and strain correlation over length scales greater than that of the shock width is achieved over later times through acoustic adjustment of the phonon spectra induced by the shock. Activation energy A at any point in the material is sensitive to the configuration of local lattice strain. As stress correlation ensues subsequent to shock passage, statistically favorable changes in the Arrhenius parameter through changes in the activation energy can occur and contribute to the condition necessary for delayed energetic reaction. Again, statistical Langevin dynamics plays a role in carrying out this shock transition.

The preceding several paragraphs regarding delayed material transformation subsequent to shock wave compression are supported by notions concerning the statistical nature of the stress and strain state within, and subsequent to, the responsible shock wave. Consider a steady structured planar shock wave of width λ propagating in a solid. Consider further the shock wave normally incident on an interior plane of the material of some thickness $\eta < \lambda$ as illustrated in figure 8.14. Conceptualize partitioning this plane into elements of dimension η^3 such that each of many elements is similarly and simultaneously carried to the current shock state as the shock wave passes. Acceleration to the shock Hugoniot state of each element is a consequence of the unique stress history experienced as the shock passes through the plane and is governed by the momentum relations

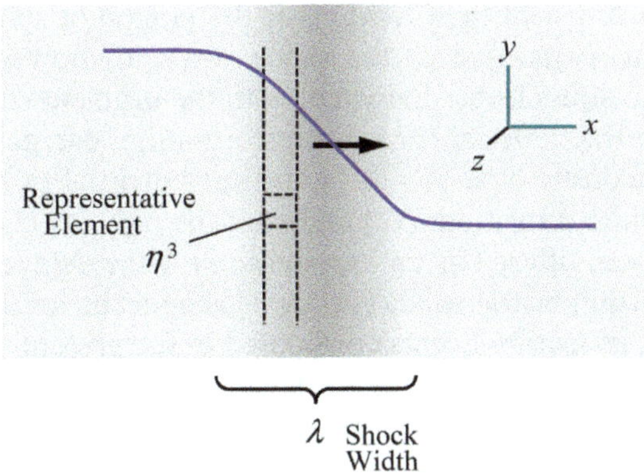

Figure 8.14. Representative element within an interior normal plane of material encompassed by a structured shock wave. The element is one of many experiencing statistically identical shock compression and acceleration.

$$\rho\frac{\partial u_x}{\partial t} = -\frac{\partial \sigma_x}{\partial h_x}, \quad \rho\frac{\partial u_y}{\partial t} = -\frac{\partial \sigma_y}{\partial h_y}, \quad \rho\frac{\partial u_z}{\partial t} = -\frac{\partial \sigma_z}{\partial h_y}. \tag{8.19}$$

Arguments are stress and particle-velocity components while indices x, y and z are Lagrangian coordinates in the direction of (x), and transverse to (y,z), the direction of shock propagation, respectively. The axial and transverse stress are written, respectively, as

$$\sigma_x(t) = \bar{\sigma}_x(t) + f_x(t), \quad \sigma_y(t) = \bar{\sigma}_y(t) + f_y(t), \quad \sigma_z(t) = \bar{\sigma}_z(t) + f_z(t), \tag{8.20}$$

where the stress components are decomposed into a slowly varying term and a random rapidly fluctuating component with an expected value of zero at any time when averaged over the many elements of dimension λ^3 over a sufficient transverse span of the interior plane experiencing the same shock acceleration. The randomly fluctuating stresses $f_x(t)$, $f_y(t)$ and $f_z(t)$ are a consequence of the unique heterogeneous microstructural character of the material in question, both inherent and that induced during the shock event. Consequently, on achieving the Hugoniot state the expected value of the axial particle velocity of any element is $\langle u_x \rangle = u_h$, the Hugoniot particle velocity. The corresponding expected value of the transverse particle velocities are $\langle u_y \rangle = 0$ and $\langle u_z \rangle = 0$.

The variance of the particle velocity at the Hugoniot state, namely $\langle (u_x - u_h)^2 \rangle$, $\langle u_y^2 \rangle$ and $\langle u_z^2 \rangle$, are, however, non-zero. This creates at the Hugoniot state a random acoustic phonon field, a Langevin field, of energy and frequency content unique to the microstructure and failure modes of the material. This field of course also evolves through the rise of the shock wave and is responsible in part for the shock viscosity that determines the structure of the shock wave (Mescheryakov and Divakov 1994, Grady 1998b) (see also section 8.5.3).

In most cases this transient field residing at the Hugoniot shock state thermalizes over time subsequent to passage of the shock wave. If, however, the shock wave carries the material to stress levels consistent with the insipient stage of onset of some critical state event in the material (phase transformation, energetic reaction, failure), then this Langevin acoustic field has the statistical potential of assisting the delayed onset of the critical state transition. Although not the only explanation put forth, this statistical basis has been offered as an explanation for the delayed failure (the failure wave) observed in certain brittle solids (see the closing technical section of chapter 6).

The several shock transition events considered in the present subsection (dynamic spall, failure waves, shock to detonation), as described, exhibit a commonality suggestive of underlying statistical causes addressed through principles of Langevin physics. Stresses imposed by the shock compression event readily decompose into constant, or slowly varying, and statistically random components. As decomposed, the stress field is governed by markedly different time scales, with that of the former markedly longer than that of the random field. The latter chaotic stress field has the potential to play an important role in a statistical determination of the history and nature of the delayed shock-induced transition event.

The present and the previous several subsections have explored the application of Langevin statistics in attempts to explain the statistical features of material change of state (dynamic failure, shock transformation) in the shock compression event. Motivation for this avenue of investigation is provided by experimentally observed systematics (such as the fourth-power law) in shock wave phenomena that demonstrate similarities to other areas of physics aptly treated within the framework of Langevin dynamics. Intriguing analyses are offered. Considerable further efforts are necessary, however, before application of this area of irreversible statistical physics to the shock transition is convincingly demonstrated.

8.5 Viscosity and the structuring of shock waves

Association of viscosity with the structuring of shock waves in the shock compression of solids has long been a part of the lore. The early papers of Band (1960), Band and Duvall (1961) and Bland (1965), provide a seminal role in linking a solid viscosity to the shaping of finite amplitude shock waves. The complementary viscous structuring of shock waves through the introduction of artificial viscosity into the numerical solution of shock waves is surveyed by Richtmyer and Morton (1967). Use of the name artificial is probably unfortunate as the underlying physics it attempts to replicate is anything but artificial. The experimentally observed fourth-power behavior is often referred to as a law of shock wave viscosity. Thus, some discussion of viscosity as a mechanism for structuring shock wave is warranted.

Any constitutive expression that relates the dissipative stress to the strain rate can be referred to as a viscosity relation. The mechanics through which strain-rate dependence or viscosity leads to the evolution of steady structured waves are reasonably well known. It is the underlying microstructure or molecular physics responsible for the viscous constitutive behavior that remains poorly understood.

Rate-dependent plasticity, with the attendant non-conservative dislocation mobility governed by the processes of thermally activated penetration of energy barriers, represents one mechanism that can account for the finite risetime of structured steady waves. Dislocation plasticity has been the most common deformation mechanism explored in explaining structured waves and the fourth-power behavior in metals (e.g. Johnson and Barker 1969, Swegle and Grady 1985, Rubin 1990, Partom 1990, Johnson 1992, Armstrong *et al* 2007).

8.5.1 Viscosity of steady shock waves

Solid viscosity as explore here is a consequence of mechanisms that accommodate the transport of momentum along the velocity gradient imposed by passage of a structured shock wave. In a kinetic model of a viscous process, a quantum of momentum separates from the local flow and is transported along the velocity gradient some characteristic length before reconnection with the flow. It adds to, or subtracts from, the momentum at the new position. In a structured shock wave, viscosity is viewed as the diffusion of momentum along the axis of wave propagation.

Viscosity is commonly considered a material constant that provides a proportionality between a dissipative stress component and the velocity gradient or strain rate. The magnitude of the viscosity coefficient η in a steady structured shock wave can be estimated through the material properties that establish the upward concavity of the Hugoniot curve, and the geometry imposed by the Rayleigh line corresponding to the steady shock wave amplitude. An estimate of the viscosity is based on the assumption of a decomposition of the dynamic pressure into equilibrium EOS and dissipative components. A first pressure derivative of the bulk modulus K_o', or more commonly in shock wave analysis the parameter S in the linear shock-velocity versus particle-velocity relation $U = C_o + Su$, provides a measure of the upward concavity of the Hugoniot and the rate at which a finite amplitude pressure wave will steepen or 'shock up' with propagation distance. On the other hand a ratio of the solid viscosity to the impulse imparted by the shock, $\eta/p_h\tau$, where p_h is the pressure jump through the wave and τ is the risetime of the wave, provides a counterbalancing dimensionless measure of the rate of spreading of the wave by the viscous forces. Steady wave conditions are achieved when these opposing forces are comparable or, $S \sim \eta/p_h\tau$. A more studied analysis of a steady structured shock wave in a linear $U = C_o + Su$ material provides for the maximum viscous pressure (Grady 1998b)

$$p_v^{\max} = \frac{1}{4}\rho_o S u_h^2, \tag{8.21}$$

and for the viscosity

$$\eta = \frac{1}{4} S p_h \tau. \tag{8.22}$$

Metals exhibit measurable risetimes in the deformation shock wave provided the shock amplitudes range from a few to perhaps a few tens times the HEL. Uranium is representative with a HEL modestly in excess of 1 GPa, exhibiting an approximately 0.1 μs steady wave risetime at a shock pressure of 10 GPa (Grady 1986). With a value of $S = 2$, equation (8.22) provides a viscosity coefficient of about $\eta = 500$ Pa s. For a constant solid viscosity the shock wave risetime is observed from equation (8.22) to decrease with the shock pressure according to $\tau \sim p_h^{-1}$, and the corresponding steady wave strain rate increases as $\dot{\varepsilon} \sim p_h^2$. The experimental fourth-power law reveals that the steady wave risetime decreases with a stronger $\tau \sim p_h^{-3}$ dependence on shock pressure. The latter dependence through equation (8.22) provides for a viscosity coefficient that decreases with strain rate according to $\eta \sim \dot{\varepsilon}^{-1/2}$. This form of strain-rate-dependent viscosity has found application in computational models (e.g. Swegle and Grady 1985, Swegle 1991).

8.5.2 Mechanisms of shock viscosity

If viscous processes as defined above are responsible for the finite thickness of steady shock waves in solids, then the underlying physical mechanisms of solid state viscosity are the basis of a fuller understanding of shock processes in solids. Gilman (1979) explored sources of solid viscosity based on a kinetics of momentum diffusion of atomistic processes active in the shock wave front. Contributions from the electron and phonon gas, and atomic interactions were considered. Estimates of mass fluxes and transport cross-sections were found to lead to negligible contributions for most of the processes examined. According to Gilman, only dislocation drag effects were found to contribute significantly to an effective shock viscosity.

In the kinetic theory of viscosity of fluids, momentum diffusion is usually synonymous with mass diffusion. Dremin and Breusov (1968, 1971) proposed a mechanism of material mixing during the inelastic shearing of neighboring planes in the shock front. Particularly interesting are the experiments reported by Alekseevskii *et al* (1989) in which radio-isotope tagged atomic species are observed to transport many numbers of atomic spacing upon passage of a strong shock wave. A clear explanation of this phenomenon is not yet available and questions concerning the experiments remain. Several possible explanations are discussed in Grady (1998a) and also in chapter 9. Sufficient mass diffusion within the shock front to account for observed wave structuring seems unlikely, however. Mass diffusion from onset of non-laminar turbulent deformation in the shock wave has not been adequately explored.

An alternative physical mechanism that may account for the observed structured shock waves in metals is suggested by ultrasonic experiments of Mason and McSkimin (1947) on metals. Their work shows that the attenuation of ultrasonic waves above a frequency of several megahertz is dominated by scattering within the grain structure of the polycrystalline metal. Acoustic scattering would lead to wave dispersion at any amplitude and could also counterbalance the shock-up tendency of nonlinear solids, thus accounting for the observed structured shock waves and corresponding viscosity.

8.5.3 Wave scattering and shock viscosity

A theory of solid viscosity in the shock compression of heterogeneous nonlinear solids based on a mechanism of momentum diffusion brought about by acoustic scattering has been explored (Grady 1998b). The theory uses the quasi-harmonic normal-mode description of matter to characterize the acoustic state caused by shock-induced scattering. A decomposition of the total shock pressure

$$p = p_{eq} + p_v \tag{8.23}$$

is assumed, where p_{eq} is the equilibrium pressure while p_v is the viscous pressure resulting from the diffusion of scattered acoustic energy. The theory is illustrated pictorially in figure 8.15 and conceptually can account for shock wave structuring in heterogeneous thermo-elastic solids.

The development leads to a rate expression for the viscous pressure of the form

$$\dot{p}_v = \dot{p}_s + \frac{1}{\tau}(p - p_{eq}). \tag{8.24}$$

The first term on the right is a stimulation pressure component brought about the diffusion of phonon momentum resulting from the shock-induced scattering. The second term accounts for relaxation of the pressure towards statistical equilibrium of acoustic energy (the Planck spectra) over the normal modes.

When equation (8.24) is specialized to the more restricted form

$$\dot{p}_v = f(\varepsilon)\dot{\varepsilon}^m + \frac{1}{\tau}(p - p_{eq}), \tag{8.25}$$

the relation encompasses the anelastic equation used by Knopoff (1965) to describe acoustic scattering in metals, as well as the expressions developed on a more intuitive basis by Barker (1971), and by Kanel *et al* (1995), to characterize the viscous structuring of shock waves. Equation (8.25) is shown, for appropriate constitutive

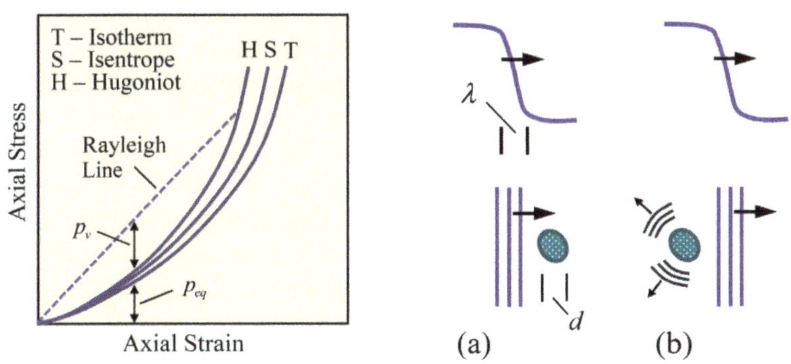

Figure 8.15. Structured shock waves as a consequence of acoustic scattering in a nonlinear heterogeneous thermo-elastic solid.

relations in the stimulation term, to readily reproduce the fourth-power behavior of steady structured shock waves.

The theory is physically appealing in that it characterizes a true solid acoustic, or phonon, viscosity brought about by shock-induced scattering in heterogeneous media. The complication is that the developments of dynamic dislocation plasticity to describe material deformation under shock compression result in very similar continuum governing equations. Thus, it is difficult for experiment to select between the two competing physical explanations of deformation plasticity versus acoustic scattering for the structuring of shock waves. Note, however, the more recent structured wave studies on composite and porous solids surveyed in the final section of this chapter. Some trends observed are consistent with shock wave structuring through wave scattering mechanisms for these materials.

In a finite amplitude shock wave a further contribution to the stimulation of wave scattering can occur. When a pressure pulse is incident on an impedance discontinuity, the resulting reflected and redirected acoustic energy is a common picture of the scattering phenomenon. Shock compression allows, however, for additional scattering processes. Shock compression in heterogeneous solids on the microscale can lead to local concentrations in elastic strain energy, followed by the sudden unstable release of wave energy due to local failure mechanisms such as microfracture or the avalanche of dislocation slip bands. The process is akin to acoustic emission brought about by the static stress loading of solid bodies and are intrinsic to the considerations of Langevin dynamics pursued in the previous section. Both elastic wave scattering and the acoustic emission of wave energy from local failure can contribute to the dispersion of large-amplitude shock waves.

8.6 Structured waves in porous and composite materials

Measurements of shock wave profiles over the past several decades have shown that a number of monolithic materials exhibit a near-universal relationship between the shock pressure and the steady wave strain rate in the shock wave. Namely, as emphasized in each of the preceding sections, the strain rate is found to display a fourth-power dependence on the step in pressure through the shock wave. It is of interest to explore whether such behavior extends to non-monolithic classes of materials. Generally, it does not, although fourth-power behavior is observed in some intriguing cases. The different behaviors are interesting and in some cases unexpected. Here, some steady wave shock profile measurements on porous and composite materials are examined. These results reveal features that may shed light on the mechanisms underlying structuring of steady shocks.

Shock compression studies on both porous and composite materials abound in the literature. A number of practical needs over the past half century have motivated many such investigations. Here only a few more recent studies are considered where reasonable efforts were undertaken to capture the wave profile shape of the measured or calculated structured shock wave.

In porous and composite solids, a material length scale is commonly apparent, and is recognized as paramount in shaping the structure of the shock wave. In porous solids, it may be the dominant grain size or the nominal spacing of pores. In composite solids, the length scale relates to the characteristic granularity of components on the microscale. The widths of steady structured shock waves are expected to replicate this length scale and be on the order of, perhaps several times, this characteristic size of the microstructure, reflecting the time required to equilibrate the requisite stress and deformation over this dimension. The shock amplitude also plays a role in defining the shock width. The extensive study of the shock compaction of granular HMX by Sheffield *et al* (1997) clearly reveals this relationship between shock width and grain size. Computational simulations of the shock compaction of granular solids similarly demonstrate the influence of the particle size scale on shock width (Vogler *et al* 2012). Experimental measurements of structured compaction waves in granular tungsten carbide are related to the particle size in the work of Vogler *et al* (2007) (see also chapter 7). Theoretical modeling of the response of composite solids to finite-amplitude wave propagation usually entrains this length scale within a representative volume element (e.g. Barker 1971). Shock wave experiments on several laminar composites further reveal the dependence of structured shock wave width on the microscale characteristic size (Zhuang *et al* 2003). Generally, the trends noted are observed in experimental studies of structured shock waves in porous and composite solids. Some unexpected results, however, have been uncovered.

As in earlier chapters, structured shock waves are characterized by a single strain-rate parameter $\dot{\varepsilon} \simeq \Delta u/\lambda$, where Δu is the jump in particle velocity through the shock wave and λ is a measure of the nominal width of the wave. Correspondingly, $\dot{\varepsilon} \simeq \Delta\varepsilon/\tau$ where $\Delta\varepsilon$ is the jump in strain and τ the wave risetime. They are the same for a steady shock wave where $\Delta u = U\Delta\varepsilon$ with U the steady wave shock velocity. Higher order detail in structured shock waves can, of course, be explored (e.g. Dunn and Grady 1986), although limitations in experimental resolution can compromise such efforts.

8.6.1 Structured shock waves in granular and porous solids

Structured shock wave profile data are available for exploring the trends in wave structure for steady wave compaction in granular media. The systematics differ markedly from comparable experimental observations in solid metals. The results offer reasons to ponder shock wave structure in porous metals.

8.6.1.1 Structured waves in granular compounds

A comparison of the shock pressure versus the strain rate of steady structured waves in the shock compaction of granular compounds within the compaction regime is shown in figure 8.16. Brown *et al* (2007) have reported this comparison of several refractory materials. The data for TiO_2 are from the study of Anderson *et al* (1994). The SiO_2 tested is a silicate beach sand. The samples tested were near pour density. The silica sand bulk density was 59% of the theoretical density, while the tungsten carbide was 56% theoretical. The titanium dioxide was presumably comparable.

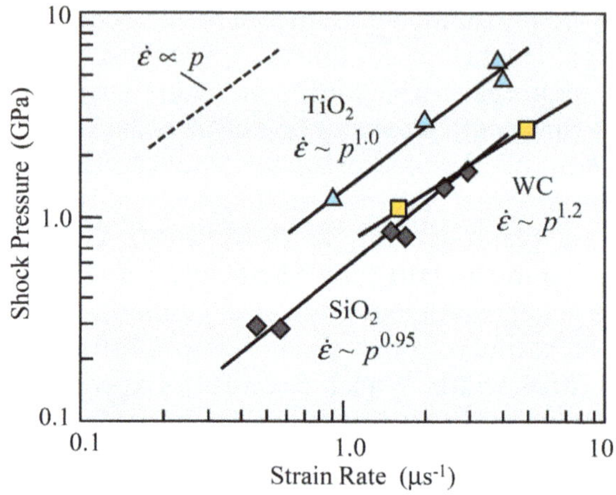

Figure 8.16. Experimental dependence between the shock pressure and the strain rate of structured steady shock waves in the compaction of several granular ceramics (Brown *et al* 2007).

Vogler *et al* (2012) have reported computational simulations of corresponding shock compaction experiments. The EMU code of Silling *et al* (2006) is applied in these calculations—a computational technique utilizing peridynamic methodology (Silling 1998). A steady wave strain-rate dependence on Hugoniot pressure to a power of about 0.88 is observed over the range of the computational simulations.

The selected shock compaction data on granular solids and the computational simulations both suggest a power-law dependence of steady wave strain rate on compaction pressure near unity. Thus, the ratio of $p/\dot{\varepsilon}$, with dimensions of viscosity, is constant, or nearly constant. It is worthwhile to examine the material properties responsible for this viscosity. The data provided in figure 8.16 imply a steady wave shock viscosity η of about 600 Pa · s for the silica sand or for the tungsten carbide granular powder.

Density and porosity should play a role and are captured by the initial distended density ρ_{oo} and the compacted or solid density ρ_o. As previously noted, grain size, denoted by the size scale a, should matter. Further, some measure of the crystal strength of the granular particulate, identified as the constant p_c, should also participate in structuring the wave. These are first-order properties of the granular medium that, taken together, may govern the wave risetime and the viscosity of the material. One is led to the functional form $\eta = f(\rho_{oo}, \rho_o, a, p_c)$. Dimensional considerations provide

$$\eta \sim (\rho_{oo}/\rho_o)^m \sqrt{\rho_o a^2 p_c}\,, \tag{8.26}$$

where the radical expression is the only relation with dimensions of viscosity that can be formed from the available properties. A self-similar power-law dependence on the distension ρ_{oo}/ρ_o is sensible. A dimensionless constant of proportionality is implicit.

Particle size for the ungraded silica sand ranges over 100–400 μm. Quartz density is 2600 kg m^{-3}, while ~3 GPa is a reasonable estimate for the crystal strength. The

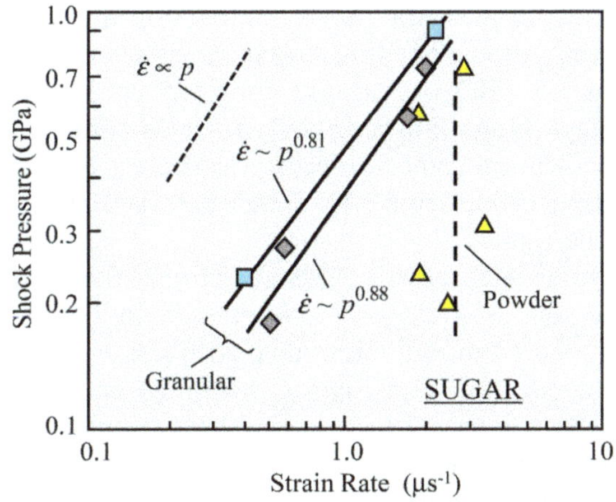

Figure 8.17. Experimental dependence between the shock pressure and the strain rate of steady structured waves for granular and powdered sugar (a surrogate for HMX). The upper and lower granular curves represent initial densities of 73% and 65% of TMD, respectively. Data from Sheffield *et al* (1997).

proportionality constant and the distension factor are not known. The radical term in equation (8.26) yields a viscosity of 250–1000 Pa · s, in reasonable agreement with the 600 Pa · s inferred from the experimental data.

Sheffield *et al* (1997) have conducted shock compaction structured wave profile measurements on granulated and fine powdered HMX and comparable studies on sugar, a mechanical surrogate for HMX. The experimental results for the two materials are comparable and only those for sugar are considered here. Plots of shock pressure versus the steady wave strain rate for the sugar are shown in figure 8.17. In addition to granular and powdered sugar media, two modestly different initial distensions of the granular materials are tested.

The granular sugar exhibits a functional dependence comparable to the ceramics shown in figure 8.16, namely, near-first-power dependence of the structured wave strain rate on shock pressure. The finely powdered sugar, on the other hand, displays a nominally constant strain rate, independent of shock amplitude over the range tested. A likely explanation is that because of the very small particle size, any dependence of wave thickness on shock amplitude is below the resolution of the instrumentation. Two different distensions of the granulated sugar suggest a modest dependence of shock viscosity on the initial distension.

8.6.1.2 Dissipative action and structured shock waves in granular solids

Further investigation of steady structured shock waves in granular solids lends support for an invariant of the dissipative action in the shock compaction event. Shock waves in granular solids notably do not exhibit the fourth-power dependence touted for a number of other solids in the present chapter. Rather, closer to a first-power behavior between the shock pressure and the steady wave strain rate is reported. This starkly different observed behavior can be considered in the context of the dissipative action. Data for three granular compounds for which a reasonably clear representation of the pressure versus compaction and pressure versus strain-rate trends have been measured are shown in figure 8.18. The materials are respectively quartz sand (SiO_2), granular tungsten carbide (WC) and granular mullite ($3Al_2O_3{\bullet}2SiO_2$). Hugoniot compaction versus compaction strain is shown on the left. Full densification (ignoring crystal compressibility) is identified on the left by the vertical dashed line segments. Shock pressure versus steady wave strain rate is provided on the right.

This structured wave data clearly fails to exhibit the fourth-power behavior characteristic of the metals and solid compounds in figure 8.1. Strong shock compaction of powders is commonly described with a snow-plow model. The model is characterized by a constant compaction strain ε_c, corresponding to complete collapse of the void volume, irrespective of the shock amplitude. Shock compaction to any Hugoniot pressure p_h achieves a constant Hugoniot strain $\varepsilon_h = \varepsilon_c$. Energy dissipation in the Hugoniot compaction event within the framework of the snow-plow model is provided by $\mathcal{E} = p_h \varepsilon_c / 2\rho_{oo}$ with ρ_{oo} the density of the distended solid. The time to compaction is given by the ratio of Hugoniot strain to the strain rate in the steady structured wave $\tau \simeq \varepsilon_h / \dot{\varepsilon}$. The energy–time product yields

$$\mathcal{E}\tau \simeq \frac{\varepsilon_c^2}{2\rho_{oo}} \frac{p_h}{\dot{\varepsilon}}. \tag{8.27}$$

Constancy of the product leads to a first-power dependence of the strain rate on the Hugoniot pressure, in better agreement with the experimental results in figure 8.18 as well as with data in the two previous figures.

The analysis can be improved. Hugoniot pressure versus compaction strain data in the region of partial compaction for the three materials are shown on the left in

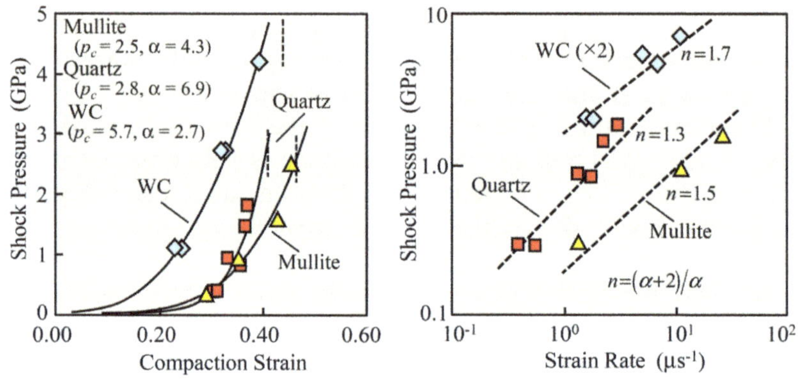

Figure 8.18. Experimental dependence between the shock pressure, compaction strain and the strain rate of structured steady waves in the shock compaction of several granular ceramics including quartz (Brown *et al* 2007), tungsten carbide (Vogler *et al* 2007) and mullite (Vogler 2015). Vertical dashed lines on the left identify full compaction of the respective materials. Note the times-two vertical offset of tungsten carbide data on the right.

figure 8.18. The Hugoniot compaction data are reasonably described with a power-law relation of the form

$$p(\varepsilon) = p_c\left(\frac{\varepsilon}{\varepsilon_c}\right)^{\alpha}. \tag{8.28}$$

A rigid-compaction model (shock compaction up a Rayleigh line and decompression down a vertical path) provides for the specific energy dissipated in the shock compaction event,

$$\mathcal{E} \simeq \frac{1}{2\rho_{oo}}p(\varepsilon)\varepsilon = \frac{1}{2\rho_{oo}}p_c\varepsilon_c\left(\frac{p}{p_c}\right)^{\frac{\alpha+1}{\alpha}}. \tag{8.29}$$

With the structured wave compaction time of $\tau \simeq \varepsilon/\dot{\varepsilon}$, the product of the energy and time yields

$$\mathcal{E}\tau \simeq \frac{p_c\varepsilon_c^2}{2\rho_{oo}}\frac{1}{\dot{\varepsilon}}\left(\frac{p}{p_c}\right)^{\frac{\alpha+2}{\alpha}}. \tag{8.30}$$

Constancy of the energy–time product requires that strain rate exhibit a power-law dependence on Hugoniot pressure according to $\dot{\varepsilon} \sim p^n$, where $n = (\alpha + 2)/\alpha$. Note that as the power $\alpha \to \infty$ the analytic model pointwise converges to the snow-plow model with fixed compaction strain ε_c.

Strain rate versus Hugoniot pressure for the shock compaction of the selected materials are shown on the right in figure 8.18. Curves base on the constancy of equation (8.31) are compared with the experimental data. Again, the value of n for each material is calculated from the power α determined from the fit to the shock compaction data as shown on the left in the figure. Shock dissipation is modestly over-predicted as the model ignores the elastic strain energy recovery on decompression.

Over the range of the data shown in figure 8.18, constancy of the energy–time product, namely the dissipative action, appears to provide a more unifying rule for steady structured shock compaction waves than does the Hugoniot pressure versus strain-rate fourth-power relation characteristic of many monolithic solids and selected full-density composites.

8.6.1.3 Steady structured waves in porous metal

Porosity and the structured wave introduces several interesting questions. The assessment of structured wave data for porous granular solids clearly suggests that constancy of the dissipative action, rather than the fourth-power trend observed for full-density solids, and particularly for the common metals, is the more appropriate underlying premise. It is worth revisiting the analysis leading to equation (8.6) and the dependence of the steady structured wave strain rate on the

Hugoniot pressure. This is done in spite of the lack of data to support or refute any observations garnered from this effort. A linear U–u Hugoniot is again assumed. This time one further term in the Hugoniot expansion is retained, yielding for the principal Hugoniot expression

$$p_{ho}(\varepsilon) = \rho_o C_o^2(\varepsilon + 2S\varepsilon^2 + 3S^2\varepsilon^3), \tag{8.31}$$

with the lattice compression volume strain $\varepsilon = 1 - \rho_o/\rho$.

Specify an initial porosity of the solid through a distention $\alpha = \rho_o/\rho_{oo}$. Hugoniot compression of the distended solid is then

$$p_h(\varepsilon, \alpha) = p_{ho}(\varepsilon)\frac{1 - \frac{1}{2}\Gamma_o\varepsilon}{1 - \frac{1}{2}\Gamma_o(\varepsilon + \alpha - 1)}, \tag{8.32}$$

where a Gruneisen EOS is applied with $\rho\Gamma = \rho_o\Gamma_o$. The corresponding Hugoniot energy is

$$E_h(\varepsilon, \alpha) = \frac{1}{2}\frac{p_h(\varepsilon, \alpha)}{\rho_o}(\varepsilon + \alpha - 1). \tag{8.33}$$

The principal isentrope (referenced to the initial solid density ρ_o) to the same order of expansion is

$$p_s(\varepsilon) = \rho_o C_o^2\left(\varepsilon + 2S\varepsilon^2 + 3S^2\left(1 - \frac{\Gamma_o}{9S}\right)\varepsilon^3\right), \tag{8.34}$$

with energy

$$E_s(\varepsilon) = \frac{1}{2}C_o^2\left(\varepsilon^2 + \frac{4}{3}S\varepsilon^3 + \frac{3}{2}S^2\left(1 - \frac{\Gamma_o}{9S}\right)\varepsilon^4\right). \tag{8.35}$$

For small levels of porosity the principle isentrope from equation (8.34), modestly translated, and the isentrope energy from equation (8.35), are adequate approximations for the decompression isentrope from states on the distended Hugoniot.

Equations (8.31)–(8.35) are used to calculated dissipation at the Hugoniot state at selected levels of initial distention. Predictions of Hugoniot pressure versus steady wave strain rate are again based on the premise of constancy of the dissipative action $\mathcal{A} = \mathcal{E}\tau$. Calculations are applied to aluminum with parameters $\rho_o = 2.70$ g cm^{-3}, $C_o = 5.39$ km s^{-1}, $S = 1.34$ and $\Gamma_o = 1.97$ (Steinberg 1996). The parameter $\mathcal{A}$ is optimized for the full-density aluminum to the data of Barker (1968) (see also figure 8.1). The parameter $\mathcal{A}$ is maintain constant for subsequent calculations for selected initial porous levels of distention. Plots of the shock pressure versus steady wave strain rate resulting from this calculation are shown in figure 8.19.

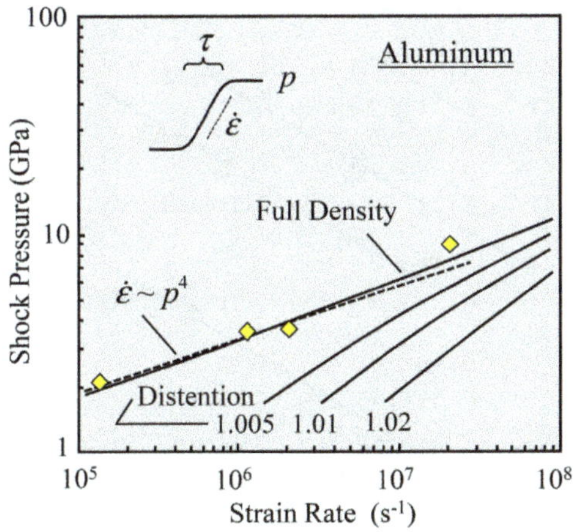

Figure 8.19. Shock pressure versus steady wave strain rate for aluminum metal based on the premise of constancy of the dissipative action. Solid curves are for full-density aluminum and selected porous distention. The dashed curve is fourth-power. The data are those of Barker (1968).

Several observations are notable. With the inclusion of an added term in the description of the Hugoniot, including a more accurate representation of the decompression isentrope, the predicted shock pressure versus strain rate for the full-density aluminum is no longer fourth-power. Although close, it increasingly diverges at higher pressures.

Second, similar curves for steady wave shock compression of modestly porous aluminum are calculated based on an assumption of constancy of the same value of dissipative action determined for the full-density material. Constancy of the action for a specified porosity is partially justified by the experiments on porous granular media. There is no justification for the assumption of constancy over the span of different porosities. Implicit also is the assumption that porosity is microscopic and homogenous on length scales that are small compared to the width of the steady wave.

The same analysis was performed using a linear-*Z* EOS (see chapter 10) description of the aluminum shock response. Differences between the two calculations were negligible.

8.6.2 Structured shock waves in composite solids

Composite solids of disparate compliance components provide a further opportunity for examining the character of the structured shock wave and dependence of the structure on the amplitude of the shock.

8.6.2.1 Particulate composites

A marked change in the functional dependence of the structured steady shock wave is observed when the void space of a granular material is filled with epoxy creating, rather than a porous solid, a several-component composite. Such shock wave data for aluminum oxide and tungsten carbide ceramic are provided in figure 8.20. With

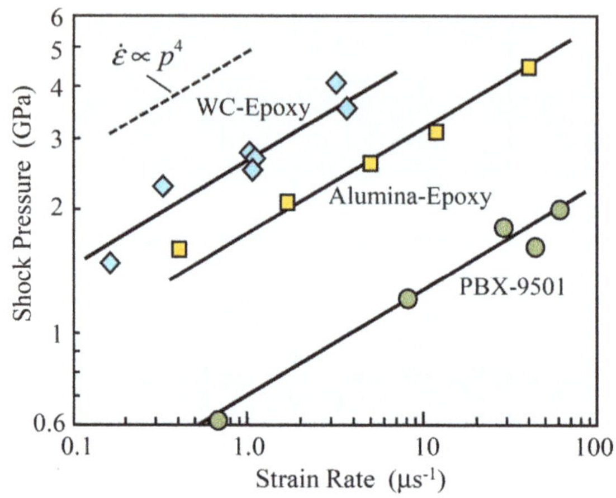

Figure 8.20. Experimental dependence between shock pressure and the strain rate of steady structured waves for composites of aluminum oxide and epoxy (Setchell and Anderson 2005), tungsten carbide and epoxy (Vogler 2009) and PBX-9501 (HMX and binder) (Dick 1998).

the addition of epoxy, the porous solid becomes a two-component composite. When subjected to steady wave shock compression these two composites are found to exhibit a near-fourth-power relation between steady wave strain rate and shock pressure; the aluminum oxide and epoxy strikingly so. More scatter is seen in the tungsten carbide and epoxy, but the trend is clearly similar.

Such dramatic differences between the shock response of the porous granular ceramic and the same ceramic imbedded in epoxy is unexpected. The response of the porous granular solid is that of compaction crushing accommodated by intense granular fracture. Shock response of the ceramic–epoxy composite is, on reflection, expected to be similar. Epoxy is markedly more compliant than the ceramic and similar crush and granular fracture is expected to accommodate the substantial difference in volume compressibility of the two components.

There is, however, an important difference between the shock compaction experiments on the porous granular media represented by the plots in figures 8.16–8.18, and the shock experiments conducted on the ceramic–epoxy composites shown in figure 8.20. The volume fraction of the ceramic in the porous medium is near the value achieved by freely pouring granular ceramic into a container (55%–60%). The volume fraction of the ceramic in the composite is notably less. Setchell and Anderson (2005) report a volume fraction of 43% for the aluminum oxide and epoxy composite tested. They also observed, from micrographs of the samples, near-isolation of ceramic granules within the epoxy matrix. A similar mixture ratio was used in the tungsten carbide and epoxy composite. A degree of diminution of the ceramic component fraction leads to markedly different mechanisms of the shock compression process on the microscale. This difference should be considered in attempting to understand the marked difference in the steady wave power-law behavior observed between the porous and the composite materials.

As noted earlier, the shock compaction of granular sugar reported by Sheffield *et al* (1997) exhibits near-first-power dependence of the shock pressure on strain rate. Similar shock compaction behavior was reported for granular HMX by the same authors. The structured wave behavior of PBX-9501, a composite of HMX and a binder, has been reported by Dick *et al* (1998) and is also shown in figure 8.20. The near-fourth-power character of this composite, comparable to the two ceramic–epoxy composites, is worth noting.

More recent structured wave measurements on a 60%–40% by volume ice–quartz composite (Kraus *et al* 2010) do not, however, exhibit the same trend. Although only two wave profiles are reported, a power closer to 1.85 between shock pressure and strain rate is observed. Whether the two profiles exhibit steady wave behavior is not known.

8.6.2.2 Laminate composites

A study of structured wave propagation in laminated composites is reported in the work of Zhuang *et al* (2003). In this investigation, laminate samples were constructed by bonding alternate plates of polycarbonate with plates of several metals (stainless steel or aluminum) or with plates of glass. These two-component composite samples were subjected to shock compression loading normal to the laminate plane. The authors report plots of shock pressure versus steady wave strain rate. A nominal power of two was collectively observed for all of the data, although a direct fit to individual data sets yields powers ranging over about 1.8–2.4. The data are shown in figure 8.21 with each set delineated with a line of power-law slope of two.

The upper-most curve identifies data for polycarbonate and stainless steel. Successive curves are for polycarbonate and glass, polycarbonate and aluminum and, again, polycarbonate and glass. The two curves for glass differ in the dimension of the unit cell. For the upper curve 0.74 mm polycarbonate and 0.55 mm glass

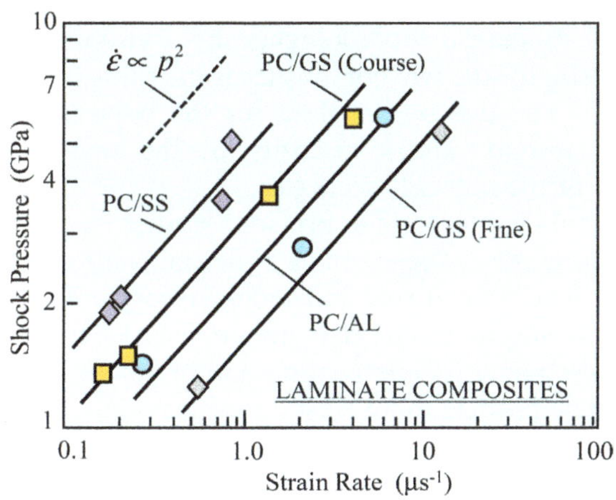

Figure 8.21. Experimental dependence between the shock pressure and the strain rate of steady structured waves for laminated composites. The curves from top to bottom are alternating plates of 0.74 mm polycarbonate and 0.37 mm stainless steel, 0.74 mm polycarbonate and 0.55 mm glass, 0.74 mm polycarbonate and 0.37 mm aluminum, and 0.37 mm polycarbonate and 0.20 mm glass. Data from Zhuang *et al* (2003).

constitute the unit cell. For the lower curve the unit cell is 0.37 mm polycarbonate and 0.2 mm glass.

8.6.2.3 Trends in structured shocks in composites

The three composites for which data are presented in figure 8.20 exhibit a nominal fourth-power relation between shock pressure and steady wave strain rate. The comparable plots for the laminar composites in figure 8.21 show behavior closer to second-power. Despite this difference in slope, several insightful observations are revealed by the two sets of data.

Shock viscosity for all of the materials displayed in figures 8.20 and 8.21 are dependent on the shock amplitude, although that for layered media approaches amplitude independence. Viscosity generally decreases as shock pressure increases. Viscosity from material to material decreases from curves at the upper left to curves at the lower right of the plots. For example, in figure 8.20 the tungsten–carbide composite exhibits higher viscosity while the PBX-9501 exhibits the lowest viscosity. Similar trends are noted in figure 8.21 for the laminate composites.

Two observations emerge from the shock viscosity data. The first relates to the difference in shock impedance of the components of the composites. The second relates to the microstructure unit cell size. The unit cell size for the granular composites shown in figure 8.20 is determined by the granular medium particle size. Although there are some differences, for the present purposes cell sizes are nominally the same for the three composites. Shock impedance differences for the components of the three composites differ markedly, however. The shock impedance difference between tungsten carbide and epoxy is substantial. That between aluminum oxide and epoxy is markedly less. That between HMX crystallites and the estane binder is even less. It is observed that steady wave shock viscosity for the granular composites decreases as the difference in component shock impedance decreases. A similar observation is noted in figure 8.21 for the layered composites, where the polycarbonate–stainless steel laminate exhibits a higher shock viscosity than polycarbonate–aluminum. Unit cell size for the two composites is the same (0.74 mm polycarbonate and 0.37 mm metal). The two sets of data for the polycarbonate–glass laminate provided additional insight. Shock viscosity of the two-component composite decreases as the size of the unit cell decreases.

Earlies in the present chapter, and in previous studies (Grady 1998b, Grady *et al* 2000), stress wave scattering is suggested as a mechanism for momentum diffusion, and the concomitant wave structuring and shock viscosity of solids. The shock data for porous and composite materials that are reviewed here lend support to that underlying cause. Rayleigh (1929) derived a relation for the elastic scattering of the wave energy of a single particle in an elastic medium,

$$\frac{A_s(r,\theta)}{A_I} \cong \frac{\pi a^3}{r\lambda^2}\left(\frac{\Delta E}{E} + \cos\theta\frac{\Delta\rho}{\rho}\right), \tag{8.36}$$

in the long wavelength limit of $\lambda \gg a$. Wavelength of the plane wave incident on the particle is λ while the size of the particle is a. A_s/A_I is the relative amplitude of the scattered energy at the position (r, θ) from the scattering body. $\Delta E/E$ is a measure of the difference of elasticity between particle and medium, while $\Delta\rho/\rho$ is the corresponding difference in density. The dependence of scattering energy in the Rayleigh relation on both the size scale of the particle and on the difference in elasticity and density between the particle and medium is consistent with the observed trends for the selected composite materials discussed here.

Perhaps more relevant to the processes governing structured shock waves in heterogeneous solids is the study of Mason and McSkimin (1947). They applied the Rayleigh relation to the scattering of acoustic waves by a single grain within a polycrystalline material. The projected impedance normal to the plane wave will statistically differ from the polycrystalline average. By integrating the scattered energy from equation (8.36) over a sphere, and accounting for the average contribution from all grains, they derive

$$\alpha \cong \frac{4\pi^2 a^3}{\lambda^4}\left\langle\left(\frac{\Delta E}{E}\right)^2\right\rangle_{\text{ave}}, \tag{8.37}$$

for the acoustic attenuation factor due to elastic scattering. The expression exhibits the well-known Rayleigh scattering fourth-power dependence on wavelength.

Although the large-amplitude structured shock waves of present concern are nonlinear, and are not elastic, extended concepts of wave scattering probably play an important role in establishing the wave structure and shock viscosity observed in heterogeneous solids. Lastly, it is of interest to note that the three-dimensional granular composite behavior shown in figure 8.20 exhibits fourth-power dependence of steady wave strain rate on shock pressure amplitude, whereas the one-dimensional laminate composites of figure 8.21 show nominal second-power dependence. It would be enlightening to perform steady shock experiments on a two-dimensional composite. A composite composed of random parallel rods in an epoxy matrix with shock propagation normal to the rod length is a possible configuration.

8.7 Physics of structured shock waves

The fourth-power law for steady structured shock waves, and the potentially more general invariant action in the shock compression event, are now long-standing observations that have not as yet received a solid physical explanation for the underlying causes. The range of materials and shock wave conditions to which these observations apply has not yet been fully established. Nonetheless, implementation into constitutive models where the formation and evolution of structured shock waves is essential to the physics, has been found to be a useful expedient of the fourth-power law and concomitant viscosity models.

The purpose of the present chapter is to place some historical perspective on the fourth-power law and to identify physics that may play a role in causing the fourth-power trend. The original experimental observations were focused on the apparent constancy of the product of an energy and a temporal duration of a structured shock wave. This potential shock invariant was observed to have dimensions of specific action in the context of classical physics. This is a connection that leads to interesting pursuits, some of which are undertaken here, that may or may not prove productive. The application of the statistical methods of Langevin dynamics to the failure mechanics underlying shock structuring is offered as one possibility.

Invariance of the energy–time product has application to unsteady structured waves and, when restricted to steady shock waves, is equivalent to the fourth-power law. Further, when cast in this form it has an apparent relationship to other energy criteria in the field of shock and impact such as that of explosive detonation (e.g. Held 1968, Walker and Wasley 1969) and dynamic fracture (e.g. Tuler and Butcher 1968).

The fourth-power behavior appears not to be unique to any specific class of materials. Although most of the steady wave risetime measurements have been on metals, limited data on brittle materials also show the fourth-power trend. Agreement with the fourth-power shock behavior of the unreacting PBX-9501, as well as an epoxy–alumina and epoxy–tungsten carbide mixtures, is particularly interesting. It should be noted, however, that the available data demonstrating fourth-power behavior are not exhaustive. Further, measurements of profile structure where fourth-power behavior is observed are near the upper limits of time resolution of the instrumentation applied. The later measurements of structured waves in aluminum by Crowhurst *et al* (2011) are particularly noteworthy in this regard. Their measurements of structured steady waves with picosecond time resolution supports a fourth-power trend to markedly higher strain rates.

Experimental observations show, however, that porous solids, and composite solids in certain configurations, do not exhibit fourth-power behavior in steady structured shock waves. That invariance of the dissipative action for structured shock compaction waves in selected compound powders predicts the observed near-first-power dependence on strain rate is profound and has not been explained.

The fourth-power law for steady wave risetimes is intimately coupled with the solid viscosity mechanisms responsible for structuring of the wave. Again, the underlying physics has not been reconciled. The dynamics of dislocation plasticity has been the favored explanation, and the governing constitutive equations with appropriate functional forms have been shown to adequately describe the experimental data. Dislocation plasticity may in fact be the responsible mechanism for certain classes of materials. Alternatively, the behavior may result from more fundamental laws of physics, such as adiabatic invariance of the dissipation field as discussed in the present chapter. As a further wave structuring mechanism, mesoscale scattering and acoustic phonon viscosity may play a role. Turbulent mass motion in the shock front may also contribute to wave structuring and is a mechanism that has received little attention.

Some systematic trends are noted for structured waves in composite solids relating to the characteristic microstructure length scales and component impedance disparities that suggest microstructure scattering may be a first-order mechanism in the structuring of shock waves in this class of materials. There is a need for more detailed structured wave experiments that are capable of discriminating among the physics models and explanations that have been pursued.

References

Alekseevskii V P *et al* 1989 Mass transfer in metals caused by a convergent cylindrical shock wave *Sov. Powder Metall. Metal Ceram.* **10** 809–13

Anderson M U, Graham R A and Holman G T 1994 Time resolved shock compression of porous rutile: wave dispersion in porous solids *High Pressure Science and Technology* (New York: AIP) pp 1111–4

Archambeau C B, Minster J B, Achenbach J D, Pao Y-H and Tiersten H F 1976 *Theory of Failure as a Transition Process in a Stressed Medium* (Washington, DC: National Science Foundation)

Armstrong R W, Arnold W and Zerilli F J 2007 *Metall. Mater. Trans.* A **38** 2605

Asay J R and Chhabildas L C 1981 Determination of the shear strength of shock compressed 6061-T6 aluminum *Shock Waves and High-Strain-Rate Phenomena in Metals* ed M A Meyers and L E Murr (New York: Plenum) pp 417–32

Band W 1960 Studies in the theory of shock propagation in solids *J. Geophys. Res.* **65** 695

Band W and Duvall G E 1961 Physical nature of shock propagation *Am. J. Phys.* **29** 780

Barrenblatt G I 2006 *Int. J. Fracture* **138** 19–35

Barker L M and Hollenbach R E 1965 Interferometer techniques for measuring the dynamic mechanical properties of materials *Rev. Sci. Instrum.* **36** 1965

Barker L M 1968 Fine structure of compressive and release wave shapes in aluminum measured by the velocity interferometer technique *Behavior of Dense Media under High Dynamic Pressures* (London: Gordon and Breach) 483 p

Barker L M 1971 A model for stress wave propagation in composite materials *J. Composite Mater.* **5** 140

Bland D R 1965 On shock structure in solids *J. Inst. Maths Appl* **1** 56

Brown J L, Vogler T J, Grady D E, Reinhart W D, Chhabildas L C and Thornhill T F 2007 Dynamic compaction of sand *AIP Conf. Proc.* **955** 1363–6

Callen H B and Welton T A 1951 Irreversibility and generalized noise *Phys. Rev.* **83** 34–40

Chhabildas L C and Asay J R 1979 *J. Appl. Phys.* **50** 2749

Chhabildas L C and Hills C R 1986 Dynamic shock studies of vanadium *Metallurgical Applications of Shock-Wave and High-Strain-Rate Phenomena* ed L E Murr, K P Staudhammer and M A Meyers MA (New York: Marcel Dekker) 763 p

Cottrell A H and Bilby B A 1949 Dislocation theory of yielding and strain aging of iron *Proc. Phys. Soc.,* A **62** 49

Crowhurst J C, Armstrong M R, Knight K B, Zaug J M and Behymer E M 2011 Invariance of the dissipative action at ultrahigh strain rates above the strong shock threshold *Phys. Rev. Letts* **107** 144302

Dick J J, Martinez A R and Hixson R S 1998 Plate impact response of PBX 9501 and its components below two GPa *Los Alamos National Laboratory Report* LA-13426-MS, April

Dremin A N and Breusov O N 1968 Processes occurring in solids under the action of strong shock waves *Russ. Chem. Rev.* **37** 396–402

Dremin A N and Breusov O N 1971 The chemistry of shock compression, *Priroda, 12, 10 (translation Sandia National Laboratories Report* SAND80-6003, 1980).

Dugdale D S 1960 Yielding of steel sheets containing slits *J. Mech. Phys. Solids* **8** 100

Dunn J E and Grady D E 1986 Strain rate dependence in steady plastic shock waves *Shock Waves in Condensed Matter* ed Y M Gupta (Plenum) 359 p

Einstein A 1905 Investigations on the theory of the Brownian movement *Ann. Phys. Leipzig* **322** 549

Falkovich G and Sreenivasan K R 2006 *Physics Today* **59** 43–9

Feymann R P, Leighton R B and Sands M 1964 *The Feymann Lectures on Physics*vol 2 (London: Addison-Wesley) 19.8

Gilman J J 1979 Resistance to shock front propagation in solids *J. Appl. Phys.* **41** 1664

Grady D E 1977 Processes occurring in shock wave compression of rocks and minerals *High Pressure Research Applications in Geophysics* (New York: Academic) pp 389–438

Grady D E 1980 Shock deformation in brittle solids *J. Geophys. Res.* **85** 913

Grady D E 1981 Strain-rate dependence of effective viscosity under steady-wave shock compression *Appl. Phys. Lett.* **38** 825

Grady D E 1986 Steady-wave rise-time and spall measurements on uranium (3–15 GPa) *Metallurgical Applications of Shock-Wave and High-Strain-Rate Phenomena* ed L E Murr, K P Staudhammer and M A Meyers (New York: Marcel Dekker) 763 p

Grady D E 1992 Properties of an adiabatic shear-band process zone *J. Mech. Phys. Solids* **40** 1197–215

Grady D E 1998a Shock wave compression of brittle solids *Mech. Mater.* **46** 181–203

Grady D E 1998b Scattering as a mechanism for structured shock waves in metals *J. Mech. Phys. Solids* **46** 2017

Grady D E 2010 Structured shock waves and the fourth-power law *J. Appl. Phys.* **107** 013506

Grady D E 2015 Unifying role of dissipative action in the dynamic failure of solids *J. Appl. Phys.* **117** 165905

Grady D E, Murri W J and DeCarli P S 1975 Hugoniot sound velocities in two silicates *J. Geophys. Res.* **80** 4857

Grady D E and Murri W J 1976 Dynamic unloading in shock compressed feldspar *Geophys. Res. Lett.* **3** 472

Grady D E and Lipkin J 1980 Criteria for impulsive rock fracture *Geophys. Res. Lett.* **7** 255–8

Grady D E and Asay J R 1982 Calculation of thermal trapping in shock deformation of aluminum *J. Appl. Phys.* **54** 7350–4

Grady D E and Kipp M E 1987 The growth of unstable thermoplastic shear with applications to steady-wave shock compression in solids *J. Mech. Phys. Solids* **35** 95–118

Grady D E, Ravichandran G and Zhuang S 2000 Continuum and subscale modeling of heterogeneous media in the dynamic environment *Proc. Int. Conf. on High Pressure Science and Technology AIRAPT-17, (Honolulu, HI, 25–29 July 1999)* ed M Manghnani, W Nellis and M NicolUniversity Press) pp 189–93

Graham R A 1981 Active measurements of defect processes in shock-compressed metals and other solids *Shock Waves and High-Strain Rate Phenomena in Metals* ed M A Meyers and L E Murr (New York: Plenum) pp 375–86

Griffith AA 1924 Theory of Rupture. *Proc. First International Congress on Applied Mechanics.* Delft, pp 55–63

Hayes D B and Grady D E 1982 A thermal-viscous model for heterogeneous yielding in aluminum *AIP Conf. Proc.* **78** 412

Hayes D B and Graham R A 1978 Yielding of shock-loaded x-cut quartz *Bull. Amer. Phys. Soc.* **23** 71

Held M 1968 Initiation of explosives: a multiple problem of the physics of detonation *Explosivstoffe* **5** 98–113

Jackson R K *et al* 1976 Initiation and Detonation Characteristics of TATB *Proc. 6th Symp. (Int.) on Detonation* pp 755–65

Johnson J N 1992 Calculation of path dependent shock wave hardening *J. Appl. Phys.* **72** 797

Johnson J N and Barker L M 1969 *J. Appl. Phys.* **40** 4321

Kanel G I, Ivanov M F and Parshikov A N 1995 Computer simulations of heterogeneous material response to impact loading *Int. J. Impact Eng.* **17** 455

Knopoff L 1965 Attenuation of elastic waves in the Earth *Physical Acoustics* vol 3 ed W P Mason (New York: Academic) 287 p

Kraus R G, Stewart S T, Seifter A and Obst A W 2010 Shock and post-shock temperatures in an ice–quartz mixture: implications for melting during planetary impact events *Earth Planetary Sci. Letts.* **289** 162–70

Kubo R 1957 General theory and simple applications to magnetic and conduction problems *J. Phys. Soc. Japan* **12** 570

Kubo R 1959 *Lectures in Theoretical Physics* vol 1 ed W Brittin (New York: Inter Science)

Maasilta I and Minnich A J 2014 Heat under the microscope *Phys. Today* **67** 27

Malygin G A, Ogarkov S L and Andriyash A V 2013 Two-wave structure of plastic relaxation waves in crystals under intense shock loading *(RU) Phys. Sol. State* **55** 780–6

Mason W P and McSkimin H J 1947 Attenuation and scattering of high frequency sound waves in metals and glasses *J. Acous. Soc. Am.* **19** 464–73

Mescheryakov Y I and Divakov A K 1994 Multiscale kinetics of microstrucure and strain rate dependence of materials *DYMAT J.* **1** 271–87

Mott NF 1948 Brittle Fracture in Mild Steel Plates, *Engineering*, 165, 16.

Nellis W J, Kanel G I, Razorenov S V, Savinykh A S and Rajendran A M 2009 Entropy-dominated dissipation in sapphire shock-compressed up to 400 GPa (4 Mbar), arXiv: 09054730

Nyquist H 1928 Thermal agitation of electric charge in conductors *Phys. Rev.* **32** 110–3

Partom Y 1990 Understanding the Swegle–Grady fourth power relation *Shock Waves in Condensed Matter* ed S C Schmidt, J N Johnson and L W Davison (Washington, DC: AIP) pp 317–20

Rayleigh Lord 1910 *Proc. R. Soc. Lond.* **84** 247

Rayleigh Lord 1929 *Theory of Sound* (New York: MacMillan)

Reif F 1965 *Fundamentals of Statistical and Thermal Physics* (New York: McGraw Hill)

Richtmyer R D and Morton K W 1967 *Difference Methods for Initial-value Problems* (Interscience)

Rubin M B 1990 Analysis of weak shocks in 6061-T6 aluminum *Shock Compression of Condensed Matter* ed S C Schmidt, J N Johnson and L W Davison (Elsevier Science) pp 621–4

Ruoff A L 1980 Estimates of uncertainty in isotherms deduced from Hugoniots resulting from shock wave generated defects *J. Appl. Phys.* **51** 6221

Setchell R E and Anderson M U 2005 Shock compression response of an alumina filled epoxy *J. Appl. Phys.* 083518

Sheffield S A, Gustavsen R L and Anderson M U 1997 Shock loading of porous high explosives *High-Pressure Shock Compression of Solids* vol 4 ed L Davison, Y Horie and M Shahinpoor (Berlin: Springer) pp 23–61

Silling S A 1998 Reformulation of elasticity theory for discontinuities and long-range forces *Sandia National Laboratories Report* SAND98-2176, October.

Silling S A, Demmie P N, Cole R A and Taylor P 2006 EMU user's manual *Sandia National Laboratories Report*

Smoluchowski M 1906 Zur kinetischen Theorie der Brownschen Molekularbewegung und der Suspensionen *Ann. Phys.* **326** 756–80

Steinberg D J 1996 equation of state and strength properties of selected materials *Lawrence Livermore National Laboratory Report* UCRL-MA-106439, February

Swegle J W 1991 Non-steady wave profiles and the fourth-power law *Shock Compression of Condensed Matter* ed S C Schmidt, R D Dick, J W Forbes and D G Tasker (New York: Elsevier Science) 249 p

Swegle J W and Grady D E 1985 Shock viscosity and the prediction of shock wave rise times *J. Appl. Phys.* **58** 692

Swegle J W and Grady D E 1986 Shock viscosity and the calculations of steady shock wave profiles *Shock Waves in Condensed Matter* ed Y M Gupta (New York: Plenum) 353 p

Taylor G I 1910 *Proc. R. Soc. Lond.* **84** 371

Ter Haar D 1964 *Elements of Hamiltonian Mechanics* (Amsterdam: North Holland)

Tuler F R and Butcher B M 1968 A criterion for the time dependence of dynamic fracture *Int. J. Fracture Mech.* **4** 431–7

Vogler T J 2009 *Society Engineering Mechanics Annual Conf. (Albuquerque, NM,* 1–4 *June* 2009) (presentation only)

Vogler T J, Lee M Y and Grady D E 2007 Static and dynamic compaction of ceramic powders *Int. J. Solids Struct.* **44** 636–58

Vogler T J, Borg J P and Grady D E 2012 On the scaling of steady structured waves in heterogeneous materials *J. Appl. Phys.* **112** 123507

Vogler T J 2015 Personal communication

Walker F E and Wasley R J 1969 Critical energy for shock initiation of heterogeneous explosives *Explosivstoffe* **17** 9

Zhuang S, Ravichandran G and Grady D E 2003 An experimental investigation of shock wave propagation in periodically layered composites *J. Mech. Phys. Solids* 245–65

IOP Publishing

Physics of Shock and Impact, Volume 2
Materials and shock response
Dennis Grady

Chapter 9

Kinetics of shocks in brittle solids

Brittle solids including ceramics, competent rock, silicate and metallic glasses, and selected single-crystal materials with inadequate plastic slip systems, store significant elastic strain energy during the shock compression process. The extent of this transient stored potential energy and the delayed release of this energy render the shock response of brittle solids unique. This dynamic plays a central role in the shock wave critical state phenomena, including transient failure, phase transformation and spall. The important role of strain energy in shaping the unique fragmentation characteristics of brittle solids is addressed in earlier chapters. The structuring of shock waves is also dependent on this character of brittle materials. In the present chapter, emphasis is on the nature of the shock compression processes in brittle solids. A number of experimental observations are described. The underlying physical mechanisms that govern the unique shock wave response of brittle solids are explored.

9.1 Introduction

The present chapter complements chapter 6, which deals with many of the shock properties of ceramics and ceramic mixtures. Additional structured shock data for ceramics are studied here. These shock data are supplemented with comparable data on hard geological materials (rocks and minerals). Emphasis is on the underlying shock physics governing structuring of the shock wave, as well as the time-dependent and strain-rate-dependent features of strength often observed in the shock and high-rate response of the hard and brittle solids.

Phase transformation, dynamic strength and solid viscosity are collectively examined in the shock structuring process. Discussions of the latter complement the more extensive discourse of chapter 8 on the viscous character of solids under shock compression. In particular, features of the silicon dioxide shock-induced quartz-to-stishovite phase transformation are explored, including a not widely recognized critical point on the shock Hugoniot. Comparisons are made with aluminum nitride ceramic, a material exhibiting phase transformation with remarkable ease under shock compression conditions. A marked contrast is provided by the carbonate mineral calcite that displays high-pressure shock phase transformation only through a distinct rarefaction shock. Wave features of calcite provided an earlier wave structuring example in chapter 1. Phase transformation under shock in calcite in turn contrasts sharply with phase transformation in the companion carbonate mineral dolomite. The disconcerting disparities in shock wave and diamond anvil cell measurements of phase transition in this mineral are visited further in chapter 11.

The elastic wave precursor, and its relationship to the dynamic strength of solids, is fundamental to shock physics. Features of the elastic precursor are examined as they relate to the shock compression of brittle solids. Observations are reported regarding elastic precursor decay in these materials. Comparisons are made between static indentation hardness and the Hugoniot elastic limit (HEL) as assessed from elastic precursor measurements. An unusual shock phenomenon, identified as super-deep penetration, is described and viable physical explanations are discussed. Novel divergent wave shock experiments are described that assess the dynamic failure of brittle solids subject to stress paths other than Hugoniot state failure.

Velocity interferometry provides detailed measurements of time-resolved shock wave structure. An extensive study of structured waves in a brittle solid using this test method has been performed on the carbonate rock dolomite. Both amplitude and propagation-distance dependence are examined within the range of pronounced shock wave structuring. These data and the supporting experimental analysis are described in some detail.

Time and strain-rate dependence of the elastic moduli and strength of brittle solids are ascribed to the micromechanics of flaws and defects. A number of workers have developed models describing dynamic elasticity, strength and failure based on the activation and growth of internal microcracks. One such effort is described towards the end of chapter 6. In the later sections of this chapter, features of microcrack models are explored as they relate to the shock and high-rate response of brittle materials, with emphasis on rocks and ceramics. The shock forming character of crack microstructure dynamics is applied to structured shock waves in brittle solids.

9.2 Shock compression of brittle solids

Solids classified as brittle are unusually rich in physical effects promoted by intense shock compression. These physical processes are in turn manifest in the structure of the affecting shock wave. The measurement of structured shock waves in brittle solids and the underlying physical causes of shock wave structure are pursued in this section.

9.2.1 Nature of the shock wave

Underlying every shock wave event is a shock wave profile that reflects the temporal history of the mechanical and thermal state at a material point. In most cases, there is not the wherewithal to measure that wave profile. It nevertheless does exist and is an important issue in the physics of the event. Over the past four or five decades, a substantial effort in the shock wave community has focused on the development of diagnostic methods for measuring wave profiles in solids. Numerous physical effects have been exploited in providing a visualization of the wave profile. Piezo-resistivity, electro-induction and optical interference are but a few of the more successful techniques that continue to be in active use today.

The discipline of measuring such waves has evolved over the past few decades into a reasonably mature technology with capabilities of measuring waves of stress and particle velocity using several techniques with quite high resolution. This technology has been driven by the expectation that such measurements would provide crucial information concerning the transient or high-strain-rate equation of state (EOS) and strength physics of condensed matter.

This expectation has only been realized in part, and it is now recognized that wave profile data alone are not usually sufficient to establish the wide range of material response behaviors sought. Such data are only part of a wider range of data from other test methods that are necessary to achieve the desired understanding of the material response. This should not diminish in any way the significance of the contributions of the many technical and scientific workers in achieving today's capabilities in the measurement of shock wave profiles.

There is, perhaps, a higher reason for making such measurements. Wave profiles, like the setting of the Sun, or the splash of a raindrop, are an interesting phenomenon in the natural world. They are rich in features that stimulate the curiosity toward the underlying responsible physics. Thus, wave profiles in condensed matter deserve, in and of themselves, a satisfactory scientific explanation. It is for these reasons that the shock wave profile has come to personify the field of shock wave physics.

In this spirit, portions of the present chapter will focus on shock wave profiles in ceramic-like materials or, more generally, wave profiles in that class of materials that we tend to think of as hard and brittle (brittle solids). The experimental methods pursued in this investigation have been those of measuring the form and evolution of large-amplitude deformation waves produced by carefully controlled impact experiments. The principal diagnostic is largely that of velocity interferometry used to measure a time history of velocity at an interface between the solid material of interest and a transparent window material. Other wave measurement methods, such as in-material piezo-resistive stress measurements and electro-inductive velocity measurements, are also included and are found in other chapters of the text.

Wave profile measurements on brittle solids have revealed an unusually rich display of nonlinear features that hint at the complex interplay of physical processes active during passage of the transient shock wave. An intermix of strength, viscosity, failure and phase transformation processes in solid matter have been put forth to explain such features. The purpose of the present section is to discuss some of the nonlinear features observed in the wave profiles of brittle solids and the underlying causative physics.

It is the compressive failure and deformation properties of brittle solids, under the unequal principle stresses brought about by the uniaxial nature of the shock process, which largely contribute to the richness in the compressive wave profile characteristics of these materials. Lattice volume compressibility for the materials of interest is usually a decreasing function of pressure in the shock stress range of concern, and would itself lead to the propagation of a single steady shock wave. The exception occurs in those materials in which phase changes brought about by lattice instabilities at critical pressure levels complicate the volumetric compressibility.

9.2.2 Experimental shock wave profiles

One test technology for probing the response of solids under uniaxial strain compression loading is illustrated in figure 9.1 (Grady 1992a, 1992b, 1995a). In the test shown, a disc of the test material such as a ceramic is mounted in the projectile and supported on the main projectile body by a disc of low-density polyurethane foam. For the target, a similar disc of the material is mounted in the stationary supporting target fixture. An optical quality disc of single-crystal lithium fluoride is intimately bonded with epoxy to the back of this ceramic sample. All critical surfaces are lapped and polished, and are typically flat to within a few bands of sodium light. The bonded lithium fluoride surface is first lightly diffused and vapor-deposited with about 100 nm of aluminum. The ceramic and lithium fluoride epoxy-bond thickness is approximately 10–20 μm.

The ceramic-on-ceramic planar impact produces a compressive wave of uniaxial strain, which propagates across the stationary ceramic specimen and through the ceramic–lithium fluoride interface. An equivalent compressive wave propagates

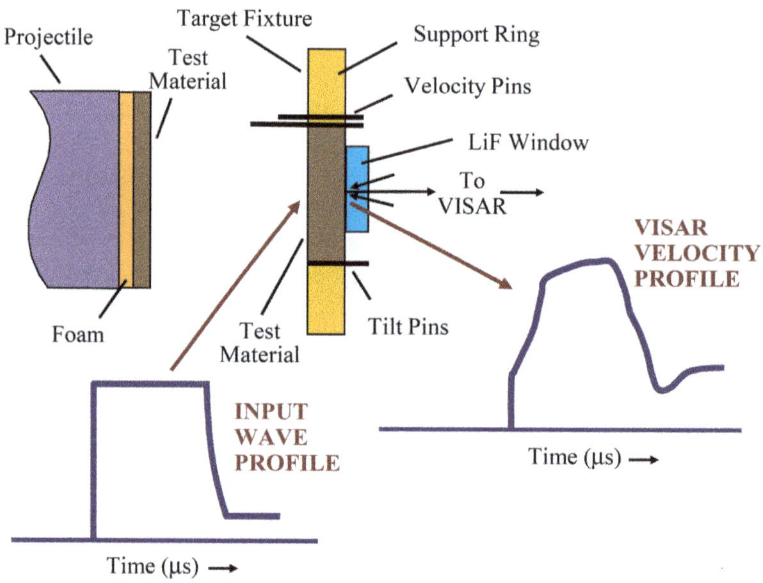

Figure 9.1. Experimental configuration for time-resolved velocity interferometry shock wave profile measurements on a test material.

through the projectile ceramic specimen. This wave reflects at the low-impedance foam interface as a release wave, which unloads the compressed ceramic. The dimensions of the ceramic discs are selected such that lateral release waves from the boundaries of the disc do not interfere with the central one-dimensional motion until after the primary experimental measurement is completed. Experimental recording is typically on the several microseconds time scale.

The compression and release wave behavior is measured by monitoring the time-resolved longitudinal motion at the center of the ceramic–lithium fluoride interface with laser velocity interferometry (VISAR) techniques (Barker and Hollenbach 1972). Measurements are recorded on transient digitizers. Lithium fluoride is used as the laser window material because, although its mechanical impedance is somewhat lower than the ceramic being tested, it is a material that has been optically calibrated, and remains transparent when subjected to the 10–50 GPa shock stresses generated in the present experiments (Wise and Chhabildas 1986).

The interference fringes, with a time resolution of approximately one nanosecond, measured with the VISAR system, are converted to time-resolved histories of the velocity of the interface using the method of Barker and Hollenbach (1972). The amplitude resolution is approximately 2% per fringe and typically two to three fringes are achieved in the interface acceleration through the compressive shock front. From these records, the dynamic axial stress and strain characteristics of the ceramics are determined through further computational and analytic techniques. The present technique, and a range of variations of the technique, have been used to explore shock wave properties of a wide range of ceramic and rock materials.

Measured wave profiles from a series of compression and release wave experiments performed on aluminum oxide ceramic (Grady 1994c, 1995a) are shown in figure 9.2. The lowest impact velocity achieves a Hugoniot state just in excess of the

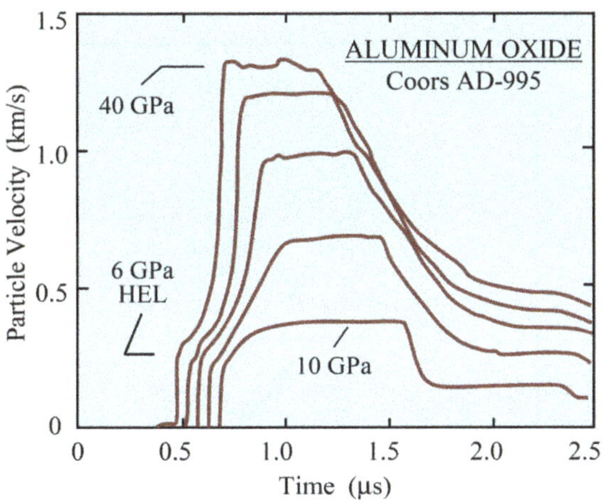

Figure 9.2. Shock compression profiles for aluminum oxide ceramic (Coors AD 995) measured with velocity interferometry (VISAR) diagnostics with the experimental configuration of figure 9.1 (Grady 1995a, 1998a). Peak pressures range from about 10 GPa for the lowest amplitude profile to just over 40 GPa for the highest. The test and impact alumina discs were approximately 10 mm and 5 mm in thickness, respectively.

HEL of the material. The remaining experiments range up to a Hugoniot state pressure of approximately 40 GPa, which is about six times the HEL of this material. This maximum pressure is achieved from a relative impact speed of the ceramic onto the same ceramic of about 2 km s^{-1}. The wave structure shown forms over a propagation distance from impact of 10 mm.

The wave profile data as shown in figure 9.2 are appropriate for determining the Hugoniot properties and some of the uniaxial strength properties of the aluminum oxide ceramic. Hugoniot properties refer to the peak-stress-versus-specific-volume states achieved in the shock compression process. The amplitude of the initial precursor corresponds to onset of dynamic shear failure (the HEL) and for the present aluminum oxide ceramic is determined to be 6.2 ± 0.4 GPa. The precursor wave velocity is 10.7 km s^{-1}.

Both the HEL and the subsequent final stress-versus-specific-volume Hugoniot states for this ceramic are shown in figure 9.3. The shock data can be compared with the static x-ray diffraction data to 12 GPa on sapphire of Sato and Akimoto (1979). The latter static data have been extrapolated to 30 GPa (approximately 15% volume strain) with a Birch–Murnaghan EOS to provide a measure of the hydrostatic compressibility of Al_2O_3 for purposes of comparisons with the shock data to comparable levels of compression. The offset between the static and shock data illustrates, and is commonly used to assess, the sustained non-hydrostatic stress state achieved in the shock compression wave. Shock temperature would account for less than 0.1 GPa of the offset. Impurities and porosity, however, can complicate a detailed evaluation of the deviator stress state under shock compression by this method. The difference in the initial specific volume of the two materials reflects initial porosity in the ceramic. Some residual porosity may remain even to the higher levels of shock pressure.

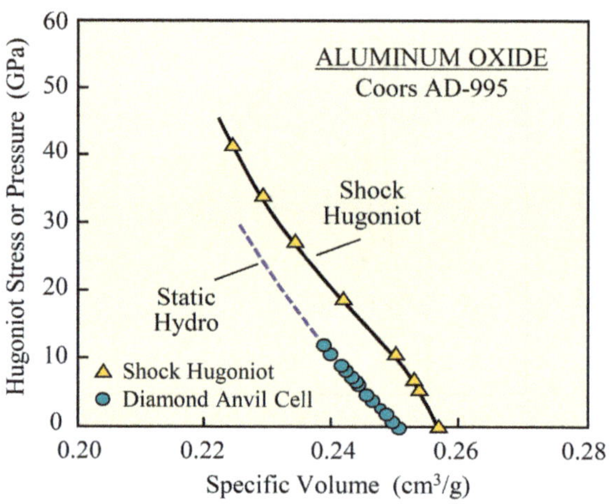

Figure 9.3. Hugoniot states for shock-compressed aluminum oxide (Grady 1995a) compared with hydrostatic x-ray diffraction assessed compression (Sato and Akimoto 1979). Note the difference in the initial starting densities of crystalline aluminum oxide versus alumina ceramic.

9.2.3 Shock- versus particle-velocity characteristics

A plot of shock velocity versus particle velocity has historically been the fundamental representation of shock wave data. The development of high-resolution time-resolved instrumentation capable of resolving the complete compression wave profile has done little to diminish use of this basic presentation of shock wave data. The seminal shock wave study of Gust and Royce (1971) provides such data for ceramics, and these are used today for the calibration of high-pressure EOSs. Shock-velocity versus particle-velocity data extracted from wave profile measurements from later work on a range of ceramics (Grady and Moody 1996) are displayed in figure 9.4. These data illustrate complexities that arise in the EOS characteristics of brittle solids.

First, some points about the plot of the data in the figure: shock velocity and particle velocity refer to the first deformation wave or shock wave following the elastic precursor wave. The data shown in the plot do not reflect third-wave characteristics in those materials that exhibit a third wave. To compare data on the same plot, the ordinate shock velocity is scaled by the ultrasonic bulk velocity $C_B = C_L - 4C_S/3$ and longitudinal velocity C_L of the material, while the particle velocity has been normalized by dividing by $C_L - C_B$. More specifically, it is a plot of $(U - C_B)/(C_L - C_B)$ as the shock-velocity ordinate and $u/(C_L - C_B)$ as the particle-velocity abscissa. A linear relation between shock velocity and particle velocity, $U_s = C_o + Su_p$ is frequently observed for many materials (Jeanloz 1989). For ceramics, $C_o = C_B$ and S = 1.0 are commonly assumed for first-order EOS characteristics and appear to sensibly describe a reasonable body of the present data shown in the figure. This curve is the dashed line in the figure. There are some interesting exceptions, however. Both TiB_2 and ZrO_2 indicate significantly higher values of S, probably hinting at the subsequent yield or transition attested to by the third wave in the compression profiles of these two materials (Grady 1992b, Grady and Mashimo 1992, Grady and Wise 1993). In contrast, B_4C and Si_3N_4 appear to

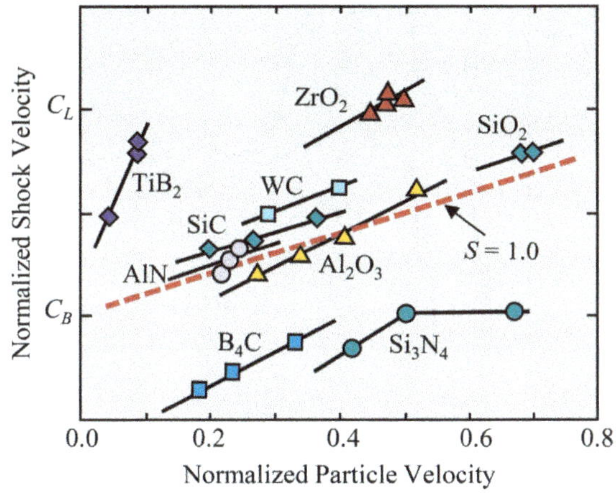

Figure 9.4. Shock velocity versus particle velocity for selected ceramics. Note normalization of both shock velocity and particle velocity (see text).

have values of C_o significantly less than C_B, perhaps indicating significant loss of strength or lattice collapse at elevated shock pressure (Grady 1998a). A phase transformation on the Hugoniot has been identified in silicon nitride at approximately 36 GPa (Hongliang *et al* 2000). Phase transformation may also play a role in the shock EOS of boron carbide (Grady 2010b, 2015).

9.2.4 Phase transformation under shock compression

The phase transformations expected when lattice instabilities occur within the range of pressures experienced in a shock wave event occur with remarkable ease in some materials during the brief shock compression process. Transformation kinetics is apparently minimal in many materials. In others, there are indications that overpressures consistent with the limits of absolute mechanical instability are achieved before transformation occurs under shock compression. The effects of phase transformation on the evolution of shock profile structure can be remarkable. Wave profile signatures of phase transformation have been observed in a number of the brittle solids. Phase transformation in a wide range of materials under shock wave compression has been reviewed (Duvall and Graham 1977, Shono 1988).

Shock wave data for aluminum nitride ceramics, shown in figure 9.5, illustrate the extent to which phase transformation can contribute to the complexity of nonlinear wave profiles (Kipp and Grady 1994). In the experiments shown, shock wave loading to approximately 35 GPa substantially exceed both the HEL (~8–10 GPa) and the wurtzite-to-rocksalt phase transformation stress (20–22 GPa) in this material. The 20% volume strain and large dissipation associated with the shock transition process leads to a marked compression wave separation and rapid attenuation of the peak wave amplitude. This transformation was first discovered by shock wave techniques (Kondo *et al* 1982), and was subsequently observed

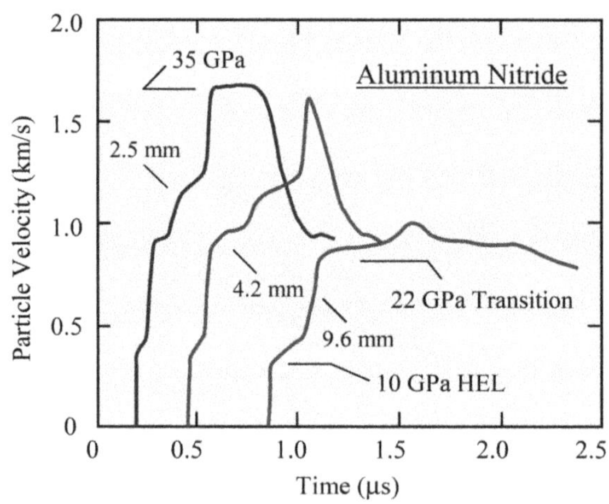

Figure 9.5. Velocity interferometry wave profiles measured in aluminum nitride ceramic. Three samples of increasing thicknesses were subjected to 1.5 mm thickness tantalum plate impact at the same 2.2 km s^{-1} impact velocity (Kipp and Grady 1994). A peak pressure of about 35 GPa is achieved in the sample before attenuation of the peak proceeds.

through static compression methods (Vollstadt *et al* 1990). The static transition pressure of 16.5 GPa reported by Vollstadt *et al* (1990) would appear to be in remarkably good agreement with the shock transition if elementary corrections for the strength at the shock state are accounted for (Jones and Graham 1968). Subsequent shock wave data on aluminum nitride from other suppliers, and with alternative shock diagnostic techniques, have been remarkably consistent in identifying the onset of the transition in the 19–22 GPa range (Kipp and Grady 1994, Nakamura and Mashimo 1994).

Dandekar *et al* (1994) noted, however, that more recent static studies of the wurtzite-to-rocksalt transition in aluminum nitride seriously cloud this comparison and that even the issue of material strength in the shock state leading to the transition remains an open question. Static transition pressures ranging over 10–23 GPa have now been reported for the aluminum nitride phase change by various authors (Dandekar *et al* 1994, Xia *et al* 1994), and suggest that the equilibrium transition pressure may even be less than 10 GPa. The wide range of reported values is a consequence of kinetics associated with the reconstructive phase change. That the shock transition proceeds with considerable ease at a stress level near the upper end of the reported static value may reflect an overstress sufficient to drive the kinetics combined with sufficient defect production in the shock deformation process. Alternatively, stress levels in the shock state may have achieved the limit of absolute stability (spinoidal) for the wurtzite structure.

Unlike aluminum nitride, certain materials have compression shock wave structures that are silent to the occurrence of phase transformation during the shock compression process. This lack of compressive wave profile signature can be due to the nature of the phase transformation and to material and EOS properties that preclude influencing the compression shock structure. Such behavior is aptly illustrated by the characteristics of the shock-induced phase transformation in calcium carbonate or calcite ($CaCO_3$), which occurs under shock compression in a pressure range of approximately 12–15 GPa (Grady 1986).

Wave profiles measured in crystalline calcite to a maximum shock stress of about 18 GPa are shown in figure 9.6. The compressive shock profile shows no indication of transformation within the 12–15 GPa range. The precursor at the substantially lower 1.5–2.0 GPa level is a consequence of yield and the low-pressure metastable calcite I–II and calcite II–III transformations (Grady *et al* 1978, Grady 1979, Aidun and Gupta 1995). The lack of a compressive shock wave signature for the higher pressure transformation is collectively due to the relatively small volume change associated with this transition (3%–5%) and to the lower pressure calcite I–II–III transitions. The collective volume strain of about 5% for this low-pressure set of transformations effectively centers the shock wave from an effectively distended state. The consequence is that the deformation shock path to the Hugoniot state (the Rayleigh line) skirts above the pressure–volume inflection point associated with the phase transformation. The conditions necessary for wave splitting due to the transformation are thus not achieved. This effect and the possible occurrence of shock structuring is explored in chapter 1. The phase transition in calcium carbonate is instead made apparent by a dramatic rarefaction shock wave that occurs during

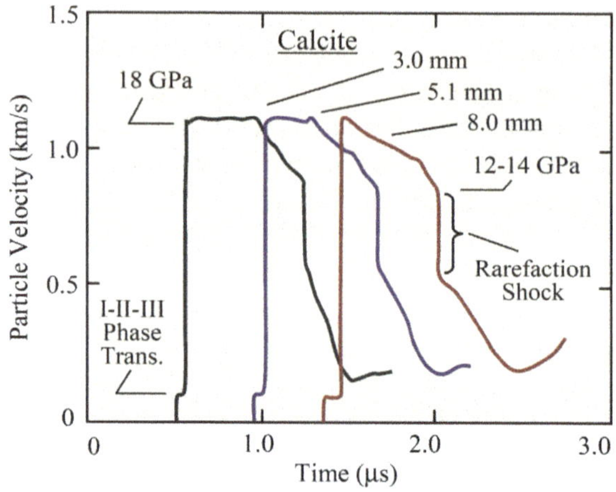

Figure 9.6. Velocity interferometer wave profiles in crystalline calcite (Grady 1986). Aluminum plate impactor 2.6 mm thickness and 2.2 km s^{-1} impact velocity. A peak pressure of approximately 18 GPa.

transformation reversal in the unloading wave structure. The sudden drop in particle velocity (<10 ns) attests to the very rapid kinetics associated with the reverse transformation.

Aragonite is the stable phase of calcium carbonate above approximately 0.3 GPa (Jamieson 1953). Transformation from calcite to aragonite is resisted by a very substantial kinetic barrier and is further thwarted by the metastable displacive or near-displacive calcite I–II and calcite II–III transitions in the 1–2 GPa range. The latter lower pressure transformations exhibit similar characteristics under both shock and static conditions. Kerley (1989, 1990) notes that EOS properties of aragonite (Vizgirda and Ahrens 1982) are consistent with the high-pressure Hugoniot and shock profile data of calcite, including those shown in figure 9.6. Static studies have, however, identified transition from calcite-III to a calcite-VI phase in the neighborhood of 15 GPa with volume change of about 3% (Merlini *et al* 2012). The latter appears a likely candidate for the distinctive transformation wave structure observed in figure 9.6.

Wave profile measurements on aluminum nitride and calcium carbonate are illustrative of the clear signatures of phase transformation provided by the shock wave profile structure. The dominant signature for the former is in wave splitting of the shock compression profile. The signature for the latter occurs in the release wave rarefaction shock structure.

Such obvious wave profile features associated with the massive tetrahedral-to-octahedral reconstructive phase transformation in quartz (SiO_2) and other silicate minerals are not observed. Extensive earlier studies did not reveal a critical transition state on the quartz Hugoniot (Wackerley 1962, Ahrens and Gregson 1964). Instead, the data suggested a rather broad pressure range (~10–40 GPa) over which the Hugoniot gradually transitioned through the mixed-phase region, fully achieving the high-density phase only after exceeding the mixed-phase pressure range. Several theories of phase transformation in the shock compression process are based on this observed 'mixed-phase Hugoniot' of quartz and the quartz-like minerals (Podurets and Trunin 1971, Grady 1980).

A critical transition state and a definitive feature in the shock compression profile of quartz associated with the transition, however, is observed in a later study (Zhugin and Krupnikov 1987). The reasons for this feature eluding the extensive earlier studies constitute an interesting story in the development of experimental shock wave technology. The Manganin gage profiles shown in figure 9.7 of Zhugin and Krupnikov (1987) measured in quartz rock identify a critical stress level, with the onset of high-pressure phase transformation, at approximately 23 GPa. Profiles at increasing impact amplitudes in the measured profiles clearly delineate the transition cusp.

Although the measurements of Zhugin and Krupnikov (1987) strongly suggest a transformation critical state, the unique characteristics of the Manganin stress gage technique warrant independent exploration of the phenomenon with an alternative experimental diagnostic. Further measurements with time-resolved VISAR wave profile techniques (Grady and Zhugin 1994, Grady 1995a) were performed on the same full-density quartzite rock tested in the earlier study. The two experiments performed used the same experimental configuration shown in figure 9.1, but with the impacting plate replaced by tungsten metal. This very high-shock impedance material was selected to achieve shock stress amplitudes that exceeded the transition stress reported by Zhugin and Krupnikov (1987) with single-stage powder gun technology. The measured wave profiles are shown on the left in figure 9.8.

The two wave profiles are arbitrarily separated in time in figure 9.8 to more clearly reveal the individual wave features. Each wave is identified by the impact velocity of the tungsten plate in the experiment. Although impact amplitudes did not significantly exceed the 23 GPa transition reported by Zhugin and Krupnikov (1987), a clear cusp at an interface velocity of 1.4 km s^{-1} identified in the figure was observed in both profiles. Shock analysis provides a stress amplitude corresponding to this interface velocity of about 23.5 GPa, in good agreement with the stress wave

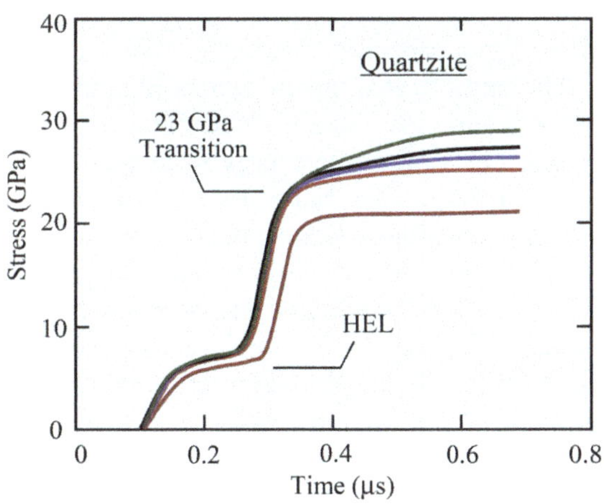

Figure 9.7. Manganin gage stress wave measurements in quartzite (Zhugin and Krupnikov 1987).

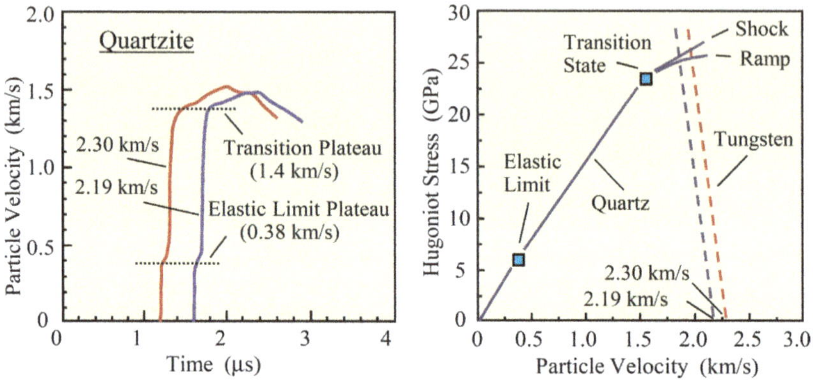

Figure 9.8. Left: VISAR shock wave profiles in quartzite (polycrystalline alpha-quartz) (Grady and Zhugin 1994, Grady 1995a, 1998a). The two experiments are identified by the velocity of the impacting tungsten plate. The elastic limit and onset of phase transformation under shock compression are noted on the profiles. Right: Hugoniot states and stress versus particle-velocity features.

measurements of the earlier authors. Stress versus particle-velocity paths identified as shock and ramp in the right-hand plot are based on different assumptions of the wave behavior above the transition cusp.

Although shock wave experiments to yet higher amplitude would provide added detail to the shock transition character of quartz, the testing shown was performed at the high end of the velocity range of the equipment available. For comparison, similar wave profiles are shown for zirconia (polycrystalline tetragonal ZrO_2) in figure 9.9, which undergoes a comparable transition under lower impact velocity conditions (Grady and Mashimo 1992). The character of the transformation cusp, particularly in the lower amplitude experiment, is clearly similar to that on quartzite.

Because of the more recent wave profile data that indicate a critical transition stress level for lattice instability in the quartz structure, both the interpretation of the earlier shock data, and theories of the mixed-phase transformation in quartz should be reconsidered. Later static compression work in which a crystal-to-amorphous transition in quartz at approximately 20 GPa was reported should be noted (Hemley *et al* 1988). The static transition in quartz has been explained in terms of free-energy and kinetics arguments (Sikka 1992). The α-quartz crystalline state is driven by elevated pressure to a high free energy state with respect to both the high-density crystalline and amorphous state, but transformation is prohibited by transition kinetics. When the mechanical stability limits are achieved (spinoidal decomposition), the material follows the easiest kinetic path to an amorphous high-density state. Earlier, De Carli and Milton (1965) suggested the possibility of a solid state transition to an amorphous phase. Molecular dynamics simulations of the α-quartz transition have been satisfactory in reproducing the expected transition pressure and volume strain (Tse and Klug 1992, Chaplot and Sikka 1992). The shock transformation mechanism warrants reconsideration considering the remarkably close agreement between critical stress levels reported in the static work of Hemley *et al* (1988), and in the dynamic studies (Zhugin and Krupnikov 1987, Grady and Zhugin 1994).

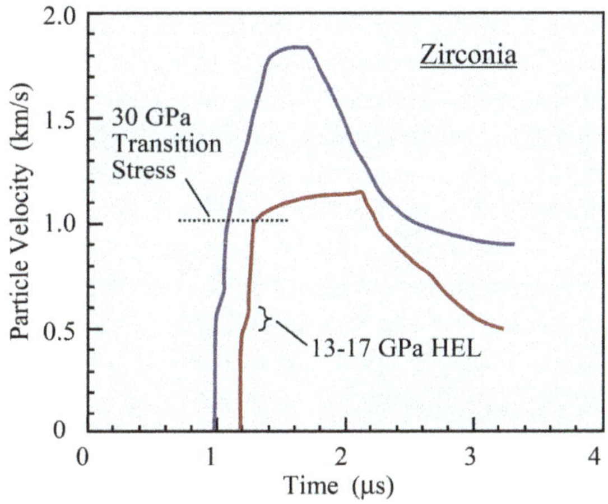

Figure 9.9. VISAR shock wave profiles in zirconia (polycrystalline tetragonal ZrO_2) revealing the elastic limit and onset of phase transformation under shock compression (Grady and Mashimo 1992).

Earlier analysis of shock wave measurements of quartz, and other silicate minerals, revealed a broad Hugoniot pressure range (approximately 20–40 GPa) where the shock velocity remained constant with increasing particle velocity. Interpretation of this data is open to question. The wave structure above the 20 GPa level is apparently subtle and requires high-fidelity wave profile diagnostics to resolve. Earlier shock wave diagnostics would have easily missed such additional structure. The consequence was to report an unchanging shock-velocity level (the transition shock velocity), and a corresponding increasing Hugoniot particle-velocity level. Shock velocities would again begin to increase with particle velocity only after the shock transition pressure was overdriven.

Similar errors in Hugoniot data interpretation have been made on ground shock data in silicate geology (Grady 1991). This experimental fidelity interpretation of the constant shock-velocity region in quartz seemed more reasonable than earlier explanations based on EOS or kinetic arguments. The later data (Zhugin and Krupnikov 1987, Grady and Zhugin 1994) imply that the dynamic compression curve of α-quartz would lie below the previously reported Hugoniot curve for quartz for pressures above the 23 GPa level over the range of a third transformation wave. It should be noted that Wackerle (1962), in his seminal study on the shock EOS of quartz, was concerned about the possible existence of a third wave in this range, but instrumentation at that time was not sufficient to resolve the details of the wave profile structure.

9.2.5 Viscosity of brittle solids and the steady shock process

A behavior characteristic to nonlinear waves is the development of shock profiles that propagate unchanged in form. Such waves of steady profile structure (steady waves) rise from the wave trough to the wave crest over a finite time determined by both material properties and the amplitude of the wave. The amplitude dependence of the risetime of such steady waves has been found to exhibit a common trend over a range of metallic materials (Grady 1981, Swegle and Grady 1985, Grady 2010a). Similar behaviors have been observed in brittle solids.

Shock profile measurements for soda-lime glass are shown in figure 9.10. Also shown in the figure is a plot of the stress jump through the structured wave against the maximum strain rate achieved in the wave. This depiction of the data again reveals the fourth-power relationship between these shock properties previously observed for metals. Additionally, shock pressure versus strain-rate data for aluminum oxide ceramic are shown in figure 9.11. The presented data are extracted from VISAR measurements on the Coors AD-995 ceramic shown previously in figure 9.2. Results for this brittle solid are again sensibly consistent with fourth-power dependence subject to the caveat that wave structuring effects due to the ceramic and lithium fluoride impedance mismatch are difficult to quantify. The profile measurement on the left illustrates where the shock particle-velocity slope (and strain rate) is assessed in each wave profile. The pressure step is based on an HEL of 6.2 GPa, while the Hugoniot pressure for the data ranges over 20–65 GPa.

This shock structuring behavior has been identified with a 'law of solid viscosity' in the development of computational visco-plastic models that reasonably reproduce the observed profile shapes in shock propagation applications (Swegle and Grady 1985, Rubin 1990, Johnson 1992, Armstrong *et al* 2007). The nature of structured waves is addressed is some detail in chapter 8. Further reflection with emphasis on brittle solids is considered here.

Although the mechanics through which rate dependence and viscosity leads to the evolution of steady waves is reasonably well known, the underlying molecular or microstructure mechanisms responsible for the dynamic behavior remain elusive. Rate-dependent plasticity, with attendant inelastic dislocation motion governed by the thermally activated penetration of energy barriers, provides a non-viscous mechanism for establishing finite risetime steady waves. The possible viscosity-based mechanisms that may be active in the structuring of shock waves in solid materials are less well understood.

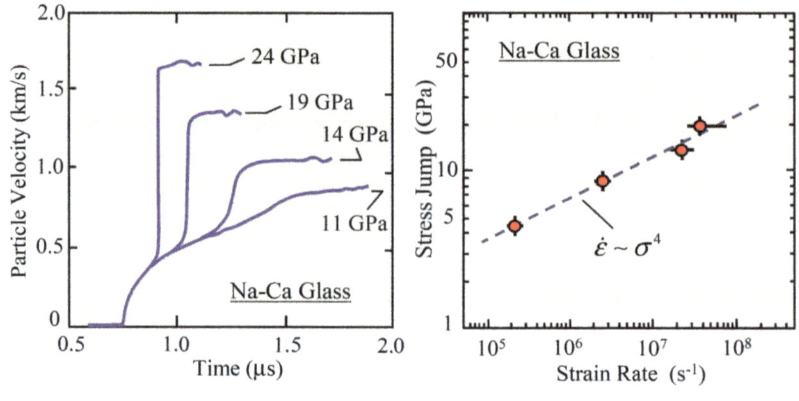

Figure 9.10. Left: Wave profiles in soda-lime glass illustrating shock formation and steady wave risetime features. Right: Steady shock wave stress jump versus strain-rate data for the soda-lime glass shock wave profiles (Grady and Chhabildas 1997, Grady 1998a).

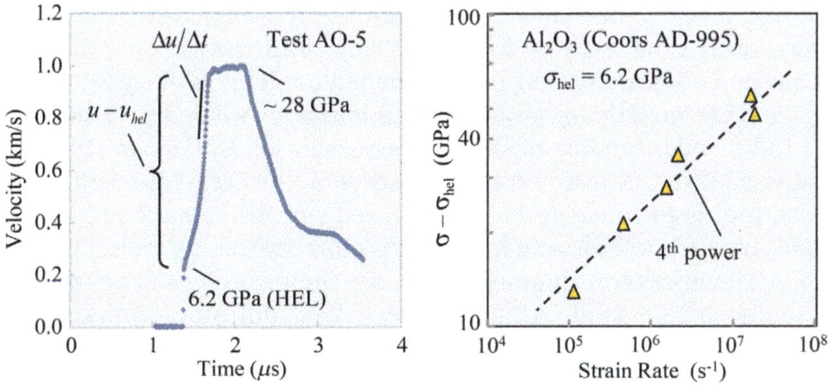

Figure 9.11. Structured wave data for aluminum oxide ceramic. Left: Point on the profile where strain rate is assessed. Right: Pressure step versus strain rate compared with fourth power.

Viscosity is commonly considered a material constant that provides proportionality between a viscous stress component and the velocity gradient or strain rate. The magnitude of a viscosity coefficient η expected in a steady shock wave process can be estimated through the material parameters that establish the upward concavity of the Hugoniot curve, and the slope of the Rayleigh line corresponding to a specific steady shock event. This estimate is based on the assumption of a decomposition of the dynamic stress into equilibrium elastic and viscous components.

The first pressure derivative of the bulk modulus K' or, more common in shock wave analysis, the slope S in a linear shock-velocity versus particle-velocity relation, provides a dimensionless measure of the concavity of the Hugoniot, and the rate at which a finite-amplitude compression wave will steepen with propagation distance. On the other hand, the ratio $\eta/\sigma\tau$, where σ and τ are the jump in stress amplitude and the risetime of the wave, respectively, provides an opposing dimensionless force determining the rate of wave spreading. A steady wave is achieved when a balance of the opposing forces $S \sim \eta/\sigma\tau$ is achieved. A slightly more rigorous analysis based on the geometry of the Hugoniot and the Rayleigh line leads to the following expression for the viscosity in a steady shock wave,

$$\eta = \frac{1}{4} S\sigma\tau. \tag{9.1}$$

Brittle solids are observed to exhibit measurable risetimes in the deformation shock wave provided the shock amplitude is of the order of or less than a few times the magnitude of the HEL (Kipp and Grady 1989, 1990). Representative values would be a 10 GPa shock jump and a corresponding 0.1 μs risetime. With a value for S of unity representative of ceramics, equation (9.1) provides $\eta \sim 250$ Pa · s for an order of magnitude estimate of the expected viscosity. If a constant viscosity coefficient governs the steady shock wave formation, equation (9.1) implies that the risetime of the steady wave decreases as the inverse first power of the jump in stress amplitude. Wave profile measurements, of which the data in figure 9.10 and figure 9.11 are representative, reveal a much more rapid decrease in risetime with shock amplitude, suggesting a decreasing viscosity with increasing shock stress.

If viscous processes are responsible for the finite thickness of steady shock waves in solids, then the underlying physical mechanisms of solid state viscosity are the basis of a fuller understanding of shock processes in solids. Gilman (1979) explored sources of viscous drag stress in metals based on a kinetics of momentum diffusion of atomistic processes active in the shock wave front. If the mechanisms within the shock front are widened to include microstructure as well as molecular processes, additional mechanisms contributing to viscosity are possible, and are perhaps more relevant to the brittle solid. Although the kinetic diffusion of mass provides a principal mechanism for viscosity in fluids, mass diffusion seems a less likely viscosity mechanism in the shock compression of solids. Evidence supporting some degree of mass diffusion in the shock front has been observed, however, possibly suggesting some viscous contribution from this process (Dremin and Breusov 1968, 1971, Alekseevskii 1989).

A microstructure analog to phonon viscosity (the diffusion of momentum through acoustic energy) seems relevant to the brittle solid. Defects on all scales (dislocations, vacancies, porosity, granularity, lamination, etc) provide sites for wave scattering. In addition, unstable failure on the microscale due to fracture or adiabatic slip can lead to acoustic emission. Both acoustic scattering and acoustic emission provide a mechanism for diffusion of momentum along the direction of the propagating wave (Grady 1998b). Coherent energy within the wave front can be transferred to incoherent energy behind the wave by such processes. That this dissipated energy may initially reside as random acoustic energy at the microstructure scale, rather than thermal kinetic energy at the molecular scale, does not bear on the structuring of the wave front profile. The consequences to the viscous character of wave propagation are the same.

In fluids, a kinetic theory of the random transport of momentum has had a reasonable measure of success in providing insight into a fuller understanding of viscosity. A relation for the viscosity coefficient based on kinetic theory can be expressed in various ways. One form developed to examine an application to wave structure in solids (Passman *et al* 1980) is

$$\eta \approx \frac{2}{3} N E_k \tau_c. \tag{9.2}$$

In this relation, N is the number of momentum transporting elements per unit volume, E_k is the kinetic energy per element and τ_c is the characteristic collision time. Equation (9.2) in the work of Passman *et al* (1980) established the wave profile structure in the spall of brittle solids. Momentum transporting elements are assumed to be the tips of propagating microcracks, and viscosity in the failure process is provided through equation (9.2).

A clear example in which elastic scattering plays an important role in the structuring of waves is through the propagation of shock waves in heterogeneous materials. The study by Barker (1971) on laminates of dissimilar materials examines the viscous character of the wave propagation process. The effective viscosity associated with propagation of the shock wave through such a material can be estimated with equation (9.2). As a representative example, consider a laminate of alternating 1 mm thick boron carbide and aluminum oxide plates. From the analysis of Barker (1971), one can estimate

$$E_k \sim (1 - F)(\rho\sigma^2/2Z) \tag{9.3}$$

and

$$\tau_c \sim (2h/U_s)\ln(1/F), \tag{9.4}$$

where $F = 4Z_1Z_2/(Z_1 + Z_2)^2$. In these relations, Z is the average laminate impedance, Z_1 and Z_2 are individual component impedances, and h is the plate thickness. For a 10 GPa shock wave equation (9.4) yields a viscosity of order $\eta \sim 20$ Pa · s. Additional shock viscosity features for laminated composites are addressed in chapter 8.

One expects a steady shock in such a laminate to span at least several cell dimensions and perhaps more. (Resonance achieved in the post-shock structure in evenly spaced laminates is not considered here.) This would require from equation (9.1) a viscosity of about 500 Pa · s—approximately one order of magnitude higher than the kinetic-theory-based prediction based on equation (9.2).

There are several possible explanations for difficulties with simple models that superposes an equilibrium elastic stress and a viscous stress to achieve a steady shock wave (9.1), and in kinetic theories of momentum diffusion to estimate viscous coefficients (9.2). First, within a heterogeneous solid (and all solids are heterogeneous to some degree), during the passage of a steady shock wave, it is highly unlikely that an equilibrium elastic state will be achieved within the rapidly rising portion of the wave. A Maxwell viscoelastic model, as was suggested by Barker (1971) for a laminated solid, in which a viscous transition from an instantaneous to an equilibrium elastic state is allowed, is probably a better representation of the underlying viscoelasticity in the physics of steady shocks in heterogeneous matter. A Maxwell model is applied to structured wave measurements for a dolomite rock in a later section.

Second, the elements of momentum diffusion responsible for viscous dissipation in steady shock propagation are fundamentally different from those considered in the more common kinetic theories of fluids. In the latter, those elements, whether particle-like or phonon-like, are a consequence of the underlying thermal motion of the body. Wave motion and velocity gradients introduced into the body would recognize this motion as viscosity, but are not responsible for them. In the steady shock waves of interest here, the elements of momentum diffusion (scattered elastic waves, acoustic emission, microfractures, etc) are stimulated, or induced, by the wave. In some cases, the mean free paths of these elements may exceed the characteristic wavelength of the steady wave profile. There may be a role for kinetic theories of viscosity in the steady wave shock process, but a direct adoption of traditional fluid theories is probably not applicable.

Third, turbulence is also a viable mechanism for shock viscosity (Kolmogorov 1941). When dissipation through laminar plasticity or viscosity becomes insufficient to dissipate the requisite energy necessitated in a steady shock wave, excess deformation through out-of-plane turbulence could provide the necessary energy dissipation.

Lastly, observations of mass diffusion in the steady shock wave within the context of shock viscosity are worth noting. In the shock loading of solids, even for relatively intense shocks, material elements (atoms and molecules) are expected to remain at, or close to, their original lattice sites. Dremin and Breusov (1968, 1971) proposed a model allowing some material mixing through the inelastic shearing of neighboring planes; but even here, relative motions are not expected to be large.

Alekseevskii *et al* (1989) experimentally investigated shock wave propagation in several materials in which radioisotopes were placed at an interface that intercepted the passing shock wave. In post-test analysis, it was found that some radioactive atoms were transported a considerable distance (many numbers of atomic spacings) into the material—the quantity of mass transfer dependent on the shock intensity and the strength of atomic bonds in the substance. A clear explanation for this anomalous diffusion effect (identified as deep penetration in some publications) was not provided, although the relatively low transport energy of interstitial atoms in the shock front was suggested. Gluzman and Psakh'e (1989) proposed that in such an effect, a low density of solitons propagating in the shock could be the carriers in the anomalous mass transport process. An alternative explanation is that preferred atoms—those whose thermal kinetic energy is favorably directed at the moment of arrival of the accelerating shock wave—could 'ride-the-wave' for some significant distance before dropping behind the wave due to arresting perturbations from lattice imperfections. The analogy with ocean surfers is clear, where an initial impulsive effort by the surfer is sufficient to join with the wave and transport some considerable distance along the path of the wave before departing. In any case, these data suggest that momentum diffusion due to mass transport in the solid state is apparently possible. Again, it is a situation where the diffusing elements are stimulated by the passing shock wave.

In either case the phenomenon probably exemplifies influences of classical chaos in the atomic process. Any molecular particle in the material, either alien or an expected member of the lattice, is in thermal motion orbiting in a cage determined by the composite atomic forces of neighboring molecules within the correlation span of the particle in question. When the shock wave subsumed this region, the same atomic forces and initial conditions determine acceleration and subsequent motion of the molecular particle. The possible initial conditions are innumerable with the vast number resulting in acceleration with the particle remaining within the original proximity of nearest neighbors. Some few sets of initial conditions have the potential of embarking the particle careening into a vastly different orbit possibly spanning many dimensions of the material lattice. Such motions certainly fall within the precepts of classical chaos where subtly different alignment of initial conditions results in vastly different subsequent motions. A perhaps relevant analogy is that of bodies in the solar asteroid belt in orbit under gravitational influence of the Sun, Jupiter and a myriad number of lesser bodies. Over sufficient spans of time, chance brings the alignment of forces that detach an asteroid from its historic orbit, rocketing through other reaches of the solar system. The 'butterfly effect' was coined to exemplify such outcomes of chaos physics. Perhaps equally appropriate is the 'nightfall effect'. Isaac Asimov (1941) was aware of chaos theory when *Nightfall* was penned, the tale of an Earth-like planet and civilization in orbit amongst six suns. Once every two thousand years unique astronomical alignment brought on the fall of night. Over that night, chaos reigned among the planet's population.

9.2.6 Hugoniot elastic limit and the shock wave elastic precursor

Failure in brittle solids under the intense shear stresses developed in the shock compression process also plays an important role in determining the structure of wave profiles. Both plasticity and fracture are viable mechanisms for inelastic deformation in brittle solids under shock compression. The relative roles of plasticity and fracture, under the range of loading conditions experienced in the shock process, are not fully understood.

The shock wave precursor, or elastic precursor, is perhaps the most prominent wave profile feature mirroring the failure processes in the shock compression of solids. (Failure in the present shock wave context refers to the abrupt softening in the dynamic uniaxial modulus of the material determined by the precursor wave amplitude.) A number of studies have focused on features of the elastic precursor waves in brittle materials and the relationship of these features to the underlying deformation mechanisms. A prominent feature that has received attention is the attenuation of precursor waves with propagation distance. Such wave attenuation would have implications on the rate-dependent nature of the inelastic failure processes. Investigations of precursor attenuation have provided indications of the importance of precursor decay in brittle solids. Studies on hard rock material including quartzite (Ahrens and Duvall 1966) and dolomite (Grady *et al* 1976, 1977) indicated modest elastic precursor decay. Rosenberg *et al* (1988) have reported elastic precursor decay in an aluminum oxide ceramic. Crystalline solids, which exhibit a post-yield stress-relaxing precursor profile, appear to have a propensity for elastic precursor decay. Examples include single-crystal magnesium oxide (Grady 1977) and forsterite (Furnish *et al* 1986). Boron carbide ceramic also exhibits precursor stress relaxation and precursor decay (Grady 1994a, 1998a). Legions of experimental and theoretical work have been undertaken on elastic precursor decay in crystalline lithium fluoride (e.g. Asay *et al* 1972).

On the other hand, the brittle solids that exhibit post-yield hardening (for example, the aluminum oxide ceramic precursor profiles in figure 9.2) apparently show less tendency toward precursor decay. Further studies on aluminum oxide ceramics (Cagnoux and Longy 1988, Longy and Cagnoux 1989) do not reveal definitive elastic precursor decay, and profile measurements on dense limestone rock indicated little or no precursor decay (Grady 1983). Investigations of elastic precursor decay on a wider suite of ceramic materials (Grady 1995b) do not support extensive elastic precursor decay. These latter HEL data and the measured dependence on strain rate are shown in figure 9.12. Strain rate is estimated from the amplitude of the elastic precursor and the precursor duration preceding arrival of the

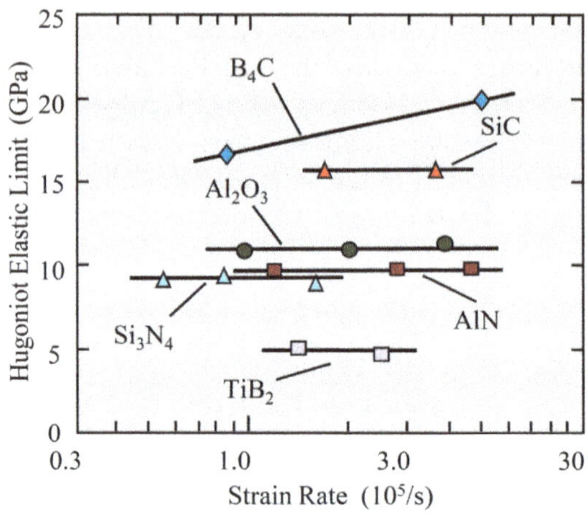

Figure 9.12. HEL stress versus strain rate for selected ceramics (Grady 1995b, 1998a).

deformation shock wave. Only boron carbide shows evidence of strain-rate dependence and precursor decay (Grady 1995b). There is concern that wave profile measurements using instrumentation with limited time resolution may falsely indicate more marked precursor decay than actually occurs.

In general, shock wave data suggest modest or negligible elastic precursor decay in most polycrystalline brittle solids. The implications are no, or very modest, rate dependence in the processes governing failure and post-failure behavior in these brittle materials.

An absence of precursor decay in some ceramics is illustrated by the wave profile data shown in figure 9.13. Elastic precursor wave profiles for aluminum oxide and aluminum nitride ceramic are shown in the plot. Precursor waves are plotted in a manner that scales with propagation distance and is examined for self-similarity of the profiles. This plotting method is a useful technique for revealing precursor decay and rate-dependent or time-dependent behavior. The profiles for aluminum oxide and aluminum nitride in figure 9.13 show no evidence of precursor decay over the range of sample thicknesses tested. Similar results were found for silicon carbide, silicon nitride and titanium diboride. Only boron carbide shows evidence of elastic precursor decay. Complexities in the precursor structures of silicon carbide and titanium diboride are discussed further in chapter 6, and suggest a more complicated two-yield mechanism in the shock-compression failure process (Grady 1992a, 1995a, 2000, 2007).

The wave profile data in figure 9.13 are used to determine both the HEL amplitude and an effective strain rate associated with the dynamic failure. The midpoint between the arrival of the elastic precursor wave and the subsequent arrival of the deformation shock wave is used to provide a consistent measure of the amplitude of the HEL. This amplitude and the time between the first and second shock waves are then used to calculate a corresponding effective strain rate for profiles at each sample thickness. This HEL stress versus strain-rate data provide the plots shown in figure 9.12. Consistent with observations of the self-similar precursor

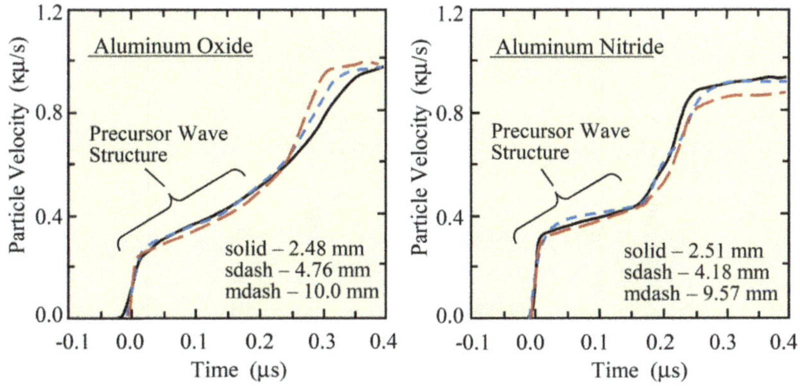

Figure 9.13. Precursor waves from VISAR profiles for aluminum oxide (left) and aluminum nitride ceramic (right). Each profile has been scaled to 10 mm sample thickness based on an assumption of centered self-similar wave propagation (Grady 1995b). Actual test sample thickness is identified in the plots.

profile characteristics in figure 9.13, there is little evidence for rate dependence over the approximately half a decade of strain rate inferred from the profile measurements. Boron carbide ceramic is the exception.

This notable lack of precursor attenuation and rate sensitivity in ceramics is surprising. Compressive failure in the shock process is believed to be a fracture-controlled mechanism and the nucleation and growth of fracture is a known time-dependent process. Such time dependence is expected to lead to rate-dependent strength and precursor decay in ceramics. Fracture failure in brittle materials under other modes of deformation (e.g. spall, compression Hopkinson bar) has shown a strength dependence on strain rate approaching a one-third-power behavior. Similar rate dependence is expected for a brittle-failure HEL mechanism (e.g. Grady 1998a). The observed lack of dependence of HEL strength on strain rate indicated by much of the data on ceramics and rocks raises the question of whether failure within the front of a compressive shock wave is indeed fracture-controlled. Cagnoux and Longy (1988) obtained similar experimental observations. From their observations of negligible precursor decay in alumina, they concluded that grain plasticity is a more likely mechanism for failure in the compressive shock wave.

The dynamic-strength measurements determined from wave profile data for a suite of high-strength ceramics are compared in figure 9.14 with corresponding static Knoop indentation hardness measurements (Grady 1994b). Dynamic failure strength is calculated from the HEL data using the relation $Y_d = ((1 - 2\nu)/(1 - \nu))\sigma_{\text{hel}}$. In some cases, strength data for several of the same materials with differing initial microstructure are shown. For TiB_2 ceramic, the two yield levels shown reflect uncertainty in the HEL brought about by the complex two-wave structure for the elastic precursor wave in this material (Grady 1992a, Grady and Wise 1993). The data compare well with earlier hardness comparisons of Cline (1989), (see Gilman 1970), on similar high-strength ceramics, although a somewhat weaker correlation of static and dynamic strength is observed. Rosenberg (1995) has proposed a Griffith failure criterion relating the HEL stress to the dynamic yield

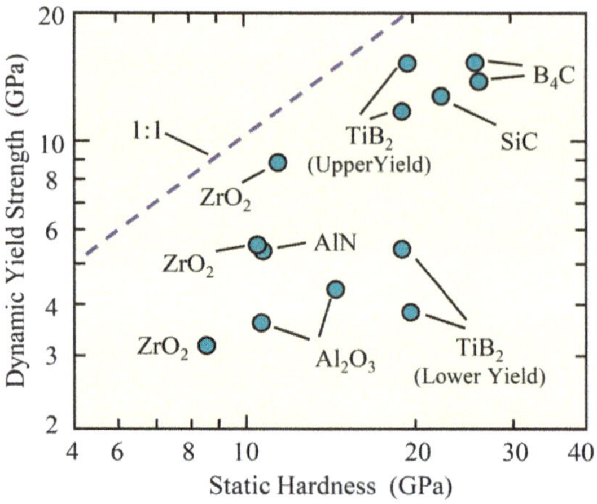

Figure 9.14. Dynamic yield strength at the HEL versus static hardness data for selected ceramics (Grady 1994b, 1998a).

stress, $Y_d = ((1 - 2\nu)^2/(1 - \nu))\sigma_{\rm hel}$. This relation is in agreement with the commonly used ratio between hardness and yield stress, $H = 3Y$ and is reasonably consistent with the data in figure 9.14.

These dynamic-strength data correspond to resolved shear stresses ranging from approximately $G/25$ to $G/40$, where G is the shear modulus of the ceramic. Various authors (e.g. Hirth and Lothe 1968) have estimated the ultimate strength of a perfect close-packed lattice to lie between $G/5$ and $G/30$. Experimental studies on single-crystal whiskers have achieved strengths as high as $G/15$. Dynamic strengths as high as $G/20$ have been obtained in shock wave studies on crystalline lithium fluoride (Gupta 1977).

9.2.7 Off-Hugoniot measurements

The preponderance of shock wave data on condensed materials has been obtained in a planar one-dimensional geometry. Explosives launching plates and gun systems accelerating flat-faced projectiles complemented emerging instrumentation and simplified analysis of the collected data. Shock wave events in application are seldom one-dimensional and it is recognized that a description of material response based strictly on planar shock data is not sufficient for predictions of more complex events.

Ground shock effects from conventional or nuclear explosion depend on the post-shock responses of the geologic materials. The attenuation of peak shock pressure with propagation distance requires attention to the subsequent shearing strength of the material. Conventional terminal ballistic events also entail multi-dimensional shock loading with the need for accurate treatment of post-shock material response.

Various experimental approaches have been undertaken to assess the post-shock strength properties of solids. The pressure–shear technique, in which gun systems launch obliquely inclined projectiles to impact compatibly oblique test-material targets, has been developed and yields dynamic-strength measurements (e.g. Abou

Sayed *et al* 1976, Chhabildas and Swegle 1980). Material testing in controlled spherical wave loading has been pursued. Laboratory scale spherical symmetry explosive testing of hard rock was performed (Swift 1971) with stress and particle-velocity measurements assessed for material strength properties through Lagrangian analysis methods. Testing from this study is considered further in chapter 11.

A unique experimental method for assessing off-Hugoniot shock strength properties of materials is presented in the study of Paris *et al* (2011). The essence of the experiment is illustrated in figure 9.15. The impact of a spherical-nose projectile onto a stationary test target inputs a divergent shock wave into the material under investigation. The initial shock carries the material to an elastic Hugoniot state along the illustrated stress path. Post-shock motion, a consequence of the imparted divergent flow, leads to reduced states of transverse stress. The impact intensity, extent of divergence and wave structuring over propagation distance determine the stress path assessed at the point of experimental measurement. Computational analysis of velocity interferometry particle-velocity records determines the shock Hugoniot state, post-shock stress path and the point of failure of the material in the shock wave event. Measurements are recorded at the interface between the test sample and an interferometer window material, or at the free surface of a backing witness plate. Through a series of such experiments, the failure surface for the material is assessed over a range of transverse stress states.

Paris *et al* (2011) used the experimental method to investigate post-shock off-Hugoniot states of silicon carbide and boron carbide ceramics. The study demonstrates the uniquely different dynamic-strength properties of these two carbide ceramics. Experimental results for both ceramics are depicted in figure 9.16. For reference, calculated curves are shown for ductile-like von Mises failure and brittle-like Griffith failure commensurate with the measured HEL. Stark differences are noted in the failure response of the two ceramics. Silicon carbide failure data fall in two sets that are nominally linear with a break at a transverse stress of about 1.5 GPa. The failure response is identified as Griffith-like or brittle-like below this

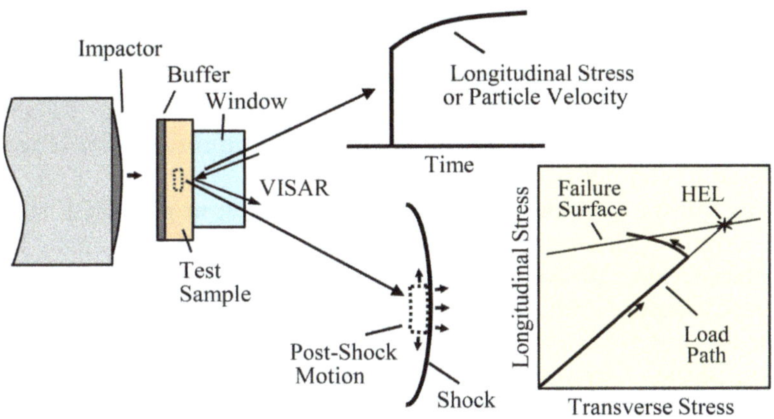

Figure 9.15. Experimental configuration and methodology of the divergent wave off-Hugoniot strength measurement (Paris *et al* 2011).

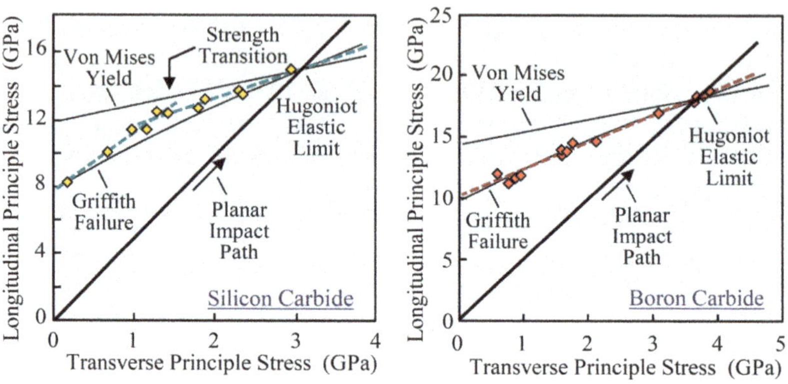

Figure 9.16. Off-Hugoniot dynamic-strength measurement for silicon carbide (left) and boron carbide ceramic (right; Paris *et al* 2011).

transition stress and von Mises-like or ductile-like above. In contrast, failure response of boron carbide is brittle-like over the full range of transverse stress up to confinement corresponding to the HEL, and perhaps above. The data are adequately described by a Griffith failure criterion. One can ponder whether the same experiment in a convergent geometry could explore the failure surface at transverse stress states in excess of the HEL, perhaps assessing the brittle-like to ductile-like transition in boron carbide.

In light of the contrasting differences in dynamic strength of silicon carbide and boron carbide depicted by the study of Paris *et al* (2011), it is instructive to examine the planar shock profiles for these two carbides. Velocity interferometer measurements provided by the experimental method described in figure 9.1 for two comparably high-quality carbide ceramics are shown in figure 9.17 (Grady 1994a). Similar sample dimensions and impact velocities allow direct comparisons of the contrasting shock measurements. Failure on the Hugoniot path (the HEL) is identified in the experimental profiles as the first break in slope subsequent to the initial steep rise in compression. The HEL for silicon carbide is about 15–16 GPa and that for boron carbide is 18–20 GPa. The post-HEL characteristics of the two ceramics differ starkly. Silicon carbide exhibits subsequent hardening. Both pressure hardening and deformation hardening probably play a role. Further details concerning the dynamic strength of silicon carbide ceramic on the Hugoniot path are considered in chapter 6.

In contrast, boron carbide exhibits a stark reduction in post-HEL stress, in line with the inferred brittle-like nature of dynamic failure in this ceramic. The chaotic nature of the measured wave profiles is attributed to the spatial coarseness of the failure microstructure relative to the spot size of the recording velocity interferometer.

There are further noteworthy features of the interferometry data in figure 9.17. Profiles for silicon carbide are smooth and regular, whereas corresponding profiles for boron carbide exhibit erratic and irregular behavior in the measured motion. The VISAR laser beam in these tests is focused to a spot of about 25–50 μm.

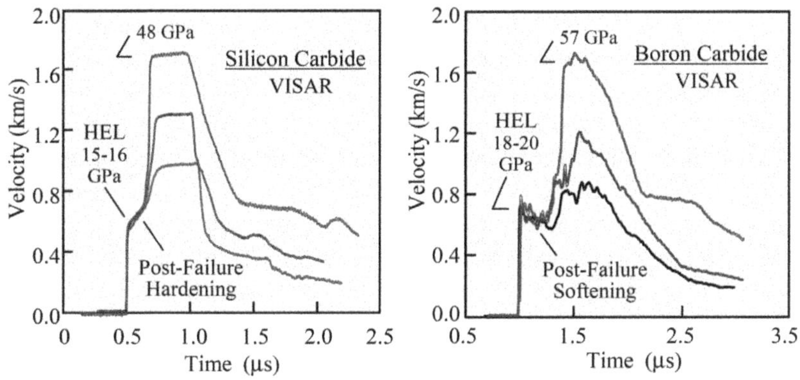

Figure 9.17. Shock wave velocity interferometer profiles spanning HEL failure, shock and subsequent release for silicon carbide (left) and boron carbide ceramic (right; Grady 1994a).

Consequently, the measured motion represents an average over this spot size. Differential interface motion at various spatial scales will influence VISAR data (Atroshenko *et al* 1990, Mescheryakov *et al* 1992). Differential motions on a scale less than the spot size will lead to reduction in VISAR contrast because interference maxima and minima at different points within the spot are achieved at different times. Differential motion on a scale larger than the spot size will not appreciably affect contrast. Random elastic wavelets from points in the material removed from the laser spot and undergoing differential motion lead to irregular motions at the recorded point. VISAR data on both the subscale (less than spot size) and mesoscale (greater than spot size) for silicon carbide indicate homogeneous motion under shock loading. VISAR data for boron carbide, on the other hand, suggests homogeneous motion on the subscale, but heterogeneous motion on the mesoscale. The latter result suggests a heterogeneous deformation process under shock failure on the 50–500 μm spatial scale in line with the brittle-like failure noted in the experiments of Paris *et al* (2011). The concentration of deformation energy into discrete deformation zones with localized softening or melting (Grady 1980) or solid state amorphization (De Carli and Milton 1965) has been proposed to accompany shock-induced failure in certain brittle solids. Chen *et al* (2003) report planar amorphization features in shock-compressed boron carbide.

The stark differences in the wave structure of the two carbides warrant consideration of the material constitutive response inferred from the shock profile data. Since measurement of the time-resolved velocity is acquired at a single station downstream from the plane of impact, assumption of simple wave behavior is necessary. Namely, characteristics corresponding to any particle-velocity amplitude propagate as straight lines, and wave speeds are inferred from the transit time from impact and the propagation distance. First, the speed and amplitude of the precursor wave is established, and from this the HEL failure strength of the material is determined. Second, the speed of the second wave and amplitude of the particle-velocity jump are assessed. From this analysis, shock-velocity and particle-velocity Hugoniot states are determined for each experiment. In the analysis, differences in impedance between test material and the lithium-fluoride window material must be accounted for. Shock-velocity versus particle-velocity Hugoniot data for the two carbides are included with other ceramic data in figure 9.4. Lastly, analysis of the full wave profile establishes the one-dimensional stress versus volume path imparted to the material by the impact experiment. This analysis is accomplished through Lagrangian analysis methods (Fowles and Williams 1970) (see also chapter 11). Alternatively, empirical response models are iteratively adjusted in computational simulations until agreement with experimental measurement is achieved. The latter method provides the experimental stress versus volume paths shown for the carbide ceramics in figure 9.18. Also shown are reasonable descriptions of the bulk hydro response to pressure of the two compounds.

Comparative plotting of dynamic compression and release along the Hugoniot path and the hydro response for silicon carbide on the left in figure 9.18 clearly support ductile-like failure and post-failure strength of this ceramic. Offset between the hydro and Hugoniot compression indicates sustained post-HEL strength. The nature of the release path supports a nominal elastic-plastic-like character to the response of silicon carbide ceramic to the shock event.

Boron carbide, in contrast, reveals substantial post-HEL loss of strength, with decompression near-fluid-like in mirroring the hydro response of the compound on stress release. Higher-pressure Hugoniot states for boron carbide that reside to the left (higher density) of the hydro response may indicate shock-induced phase transformation (Grady 2015).

Later Hugoniot path dynamic-strength studies on boron carbide (Vogler *et al* 2004) and silicon carbide (Vogler *et al* 2006) generally support the results inferred from the shock data and analysis of figures 9.17 and 9.18. The later studies use a method in which the material is subjected both to shock and release, and to shock and reshock, at a specified Hugoniot state. These studies support sustained post-HEL strength in silicon carbide and near total loss of strength in boron carbide. The test method, however, reveals added complexity to the Hugoniot path dynamic strength. Shock–reshock experiments reveal strength structure in the second shock in

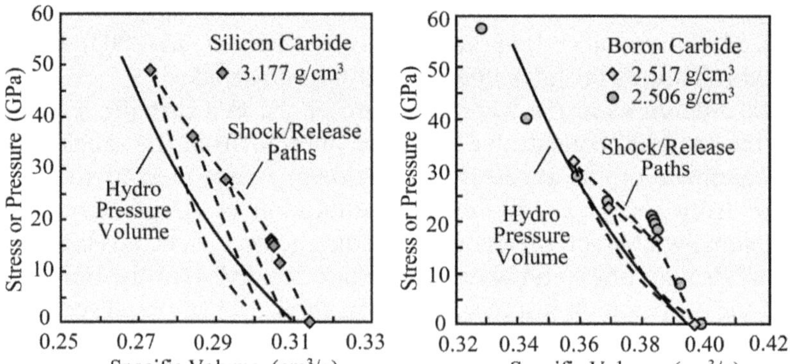

Figure 9.18. Hugoniot states with shock and release paths from shock profile measurements on silicon carbide (left) and boron carbide ceramic (right) (Grady 1994a).

both ceramics that suggest both time and path dependence to the dynamic failure surface. Discussion of these complexities is provided in the earlier works and later analysis (Vogler *et al* 2004, Vogler *et al* 2006, Grady 2015).

9.3 Shock compression of dolomite rock

Dolomite is a hard, brittle carbonate rock that has been tested extensively under both static and shock compression loading (e.g. Heard *et al* 1973, Grady *et al* 1976, 1977). It is selected for discussion here because it exhibits dynamic-strength and failure characteristics that are representative of many ceramic and ceramic-like materials. It is also one of the few brittle materials for which a large number of high-fidelity shock wave profiles have been measured that explored both the evolution of waves with propagation distance, and the dependence on the amplitude of the shock wave input. In particular, features of the measured wave profiles lend insight into the dynamic processes controlling the kinetics of both ductile and brittle aspects of failure in this class of materials. The experimental wave profiles access the shock response of dolomite in the stress range up to a few times the HEL strength of the material. Shock EOS issues of dolomite mineral, including phase transformation in the more extreme pressure regime, are explored further in chapter 11.

9.3.1 Experimental methods and materials

The dolomite selected for investigation in this study has received considerable attention in both the static (Heard *et al* 1973) and earlier shock wave (Petersen 1969, Larsen *et al* 1971) literature. Blair dolomite is a fairly homogeneous fine-grained equigranular dolomite. Dolomite rock, obtained from Martinsburg, West Virginia, is greater than 98% crystalline dolomite, $Ca,Mg(CO_3)_2$. The measured dry density is 2840 kg m^{-3} and the calculated porosity is 0.9%, based on a crystal density of 2866 kg m^{-3}. Petrographic analysis reveals a dolomitic, micritic limestone containing about 10% large (1.0–3.5 mm) dolomite crystals in a fine-grained (0.01–0.02 mm) micritic matrix. Large crystals show some local fracturing and are commonly twinned. The micritic matrix is generally unfractured and is defined by optically discontinuous equidimensional carbonate grains with local rhombohedral cleavage. Fractures and grain boundaries are sharply defined; grain boundaries are locally sutured and may be lined with organic material. Fracturing is relatively rare in the undeformed dolomite, and no preferred fabric has been demonstrated. Cylindrical samples 75 mm in diameter and with thicknesses ranging over 5–25 mm were prepared from cores of the dolomite. Impact and recording faces of the samples were lapped to a smooth polish. Both longitudinal and shear ultrasonic velocities were measured on all samples. The average values obtained were 6.83 km s^{-1} and 4.03 km s^{-1} for the longitudinal and shear speed, respectively.

The experimental study consisted of selecting five impact velocities consistent with the stress range of interest and conducting a series of several experiments at each velocity level. Sample thickness in a given series was varied so that the evolution of the wave profiles from the initial input pulse could be monitored.

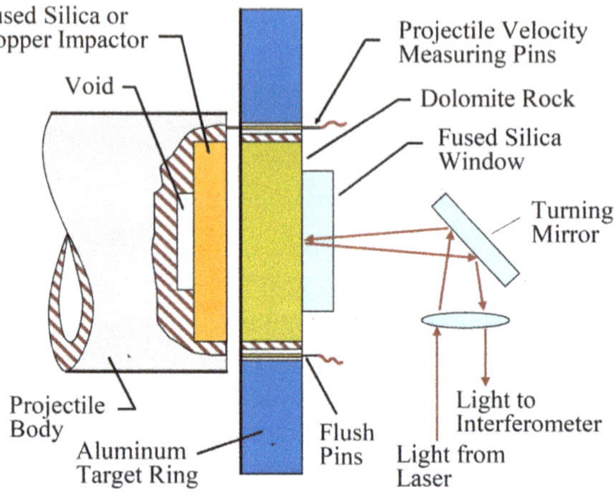

Figure 9.19. Impact target assembly for dolomite rock shock wave tests including projectile, target material, laser window and near-target laser optics (Grady *et al* 1976, 1977).

The specimens were subjected to shock wave loading resulting from the impact of a projectile fired by a gas gun. The target configuration is illustrated in figure 9.19 and is a particular application of the experimental geometry shown in figure 9.1. The impacting projectile carries a fused silica impactor disk except for the highest impact stress level experiments, in which copper impactors were used. A void was machined behind the central region of the impactor disk, and when the compressive wave produced on impact reached this free surface, it reflects as an unloading wave, propagates back through the impactor, and unloads the test specimen. Thus, the particle-velocity history at the impact interface is a square pulse of amplitude and duration determined by the projectile velocity and the impactor thickness, respectively.

The particle-velocity history, subsequent to propagation of the pressure wave through the thickness of the material, at the interface between the dolomite and a fused silica window material is monitored with diffuse-surface velocity laser interferometry (Barker and Hollenbach 1972). Nunziato *et al* (1974) provides an excellent review of this experimental technique and its applications.

9.3.2 Features of the shock wave profiles

Each impact experiment provides one complete compression loading and unloading particle-velocity history that corresponds to a specific impact velocity and specific propagation distance from the impact interface. The detailed test conditions of each experiment are provided in earlier reports (Grady *et al* 1976, 1977). However, in all tests, with the exception of the highest stress amplitude, approximately 1.5 mm thick fused silica plates impact dolomite samples ranging from about 5–25 mm in thickness. Copper plates 1.5 mm in thickness provided the highest stress experiments. Two profiles were obtained in series 1 and series 4. Three profiles were obtained in series 2, 3 and 5. Series 1–5 correspond to peak impact stress amplitudes of approximately 0.5, 1.4, 2.4, 3.9 and 6.5 GPa, respectively. The measured particle-velocity profiles are only indirectly related to the dynamic stress–strain response and further analysis is necessary to extract the experiment-independent dynamic stress–strain properties of the material.

Two issues must be addressed. First, with the velocity interferometer technique a laser window material is used to back the test sample. Since the window material is not an exact mechanical impedance match for the test material, a region of wave interaction is created in the test material near the sample–window interface. The measured interface velocity is thus a distorted version of the profile originally propagating in the undisturbed test medium. Second, particle velocity versus time is measured, and this measurement must be coupled with the equations governing conservation of mass and momentum, to determine the dynamic stress–strain response inferred by the measurement. Analysis methods have been developed to account for both issues (Grady *et al* 1977) (see also chapter 11). The analysis methods are used with the present data to determine the dynamic stress–strain characteristics of the dolomite rock under the shock loading conditions applied.

After correcting for the wave profile distortion caused by the sample and window material impedance mismatch, the experimental velocity profiles determined at differing thickness samples provide a measure of the impact imparted mass velocity as a function of time and propagation distance. The resulting set of experimentally measured wave profiles are depicted in figure 9.20. Through Lagrangian analysis methods developed initially by Fowles and Williams (1970), the stress versus particle velocity and stress versus strain response of the material through the shock compression and release process is determined from the multiple wave profiles.

Two markedly different regions of behavior are observed in the shock response of Blair dolomite. The first region corresponds to peak impact stresses below approximately 2.5 GPa, which is achieved in the lower impact amplitude experiments identified as series 1, 2 and 3. The second region occurs at higher impact amplitudes and corresponds to peak stress levels in excess of 2.5 GPa. This region of behavior is captured by the series 4 and 5 experiments.

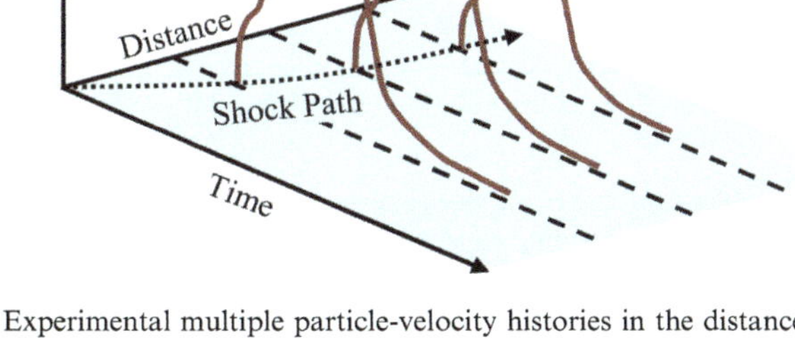

Figure 9.20. Experimental multiple particle-velocity histories in the distance versus time plane.

Particle-velocity profiles for the series 1 tests are shown on the left in figure 9.21 and are representative of the wave propagation behavior observed in all experiments below 2.5 GPa peak stress. The initial response is a sharp shock loading to a particle-velocity amplitude of about 0.015 km s^{-1}. Subsequent loading to peak particle-velocity amplitude occurs by a more gradual rise over about 0.2 μs. This inflection in the compressive shock is noted in the figure.

A similar loading response is observed in the series 2 and series 3 profiles, although the amplitude of the sharp shock region is different for each series. Similar test observations have been reported by Petersen (1969) and by Larsen *et al* (1971). From successive profiles, it is observed that the shape of the loading wave does not change with propagation distance. In figure 9.21 (left) the two profiles are displayed coincident in time to illustrate this steady wave structure of the loading wave. No attenuation of the wave peak is observed for profiles below 2.5 GPa. The unloading wave is slightly dispersive and does not differ significantly in velocity from the loading wave velocity.

Particle-velocity profiles for the series 5 experiments are shown on the right in figure 9.21, and are representative of wave propagation observed for shock amplitudes above 2.5 GPa peak stress. Initial loading occurs by a sharp shock to a stress level of about 2.5 GPa (corresponding to a particle velocity of 0.16 km s^{-1}). This stress level is no longer dependent on the peak amplitude, and is identified as the HEL stress amplitude of the Blair dolomite material. The unloading wave is observed to propagate at a velocity significantly faster than that of the loading wave. The unloading wave overtakes the loading wave over the propagation distance leading to attenuation of the peak amplitude.

In figure 9.22 the left-hand plot shows the axial stress versus strain response calculated from the experimental profiles by the methods of Lagrangian analysis achieving peak stresses below 2.5 GPa. The peak stresses achieved for the series 1, 2 and 3 experiments are approximately 0.5, 1.4 and 2.4 GPa, respectively. The

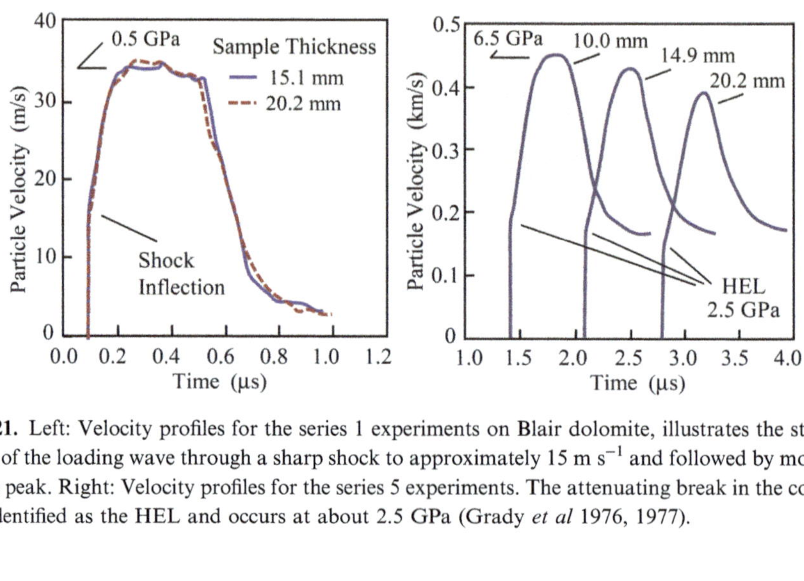

Figure 9.21. Left: Velocity profiles for the series 1 experiments on Blair dolomite, illustrates the steady wave character of the loading wave through a sharp shock to approximately 15 m s^{-1} and followed by more gradual rise to the peak. Right: Velocity profiles for the series 5 experiments. The attenuating break in the compression wave is identified as the HEL and occurs at about 2.5 GPa (Grady *et al* 1976, 1977).

compression response is near linear and the loading modulus increases slightly with increasing peak stress. The unloading response is slightly stiffer, and provides for a small amount of hysteresis and shock dissipation in the compressive load and unload cycle.

On the right in figure 9.22 the axial stress versus strain response including the higher shock amplitude tests is shown. In contrast to response below 2.5 GPa, the loading stress–strain paths for the two higher series show a cessation of proportional loading at about 2.5 GPa (the HEL), with curvature concave toward the strain axis. Unloading occurs with a markedly larger hysteresis and shock dissipation.

In figure 9.23, dependence of the loading wave velocity on stress amplitude, as determined from the time between equal amplitude points on the corresponding profiles, is shown for all five series of experiments in the left-hand plot. For instance, the series 1 curve shows that the loading wave velocity was essentially a constant value of about 6.94 km s^{-1} from zero stress to the wave peak of 0.5 GPa. Accuracy in the loading wave velocity data is about 1.5%. Wave velocities for the series 2 and 3 experiments are also nominally constant from zero to peak stress, indicating steady wave propagation of the loading wave. Note also that the wave velocity increases with peak driving stress (i.e. the loading wave velocity for the series 3 experiments is larger than that for the series 2 experiments and so forth). For the higher stress experiments (series 4 and 5), however, the character of the loading wave velocity changes above about 2.5 GPa. The velocity decreases above this stress level, indicating a lower modulus and dispersive wave propagation.

Lastly, the right-hand plot of figure 9.23 shows the corresponding unloading wave velocity dependence on stress amplitude for all five series. Velocity accuracy in the unloading data is about 3%. The velocity data indicate a dispersive unloading modulus; namely, a modulus that increases with increasing stress level. Note also that the dispersion increases more sharply in the stress region below about 1.0–1.5 GPa.

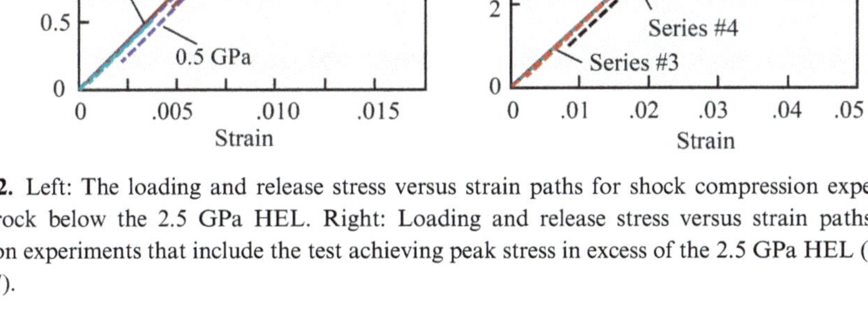

Figure 9.22. Left: The loading and release stress versus strain paths for shock compression experiments on dolomite rock below the 2.5 GPa HEL. Right: Loading and release stress versus strain paths for shock compression experiments that include the test achieving peak stress in excess of the 2.5 GPa HEL (Grady *et al* 1976, 1977).

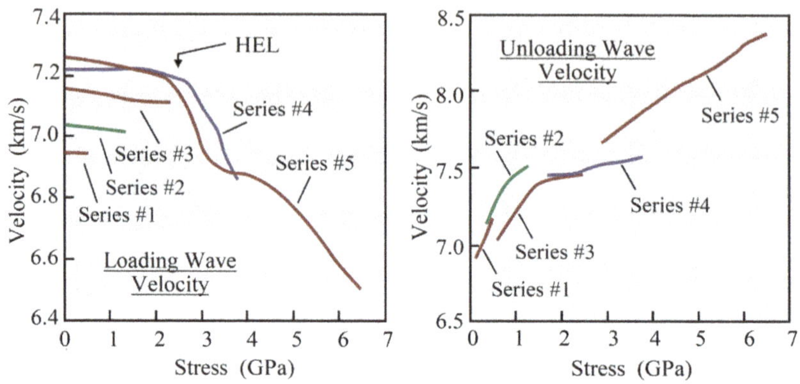

Figure 9.23. Left: Each curve is the loading wave velocity dependence on stress amplitude determined from two or three wave profiles from the same peak impact amplitude experimental series (Grady *et al* 1976, 1977). Right: Similar analysis provides the unloading wave velocity dependence on stress amplitude.

From these experiments, it is concluded that Blair dolomite undergoes pervasive dynamic failure in the shock compression event at a stress level of about 2.5 GPa. Failure and subsequent inelastic processes account for the substantial hysteresis (energy dissipation) and wave peak attenuation observed above 2.5 GPa. Compression wave propagation below 2.5 GPa is complex. The increase in loading wave velocity with increased peak stress below 2.5 GPa and the dispersion in the unloading wave behavior below 1.0–1.5 GPa are possibly explained by micro-structure features of the dolomite rock associated with pre-existing cracks in the material as proposed by Walsh and Brace (1972). This wave structure of dolomite rock is explored further in section 9.4.3 with a viscoelastic response model of the Maxwell type.

9.3.3 Ramp wave compression of dolomite rock

The shock compression data obtained on dolomite rock exhibit features of rate-dependent deformation in several ways (Grady *et al* 1977). First, steady structured elastic waves to stress amplitudes below the HEL are a consequence of rate-sensitive deformation characteristics of the intrinsic microcrack population that pervades the rock material. Second, compression under shock wave loading differs starkly from the same uniaxial loading under static compression conditions. This difference in static and dynamic compressibility indicate rate-dependent behavior in the compression loading and failure of dolomite rock.

In the static strength studies of Blair dolomite (Heard *et al* 1973), onset of failure under static uniaxial strain loading is observed to occur at about 0.3 GPa, nearly a factor of ten lower than observed under shock wave compression. This stark difference between shock wave and static deformation suggests testing of dolomite rock using ramp wave loading techniques (Grady and Hollenbach, 1977). The objective is to investigate the deformation properties of the material at intermediate rates of loading.

Experimental techniques for generating ramp waves (waves with finite risetime) were under investigation at the time. Plane ramp waves generated with impact facilities can achieve strain rates down to about 10^4 s^{-1}, approximately two orders of magnitude below the nominal 10^6 s^{-1} strain rates of the corresponding shock wave test.

Fused silica exhibits wave propagation properties that make this material useful in the study of finite amplitude stress waves. First, it is nonlinear elastic to approximately 9 GPa (Barker and Hollenbach 1970). Second, the stress–strain curve to approximately 4 GPa is concave toward the strain axis. The latter property allows control of the risetime of the loading pulse applied to a test specimen. If a fused silica buffer is placed between the impacting projectile and the test specimen, then an initial step wave input will spread into a ramp wave as it propagates through the fused silica. The elastic properties of fused silica are such that an initial step wave will acquire a risetime of about 1 μs over a propagation distance of 25 mm. This provides a capability of controlling the strain rate of the input pulse applied to a test specimen over a range of about 10^4 s^{-1} to 10^6 s^{-1}.

Four impact experiments were conducted to determine how ramp wave loading applied to dolomite rock evolved during propagation through the material. In each experiment, an impact plate of fused silica 6.36 mm in thickness is accelerated with a light gas gun to a velocity of approximately 0.58 km s^{-1}. The accelerated plate undergoes planar impact on a buffer disc of fused silica 19 mm in thickness. The impact-generated compression wave propagates through the buffer and achieves a risetime of about 0.7 μs (corresponding to a strain rate of 3×10^4 s^{-1}) before transmission into the dolomite rock. Blair dolomite specimens of thicknesses 2, 5, 10 and 15 mm are used in the successive tests to determine the wave propagation characteristics of the initial ramp wave input. The fused silica projectile plate is backed with light foam so that a subsequent release wave is produced.

The sample is backed with a plate of fused silica (the laser window material), and a diffuse-surface laser interferometer (Barker and Hollenbach 1972) is used to measure the transmitted particle velocity profile at the interface between the sample and window. The wave profiles measured in the four experiments are shown in figure 9.24.

There are several features of the present ramp wave profiles that are not observed in the shock wave measurements. Focus on the loading wave profiles shown in detail in the right-hand plot of figure 9.24. The profile at 2 mm specimen thickness nearly duplicates the fused silica input profile since little wave dispersion has occurred over this distance. Profiles measured at increasing specimen thickness show how the wave evolves with propagation distance. The wave amplitude from zero to approximately 0.15 km s^{-1} (corresponding to about 2.0 GPa) steepens with propagation distance, indicating an increasing modulus with stress amplitude. The modulus is equal to $\rho_o c^2$ where ρ_o is the initial density and c is the Lagrangian wave velocity. In contrast, above about 0.15 km s^{-1} particle-velocity amplitude, the wave spreads with propagation distance, implying that the longitudinal modulus is decreasing with further increase in stress amplitude. This response is attributed first to an initial elastic response during which closure of an intrinsic crack porosity leads to an

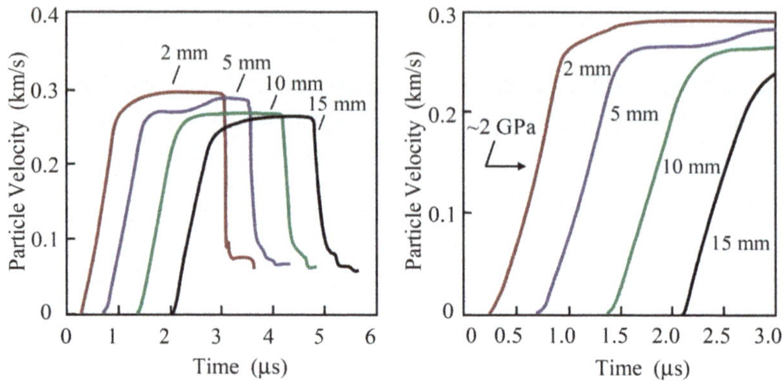

Figure 9.24. Particle-velocity profiles that evolve over the indicated propagation distance from an initial ramp wave input into the dolomite rock. Left: Total profiles including loading and release. Right: A magnified view of the loading profiles only (Grady and Hollenbach, 1977).

increasing elastic stiffening of the material. The subsequent response is ascribed to onset of inelastic yielding above about 2.0 GPa. Wave structure at the top is a consequence of subsequent wave reflection in the dolomite sample and does not partake in the analysis.

The character of the loading modulus in the ramp wave experiment is clarified further in a plot of the wave velocity versus the stress amplitude shown on the left in figure 9.25. In this figure, the dependence of the loading wave velocity on the stress amplitude (dashed curve) determined from the four ramp wave profiles in figure 9.24 is compared with similar wave speed plots from the shock wave experiments. Each solid curve is determined from a series of shock wave profiles as described earlier. Below about 2.5 GPa, all shock wave experiments show an essentially constant dependence of wave velocity on stress amplitude for a given series. However, the wave velocity (the shock velocity) increases with increasing peak stress. This trend results from shock loading along a Rayleigh line to increasing points on a positive curvature Hugoniot of a nonlinear elastic material. For experiments above 2.5 GPa (series 4 and 5), the wave-velocity plots show that failure and inelastic deformation occurs under shock compression. In contrast, the ramp wave compression indicates a continuously increasing wave velocity with stress amplitude to about 2.0 GPa followed by marked reduction in wave velocity with further stress increase.

The best estimate of the dynamic yield stress for Blair dolomite in the shock wave experiments at strain rates of order 10^6 s^{-1} was obtained from the series 5 shock wave profiles. In these tests, a HEL of 3.0 GPa at about 10 mm propagation distance decayed to about 2.5 GPa at a distance of 20 mm. In the ramp wave experiment, the onset of yielding is not abrupt and a definite yield point is not readily identified. If the point at which the loading modulus starts to decrease in figure 9.25 is selected, a yield stress of approximately 2.0 GPa is obtained. If consistency with the shock wave data is attempted by selecting the point at which a Rayleigh line lies tangent to the stress–strain response curve determined from the ramp wave data, then a yield stress of about 2.6 GPa is determined.

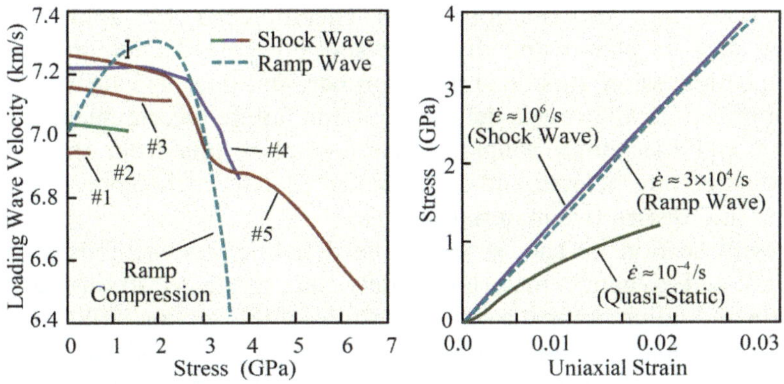

Figure 9.25. Left: Loading wave-velocity dependence on the stress amplitude for both shock wave and ramp wave experiments on dolomite rock. The accuracy in the shock wave data is about 1.5%. The accuracy in the ramp wave data is indicated by the error bar. Right: Stress versus uniaxial strain response for Blair dolomite at different strain rates. Shock wave data are compared with both quasi-static compression from Heard *et al* (1973) and compression assessed from the ramp wave tests (Grady and Hollenbach, 1977).

The exact value is not critical. The important point is that the yield stress at a strain rate of approximately 3×10^4 s^{-1} is in reasonable agreement with the shock compression value (~10^6 s^{-1}). It is markedly higher than the quasi-static uniaxial onset of failure of about 0.3 GPa (Heard *et al* 1973). On the right in figure 9.25, the stress–strain responses at the three strain rates are compared. It is clear that both the yield and subsequent deformation at rates of 3×10^4 s^{-1} are in better accord with the shock compression response than with the quasi-static response.

The ramp wave experiments show that the substantial difference in yield stress, stress–strain response and energy dissipation inferred from the comparison of shock wave and quasi-static data on dolomite rock is not simply a consequence of the extreme strain rates achieved in shock compression. Nominally, the same behavior persists down to strain rates of order 10^4 s^{-1}, nearly two decades of strain rate below the shock wave tests. The material physics underlying the observed deformation dynamics in Blair dolomite and other similar rock types has yet to be fully explained (e.g. Grady *et al* 1977). Many of these issues are touched upon in the following section.

9.4 Rate dependence in brittle deformation and failure

The processes of deformation of brittle solids such as ceramics, glass and hard rock media depend strongly on the rate of the deformation impressed on the material. The deformation is elastic, and then brittle fracture and fragmentation, except perhaps at the highest levels of strain rate and confining pressure. The off-Hugoniot experiments of Paris *et al* (2011) on silicon carbide and boron carbide ceramics lend insights into this transition. Although strain-rate dependence is clearly recognized, the assessment of the governing physics, development of descriptive constitutive models and measurement of appropriate material properties is difficult. This difficulty is because no single experimental method or apparatus spans the range of strain rates and deformation paths encompassed by application needs. Techniques such as planar and divergent shock waves, Taylor and rod impact, Hopkinson compression, torsion and tension bar, and intermediate and quasi-static triaxial compression, all provide deformation and failure measurements pertinent to the physics and mechanics sought. Experimental results from the various test methods must be interpolated and extrapolated to extract the underlying deformation physics and descriptive material models.

The present section pursues some of the dynamic test methods, experimental observations and dynamic material models that relate to strain-rate-dependent deformation and failure in brittle solids. Portions of this section build on discussions and analysis of topics provided in an early report (Grady and Hollenbach 1977). Aspects of some of the same issues have been treated more thoroughly in later studies (e.g. Ravichandran and Subhash 1995, Kimberleya *et al* 2012).

9.4.1 Observations of rate dependence

Test methods other than shock wave studies show evidence for rate dependence in the dynamic compression and failure of brittle solids. A large body of experimental data is provided by Hopkinson bar studies of the deformation of brittle materials. Impact loading of rods of the test material is an alternative shock technique that has revealed useful data on rate-dependent dynamic properties, as have several other complementary methods. Much of the earlier Hopkinson bar studies were on natural rock material. Later, higher quality manufactured ceramics were subjected to similar test methods.

Nonetheless, studies of the strain-rate dependence of failure and subsequent deformation of brittle solids is still scant, particularly at the higher strain rates, due primarily to experimental complexities. Investigations in the lower strain-rate regime (about 10^{-7} s^{-1} to 10^{-1} s^{-1}) have generally indicated a linear increase in strength with increase in log strain rate or reciprocal temperature for a number of geological materials (Robertson 1955, Turner *et al* 1956, Serdengecti and Boozer 1961, Heard 1963). Heard (1963) obtained reasonable agreement by fitting his data to a form of the Eyring equation, verifying a thermal activation rate-controlling process.

Kumar (1968) was possibly the first to note a deviation from a linear behavior between strength and log strain rate. In that work, unconfined split-Hopkinson bar experiments were conducted on granite and basalt from the Nevada Test Site to strain rates of about 2×10^3 s^{-1} at two different temperatures of the samples. An upturn in the ultimate strength was observed at strain rates of about 7–8 $\times 10^2$ s^{-1}, and ultimate strengths greater than twice that expected for a linear dependence were achieved at the highest strain rates. Kumar (1968) suggested a transition from a thermal activation controlled process to an inertial governed mechanism. He proposed crack propagation dynamics as responsible for the observed upturn in strength.

Green and Perkins (1968) in further split-Hopkinson bar studies observed a similar upturn in the ultimate strength of Solenhofen limestone at a strain rate of about 10^3 s^{-1}. Experiments on Westerly granite and a volcanic tuff did not reveal a deviation from linearity, although the higher strain-rate regime was not investigated as thoroughly. A further significant observation of Green and Perkins was a substantial increase in the loading modulus with strain rate in Westerly granite.

Lindholm *et al* (1974) conducted an extensive investigation of the strain rate and temperature dependence of failure in Dresser basalt using a split-Hopkinson bar apparatus with a fluid confining pressure. Maximum strain rates of 10^3 s^{-1} failed to reveal anything other than linear behavior in this material. Lankford (1976) noted an upturn in strength in several grades of oil shale at strain rates of about 10^3 s^{-1} with a split Hopkinson bar apparatus, and attributes the observation to a transition in the mechanism of crack nucleation and growth.

Janach (1976, 1977) considered the deviation from linearity observed in split-Hopkinson bar experiments and proposed a model that probably focuses reasonably closely on the responsible mechanisms, namely inertial effects associated with radial cracking (dilatancy) during dynamic compression. He concludes that this kind of dynamic loading is not controlled by rate-dependent material properties. Rather, that the strain-rate sensitivity is a consequence of a transition from a uniaxial stress to near-uniaxial strain loading conditions. This may be misleading since a state of elastic uniaxial stress is achieved within the specimen in times of the order suggested by Green and Perkins (1968). Subsequent deformation is governed by the rate of opening and propagation of radial cracks, which will necessarily be controlled by material inertia, as Janach suggests. A mechanism of this type is also expected to be important in a dynamic tensile experiment. Thus, Kumer's suggestion of inertial effects associated with finite crack propagation velocities (Kumar 1968), as responsible for the observed rate dependence, would also appear to be correct, although the form may be somewhat more complex than envisioned.

Strain rates of 3–4 × 10^3 represent an upper limit for the split Hopkinson bar apparatus before concerns with stress equilibrium arise. To reach greater strain rates, workers have resorted to shock wave techniques. In a typical shock wave experiment, an impact or explosive is used to achieve material loading, and the strain rates are not typically well defined. The loading wave immediately after impact is essentially discontinuous (principally determined by the surface finish of the respective materials). However, dispersive properties of the test material produce a finite risetime to the loading wave after a short propagation distance. Strain rates under shock conditions are typically 10^6 s^{-1} or higher. Ramp waves produced by appropriately designed planar impact methods, discussed earlier on the study of Blair dolomite, provide a carefully controlled uniaxial-strain strain-rate experiment in the domain between about 10^4 s^{-1} to 10^6 s^{-1}.

Strain-rate dependence in the dynamic loading of materials has been inferred from comparison of the shock wave stress versus strain compression with results achieved by other methods at much lower strain rates. Brace and Jones (1971) were apparently the first to compare static uniaxial strain and shock wave data for rocks at moderate stresses (<2.5 GPa). They concluded from investigation of three rocks—a granite, a tonalite and a limestone—that the strain-rate dependence of permanent deformation was approximately the same as earlier observations of fracture of rock under triaxial loading conditions. Comparable studies were extended to a broader suite of rocks (Brace and Riley 1972). Schock *et al* (1973) and Schock and Heard (1974) compare static uniaxial strain and shock wave data, and conclude that the shock wave results cannot be reconciled with low-strain-rate data. They suggest that both the uniaxial strain path and the failure envelope are sensitive to the loading strain rates achieved. Comparisons of static and dynamic deformation of Blair dolomite (Grady *et al* 1977) also shows marked strain-rate sensitivity, as noted in the previous section.

In figure 9.26 the failure stress dependence on strain rate are shown for various rock types from several of the sources discussed (Grady 2003). There are serious arguments for questioning the implications of this plot. First, conditions of uniaxial stress were maintained in the Hopkinson bar experiments of Kumar (1968) and Green and Perkins (1968) while uniaxial strain was achieved in the tests of Heard *et al* (1973) and Grady *et al* (1977) on Blair dolomite. Second, definitions of the failure stress are commonly not consistent over the different experiments performed. Nevertheless, the observations of a significant increase in failure stress starting at strain rates of about 10^3 s^{-1} provide impetus for further reflection on the underlying causes.

9.4.2 Analysis of strain-rate-dependent deformation and failure

In this subsection, some of the physical mechanisms of deformation that can contribute to rate sensitive yield and failure in brittle solids are pursued with an emphasis on hard rock and ceramics. Fracture mechanisms, rather than plastic processes, are the principle focus. In the closure of this section, some remarks are offered on the brittle-to-ductile transition. In this context, brittle-to-ductile implies a transition from a process of failure by fracture to a process of failure by plastic flow.

Mechanisms such as thermal activation and inertial effects associated with fracture initiation and growth introduce strain-rate dependence to the processes of

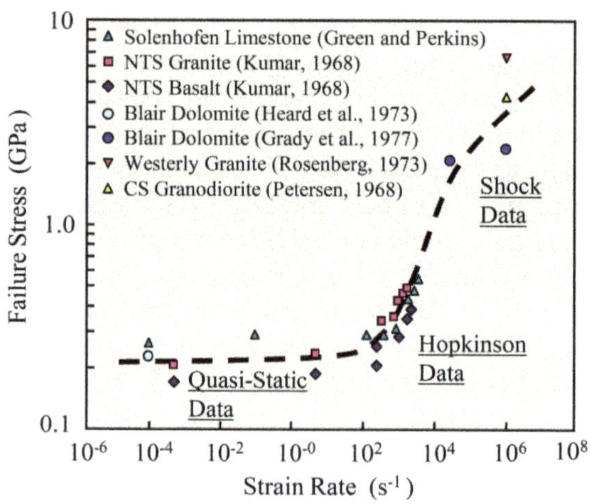

Figure 9.26. Comparison of intermediate rate Hopkinson bar data with lower quasi-static and higher shock-rate strength data in selected rocks (Grady 2003).

deformation and failure in brittle solids. These mechanisms are pursued through models of the fracture initiation and growth process as they apply to shock and high-rate deformation.

9.4.2.1 A model for brittle fracture

Ideas of fracture proposed by Shockey *et al* (1974) are exploited in the present analysis. Namely, (1) an initial distribution of flaws (or cracks) exists in the specimen that can grow according to specified growth laws and the stress history impressed on the sample; (2) the void volume associated with an individual crack can be approximated by an ellipsoidal void region in the form of a penny-shaped crack; and (3) the total strain can be represented as a superposition of the elastic strain and the crack void volume. This analysis is restricted to mode I tensile crack failure although comparisons are made with more general experimental results such as the failure data in figure 9.26.

Shockey *et al* (1974) examined in detail the distribution of inherent flaws in the quartzite rock Arkansas novaculite, and Hadley (1976) conducted a comparable study of the microcrack fabric in Westerly granite. The general observation in both studies was the existence of a flaw distribution in these two competent rocks that decreased rapidly in number with increasing flaw size, and with no flaws larger than about 1 mm. The flaw distribution was reasonably approximated by an exponential distribution function of the form

$$n(r_o) = \frac{N_o}{r_o} e^{-r_o/r_o^*}, \tag{9.5}$$

where $n(r_o)$ has the dimension of number of flaws per unit volume per unit crack length. Integration provides the cumulative distribution function considered by Shockey *et al* (1974), where N_o is the total number of flaws per unit volume, r_o is the initial half-width of a penny-shaped flaw, and r_o^* is a characteristic half-width.

The model assumes that the total strain-rate is provided by

$$\dot{\varepsilon} = \dot{\varepsilon}_e + \dot{\varepsilon}_c, \tag{9.6}$$

where $\dot{\varepsilon}_e$ and $\dot{\varepsilon}_c$ correspond to the strain rates associated with elastic strain of the solid material and the fracture void volume, respectively. The elastic strain, depending on the boundary constraints, is related to the stress through the intrinsic elastic modulus of the solid material. For example, $\dot{\varepsilon}_e = \dot{\sigma}/E$, where E is Young's modulus, for the case of loading under conditions of uniaxial stress. The strain rate associated with the fracture void volume is

$$\dot{\varepsilon}_c = \int_0^\infty n(r_o)\dot{V}_c(r_o)dr_o, \tag{9.7}$$

where $n(r_o)$ is an appropriate initial flaw distribution function such as provided in equation (9.5), and $\dot{V}_c(r_o)$ is the rate of volume growth for cracks of initial radius r_o. The volume of an ellipsoidal penny-shaped crack is

$$V_c = 4\pi r^2 b/3, \tag{9.8}$$

where b is the crack half-thickness and r is the current radius of the crack. The growth rate is

$$\dot{V}_c = \frac{4\pi}{3} r^2 \dot{b} + \frac{8\pi}{3} b r \dot{r}. \tag{9.9}$$

To complete the constitutive description, models that describe the crack expansion $\dot{b}$, and crack growth $\dot{r}$, are needed in terms of the deformation history of the material.

9.4.2.2 Elastic response of the model

Consider first the simpler problem of strictly elastic tensile loading where crack growth $\dot{r} = 0$, and the void expansion, according to equation (9.9), reduces to

$$\dot{V}_c(r_o) = 4\pi r_o^2 \dot{b}/3. \tag{9.10}$$

The present discussion is restricted to a distribution of penny-shaped cracks, closed at zero tensile stress, and oriented normal to the direction of wave propagation. Walsh and Brace (1972) provide a general analysis of elastic crack response under quasi-static conditions. The present model is sufficient to point out wave propagation features that occur during rapid dynamic loading.

When wavelengths associated with the applied pressure load approach the dimensions of the crack, inertial effects become important and the static solution for crack expansion is no longer a valid. At wavelengths near the resonant frequency, the crack will oscillate and reradiate elastic energy and the effect on the bulk material response is to create a frequency region of anomalous dispersion, or blocking band, within which waves will not propagate (Drumheller and Sutherland 1973).

Barker (1971) has shown that a viscoelastic model with normal dispersion can describe wave propagation in composite materials, whereas a strictly elastic analysis of the composite predicts anomalous dispersions. The implication is that dissipative effects are present that are not accounted for in the elastic treatment.

Similar behavior is expected in bulk rock with an inherent crack distribution. Shock wave effects observed in dolomite (Grady *et al* 1977) and in oil shale (Schuler *et al* 1976) show wave phenomena comparable to dynamic viscoelastic material response. These observations suggest the following damped response model for an individual crack,

$$\dot{b} = \frac{1}{\tau}(b - b_{\text{eq}}), \tag{9.11}$$

where b_{eq} is the equilibrium static value for the crack expansion determined by the current far field stress level, and $\tau = r_o/c$ (with c the elastic wave speed) is a measure of the dynamic relaxation time. Sneddon's expression (see Shockey *et al* 1974) provides

$$b_{\text{eq}} = \frac{4(1-\nu^2)r_o}{\pi E}\sigma, \tag{9.12}$$

for the expansion of a penny-shaped crack subject to a uniform positive tension σ, with ν and E Poisson's ratio and the elastic modulus, respectively. Equation (9.10), for the rate of volume opening of an individual crack, combined with the expansion model from equation (9.11) and Sneddon's expression in equation (9.12) yields

$$\dot{V}_c = -\frac{4r_o^2}{3\tau}\left(b - \frac{4}{\pi}\frac{(1-\nu^2)}{E}r_o\sigma\right), \tag{9.13}$$

and since $V_c = 4\pi r_o^2 b/3$,

$$\dot{V}_c = -\frac{1}{\tau}\left(V_c - \frac{16}{3}\frac{(1-\nu^2)}{E}r_o^3\sigma\right). \tag{9.14}$$

Equation (9.14) along with equation (9.6) and equation (9.7) provides

$$\dot{\varepsilon}_c = \int_0^\infty n(r_o)\dot{V}_c\,dr_o, \tag{9.15}$$

and

$$\dot{\varepsilon} = \dot{\sigma}/M + \dot{\varepsilon}_c, \tag{9.16}$$

where $M = E(1-\nu)/(1+\nu)(1-2\nu)$ is the longitudinal modulus, and a general constitutive description for the elastic response of the material under dynamic tensile loading. For a single harmonic frequency ω, equation (9.14) yields

$$V_c^\omega = \frac{16}{3}\frac{(1-\nu^2)}{E}r_o^3\left(\frac{1}{1+i\omega\tau}\right)\sigma^\omega. \tag{9.17}$$

Equation (9.16) becomes

$$\varepsilon^\omega = \left(\frac{1}{M} + \frac{16}{3}\frac{(1-\nu^2)}{E}\int_0^\infty \frac{n(r_o)r_o^3}{(1+i\omega\tau)}dr_o\right)\sigma^\omega. \tag{9.18}$$

For a delta function distribution in crack size

$$n(r_o) = \frac{N_o}{r_o^*}\delta\left(r_o^* - r_o\right), \tag{9.19}$$

which constitutes N_o cracks of equal half width r_o^*, substitution into equation (9.18) yields

$$\varepsilon^\omega = \left(\frac{1}{M} + \frac{16}{3}\frac{(1-\nu^2)}{E}\left(r_o^*\right)^3\left(\frac{1}{1+\omega^2\tau^2}\right)\right)\sigma^\omega. \tag{9.20}$$

Equation (9.20) predicts normal dispersion for a propagating periodic pressure wave with a compliance of $1/M$ for $\omega \gg \tau$ and $1/M + 16(1 - \nu^2)(r_o^*)^3/3E$ for $\omega \ll \tau$. The frequency band for the dispersion region has a width of approximately $2.7/\tau$.

The character of equation (9.18) for more general crack distribution functions (such as the exponential distribution in equation (9.5)) is normal dispersion with limiting behaviors for the low and high frequencies the same as the delta function distribution, but with a skewed and considerably broader dispersion region. The width and shape of the dispersion curve is dependent on the functional form of the distribution function.

9.4.2.3 Large-strain elasticity

The shock compression wave profiles measured on Blair dolomite below the HEL, illustrated by the waves displayed on the left in figure 9.21, exhibit a dependence on the impact stress amplitude. In figure 9.27 the arrival velocities of either the elastic shock wave, or the elastic precursor wave, is plotted for all of the experiments performed in the study described in the previous section. These data clearly illustrate the dependence of the finite amplitude elastic shock wave on the stress amplitude. For comparison, the zero-stress data point in the plot identifies the ambient pressure longitudinal ultrasonic velocity for this dolomite rock.

The behavior of dolomite rock subjected to shock compression can be compared with measurements of ultrasonic velocities in rocks with microcrack porosity when subjected to increasing levels of static confining pressure (Isenberg 1972). The longitudinal sound speed as a function of confining pressure for selected rocks is shown in figure 9.28. Closure of an initial microcrack porosity is thought to explain the increase in ultrasonic stiffness with confining pressure. There is qualitative agreement between the ultrasonic measurements in figure 9.28 and the shock wave velocities in figure 9.27. Differences in the pressure range over which changes are observed indicated added complexity, however.

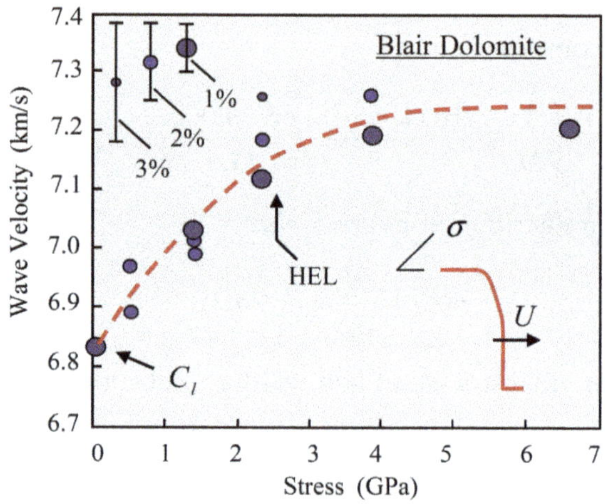

Figure 9.27. First arrival velocity measurements in shock experiments in Blair dolomite (Grady *et al* 1976, 1977). The abscissa is the peak stress achieved in the impact experiment. The approximate error in individual experiments is indicated by the size of the symbol used.

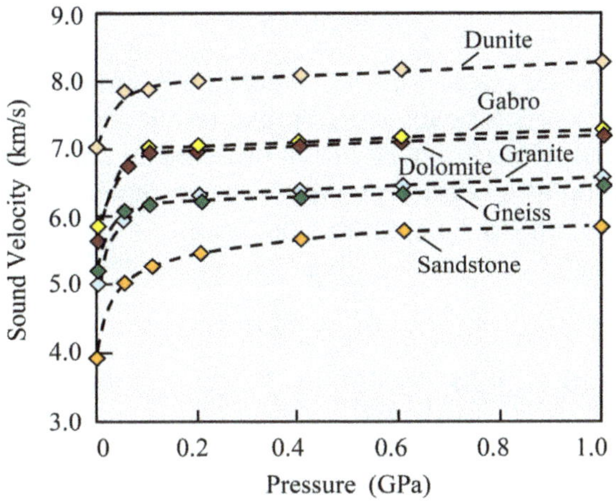

Figure 9.28. Longitudinal sound speed as a function of confinement pressure for selected rocks (Isenberg 1972).

A constitutive law governing elastic crack expansion from equation (9.11),

$$\dot{b} = -\frac{1}{\tau}(b - b_{\text{eq}}), \tag{9.21}$$

models the elastic dynamics of a penny-shaped crack. The static crack opening b_{eq} is a linear function of the applied stress provided by the Sneddon solution in equation (9.12) modified to account for an ambient (open) crack porosity. For modest compressive stress levels, this linear theory applies to the present shock wave application. For larger amplitude stress waves, however, crack closure occurs due to the higher stress levels and a linear theory is not adequate. Ultrasonic studies on rock show that the acoustic wave velocity approaches the intrinsic velocity of the solid material exponentially with increasing confining pressure. To account for crack closure in the shock or static high-pressure application, an equilibrium crack opening relation of the form,

$$b_{\text{eq}} = b_o \mathrm{e}^{-\sigma/\sigma_o}, \tag{9.22}$$

is assumed to apply where b_o is a nominal crack width and σ_o is a characteristic closure stress. From the preceding section, the rate of volume change of an individual crack is

$$\dot{V}_c = 4\pi r_o^{\,2}\dot{b}/3, \tag{9.23}$$

and the total crack-opening rate of strain due to N_o similar cracks is

$$\dot{\varepsilon}_c = -\frac{1}{\tau}[\varepsilon_c - p(1 - \mathrm{e}^{-\sigma/\sigma_o})], \tag{9.24}$$

where

$$p = 4\pi N_o r_o^2 b_o \tag{9.25}$$

is a measure of the initial crack porosity. Since higher amplitude stress levels are of interest, a linear stress versus strain expression as developed in the previous section is no longer adequate. A Murnaghan equation provides an appropriate measure of the nonlinear compression at modest pressures (a linear dependence of the modulus on stress). Consequently, an elastic compression strain,

$$\dot{\varepsilon}_e = \frac{\dot{\sigma}}{M_o + M_o'\sigma}, \tag{9.26}$$

combined with equation (9.24) to obtain the total strain, yields the constitutive relation

$$\dot{\varepsilon} - \frac{\dot{\sigma}}{M_o + M_o'\sigma} = -\frac{1}{\tau}\left[\varepsilon - \frac{1}{M_o'}\ln\left(M_o + M_o'\sigma\right) - p(1 - \exp(-\sigma/\sigma_o))\right]. \tag{9.27}$$

This constitutive relation is of the form of a generalized Maxwell equation characteristic of nonlinear viscoelastic response. The expression predicts an instantaneous stress–strain response at rapid loading rates,

$$\varepsilon_{\text{inst}} = \frac{1}{M_o'}\ln\left(M_o + M_o'\sigma\right), \tag{9.28}$$

but relaxes with a characteristic time $\tau \sim r_o/c$ to an equilibrium response,

$$\varepsilon_{\text{eq}} = \frac{1}{M_o'}\ln\left(M_o + M_o'\sigma\right) + p(1 - \exp(-\sigma/\sigma_o)). \tag{9.29}$$

This nonlinear rate-dependent viscoelastic response sensibly models the nonlinear elastic shock response of Blair dolomite.

9.4.3 Steady elastic shock waves in dolomite rock

Compression wave profiles to peak shock stresses in the dolomite rock below about 2.5 GPa show a sharp shock to a breakpoint followed by a gradual rise to peak amplitude. Wave structure of this type is illustrated in figure 9.21. Dynamic failure in the material response under shock compression results in a two-wave structure in the loading compression wave of numerous materials (Fowles 1961). There are several reasons that indicate that onset of failure is not responsible for the present observed wave structure. First, the stress at which the break occurs is a function of the peak driving stress. Second, the loading profiles do not show an unsteady wave separation of the first wave from the second wave. Rather, the wave propagates as a single steady structured wave. Lastly, the velocity of the observed precursor waves depends on the stress amplitude.

This type of structured wave propagation has been studied in materials such as plastics and composites. The behavior commonly suggests a rate-dependent viscoelastic response of the material during dynamic loading. A notable study is the experimental and theoretical investigation of nonlinear viscoelastic wave propagation in polymethyl-methacrylate (PMMA) (Nunziato *et al* 1974).

In the previous subsection, inertia of a microcrack system in rock media is explored as a deformation mechanism that can lead to an effective viscoelastic response of the Maxwell type. A further possible mechanism for the observed response in dolomite is rate-dependent frictional sliding in preexisting cracks. This phenomenon as a mechanism effecting dynamic deformation of rock materials has been studied by Brace and Jones (1971). A frictional mechanism is proposed as being responsible for the rate dependence observed in rock deformation by Schotz and Kranz (1974). The variation in acoustic velocity with increasing confining pressure observed in Blair dolomite (Heard *et al* 1973) is consistent with initial crack porosity. The shock data and microstructure analysis in the study on dolomite rock of Grady *et al* (1977) similarly indicate influences on the wave propagation of crack porosity, or effective crack porosity (cracks filled with lower impedance media such as organic materials).

Here a physical model of the Maxwell type is explored that is capable of describing features of the wave propagation data. The model determines the dynamic stress–strain response and viscous-like relaxation of the dolomite rock when subjected to shock compression. The character of particle-velocity profiles illustrated in figure 9.21 follow from solutions of a generalized nonlinear Maxwell constitutive equation of the form (Herrmann and Nunziato 1973)

$$\frac{\partial \varepsilon}{\partial t} = f(\sigma,\varepsilon)\frac{\partial \sigma}{\partial t} + g(\sigma,\varepsilon), \tag{9.30}$$

where σ and ε are the axial stress and strain, respectively, while $f(\sigma,\varepsilon)$ determines the elastic response, and $g(\sigma,\varepsilon)$ is a relaxation function that determines the viscous response. A constitutive relation of this type provides for an equilibrium stress versus strain, or stress versus particle-velocity, response at low rates of loading and an instantaneous stress–strain or stress–particle-velocity response at rapid rates of loading. The constitutive equation also accounts for relaxation from one response to the other under more complex dynamic loading. Equilibrium and instantaneous response curves of the form provided by equation (9.30) in materials supporting steady structured viscoelastic shock propagation are illustrated on the right in figure 9.29.

Steady waves result from a balancing of elastic and viscous stresses that constrain the loading path to a straight line in stress–strain or stress–particle-velocity space (Band and Duvall 1961). This straight line is the Rayleigh line illustrated on the right in figure 9.29. A single experimental steady wave profile to a specific peak amplitude will achieve one particle-velocity point on both the instantaneous response curve and the equilibrium response curve.

The instantaneous and the equilibrium responses for the three series of shock wave experiments on dolomite rock to stress levels below the 2.5 GPa HEL are plotted in figure 9.30 in shock-velocity versus particle-velocity space. Linear relations readily describe the shock wave data. Namely,

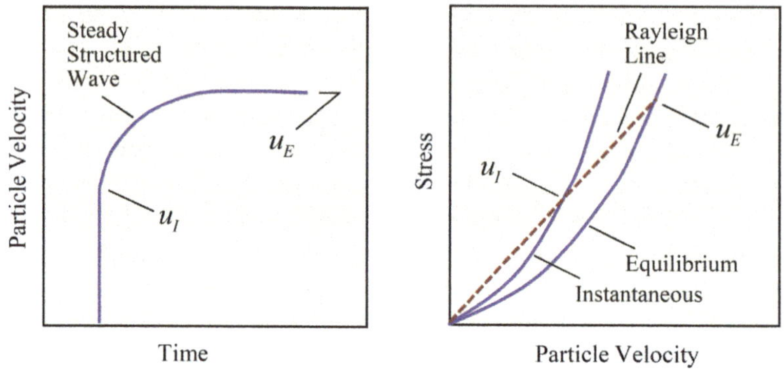

Figure 9.29. Steady wave properties of a nonlinear Maxwell constitutive equation. Right: The instantaneous and equilibrium response paths and the Rayleigh line loading path are illustrated. Left: A steady structured wave, indicating instantaneous and equilibrium particle-velocity levels.

$$U = a_I + b_I u, \tag{9.31}$$

for the instantaneous response and

$$U = a_E + b_E u \tag{9.32}$$

for the equilibrium response. U and u are the shock velocity and particle velocity, respectively.

The methods required to fit a nonlinear Maxwell equation to steady wave profiles are discussed in Nunziato *et al* (1974) and applied here. Equation (9.30) is specialized to the form

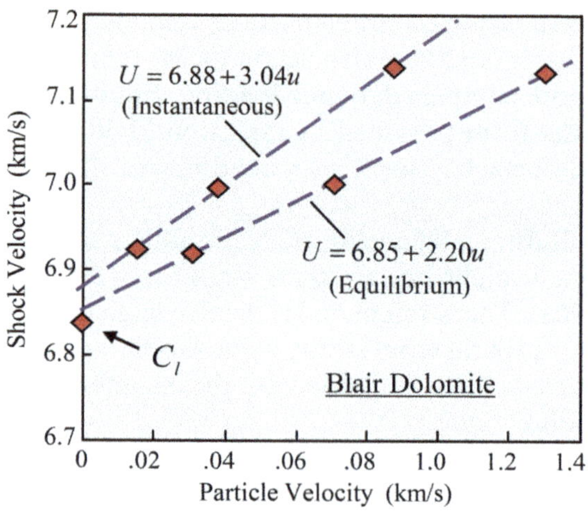

Figure 9.30. Linear shock-velocity versus particle-velocity structured shock compression data for Blair dolomite (Grady *et al* 1977). Measured states are the instantaneous and equilibrium response of the compression wave data below the 2.5 GPa HEL.

$$\frac{\partial \sigma}{\partial t} - E_I(\varepsilon)\frac{\partial \varepsilon}{\partial t} = -\frac{\sigma - \varphi_E(\varepsilon)}{\tau}, \tag{9.33}$$

which is found to be sufficiently general for structured shock waves in other studies. When combined with the equations of continuity,

$$\frac{\partial \varepsilon}{\partial t} + \frac{\partial u}{\partial h} = 0, \tag{9.34}$$

and momentum,

$$\frac{\partial u}{\partial t} + \frac{1}{\rho_o}\frac{\partial \sigma}{\partial h} = 0, \tag{9.35}$$

the set of three equations constitutes a complete system of hyperbolic differential equations that describe the evolution of nonlinear one-dimensional waves. The function $\varphi_E(\varepsilon)$ describes the equilibrium stress–strain response of the material. $E_I(\varepsilon)$ is the modulus that determines the instantaneous stress–strain response. The time constant τ characterizes material relaxation in the constitutive model.

Since, for steady wave propagation, material compression proceeds along a straight Rayleigh line, the Hugoniot jump conditions apply from the foot of the wave to any point on the wave. In conjunction with the linear U–u expressions in equations (9.31) and (9.32), the experimental wave profile data determine the functional forms for $E_I(\varepsilon)$ and $\varphi_E(\varepsilon)$. The expressions obtained are

$$\varphi_E(\varepsilon) = \rho_o a_E^2 \varepsilon/(1 - b_E\varepsilon)^2 \tag{9.36}$$

for the equilibrium response and

$$E_I(\varepsilon) = \rho_o a_1(1 + b_1\varepsilon)/(1 - b_1\varepsilon)^3 \tag{9.37}$$

for the instantaneous modulus.

It remains to determine from the data the time constant τ. This is accomplished by assuming a steady wave solution for the system of equations (9.33)–(9.35) in terms of a single scaling variable

$$\xi = 1 - h/U, \tag{9.38}$$

where U is the shock velocity of the steady profile. Equations (9.33)–(9.35) then reduce to the single ordinary differential equation

$$\frac{\rho_o U^2 - E_I(u)}{\rho_o U^2 u - \varphi_E(u)}\frac{du}{d\xi} = -\tau. \tag{9.39}$$

The functional dependence of $E_I(\varepsilon)$ and $\varphi_E(\varepsilon)$ on u follows from equations (9.36) and (9.37) and the Hugoniot conservation relations. The approximation $E_I(u) = a_I + 2b_I u$ is used, which is a reasonable simplification within the particle-velocity range of interest. Equation (9.39) integrates to the expression

$$-\frac{\xi}{\tau} = a(u - u_I) + \beta \ln \frac{u(1 - u/u_E)}{u_I(1 - u/u_E)} + \frac{\gamma}{2} \ln \frac{u_I(1 - u/u_E)}{u(1 - u/u_E)}, \tag{9.40}$$

where

$$a = 4b_I^2/Ub_E, \tag{9.41}$$

$$\beta = (2b_I/Ub_E)[a_I + (b_I/b_E)(U - a_E)] \tag{9.42}$$

and

$$\gamma = -\left[2\left(U^2 - Ua_I^2\right)/(U^2 - a_E)\right] + (4b_I/Ub_E)[a_I + (b_I/b_E)(U - a_E)]. \tag{9.43}$$

An optimum fit for the experimental steady wave profiles below 2.5 GPa is obtained with approximately $\tau = 0.01$ μs.

The question remains as to the underlying physics of the relaxation time τ in the present application to the shock deformation of rock. Within the framework of the viscoelastic model adjusted to the loading wave propagation data for dolomite rock below 2.5 GPa, it represents a characteristic relaxation time associated with the viscous or quasi-viscous mechanisms active in this strain-rate regime. The rapid increase in the modulus with stress, identified in the wave propagation data by both the increasing shock velocity, and by the dispersive behavior of the unloading wave is similar to the effects of preexisting cracking in rock material on acoustic data.

Analysis in the earlier section suggested that $\tau = r_o/c$ is a consequence of inertia in the dynamic response of the microcrack system. The value of $\tau = 0.01$ μs from a Maxwell model description of the structured wave data suggests a crack size r_o of order 50 μm, not inconsistent with crack flaw distributions suggested by Shockey *et al* (1974) in other rock materials. Alternatively, however, Brace and Jones (1971) propose that interfacial friction on closed cracks may provide the rate-dependent mechanism.

The generalized Maxwell model nonetheless adds insight into the underlying physics and dynamic deformation mechanisms underlying structured shock wave propagation in rock and perhaps other brittle media. The model captures the main features of the structured shock wave propagation in dolomite rock in the near-elastic shock regime below the HEL.

9.4.4 Strength at Hopkinson bar strain rates

A substantial body of experimental work has focus on the compressibility and strength of rock and ceramic materials using Hopkinson bar techniques to investigate strain rate response several decades below that of shock wave methods. Trends are observed in the studies that make interpretation of data less straightforward than with comparable data on metals. The present section explores the physics and mechanics underlying such testing and the resulting material response. The background for the topic follows from an earlier report (Grady and Hollenbach 1977). Similar issues have been addressed in more recent assessments (e.g. Ravichandran and Subhash 1995, Kimberleya *et al* 2012).

9.4.4.1 Single-crack dynamics

When the stress intensity factor exceeds the fracture toughness of the material in the neighborhood of a critical fracture flaw in a brittle material, fracture initiates and failure occurs. Such fracture must necessarily be accounted for in a constitutive description of brittle material dynamic deformation and failure. In the following analysis, a simple experiment is considered where a cylindrical test specimen of radius R is loaded in tension at a uniform stress rate (elastic strain rate), as illustrated in figure 9.31.

Strain rates in the neighborhood of the upturn observed in the data in figure 9.26 of about $10^3\ \mathrm{s}^{-1}$ to $10^4\ \mathrm{s}^{-1}$ are of particular interest. The present analysis explores mechanisms that account for the strain-rate dependence of deformation and failure observed in Hopkinson bar experiments on brittle materials. Although the present analysis focuses on tensile loading, and such experiments are commonly conducted under compressive loading, it is evident that the fracture mechanisms considered should occur under either loading condition.

A single ellipsoidal crack is approximately centered in the test specimen as illustrated in figure 9.31. Analysis of the test specimen response applies the model of crack growth resulting in equation (9.9),

$$\dot{V}_c = \frac{4\pi}{3} r^2 \dot{b} + \frac{8\pi}{3} b r \dot{r}. \tag{9.44}$$

Simple laws are used for the crack expansion $\dot{b}$ and crack growth $\dot{r}$ that are appropriate to the application. Brittle crack growth is considered and crack propagation velocities should rapidly approach a reasonable fraction of the sonic Rayleigh wave or shear wave speed c_s. The exact value is not critical. The critical crack in figure 9.31 is initially much smaller than the sample size $r_o \ll R$. For most of the crack-growth history $r \gg r_o$ and initial acceleration of the crack is ignored.

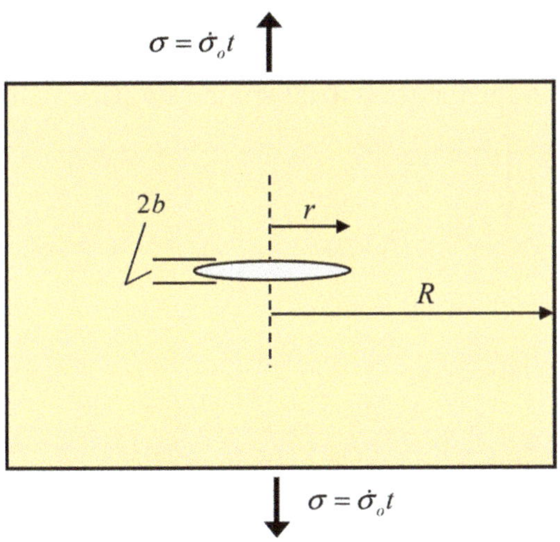

Figure 9.31. Hopkinson bar specimen with idealized fracture flaw subject to dynamic loading in tension. Failure of the specimen is through onset of fracture of the interior ellipsoidal flaw.

A constant rate of growth of the crack is assumed,

$$\dot{r} = c_s. \tag{9.45}$$

If a discontinuous step wave of tensile stress amplitude σ is normally incident on the crack face, then the wave reflection problem requires that the crack expansion is initially governed by

$$\dot{b} = \frac{\sigma}{\rho_o c_l}, \tag{9.46}$$

with ρ_o the specimen density and c_l the longitudinal wave speed. Since crack growth is governed by the crack speed in equation (9.45), the crack volume cannot immediately satisfy Sneddon's solution in equation (9.12) since expansion is constrained through inertial conditions that are bounded by the expansion speed from equation (9.46). Since the crack growth $\dot{r}$ is at or near the shear wave velocity, only modest stress resistance will propagate back to the crack center. Thus, crack expansion $\dot{b}$ will not vary significantly over the problem time and equation (9.46) is assumed as the law governing crack expansion. At a constant stress rate $\dot{\sigma}_o$ equation (9.46) provides

$$\dot{b} = \frac{\dot{\sigma}_o}{\rho_o c_l} t. \tag{9.47}$$

Incorporating the crack-growth and crack-expansion laws from equations (9.45) and (9.46) into equation (9.44) leads to a relation for the rate of growth of the penny-shaped crack volume,

$$\dot{V}_c \simeq \frac{8}{3}\pi \frac{c_s^2}{\rho_o c_l} \dot{\sigma}_o (t - t_o)^3, \tag{9.48}$$

where $t_o = \sigma_c/\dot{\sigma}_o$ is the time at fracture initiation and $\sigma_c \sim K_c/\sqrt{r_o}$ is of the order of the Griffith fracture stress.

Assuming N_o similar cracks per unit volume, and incorporating the elastic strain, provides an expression for the total axial strain rate per equation (9.6) of

$$\dot{\varepsilon} = \begin{cases} \dot{\sigma}/E & t < t_o \\ \dot{\sigma}/E + \dfrac{8}{3}\pi \dfrac{N_o c_s^2}{\rho_o c_l} \dot{\sigma}_o (t - t_o)^3 & t \geqslant t_o \end{cases}, \tag{9.49}$$

and, since $\sigma = \dot{\sigma}_o t$, the following relation is obtained,

$$\dot{\varepsilon} = \begin{cases} \dot{\sigma}/E & \sigma < \sigma_c \\ \dot{\sigma}/E + \dfrac{8}{3}\pi \dfrac{N_o c_s^2}{\rho_o c_l} \dfrac{1}{\dot{\sigma}_o^2} (\sigma - \sigma_c)^3 & \sigma \geqslant \sigma_c \end{cases}. \tag{9.50}$$

A more detailed analysis of equations (9.44)–(9.50) results in an initial $(\sigma - \sigma_c)^2$ behavior that transitions smoothly to a $(\sigma - \sigma_c)^3$ dependence for σ appreciably larger than σ_c. For the present analysis equation (9.50) is adequate.

9.4.4.2 Multiple crack behavior

An individual crack will not continue to propagate, but will eventually interact with other cracks, or with a free boundary of the test specimen. As this event occurs, the specimen loses the ability to support stress and disintegrates through the brittle fracture process. If the population of active cracks is sufficiently dilute such that the sample size provides the characteristic length scale of crack propagation, then failure subsequent to fracture onset will occur in a time $t_f \simeq R/c_s$. The sample size is R, as illustrated in figure 9.31 while c_s is the nominal crack speed. The stress at ultimate failure is then

$$\sigma_f = \sigma_c + \dot{\sigma}_o \frac{R}{c_s}. \tag{9.51}$$

If crack interaction rather than the boundary determines failure, then failure becomes independent of sample size. Sample size dependence is assumed in the subsequent calculation.

Equation (9.51) is integrated to obtain the final stress, strain and strain-rate response of the material,

$$\varepsilon = \begin{cases} \sigma/E & \sigma < \sigma_c \\ \sigma/E + \dfrac{2}{3}\pi \dfrac{N_o c_s^2}{\rho_o c_l} \dfrac{(\sigma - \sigma_c)^4}{\dot{\sigma}_o^3} & \sigma_c \leqslant \sigma \leqslant \sigma_f \end{cases}. \tag{9.52}$$

Experimental dimensions and material properties for the Hopkinson bar tests of Green and Perkins (1968) on Solenhofen limestone are used in equations (9.51) and (9.52). On the left in figure 9.32 the resulting stress versus strain responses to failure at several strain rates are shown. The curves compare closely to the experimental result reported by Green and Perkins.

In the right-hand plot of figure 9.32 the failure stress versus strain-rate relation from equation (9.51) is in the form

$$\sigma_f/\sigma_c = 1 + e^{\ln\left(\dot{\varepsilon}_o/\dot{\varepsilon}_o^*\right)}, \tag{9.53}$$

where

$$\dot{\varepsilon}_o^* = \sigma_c c_s / ER \tag{9.54}$$

is a characteristic strain rate at which failure stress has doubled in the upturn region. For Solenhofen limestone, the calculated value of $\dot{\varepsilon}_o^* = 2.8 \times 10^3\ \mathrm{s}^{-1}$, based on equation (9.54), is in sensible agreement with the data of Green and Perkins (1968).

The present rather basic model is capable of providing a qualitative description of the stress versus strain and failure stress versus strain-rate response qualitatively observed in brittle materials, and more specifically for the data of Green and Perkins (1968) on rock materials. The model suggests that stiffening of the material at increasing strain rate is due to inertia of the elastic crack expansion (the $\dot{b}$ term). The increasing failure stress with strain rate is due to finite crack growth rate (the $\dot{r}$ term). The model reveals that the rate dependence of strength is size-dependent; that is, if

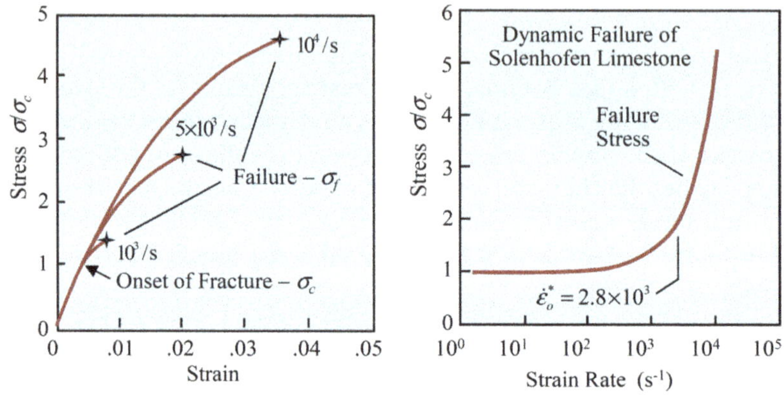

Figure 9.32. Left: Calculated tensile stress versus strain response is plotted for brittle fracture onset and failure using properties for Solenhofen limestone (Grady and Hollenbach 1977). Right: Failure stress versus strain rate based on the model for brittle fracture and failure of Solenhofen limestone subject to a uniform rate tensile loading. Stress plots are normalized by the stress at fracture onset.

the sample size is increased, the characteristic upturn strain rate (9.54) correspondingly decreases.

For a given crack distribution this size dependence would not continue unbounded. As the free boundaries become increasingly removed, crack coalescence rather than the free boundary interaction would become the mechanism for loss of the stress supporting strength. The characteristic crack coalescence length L of order $N^{-1/3}$, where N is the number of actively participating cracks per unit volume and would replace R in equation (9.54) yielding

$$\dot{\varepsilon}_o^* \approx \sigma_c c_s N^{1/3}/E. \tag{9.55}$$

It is important to emphasize that, although the rate effects observed by Green and Perkins (1968) are undoubtedly size-dependent, they are nonetheless important indicators of inertial-rate-dependent material property influences that can be expected under any geometry at these strain rates.

9.4.5 Intermediate strain rates (10^4 s^{-1} to 10^5 s^{-1})

Intermediate strain rates refer here to rates just above those typically achieved in Hopkinson bar experiments ($\sim 10^3$ s^{-1}), but below those commonly accessed in shock wave experiments ($\sim 10^6$ s^{-1} and higher). In principal, impact-induced ramp wave experiments (e.g. Grady and Hollenbach 1977) and tailored rod experiments (e.g. Chhabildas *et al* 1998) provide experimental methods that can access this strain-rate regime. The interplay of a rapidly rising stress load, and the microcrack fabric of a brittle solid, introduces new effects in this range.

When stress pulse risetimes approach wave transit times across the dimensions of initial flaws, additional strain-rate-dependent phenomena come into play. These include finite incubation times associated with fracture initiation and the early time acceleration of fracture propagation. These mechanisms, coupled with continued retardation of crack expansion due to inertial effects, further increase material dynamic stiffness during the macroscopic processes of initial inelastic yield, deformation and ultimate failure.

Under quasi-static loading, fracture initiation in a penny-shaped crack of radius r_o proceeds when the applied load satisfies the classic Griffith criterion

$$\sigma_c \simeq K_{IC}/\sqrt{r_o}, \tag{9.56}$$

where K_{IC} is the critical stress intensity factor (fracture toughness) under mode I stress loading. For rapid dynamic loading (9.56) is no longer generally valid. Various authors (e.g. Tuler and Butcher 1968, Steverding and Lehnigk 1976, Kalthoff and Shockey 1977) have addressed the question of fracture initiation under short pulse loading. The criterion of Tuler and Butcher is addressed in some detail in other applications in the present text, and is not considered here. Steverding and Lehnigk formulated a theory based on a criterion of critical action to initiate fracture. Kalthoff and Shockey used solutions of the stress-intensity history under sharp shock loading to formulate their fracture criterion. These criteria are similar, although Kalthoff and Shockey clearly illustrate how the criteria becomes independent of crack length for short pulse duration, while Steverding and Lehnigk failed to separate their criterion from the crack length. The efforts of Kalthoff and Shockey are pursued here to estimate the strain-rate dependence of fracture initiation under dynamic loading based on this facet of fracture mechanics.

The dynamic stress intensity factor K_I for a crack of radius r_o, subject to a tensile stress step loading of amplitude σ, exhibits a time dependence as illustrated in figure 9.33 (e.g. Kalthoff and Shockey 1977). For early time the growth of K_I is proportional to $\sqrt{ct}$ due to material inertia near the crack tip (c is the acoustic wave speed). K_I over-shoots the static value by an amount of about $4/\pi - 1$ (Achenbach and Brock 1973) and then undergoes oscillatory convergence to the static value. Thus, the limiting values are

$$K_I \simeq \begin{cases} \sigma\sqrt{ct} & t \ll r_o/c \\ \sigma\sqrt{r_o} & t \gg r_o/c \end{cases}. \tag{9.57}$$

Fracture initiation requires $K_I \geqslant K_{IC}$. For large times (slow loading rates) the Griffith criterion is recovered,

$$\sigma_c \simeq K_{IC}/\sqrt{r_o}, \tag{9.58}$$

while for short times (rapid loading rates) fracture stress and time are related through

$$\sigma_c \simeq K_{IC}/\sqrt{ct_c}. \tag{9.59}$$

Equation (9.58) is independent of the loading strain rate while equation (9.59) is not. The strain-rate dependence implied by equation (9.59) is of the order

$$\dot{\varepsilon} \sim \sigma_c/Et_c. \tag{9.60}$$

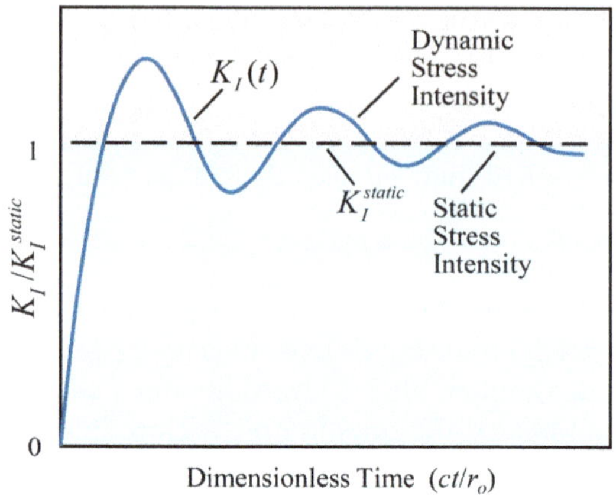

Figure 9.33. Dependence of the dynamic stress intensity factor on dimensionless time where c is the acoustic wave velocity and r_o is the initial crack half radius.

Eliminating t_c between equation (9.59) and equation (9.60) results in a cube root dependence,

$$\sigma_c \sim \sqrt[3]{\rho_o c K_{IC}^2}\,\dot{\varepsilon}^{1/3}, \tag{9.61}$$

for the fracture initiation stress dependence on the strain rate. A more detailed analysis for a penny-shaped crack (Kipp *et al* 1980) results in

$$\sigma_c \simeq \sqrt[3]{\frac{9\pi E K_{IC}^2}{16N^2 c_s}}\,\dot{\varepsilon}^{1/3}, \tag{9.62}$$

where $N = 1.12$ is determined numerically. The energy-based approaches to dynamic fracture and fragmentation discussed in chapter 2 lead to similar functional dependence of the tensile fracture stress on strain rate.

Based on these arguments, fracture initiation stress is expected to be strain-rate-dependent and have a cube root dependence for strain rates in excess of about $c\sigma_c/r_o E$, and tend to a rate-independent Griffith criterion for lower strain rates. A transition strain rate will depend on the characteristic flaw size governing fracture. For a flaw size of about 1 mm a strain-rate transition based on the above reasoning should occur at about $10^3\ \mathrm{s}^{-1}$.

After fracture initiation, there is a finite acceleration time during which brittle crack propagation transitions to a limiting crack velocity. An accurate growth law for $\dot{r}$ in the earlier model would account for this acceleration. When the strain rate is on the order of or smaller than about c/r_o, an acceleration growth law such as that proposed by Mott (1948), or more complex forms derived by Dulaney and Brace (1960) and Berry (1960), provide reasonable descriptions of fracture acceleration. (Further discussion is provided in chapter 3.) At higher strain rates, the expression for accelerated growth must become independent of crack length, as does the expression for fracture initiation. Expressions for the crack-velocity history subsequent to very rapid loading, such as that derived by Achenbach and Brock (1973), suggests that this is the case and indicate that an appreciable fraction of terminal velocity is reached in a time on the same order of the characteristic time in equation (9.59) required to achieve fracture initiation. A number of these aspects of dynamic fracture mechanics receive a treatment in the text of Freund (1990).

9.4.6 Strain-rate dependence in brittle materials

At strain rate approaching about 10^6 s^{-1} (shock wave loading) stress levels for initiation of fracture are a strong function of strain rate as anticipated by equation (9.62). Although stress concentrators (crack tips) are present, material inertia inhibits the buildup of stress concentration at high rates of strain. Stress levels consistent with plastic deformation processes such as inelastic dislocation nucleation and growth are achieved before stress relief by fracture can occur. (Shock recovery experiments reveal evidence of plastic deformation in many brittle solids (e.g. Cagnoux and Longy 1988).) It is reasonable that at strain rates of this order a transition from failure through fracture to failure through plastic flow will occur. Such a transition seems even more likely under compressive loading, where closure of pre-existing flaws will introduce additional resistance due to interfacial friction and pressure confinement, further inhibiting fracture and enhancing plastic processes (Grady *et al* 1977). Exceptions occur as demonstrated by the experiments on boron carbide of Paris *et al* (2011) described in section 9.2.7.

At the lower end of the strain-rate spectrum (below about 10^2 s^{-1}), rate-controlling processes associated with inertia are not an issue provided competent materials with intrinsic flaw sizes less than about 1 mm are the concern. In this region, a linear dependence of failure stress on the log strain rate is observed experimentally and such behavior has been identified with kinetic processes of thermal activation. (It should be noted that *in situ* geologic media, where flaws on all length scales up to those associated with joints and faults occur, will exhibit rate dependence in the dynamics of deformation and failure over a comparably wider range of strain rates. An accounting of the inherently more complex flaw and defect structure in massive geologic media is required in modeling the material response).

To identify thermally activated mechanisms in the process of brittle fracture, it is necessary to view failure within the crack tip on a microscopic and atomistic level. On this scale, processes such as molecular bond rupture and dislocation motion are thermally activated. Crack surface energy is periodic on the order of atomic dimensions and chaotic on the microscale. Local variations will differ from the continuum strain energy release rate, $G \simeq K_I^2/2E$. The periodic potential or heterogeneous potential will lead to lattice trapping of the crack analogous to the Peierls resistance in dislocation motion (Lawn and Wilshaw 1975). Thermal activation provides the means for crack progression. The growth law can be written as

$$\dot{r} = A\mathrm{e}^{-V(G)/kT}, \tag{9.63}$$

with $V(G)$ an activation potential, while A includes the activation attempt frequency, the activation entropy which relates to the directionality of attempt, and the number of sites (dislocations, bond ruptures, etc,) participating in the process. For small deviations of the strain energy release from Griffith equilibrium, the potential can be linearized in terms of the stress state, and through equation (9.63) the required relation between stress, strain rate and temperature at fracture initiation can be obtained. For large deformation, the nonlinear dependence of $V(G)$ on G must be accounted for. At strain rates below those at which inertial effects become controlling, thermal activation processes provide the dominant rate controlling mechanism. At higher strain rates, thermally activation rate dependence is second-order and superposes on the inertial rate dependence. This superposition is observed in the temperature-dependent Hopkinson bar experimental results of Kumar (1968), and would probably continue to be observed up to strain rates at which viscous drag effects become important in dislocation motion.

The previous several sections provided a heuristic model for examining the role of rate dependence in the failure of brittle solids under shock and high-strain-rate compression. The dolomite shock and ramp wave data, and the collection of failure data in figure 9.26 provide a display of experimental data that show evidence of rate dependence in compression and failure at high rates of strain. The fracture model of Shockey *et al* (1974), along with some simplifying constitutive assumptions, provides and effective analytic means for identifying underlying physics of rate-dependence in brittle solids, and in indicating the strain-rate regions and functional dependence of some of these mechanisms.

Figure 9.34 shows a strength curve for a brittle solid exhibiting a qualitative dependence on strain rate comparable to the experimental plot provided in figure 9.26. Competent laboratory samples of brittle materials are considered. Transition boundaries identified in the figure are nominal and will depend on the brittle material. Below strain rates of about 10^2 s^{-1} to 10^3 s^{-1}, processes of thermal activation provide the dominant rate-controlling mechanism governing material strength. At strain rates approaching about 10^3 s^{-1}, inertial features associated with

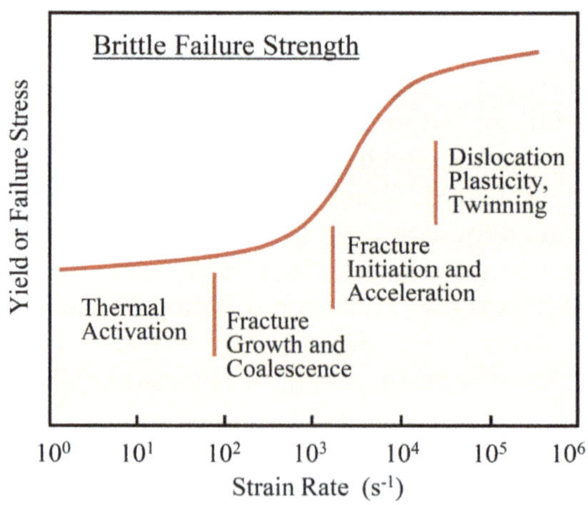

Figure 9.34. Dependence of the yield or failure stress of brittle solids on strain rate, and the associated rate controlling processes (Grady and Hollenbach 1977).

fracture growth and coalescence provide for a rapid increase in the ultimate failure stress at about this strain rate and higher. Coalescence can include either interaction with other active cracks or with free boundaries of the system subject to dynamic loading. When pulse widths approach the dimension of initial flaws at about the millimeter length scale corresponding to strain rates of about $\dot{\varepsilon} \simeq 10^4\ \mathrm{s}^{-1}$ to $10^5\ \mathrm{s}^{-1}$, inertial effects associated with fracture activation and acceleration further contribute to rate dependence of the dynamic strength and stiffness. At higher strain rates, $10^6\ \mathrm{s}^{-1}$ and greater, stresses for fracture activation and growth become excessively high, and a transition to deformation by plastic processes will occur in many brittle materials. In this region, rate-controlling plasticity mechanisms such as dislocation viscous drag come into play. In excessively brittle materials for which slip systems for plasticity processes do not occur at any realistic stress level, fracture on increasingly finer length scales continue to be the mechanism of inelastic deformation. Glasses and certain ceramics such as boron carbide are sensible examples of such materials.

Figure 9.35, illustrates further features associated with strain-rate dependence of the strength of brittle solids. At strain rates of about $10^2\ \mathrm{s}^{-1}$, the processes of initial yield by fracture and ultimately failure are quasi-static, and governed by the fracture toughness and hardening characteristics of the brittle material of interest. Rate dependence is due to thermal activation on atomistic and microscopic scales, and is manifest as a temperature dependence and modest strain-rate dependence of the deformation path and strength states at failure. At strain rates in the neighborhood of $10^3\ \mathrm{s}^{-1}$, the effects associated with crack expansion, growth and coalescence will cause a rapid increase in the stress at ultimate failure. According to equation (9.53) a nominally linear dependence on strain rate is expected in this range. Only modest increases in the yield stress (fracture initiation) are expected at this strain-rate range. At strain rates of about $10^4\ \mathrm{s}^{-1}$ failure stress continues to increase with strain rate. In addition, rate dependence of elastic crack expansion will stiffen the material modulus and the yield stress will increase due to rate-dependent fracture initiation. A trend toward a cube root dependence of the initial yield stress by fracture on strain

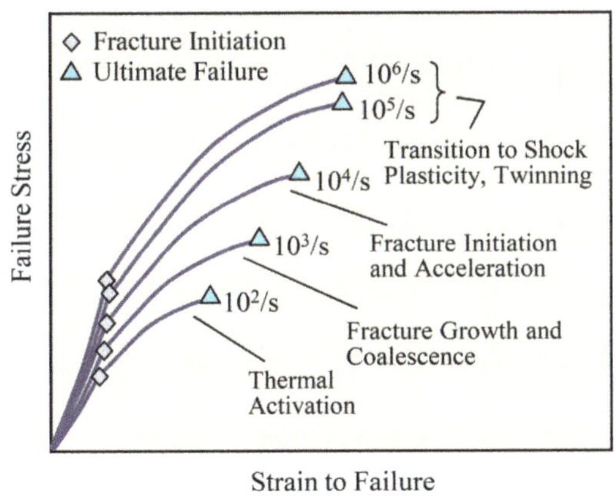

Figure 9.35. Strain-rate features of the stress versus strain load paths (Grady and Hollenbach 1977).

rate will come into play in this region. At strain rates in the neighborhood of 10^5 s^{-1} to 10^6 s^{-1}, compliance and features of failure initiation and pervasive deformation characteristic of shock wave compression will emerge. The range of transition strain rates will depend, of course, on the defect microstructure brittle-fracture characteristics of the material.

Absent in much of the discussion is the role of confinement in influencing transitions in the mechanisms of failure in brittle solids. In the shock regime the study of Paris *et al* (2011) on carbide ceramics suitably displays the contributions of confinement. Such issues are pursued further in the following section. Further features of rate-dependent deformation and failure in brittle solids are addressed in the next section. The focus is on hard rock materials where motivation for much of the studies was provided by ground shock effects (e.g. Isenberg 1972).

9.5 Testing and failure kinetics of rocks and ceramics

The study of Kanel *et al* (1992) was a seminal investigation in reinvigorating awareness of the kinetic effects in the strength of brittle solids subjected to impact loading. This study is discussed in chapter 6 in relation to the strength of ceramics under shock compression. A kinetic model of fracture specific to failure waves is explored in the final section of that chapter. The development of combined longitudinal and transverse stress gages, and application of the test methods to shock compression and failure in glass, was paramount in revealing rate-dependent behavior in the dynamic strength of brittle materials. The measured longitudinal and transverse stress in shock loaded K-2 glass, reported by Kanel and co-workers, is shown in figure 9.36, and is considered a pivotal experimental observation of kinetics-induced failure waves in brittle solids.

Kanel and co-workers were aware of concepts of failure waves preceding the landmark study undertaken by them. Substantial earlier Soviet work addressed the issue. Galanov *et al* (1989) proposed a limiting speed of a brittle failure front in an

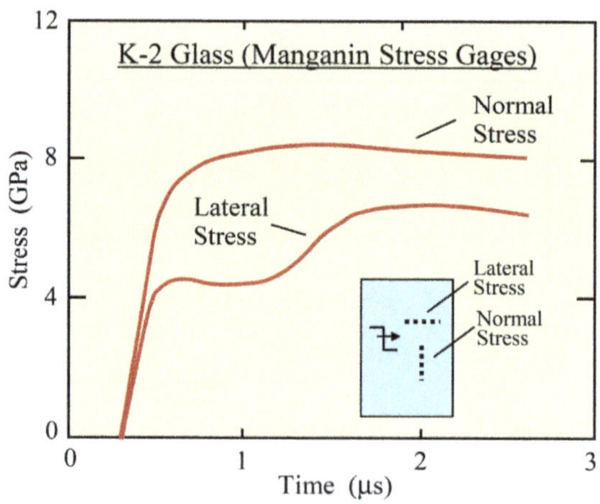

Figure 9.36. Manganin gage stress wave profile observation of failure waves in glass through measurements of axial and transverse stress in a compressive shock wave (Kanel *et al* 1992).

analysis of rod penetration in ceramics. Balankin (1988) included a limiting fracture wave velocity in his analysis of penetration in brittle materials. Zlatin *et al* (1988) discussed a transition from compressive fracture to plastic yield in penetration experiments in glass as velocities passed through the maximum velocity of crack growth. Bakun *et al* (1986) related the ability to sustain fracture waves in brittle materials to the storage of elastic shear energy in the competent microstructure. These studies referred to even earlier Soviet work that in turn reference previous studies on the kinetics of brittle failure published in the West.

Since the seminal paper of Kanel *et al* (1992), further experimental and theoretical progress on the nature of failure waves has been made (e.g. Brar and Bless 1992, Raiser *et al* 1994, Espinoza *et al* 1997, Bourne *et al* 1998). The present discussion does not address these later efforts. Rather, the purpose here is to describe earlier studies (principally within the 1970s period) that bear on the processes of dynamic failure kinetics in brittle solids.

The flourish of research on the mechanical constitutive properties of brittle solids through the decade of the 1970s was encouraged by several sources: the then US Defense Nuclear Agency (DNA, and now DTRA) and the US Department of Energy (DOE) were concerned with ground shock effects. The US Geological Survey (USGS) was actively pursuing earthquake research. A push to energy independence stimulated the DOE and university research into enhanced oil, gas, coal and mineral recovery technologies.

A large portion of the strength-of-material studies was focused on rock and rock-like materials. Natural materials with colorful names such as Arkansas novaculite, Solenhofen limestone, Wagon Wheel dolomite, Cedar City tonalite and Climax Stock granodiorite became familiar through the various testing laboratories. These hard rock materials are ceramic-like brittle solids in every sense. Arkansas novaculite, for example, is a fine-grain homogeneous quartzite, then used as a surgical sharpening stone. It has a HEL of approximately 8 GPa, comparable to the familiar Coors AD995 aluminum oxide ceramic.

These materials were subjected to an array of static strength-of-materials testing on equipment developed for this purpose within the same period. Hopkinson bar facilities, used up to that time principally for metals testing, turned toward the investigation of rocks and rock-like materials. Shock wave studies at a number of laboratories focused efforts on the dynamic strength and transient stress load characteristics of these same materials. From these multiple research efforts, a better understanding of the unique strength-of-materials characteristics of brittle solids emerged.

9.5.1 Static testing

The relevance of static strength-of-materials testing to failure kinetics in brittle materials is not immediately apparent. However, the phenomenology of failure under static load, and comparisons with corresponding dynamic testing, provided much insight into the characteristics of brittle solids subjected to impact and shock wave loading (Grady 2003).

Under the stimulus of research needs, machines capable of controlled testing of hard materials were developed at a number of laboratories. Some were massive systems capable of subjecting fist-size specimens to fluid pressures in excess of 1 GPa. The mainstream experiment was the triaxial compression test, in which a cylindrical specimen is first subjected to confining fluid pressure σ_2 and then the axial pressure σ_1 is increased until yield, and ultimately pervasive failure, of the sample is achieved. Axial strain ε_1 and diametral strain ε_2 were additionally monitored through the test. A substantial amount of testing of this type was conducted within this time period (e.g. Jaeger and Cook 1969, Franklin 1971).

Plots in figure 9.37 are representative of the mechanical response of hard rock materials under compressive loading in the triaxial tests. On the left in figure 9.37 stress difference (proportional to shear stress) is plotted as a function of axial strain. At the lower confining pressures, a yield stress is achieved with subsequent softening of the tangent modulus due to inelastic deformation. An ultimate failure stress is achieved followed by structural failure of the test specimen with an abrupt loss of stored elastic energy. With increasing confining pressure, the extent of the abrupt unstable failure reduces. At a critical confining pressure (between $\sigma_2 = 3$ and $\sigma_2 = 4$ in the example shown in the figure), a brittle-to-ductile transition is achieved above which inelastic deformation is a stable homogeneous cataclastic flow (large-strain controlled deformation of fractured and fragmented material).

On the right in figure 9.37 the means stress is plotted against the volume strain for the same triaxial tests. A reference is provided by the hydrostatic ($\sigma_1 = \sigma_2$) compression curve of the material. Deviation of the triaxial compression path from the hydro-compression curve corresponds to dilatancy (the opening of voids within the deforming microstructure) and initiates at about one-half to one-third of the ultimate failure stress largely independent of rock type (Brace and Riley 1972).

In media subjected to a planar shock wave, continuum deformation is constrained to that of uniaxial strain. Static uniaxial strain deformation is performed in a triaxial compression apparatus through manual or servo-control of the confine pressure σ_2 to maintain a condition of zero diametral strain. Tests of static uniaxial strain have

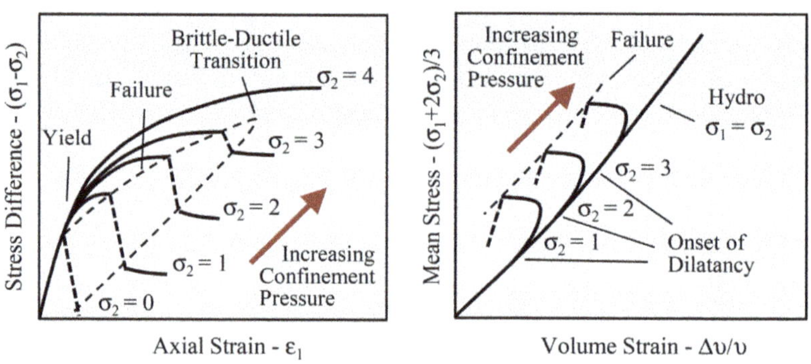

Figure 9.37. Stress difference versus axial strain (left) and mean stress versus volume strain (right) characteristics in a static-confinement triaxial compression experiment (Grady 2003).

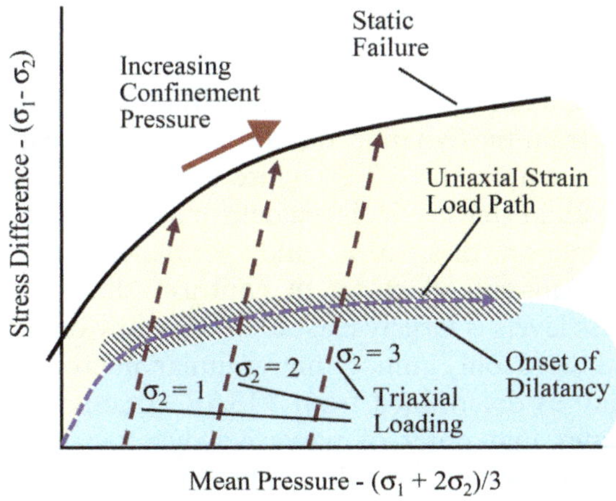

Figure 9.38. Stress path of static uniaxial strain compression is experimentally observed to coincide with the onset of dilatant states determined from triaxial compression experiments performed at constant confinement pressures (Grady 2003).

been performed in high-strength rock materials (e.g. Brace and Riley 1972). The consequences of these studies are illustrated in figure 9.38.

The tests of Brace and Riley show that the stress path for static uniaxial-strain compression resided on or near the onset of the dilatancy curve identified in the series of triaxial compression (σ_2 = constant) tests illustrated in figure 9.37 (see also Edmond and Paterson (1972)). The tests suggested that the uniaxial strain constraint of zero lateral strain limits growth of dilatant volume to at or near zero. Tests of uniaxial strain were performed which could be compared directly with dynamic uniaxial strain compression achieved in the shock wave environment.

9.5.2 Shock wave testing

The standard explosive loading or gun-driven plate impact experiment produces dynamic compression states under the constraint of uniaxial strain compression. Although geometries have been devised to explore other dynamic compression paths (pressure-shear, rod impact and divergent or convergent shock loading, among others), the preponderance of testing, including both Hugoniot and unsteady structured shock wave measurements, has been that of uniaxial strain. One objective of the early experimental work was to compare shock and static uniaxial strain test results. Other workers also tested a number of the fifteen rocks tested under static uniaxial strain loading by Brace and Riley (1972) under shock conditions.

Discussions here focus on static and dynamic testing of one specific model rock that was subjected to extensive testing. Namely, the material is Blair dolomite, the hard, near-full-density calcium-magnesium carbonate rock $(Ca,Mg)CO_3$ already addressed in several earlier sections. Shock testing was conducted (Grady *et al* 1977). Static testing on the same rock was performed at Lawrence Livermore National Laboratory (Heard *et al* 1973). Comparable dynamic and static test results for other rock materials (e.g. Schock *et al* 1973, Schock and Heard 1974) indicate that the observations are generally representative.

A comparison of the axial stress versus axial strain uniaxial strain compression data for this dolomite from the dynamic and the static testing carried to comparable axial stress levels is shown on the left in figure 9.39. Under static uniaxial strain compression, yielding and extensive inelastic deformation initiates at approximately 0.2–0.3 GPa. Deformation dissipation and permanent deformation offset is observed in the full load–unload cycle. In contrast, shock load and unload to comparable axial stress levels is largely elastic. Microscopy on recovered samples of shocked material indicated only minor intergranular microfracture damage. The HEL for this material, as determined from VISAR structured shock wave profile measurements, indicates that onset of pervasive dynamic yield for this material occurs at an axial stress of about 2.5 GPa, approximately one order of magnitude higher than the static yield. Dynamic and static loading paths are compared in a stress difference versus mean pressure plot on the right in figure 9.39. As observed in other competent rock materials, the static uniaxial-strain loading path appears bounded by the onset-of-dilatancy curve as determined from triaxial compression data and illustrated in figure 9.38. Dynamic uniaxial strain compression is not constrained by this curve. The dynamic yield (the HEL), in contrast, appears more consistent with the static triaxial ultimate failure curve above the static brittle–ductile transition, although the intersection is somewhat beyond the pressure range of the static data.

Thus, static uniaxial strain loading is apparently sensitive to and constrained by the onset of dilatancy curve as determined by confined static triaxial compression tests. Dynamic uniaxial strain loading is not aware of this static critical state and continues to load elastically until shear stresses are achieved that approach the static ultimate failure strength of the material. It has been theorized that the static and shock uniaxial strain compression paths bound a regime of rate-dependent behavior that can be probed through intermediate rate testing methods (Grady *et al* 1977).

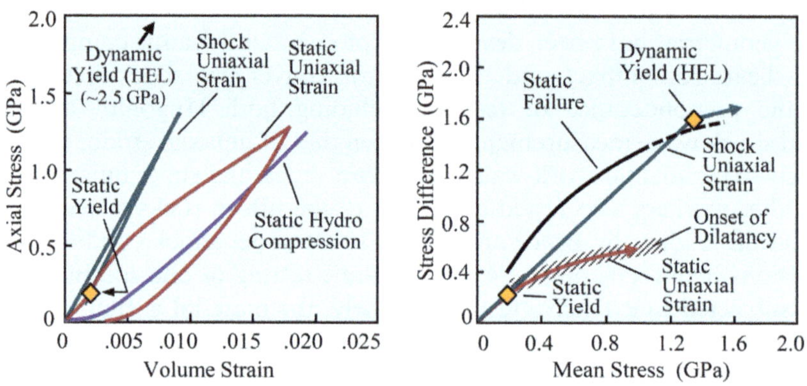

Figure 9.39. Comparison of dynamic (shock) uniaxial strain and static uniaxial strain loading in dolomite rock (Grady 2003).

Rate dependence in the deforming microstructure, as discussed in the previous section, underlies the observed deformation measurements.

9.5.3 Intermediate rate testing

It is suggested that the static and dynamic-stress versus strain compression curves on the left in figure 9.39, measured by quasi-static and shock wave uniaxial-strain compression testing, bound a range of material responses that can be accessed by appropriate intermediate rate testing methods. On the right, the onset-of-dilatancy curve and the failure envelope bound a similar rate-dependent region in stress space.

The uniaxial strain response of dolomite and other competent rocks could be productively investigated with the longitudinal and lateral stress gage technique employed on glass by Kanel and co-worker. Intuitively, shock loading to stress amplitudes in the region between the onset of dilatancy and failure envelope are expected to lead to shear stress relaxation, not unlike the behavior observed in glass. Such longitudinal and lateral stress gage studies have been pursued on engineering ceramics (e.g. Bourne *et al* 1998). Prior to the developments of Kanel, however, other techniques were pursued to explore intermediate rate properties of brittle solids.

In the 1970's and later, alternatives to hard shock loading as produced by explosives and direct impact were being developed. The high-rate ramp loading technique emerged as a method to achieve intermediate strain rates under constraints of uniaxial strain. Initially, materials with anomalous compressibility such as fused silica and pyroceram were sought which served as a buffer and ramp wave generator between the impact and test materials (Chhabildas and Graham 1987). This buffer dispersed (ramped) the shock wave before coupling the wave into the test sample. Later, graded density (Barker 1984) and layered impedance (Chhabildas and Barker 1988) impactor materials were developed which provided a direct and more versatile ramp loading technique.

Fused silica, which ramps effectively up to about 3 GPa shock pressure, was appropriate for testing the dolomite discussed in section 9.3.3. Compression strain rates approaching 3×10^4 s^{-1} were achieved in the study, nearly two orders of magnitude lower than shock loading strain rates. Compression paths from the lower rate tests were found to approximate the shock data. The tests provided no evidence of relaxation into the inferred rate-dependent regime and toward the measured static uniaxial strain curve. One further shock relaxation test on dolomite indicated a similar dynamic response down to an effective strain rate of approximately 2×10^3 s^{-1} (Grady 1983).

Other techniques for investigating the intermediate rate behavior of brittle solids include the rod impact method (Chhabildas *et al* 1998) and the Hopkinson bar technique. Traditionally developed to measure dynamic properties of ductile metals, early researchers including Kumar (1968), Green and Perkins (1968) and Lindholm (1974) explored the applicability of Hopkinson bar methods to rocks. While strain-rate sensitivity at lower rates is fully consistent with expected thermal activation processes, the upturn of rate sensitivity revealed by the Hopkinson bar test in the neighborhood of about 10^3 s^{-1} is generally considered an athermal fracture-kinetics governed process. Concerns have been raised regarding lack of stress equilibrium in the Hopkinson bar specimen at these rates, and a transition from uniaxial-stress-like to uniaxial-strain-like deformation. This issue is probably immaterial. The tests in this range are more reasonably viewed as unequilibrated brittle structural failure experiments. Considerations suggested that the observed elevated failure stresses are a consequence of achieving loading rates in which elastic strain energy in the sample is being fueled faster than fracture and deformation kinetics can accommodate failure and dissipation.

Various authors have attempted to model or otherwise analytically quantify this fracture-kinetics governed failure process. Further discussion is provided in section 9.4.5. Birkimer (1971) found that a critical work criterion adequately characterized his Hopkinson bar data on quartzite and concrete, while Steverding and Lehnigk (1976) argued that an action principle captures the dynamic strength due to fracture in brittle solids. Grady and Lipkin (1980) offered an energy-based prediction of the fracture strength. The above theories, in fact, are all founded on an underlying energy principle and lead to a predicted dependence of the fracture stress on the strain rate of the power-law form $\sigma_f \sim \dot{\varepsilon}^{1/3}$, if a dynamic energy balance is achieved in the failure process. The more recent efforts of Ravichandran and Subhash (1995) and Kimberleya *et al* (2012) provide further studies of this issue.

Intermediate rate data on high-strength commercial aluminum oxide ceramics, including both Hopkinson bar (Lankford 1981) and rod impact (Wise and Grady 1993, Chhabildas *et al* 1998) data, are shown in figure 9.40. The results mirror the earlier data on rock in figure 9.26. Both the transition in rate sensitivity and the magnitude are consistent with the failure data on rock. Shock strengths in the figure

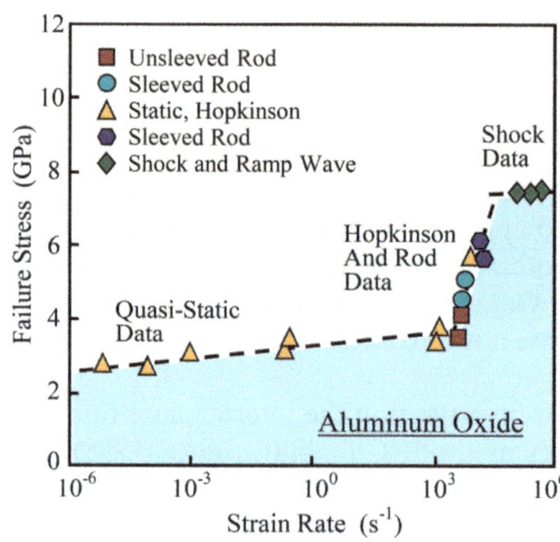

Figure 9.40. Strength versus strain-rate data for aluminum oxide ceramic from multiple test methods including quasi-static and Hopkinson bar (Lankford 1981), sleeved and unsleeved rod (Wise and Grady 1993, Chhabildas *et al* 1998), and shock and ramp wave (Grady 1995a, 1995b).

are unconfined compressive strength for aluminum oxide calculated from the data in figure 9.12.

It is reasonable that the high-rate Hopkinson bar behavior and failure waves in ceramic materials are different manifestations of the same phenomenon (Grady 1998a). Namely, distortional elastic strain energy is introduced into the material at a rate faster than this energy can be dissipated by fracture and deformation. In the failure wave event, elastic loading is a Heaviside (shock) process. A high nominally uniform strain-rate loading introduces elastic strain energy in the upper range of the Hopkinson bar experiment.

Other evidence speaks for a brittle fracture failure in the two loading methods. In the Hopkinson bar studies, collection and correlation of resulting fragment debris are consistent with a brittle fracture kinetics modeling of the event (e.g. Shih *et al* 1998). High-speed photography of shock-induced fracture in transparent brittle solids clearly illustrates the character of fracture in the failure wave (Brannon *et al* 1983, Bourne *et al* 1994). The same data have been use to extract kinetic material properties for selected brittle solids (Grady 1995b).

9.5.4 Kinetic effects in brittle solids

A competent full-density brittle solid that does not undergo polymorphic phase transformation fails at the HEL through inelastic shear deformation. Inelastic deformation may include shear fracture and/or plasticity (dislocation mobility, twinning, adiabatic shear). The evidence outlined in section 9.2.6 supports HEL failure in most ceramics and rocks through processes that exhibit no, or very modest, rate dependence. Most polycrystalline ceramics do not show evidence of elastic precursor attenuation. (Single-crystal compounds, however, generally do exhibit precursor decay with propagation distance.) Ramp wave experiments on dolomite rock described in section 9.3.3 reveal uniaxial failure strength comparable to the HEL strength under loading conditions that are approximately two orders of magnitude lower strain rate. Microstructure examination of selected ceramics shows evidence of plastic deformation under HEL failure. HEL strengths for many ceramics are an appreciable fraction of the theoretical strength.

There are probably exceptions. HEL failure in boron carbide ceramics shows evidence of brittle failure. Plasticity in silicate and metallic glasses under shock and high-rate failure is conceptually difficult, although adiabatic shear plasticity may be a plausible mechanism. Further, any near-full-density brittle solid can exhibit a range of HEL strengths determined by the preparation microstructure. Aluminum oxide ceramic, for example, depending on microstructure, exhibits HEL strengths that range over about 6–12 GPa.

Evidence leans toward plastic processes governing the high-rate high-confinement failure and post-failure deformation of brittle solids under shock compression conditions. The evidence is not conclusive, however, and there are probably exceptions in both material and loading conditions. The study of Paris *et al* (2011) indicates that boron carbide ceramic is such an exception. Nevertheless, it is useful to examine the HEL failure process in detail, assuming a mechanism of brittle shear fracture. Such analysis reveals conditions under which shear fracture becomes physically implausible. Brittle shear fracture in the shock failure event is explored in the following subsections. Analysis of shear fracture leads to shock-loading conditions in which delayed fracture failure (the failure wave) is predicted. Combined with material strength phenomena from quasi-static triaxial testing, as well as intermediate rate testing, a plausible physical picture of strength kinetics in brittle solids emerges.

9.5.4.1 Shear fracture in the shock front

As elastic compression within the shock front increases in a solid with strength, shear stresses ultimately achieve levels in excess of the critical strength. Stress can be concentrated at defects within the material. At these stress concentrations, inelastic deformation accompanied by stress relaxation will lead to pervasive shear failure in the material. The character of structured shock waves observed in brittle solids is a consequence of the microscopic and mesoscopic nature of shear failure and deformation.

Inelastic shear within the shock front is either ductile or brittle (or a mixture of both). Ductile deformation is associated with shear deformation in which material displacements normal to the shear planes do not exceed atomic dimensions and material cohesion is retained. Brittle shear deformation, or shear fracture, in contrast, involves the normal separation of material points over a sufficient number of atomic dimensions such that cohesion is lost and internal stress-free surfaces are created.

Metals are generally associated with ductile failure, while ceramics are commonly considered brittle. This intuitive classification becomes less certain when a substantial confining pressure component of the stress field plays a prominent role in the shear failure and deformation process. In principle, sufficient confining pressure can cause any material to undergo a brittle-to-ductile deformation transition. Consequently, ductile shear within the shock front in ceramics must necessarily be entertained.

Spall experiments, in which shock wave tests are designed to first dynamically compress the specimen, and then subsequently test the tensile strength, have been used to examine the residual cohesion of a material after shock compression. It is suggested that the loss of spall strength observed in ceramics, when first shock compressed above the HEL, argues for a brittle failure process at the HEL state of ceramics. This proposition is arguable. Reverse shear yield is usually experienced on decompression from the Hugoniot state before tensile axial stress levels are achieved. The significantly lower confining pressure associated with reverse yielding would much more readily accommodate brittle failure and loss of cohesion than would initial failure during shock compression.

Nonetheless, shear fracture during shock compression does appear to be a viable mechanism for inelastic deformation, and speculation on physical details of the fracture processes in the shock front is a worthy pursuit (Grady 1994c). The shock compression profile shown in figure 9.41 for aluminum oxide is representative of the wave structure observed in most ceramic and rock materials. There are a few notable

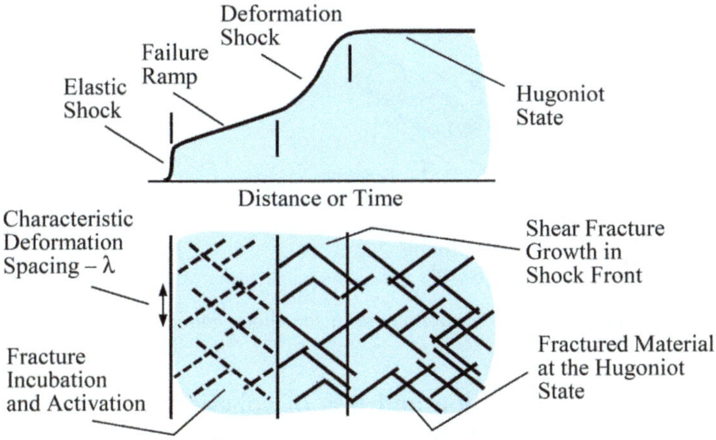

Figure 9.41. A structured shock compression wave in a brittle solid (aluminum oxide ceramic for example) and the sequence of deformation processes leading to brittle shear fracture at the shock state (Grady 1998a).

exceptions, however. Boron carbide, for example, exhibits marked stress softening in post-HEL deformation.

The structured shock wave profile is characterized by three distinct regions. First is the sudden elastic rise to the failure limit (HEL). For high-strength ceramics, elastic stresses over the range of 5–20 GPa are achieved at the HEL with corresponding elastic lattice volume and shear strains of about 2%–5%. The failure ramp following the HEL state identifies a region of rapidly changing compressibility. The material accommodates only modest additional strain during passage of the post-HEL ramp. The small strain and changing compressibility suggest that precursors to the inelastic deformation process including fracture incubation occur within this region. Finally, formation of a steady, rapidly rising deformation shock is required by the intrinsic nonlinearity and upward concavity of the volumetric elastic compressibility of the solid. Polymorphic phase transformation in the shock will profoundly alter the response, however. The preponderance of inelastic shear deformation due to shear fracture growth is accommodated within the rapidly rising shock wave. The spatial width of this wave is governed by the deformation kinetics and viscous mechanisms intrinsic to the material.

A physical model of the shear fracture process within the shock front is shown in figure 9.41. Illustrated in the figure is the stress (or particle velocity) wave profile propagating toward the left through the solid. Material to the left of the elastic front is undisturbed. Material to the right is fully fractured due to passage of the shock wave. Shear fracture incubation and activation occurs within the failure ramp. Completion of shear fracture and inelastic fracture deformation follows within the deformation shock wave.

In the failure ramp and early in the deformation shock wave, confining pressures are of order 3–10 GPa and can rise to values of 20–60 GPa or more as the Hugoniot state is approached. To accommodate brittle shear fracture the material must be capable of locally supporting stress gradients of this order if material displacements

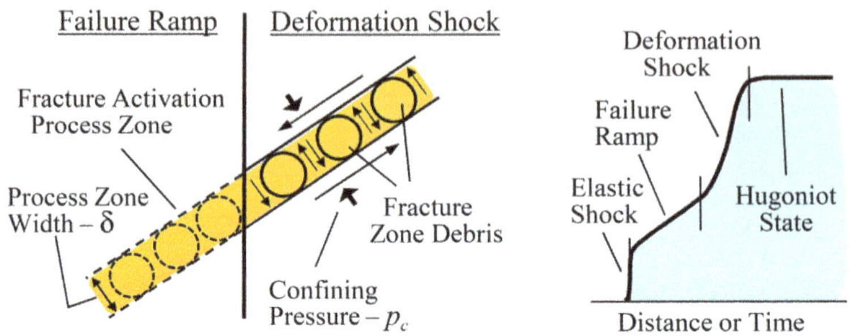

Figure 9.42. Activation and growth of shear fracture in the shock compression zone under high dynamic confining pressure (Grady 1998a).

consistent with brittle deformation and loss of cohesion are to occur in zones of fracture. This action could readily be accommodated in microscopic shear fracture if regions of ultra-fine debris are generated within the fracture process zone and persist to support the walls of the shear fracture during mature growth and subsequent arrest, as illustrated in figure 9.42.

The size of the fracture zone debris and dimension of the shear fracture can be estimated from the size dependence of the strength of brittle bodies based on energy arguments (Grady 1994c),

$$\delta \sim \frac{2E\gamma}{p_c^2}. \tag{9.64}$$

In equation (9.64) δ is the characteristic particle size, E is an elastic modulus, γ is the cohesive surface energy and p_c is measure of the confining stress that determines the elastic strain energy within the supporting particle. Assuming $E = 400$ GPa and $\gamma = 5\ \text{J m}^{-2}$, appropriate properties for aluminum oxide, and p_c equal to the local confining pressure of about 5–10 GPa, a characteristic particle size and fracture zone dimension of $\delta \sim 40$–160 nm is calculated. Values for p_c somewhat higher than the confining pressure are expected due to stress concentration near the environment of the supporting particles. Consequently, estimates of δ are probably an upper bound. A small measure of dilatancy would accommodate the shear fracture free-volume. The parameter δ will decrease through the deformation shock as confining pressures approach the Hugoniot state. The corresponding crushing of fracture zone debris would lead to local energy dissipation and elevated temperatures. In certain materials, and under certain shock conditions, localized melting and transition to near-fluid shear zones within the fracture region late in the deformation shock may occur (Grady 1980).

The characteristic spacing of fractures λ in the shock process zone (see figure 9.41) is estimated through the energy-based methods described in chapters 2. A sensible prediction of fracture spacing is

$$\lambda \sim \left(\frac{48\gamma}{\rho\dot{\varepsilon}^2}\right)^{1/3}, \tag{9.65}$$

based on the shearing strain rate $\dot{\varepsilon}$ in the shock wave and the fracture energy γ. The strain rate is calculated from $\dot{\varepsilon} \sim 3\sigma/2\eta$, where η is the solid viscosity within the shock front and σ is the stress jump through the deformation shock wave. The properties in the shock wave that determined this viscosity are not fully understood (see the previous sections and chapter 8). However, the experimental measurements of shock widths in brittle materials can be used to estimate the necessary deformation strain rates in equation (9.65).

For the aluminum oxide for which profiles are shown in figure 9.2, a shock Hugoniot state of 19 GPa leads to a strain rate in the shock front of about $\dot{\varepsilon} \sim 1.7 \times 10^5\ \mathrm{s}^{-1}$ and a fracture spacing from equation (9.65) of about 130 μm. In contrast, the highest shock stress of 42 GPa yields a shock front strain rate in excess of $\dot{\varepsilon} \sim 1.0 \times 10^7\ \mathrm{s}^{-1}$ and a fracture spacing of less than 8 μm. Material parameters of $\rho = 3890\ \mathrm{kg\ m^{-3}}$ and $\gamma = 5\ \mathrm{J\ m^{-2}}$ are used in equation (9.65). Such predictions of fracture spacing exceed the fracture process zone dimension δ by about two to three orders of magnitude, and at the lower strain rates are well above the characteristic grain size of this material.

Rosenberg (1995) has argued for a fracture mechanism in the compressive shock wave and proposed that a Griffith failure criterion would be appropriate for relating compressive fracture (HEL) and tensile fracture (spall) for brittle solids. The criterion developed by Griffith (1924) is based on stress concentrations at the tips of a population of microcracks assumed to pervade the material subjected to a stress load. The theory identifies a critical stress state at which favorably oriented cracks achieve a value characteristic of the material at which crack growth initiates. The analysis carried out by Griffith was strictly two-dimensional. Rosenberg (1995) proposed an extension of the Griffith analysis to the shock compression environment resulting in the criterion

$$(\sigma_x - \sigma_y)^2 = 8T_o(\sigma_x + \sigma_y), \tag{9.66}$$

where σ_x and σ_y are the axial and transverse stresses, respectively, and T_o is the unconfined tensile fracture strength of the material. A modification of the Griffith criterion proposed earlier by Murrel (1963) leads to a modestly different expression,

$$(\sigma_x - \sigma_y)^2 = 12T_o(\sigma_x + 2\sigma_y). \tag{9.67}$$

Figure 9.43 provides a plot of the ratio of the experimental spall strength to the HEL for selected ceramics (Grady 1994c). Overlaid on the plot are predictions of the ratio based on the von Mises failure criterion, and the Griffith failure criteria as modified by Rosenberg and by Murrel. Although agreement in detail is not observed, the Griffith criteria are consistent in trend with low values of this ratio for most ceramics. The ZrO_2 tested was a transformation-toughened material, possibly accounting for the higher spall strength and higher value of the ratio plotted in figure 9.43. The low value for SiO_2, the one geological material included, is consistent with comparisons of the Griffith criterion with strength of other rock materials (e.g. Januch 1977).

In chapter 2, a theory based on energy balance principles is described from which predictions can be made for both the fragment size and the spall strength of a solid in a dynamic tensile failure event. When the dynamics of tensile failure is characterized

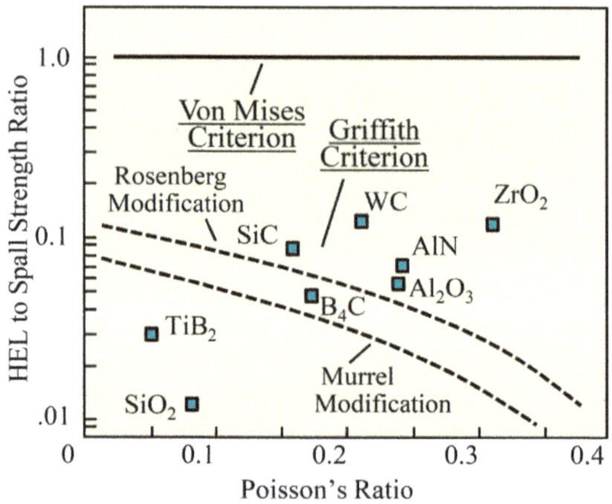

Figure 9.43. Comparison of the spall versus HEL ratio for selected ceramics with predictions based on the von Mises failure criterion and on the Griffith failure criterion as modified by Murrel and by Rosenberg (Grady 1998a).

by a strain rate $\dot{\varepsilon}$, the nominal fragment size is provided by equation (9.65), while the spall strength is given by

$$P_s = (6\gamma\rho^2 c^3 \dot{\varepsilon})^{1/3}. \tag{9.68}$$

The Griffith failure criterion can be joined with the spall relation in equation (9.68) to provide an expression for the HEL of a brittle solid based on intrinsic material properties. Working with Murrell's modification from equation (9.67), a prediction for the HEL based on fundamental properties of the material and an assumption of brittle fracture in the shock compression process is

$$\sigma_{\mathrm{hel}} \simeq 12\frac{1-\nu^2}{(1-2\nu)^2}(6\rho^2 c^3 \gamma\dot{\varepsilon})^{1/3}. \tag{9.69}$$

Using reasonable properties for aluminum oxide ($\nu = 0.23$, $\rho = 4$ g cm^{-3}, $c = 10^4$ m s^{-1}, $\gamma = 5$ J m^{-2}), and assuming a strain-rate range of 10^5 s^{-1}–10^6 s^{-1}, yields a HEL of about σ_{hel}= 14–30 GPa, which can be compared with experimental values of σ_{hel} = 6–12 GPa for reasonably competent alumina ceramics.

Equation (9.69) is not expected to yield an accurate estimate of the HEL strength of a brittle material. Brittle fracture has not been definitively shown as the dominant failure mechanism under shock compression. Numerous microstructural issues known to influence dynamic strength (i.e. grain structure, porosity) are not accounted for. Nonetheless, the reasonable agreement with measured dynamic strengths of ceramics (within a factor of two) offers some support for the mechanics leading to equation (9.69). This agreement may provide some argument for a fracture-dominated process controlling failure in the compressive shock wave.

The strain-rate dependence predicted by equation (9.69) is not, however, consistent with that inferred from the analysis of experimental data for several ceramics leading to the lack of strain-rate dependence plotted in figure 9.12. The algorithm for assigning strain rate to the experimental data plotted in figure 9.12 may, however, be at odds with logic leading to equation (9.69).

Other authors have argued for a dislocation plasticity mechanism at the HEL in competent brittle solids (Cagnoux and Longy 1988). Later studies of Cagnoux (1994) continue to point toward intra-crystal mechanisms in selected ceramics dominating deformation in the shock compression process. Similar conclusions supporting intra-granular plasticity mechanisms within the shock-wave-failure and post-failure deformation have been suggested in studies on polycrystalline aluminum oxide by Munson and Lawrence (1979) and Longy and Cagnoux (1989), and on dolomitic rock by Grady *et al* (1977). The study of Paris *et al* (2011) illustrated in figures 9.15 and 9.16 argue for plasticity at the HEL failure, at least for silicon carbide.

Arguments leading to estimates of a characteristic particle size within the zone of a brittle shear fracture in equation (9.64) are not inconsistent with brittle fracture. Confining pressures at HEL stress levels of 5–10 GPa predict particle dimensions of order 40–160 nm, sufficient to accommodate the crack opening and loss of cohesion necessary for brittle fracture. As stresses approach the Hugoniot state with confining pressures of 30 GPa and higher, the characteristic dimension required to support pressure differences sufficient for decohesion become increasingly unlikely. A reasonable description of failure in a brittle solid is that of the activation and progression of inelastic deformation at the HEL through brittle fracture with the attendant loss of cohesion and formation of fracture free-volume when confining pressures are relatively modest (5–10 GPa).

Late in the deformation process, as confining pressures approach the Hugoniot state (about 30 GPa and higher), molecular cohesion is re-established and deformation within the shear zone approaches that of elevated-temperature viscous-like flow of molecular-bonded near-amorphous matter. This model of the inelastic shear deformation process in the shock wave for a brittle solid is sketched in figure 9.44. Thermo-mechanical properties would dictate whether the deforming material is in the solid–amorphous or liquid state. Issues of adiabatic shear deformation explored in chapter 5 would potentially play a significant role.

9.5.4.2 Features of failure waves

The emergence and evolution of structure in a compressive shock wave such as illustrated in figure 9.41 or figure 9.44 is dependent on the amplitude of the input stress load. If the input stress level is less than the HEL, a finite amplitude elastic wave is sustained and propagates through the brittle solid. This response is brought into question by the experimental shock wave studies on K-2 glass (Kanel *et al* 1992). This work provided evidence for the propagation of a delayed front of fracture following the initial elastic compression wave as is depicted in figure 9.45. The phenomenon is termed a failure wave, and is presumed to be a delayed brittle shear fracture process that is driven by the large elastic shear strain energy residing in the body behind the elastic uniaxial-strain compressive shock wave.

Further evidence for failure waves is provided by the subsequent studies of Brar *et al* (1991) on a soda-lime glass, and Raiser *et al* (1994) on an alumina-silicate glass.

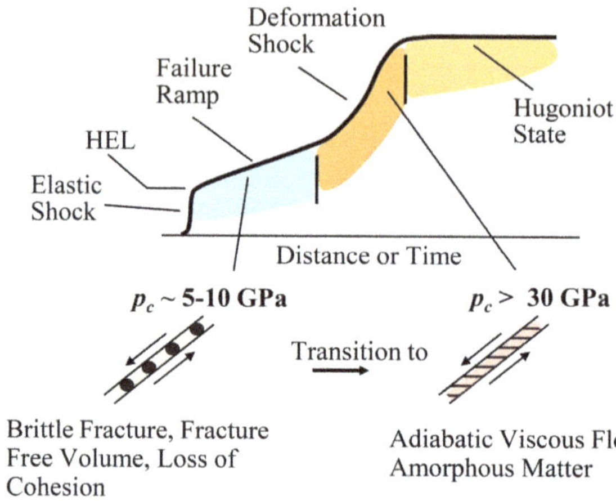

Figure 9.44. Illustrates activation of brittle fracture with attendant loss of cohesion and fracture free-volume. Deformation transitions to near-adiabatic solid state flow at near-Hugoniot state confining pressures that preclude decohesion.

In the first of these studies, transverse stress gages provide evidence for a substantial increase in the transverse stress component σ_y following passage of the failure wave, significantly reducing the shear stress $\tau = (\sigma_x - \sigma_y)/2$. In the same study, the tensile strength of glass behind the initial elastic shock wave was measured both in front of and behind the failure wave. The tests indicated a tensile strength in excess of 3 GPa in front of the failure wave and close to zero strength behind the failure wave, indicating a transition from competent material to fully fractured material upon passage through the failure wave.

High-speed photography has been used to examine failure wave propagation in glass. Bourne *et al* (1994) used orthogonal shadowgraph photography to track the fracture front in uniaxial shock impact experiments on glass, revealing a coarse propagating fracture interface behind the shock front. Senf *et al* (1994), using similar photographic methods, identified delayed shear fracture processes behind spherical shock fronts in glass. Earlier photographic studies of Brannon *et al* (1983) on crystalline quartz and glass, imaging triboluminescent failure planes behind a compressive shock front, may also have identified features of a failure wave process.

These studies suggest that the failure wave may be a propagating fracture wave trailing the initial elastic shock wave at a velocity substantially less than the shock velocity (Balankin 1988, Galanov *et al* 1989). A failure wave velocity close to the Rayleigh wave speed was proposed by Raiser and Clifton (1993). An alternative explanation is that the delay time between the shock wave and failure wave is governed by the kinetics of shear fracture activation and growth. A quantitative fracture model leading to a failure-wave-based activation and growth kinetics is explored in chapter 6.

The reported changes in the state of the material upon passage of a failure wave offer constraints on the material response through the failure wave process. Several noteworthy observations in the shock wave studies on glass are as follows.

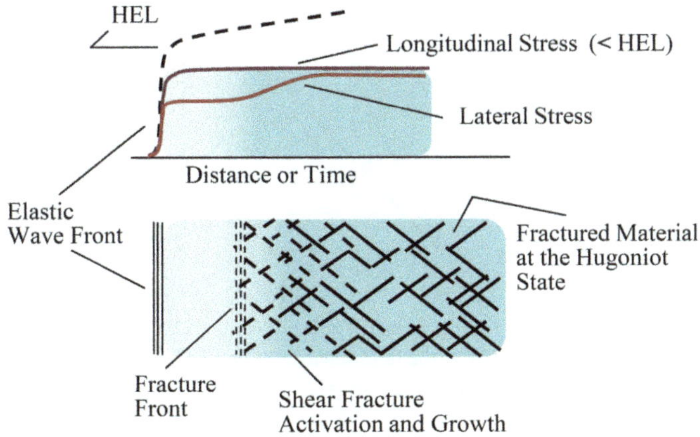

Figure 9.45. Depicts shock compression and accompanying stress waves to stress levels below the HEL in a brittle solid with delayed activation and growth of brittle shear fracture.

(1) Velocity interferometry measurements indicate small or negligible change in particle velocity through the failure wave. (2) Transverse and normal stress measurements indicate that the stress difference reduces substantially during passage of the failure wave, but most of the change is due to an increase in the lateral stress σ_y with little change in the longitudinal stress σ_x. This observation would also be inferred from the first observation and the momentum jump condition across the failure wave. (3) The failure wave is reported to propagate at a velocity substantially less than the longitudinal elastic wave velocity in glass. Failure wave velocities of order 2 km s^{-1} have been reported from the experimental data. The evidence is weak, however, and any velocity up to the longitudinal sound velocity (~6 km s^{-1}), including time and/or amplitude dependence, is possible.

Consider a material response in a brittle solid (glass) that leads to a failure wave with a characteristic velocity V_{failure}, and with an accompanying loss of the stress deviator. If the volumetric compressibility behind the failure wave corresponds to that of full-density glass, then the situation in figure 9.46(a) applies. State 1 precedes the failure wave and state 2 is behind the wave. The velocity of the failure wave is determined by the chord connecting state 1 and state 2,

$$V_{\text{failure}} = \sqrt{\frac{\Delta\sigma}{\rho\Delta\varepsilon}}. \tag{9.70}$$

The difficulty with this explanation is that a substantial increase in particle velocity ($\Delta u = \Delta\varepsilon V_{\text{failure}}$) is not consistent with the first observation above.

A second possibility is that material behavior through the failure wave represents catastrophic relaxation of the shear stress due to time-dependent shear fracture. This model of the process is illustrated by 9.46(b) where complete relaxation of the stress deviator at constant strain occurs through the failure wave. To achieve the volumetric compression state of fully dense glass, identified by state 2 in (b), a substantial drop in σ_x through the failure wave is required. This decrease in the

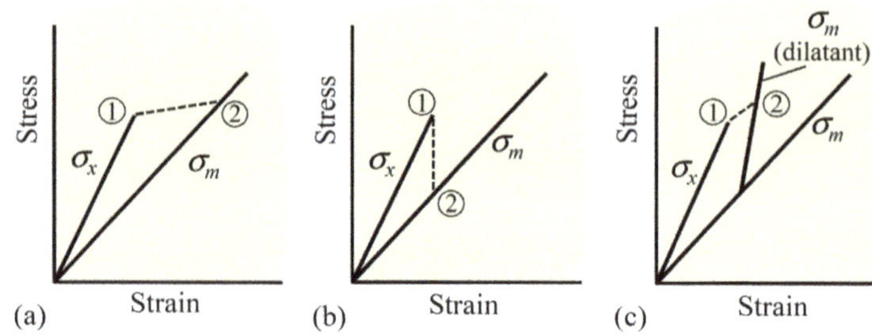

Figure 9.46. Transition states for failure waves in brittle solids (Grady 1995b). In each example the shock state at point one is below the HEL and in front of the failure wave. State two is behind the failure wave. The stress paths σ_x and σ_m refer to the elastic longitudinal stress and mean stress response under uniaxial strain compression.

longitudinal stress is approximately twice the corresponding increase in the lateral stress σ_y and is not consistent with the second observation above.

A third possible material behavior, which would be consistent with the three experimental observations is illustrated in figure 9.46(c). An alternative pressure (mean stress) versus volume curve characterizes the fractured material behind the failure wave due to shear-fracture-induced dilatancy, or bulking, in the failure wave. The process could lead to a constant failure wave velocity, with small changes in both particle velocity and longitudinal stress through the failure wave. A complete loss of shear strength is illustrated in figure 9.46(c), although a partial strength loss accompanied with a dilatancy process is consistent with the same physical model. The dilatant void volume required to accommodate the behavior can be estimated. A pressure change is related to volumetric strain of the lattice v_{lat} through the bulk modulus K according to

$$dp = -K\frac{dv_{\text{lat}}}{v_{\text{lat}}}, \tag{9.71}$$

If a dilatant void volume dv_{dil} occurs within the failure wave, then the total specific volume change is $dv = dv_{\text{lat}} + dv_{\text{dil}}$. Assume for simplicity a constant volume strain process through the failure wave, $\Delta v = 0$ (coincidence of state 1 and state 2 in figure 9.46(c)), then approximately

$$\frac{\Delta v_{\text{dil}}}{v} = \frac{\Delta p}{K}. \tag{9.72}$$

Assume further that $\Delta\sigma_x = 0$ and that $\sigma_x = \sigma_y$ after passage of the failure wave. Then $\Delta p = 4\tau/3$ where τ is the shear stress in the elastic state preceding the failure wave and

$$\frac{\Delta v_{\text{dil}}}{v} = \frac{4}{3}\frac{\tau}{K}. \tag{9.73}$$

For glass shocked to about 5 GPa on the elastic Hugoniot with $\tau = 1.7$ GPa and $K =$ 56 GPa, equation (9.73) predicts a dilatant strain of about 4%. The complete transfer of elastic shear strain energy into dilatant strain energy yields a comparable result. Further aspects of the failure wave event are explored in chapter 6.

9.5.4.3 High-rate and high-confinement failure

Although a consensus has not been achieved, evidence favors failure at the HEL in the shock compression of many brittle solids as governed by crystal plasticity, or perhaps shear fracture that transitions to plastic-like failure as confinement pressure increases in the following shock wave. Failure in the shock wave of this nature is illustrated by the sketch in figure 9.44. Porosity in the material is a complicating matter that is not considered here, but when present commonly dominates failure at the HEL. Strength in the presence of porosity is discussed in both chapters 6 and 7.

In any case, the deformation behavior and weak strain-rate dependence associated with the failure of brittle solids at the HEL differs markedly from the strong strain-rate-dependent brittle fracture processes characterizing failure in unconfined Hopkinson bar (Lankford 1981) and dynamic spall experiments (Grady and Kipp 1989). In addition, quasi-static triaxial and uniaxial compression studies on hard rock materials (Schock and Heard 1974) contribute to a picture of strain-rate dependence in brittle solids that is not reconciled within the more conventional models of failure and deformation.

A reasonable physical model accounts for the range of phenomena observed in the dynamic and static failure of brittle solids (Grady *et al* 1977). The model accommodates features of both time-dependent fracture and rate-independent (or weakly rate-dependent) plasticity. The essential features of the model are depicted in figure 9.47. On the left, within a plot of mean pressure versus shear stress, two uniquely different failure curves are identified. Stress loading paths consistent with the uniaxial strain compression are also shown, although other loading paths such as uniaxial stress can equally well be considered. The lower curve identifies the quasi-static fracture limit at which a critical flaw or a system of flaws activate onset of fracture by the stress load. Stress loading of the brittle material at any strain rate from quasi-static to shock below this curve is elastic. This failure curve is delineated by the onset-of-dilatancy states from the static triaxial tests and by

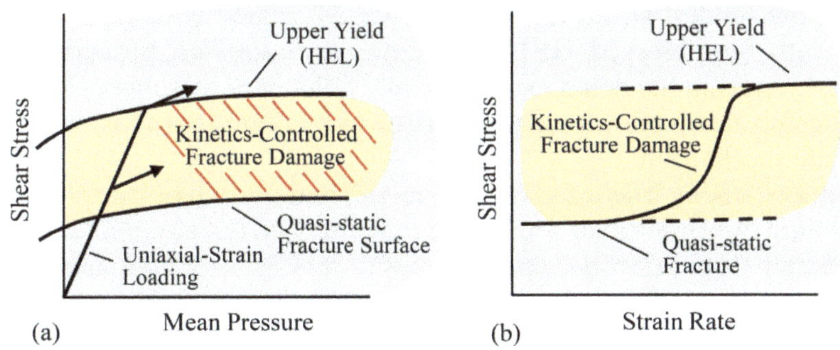

Figure 9.47. Dynamic failure in brittle solids based on fracture-kinetics and mechanism-transition model (Grady 1998a).

the inelastic portion of the static uniaxial strain experiment as illustrated by the plot in figure 9.38.

Dynamic or transient stress loading, however, can exceed this quasi-static fracture curve because of inertia of the kinetics-controlled microscopic and mesoscopic fracture activation and growth processes. As a cause of the very high strain rates associated with the shock compression process, stress levels consistent with the upper failure curve can be achieved with negligible fracture damage while passing between the two curves. When the upper failure is reached in the shock compression process (the HEL), a transition in shear failure mechanisms is achieved. At these highest of loading rates, because of the inertia associated with brittle fracture, stress levels are achieved that activate inelastic failure through alternative plasticity-like processes. The shock wave HEL delineates one point on this upper failure curve. Whether loading at comparable rates that achieve this upper failure curve at lower levels of confining pressure would fail through comparable plasticity would be material-dependent. The divergent wave experiments of Paris *et al* (2011) on silicon carbide, for example, suggest a transition to brittle-like failure at lower levels of confinement. There is some evidence that this upper failure is also consistent with the ultimate material strength determined from the highly confined triaxial static compression tests.

The dependence of the shear strength of the material as loading strain rate is systematically increased is illustrated on the right of figure 9.47. Such a plot would represent, for example, strength data under either uniaxial stress (Hopkinson bar), or uniaxial strain (shock wave or ramp wave), loading. Occasionally such data are mixed by making reasonable corrections for differences in stress loading paths. At lower strain rates, strength is governed by onset of fracture with a relatively modest strain-rate dependence due to, in certain brittle solids, thermally activated subcritical crack growth (Lankford 1981). This strength level is identified on the left in figure 9.47 by the lower quasi-static fracture curve.

At a characteristic transition strain rate, a regime is entered in which athermal, inertia-dominated fracture damage kinetics lead to a rapidly increasing failure stress with strain rate. At the highest strain rates, the delay due to damage kinetics precludes failure by brittle fracture and an alternative failure mechanism is activated —reflected by the upper failure or yield surface (the HEL) in figure 9.47. In this regime, a more modest strain-rate dependence associated with this upper yield is consistent with the observed HEL and ramp wave strength properties of brittle solids. These trends in the behavior of dynamic failure are reasonably interpreted as a strain-rate-dependent transition from brittle-like failure to a more ductile-like or plastic failure mechanism.

Some data reflecting fracture-damage kinetics and a mechanism transition in the dynamic failure of brittle solids are shown in figure 9.48 (Grady 1998a). On the left, quasi-static and Hopkinson bar data of Lankford (1981) are combined with shock wave data (figure 9.16) for aluminum oxide and silicon carbide (Grady 1995a, 1995b).The shock data on the left have been corrected to a von Mises unconfined compressive strength. The data on the right are strictly static and dynamic uniaxial strain. These data indicate first a transition strain rate to a fracture-kinetics

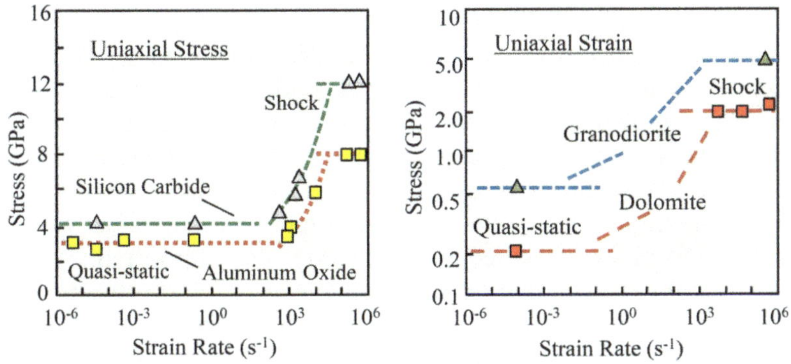

Figure 9.48. Dynamic failure properties of selected brittle solids under quasi-static (Schock and Heard 1974, Lankford 1981) and dynamic compressive loading (Petersen 1969, Grady *et al* 1977, Grady 1995b).

controlled failure strength, and a subsequent transition to an alternative rate-insensitive failure mechanism at the higher ramp wave and shock wave strain rates. Of note is the difference in transition to a rapidly increasing strength regime between aluminum oxide and silicon carbide, possible speaking to differences in fracture damage kinetics between to two ceramics. The controversial nature of Hopkinson bar data in the high-strain-rate regime, and the mixing of uniaxial-stress and uniaxial-strain data in the plot on the left, make conclusions from these data tentative. Loss of uniaxial stress loading and the nature of the structural strength of Hopkinson bar samples are discussed earlier in this section.

The plot on the right includes both static (Schock and Heard 1974) and shock (Petersen 1969, Grady *et al* 1977) uniaxial-strain strength-at-failure data for both Blair dolomite and Climax Stock granodiorite. For the dolomite, impact experiments involving shock wave (Grady *et al* 1977), ramp wave (Grady and Hollenbach 1977) and shock dwell (Grady 1983) techniques, fail to reveal a transition to the strength implied by the markedly lower quasi-static strength. Ramp wave and shock wave studies on aluminum oxide (Cagnoux and Longy 1988) similarly failed to identify significant strain-rate-dependent strength in the high strain-rate region accessed by impact techniques. The above model would predict a rapidly rising damage-kinetics controlled region joining the high-strain-rate and quasi-static data.

The present physical model based on fracture kinetics of the failure process in brittle solids is also contemplated by considering the shock wave (or Heaviside) compressive loading to states below the HEL, but above the quasi-static fracture curve on the left in figure 9.47. The initial near-instantaneous response of the material to the shock load is expected to be elastic as a consequence of fracture kinetics inertia. Delayed failure governed by shear-fracture damage kinetics would occur after sufficient time at stress following the propagating shock wave as illustrated in figure 9.45. Characteristic delay times of microseconds to tens of microseconds have been predicted for silicon carbide and aluminum oxide based on the data of Lankford (1981) and a damage kinetics model (Grady 1995a, 1995b).

Alternatively, under such loading conditions fracture may proceed through a following failure wave at a characteristic speed governed by a limiting fracture velocity (e.g. Nickolaevskii 1981). Evidence is increasing for delayed shear fracture failure and failure waves in the shock compression of brittle solids. The seminal study of Kanel *et al* (1992) on glass reveal evidence of wave reflections from propagating damage fronts, while measurements of lateral and longitudinal stress (figure 9.36), indicate a delayed collapse of the stress difference associated with the failure process. Work summarized by Brar *et al* (1991) supports the results of Kanel and describe spall studies demonstrating loss of tensile strength behind the damage front. The stress profiles measured in that study also support the occurrence of dilatancy or fracture free-volume in the damage wave. Dandekar and Mole (1995) note a tendency toward a constant failure wave velocity in glass, at least for a fixed impact amplitude. The photographic studies of Brannon *et al* (1983) on crystalline quartz, Barker *et al* (1978) on brittle plastics, and Bourne *et al* (1994), Senf *et al* (1994) and Bless *et al* (1992) on glass, all provide visual evidence for failure waves in brittle solids. Some efforts in modeling failure wave kinetics are described by Espinosa (1996) and by Grady (2006). The latter model is discussed in a later section of chapter 6.

References

Abou Sayed A S, Clifton R J and Herman L 1976 The oblique-plate impact experiment *Exp. Mech.* **16** 127–32

Achenbach J D and Brock L M 1973 *On Quasistatic and Dynamic Fracture in Dynamic Crack Propagation* ed G C Sih (Noordholf International) pp 529–42

Ahrens T J and Duvall G E 1966 Stress relaxation behind elastic shock waves in rocks *J. Geophys. Res.* **71** 4349–60

Ahrens T J and Gregson V G 1964 Shock compression of crustal rocks: data for quartz, calcite and plagioclase rocks *J. Geophys. Res.* **69** 4839–74

Aidun J B and Gupta Y M 1995 Shear and compression waves in shocked calcium carbonate *J. Geophys. Res.* **100** 1955–80

Alekseevskii V P *et al* 1989 Mass transfer in metals caused by a convergent cylindrical shock wave *Sov. Powder Metall. Metal Ceram.* **10** 809–13

Armstrong R W, Arnold W and Zerilli F J 2007 *Metall. Mater. Trans.* A **38** 2605

Asay J R, Fowles G R, Duvall G E, Miles M H and Tinder R F 1972 Effects of point defects on elastic precursor decay in LiF *J. Appl. Phys.* **43** 2132–45

Asay J R, Fowles G R and Gupta Y M 1972 Determination of material relaxation properties from measurements on decaying elastic shock fronts *J. Appl. Phys.* **43** 744–6

Asay J R and Gupta Y M 1972 Effect of impurity clustering on elastic precursor decay in LiF *J. Appl. Phys.* **43** 2220–3

Asimov I 1941 *Nightfall (Astounding Science Fiction)*

Atroshenko S A, Vasilkov V B, Meshcherykov Y I, Savenkov G G and Chernyshenko A I 1990 *Sov. Phys. Tech. Phys.* **35** 336–41

Bakun O V, Grigor'yev O N, Kartuzov V V and Trefilov V I 1986 Fracture of heterogeneous polycrystals based on dense modifications of boron nitride *Sov. Phys. Dokl.* **31** 513

Balankin A S 1988 Penetration of a plastic object into obstacles made of brittle materials *Sov. Phys. Dokl* **31** 31

Band W and Duvall G E 1961 Physical nature of shock propagation *Am. J. Phys.* **29** 780

Barker D B, Fourney W L and Dally J W 1978 Photoelastic investigations of fragmentation mechanisms *University of Maryland Technical Report*

Barker L M 1971 A model for stress wave propagation in composite materials *J. Composite Mater.* **5** 140–62

Barker L M 1984 High pressure quasi-isentropic impact experiments *Shock Waves in Condensed Matter-1983* (Amsterdam: Elsevier) pp 217–20

Barker L M and Hollenbach R E 1970 Shock wave studies of PMMA, fused silica and sapphire *J. Appl. Phys.* **41** 4208–26

Barker L M and Hollenbach R E 1972 Velocity interferometer for measuring the velocity of any reflecting surface *J. Appl. Phys.* **43** 4669–80

Berry J P 1960 Some kinetic considerations of the Griffith criterion for fracture *J. Mech. Phys. Solids* **8** 194–216

Birkimer D L 1971 A possible fracture criterion for the dynamic tensile strength of rock *Proc. 12th Symp. on Rock Mechanics* ed G B Clark p 573

Bless S J, Brar N S, Kanel G and Rosenberg Z 1992 Failure waves in glass *J. Am. Ceram. Soc.* **75** 1002–4

Bourne N K, Rosenberg Z, Mebr Y, Obara T and Field J E 1994 A high-speed photographic study of fracture wave propagation in glasses, *Proc. EuroDYMAT 94 Int. Conf. of Mechanical and Physical Behavior of Materials under Dynamic Loading, (Oxford, 26–30 September*) pp 635–40

Bourne N K, Millet J C F, Rosenberg Z and Murray N H 1998 On the shock induced failure of brittle solids *J. Mech. Phys. Solids* **46** 1887–908

Brace W F and Jones A H 1971 Comparison of uniaxial deformation in shock and static loading of three rocks *J. Geophys. Res.* **76** 4913–21

Brace W F and Riley D K 1972 Static uniaxial deformation of 15 rocks to 30 kb *Int. J. Rock. Mech. Min. Sci.* **9** 271–88

Brannon P J, Konrad C, Morris R W, Jones E D and Asay J R 1983 Studies of the spectral and spatial characteristics of shock-induced luminescence from x-cut quartz *J. Appl. Phys.* **54** 6374–81

Brar N S, Bless S J and Rosenberg Z 1991 Impact-induced failure waves in glass bars and plates *Appl. Phys. Lett.* **59** 3396–8

Brar N S and Bless S J 1992 Failure waves in glass under dynamic compression *High Press. Res.* **10** 773–84

Cagnoux J 1994 Analysis d'Echantillons d'Al_2O_3, TiB_2 et SiC Recuperes Apres des Chocs Allamt Jusqu'a 20 GPa *J. Physique* 257–62

Cagnoux J and Longy F 1988 Is the dynamic strength of alumina rate dependent *Shock Waves in Condensed Matter* (Amsterdam: Elsevier) pp 293–6

Chaplot S L and Sikka S K 1992 Molecular dynamics simulations of pressure induced crystal to amorphous transitions in some corner linked polyhedral compounds *Recent Trends in High Pressure Physics* (Oxford) pp 259–61

Chen M W, McCauley J W and Hemke K J 2003 Shock induced localized amorphization in boron carbide *Science* **299** 1563

Chhabildas L C and Barker L M 1988 Dynamic quasi-isentropic compression of tungsten *Shock Waves in Condensed Matter* (Amsterdam: Elsevier) pp 111–4

Chhabildas L C and Graham R A 1987 *Developments in Measurement Techniques for Shock-Loaded Solids*vol 83 (New York: ASME)

Chhabildas L C, Furnish M D, Reinhart W D and Grady D E 1998 Impact of AD995 alumina rods *AIP Conf. Proc.* pp 505–8
Chhabildas L C and Swegle J W 1980 Dynamic pressure-shear loading using anisotropic crystals *J. Appl. Phys.* **51** 4799
Cline C F 1989 personal communication.
Dandekar D P, Abbate A and Frankel J 1994 Equation of state of aluminum nitride and its shock response *J. Appl. Phys.* **76** 4077–85
Dandekar D P and Mole P 1995 Failure waves under shock wave compression in soda lime glass *Metallurgical and Materials Applications of Shock-Wave and High-Strain-Rate Phenomena* ed L Murr *et al* (Amsterdam: Elsevier) 211–8
De Carli P S and Milton D J 1965 Stishovite: synthesis by shock wave *Science* **147** 144–5
Dremin A N and Breusov G A 1968 Processes occurring in solids under the action of powerful shock waves *Russian Chem. Rev.* **37** 396–402
Dremin A N and Breusov O N 1971 The chemistry of shock compression, *Priroda*, 12, 10, (translation *Sandia National Laboratories Report* SAND80-6003, 1980)
Drumheller D S and Sutherland H J 1973 A lattice model for stress-wave propagation in composite materials *J. Appl. Mech.* 149–54
Dulaney E N and Brace W F 1960 Velocity behavior of a growing crack *J. Appl. Phys.* **31** 2233
Duvall G E and Graham R A 1977 Phase transitions under shock wave loading *Rev. Modern Phys.* **49** 523–79
Edmond J M and Paterson M S 1972 Volume changes during the deformation of rocks at high pressures *Int. J. Rock. Mech. Min. Sci.* **9** 161–82
Espinosa H D 1996 *Modeling failure waves in brittle materials* Structures under Shock and Impactvol 4 ed N Jones *et al* (Boston, MA: Computational Mechanics) pp 449–58
Espinoza H D, Xu Y and Brar N S 1997 Micromechanics of failure waves in glass *J. Amer. Ceram. Soc.* **80** 2061–85
Franklin J A 1971 Triaxial strength of rock *Rock Mech.* **3** 86–98
Freund L B 1990 *Dynamic Fracture Mechanics* (New York: Cambridge University Press)
Fowles G R 1961 Shock wave compression in hardened and annealed 2024 aluminum *J. Appl. Phys.* 1475–87
Fowles R and Williams R F 1970 Plane stress wave propagation in solids *J. Appl. Phys.* **41** 460
Furnish M D, Grady D E and Brown J M 1986 Analysis of shock wave structure in single-crystal olivine using VISAR *Shock Waves in Condensed Matter* (New York: Plenum) pp 595–600
Galanov B A, Grigor'yev O N, Kartuzov V V, Ostapenko V N and Trefilov V I 1989 Impact failure of ceramic plates *Sov. Powder Metall. Metal Ceram.* **4** 299
Gilman J J 1970 Strength of solids *J. Appl. Phys.* **41** 1644–74
Gilman J J 1979 Resistance to shock-front propagation in solids *J. Appl. Phys.* **50** 4059–64
Gluzman S L and Psakh'e S G 1989 Hydrodynamic mode in a solid under conditions of a strong external action *Sov. Tech. Phys. Lett.* **15** 764–8
Grady D E 1977 Processes occurring in shock wave compression of rocks and minerals *High Pressure Research: Applications in Geophysics* (New York: Academic) 389–438
Grady D E 1979 Interrelation of flow or fracture and phase transition in the deformation of carbonate rock *J. Geophys. Res.* **84** 7549–55
Grady D E 1980 Shock deformation of brittle solids *J. Geophys. Res.* **82** 913–24 .
Grady D E 1981 Strain-rate dependence of the effective viscosity under steady-wave shock compression *Appl. Phys. Lett.* **38** 825–6

Grady D E 1983 Stress wave studies on selected rocks and minerals, *Sandia National Laboratories Technical Report* SAND83-1898, September.
Grady D E 1986 High-pressure release-wave measurements and phase transformation in $CaCO_3$ *Shock Waves in Condensed Matter* (New York: Plenum) pp 589–93
Grady D E 1991 On the determination of a Hugoniot equation of state from ground-shock measurements *Sandia National Laboratories Technical Report* SAND91-2000, November.
Grady D E 1992a Shock-compression properties of ceramics *Recent Trends in High-Pressure Research* ed A K Singh (Oxford and IBH) pp 641–50
Grady D E 1992b Shock-wave properties of high strength ceramics *Shock Waves in Condensed Matter* (Amsterdam: Elsevier) pp 455–8
Grady D E 1994a Shock wave strength properties of boron carbide and silicon carbide *J. Physique* 385–91
Grady D E 1994b Impact strength and indentation hardness of high-strength ceramics *Shock Waves in condensed Matter* (Amsterdam: Elsevier) pp 741–4
Grady D E 1994c Dynamic failure of brittle solids *Fracture and Damage in Quasibrittle Structures* ed Z P Bazant *et al* (E and FN Spon) pp 259–72
Grady D E 1995a Dynamic properties of ceramic material *Sandia National Laboratories Report* SAND94-3266, February.
Grady D E 1995b Shock properties of high-strength ceramics *Constitutive Laws: Experiment and Numerical Implementation* (Barcelona: CIMNE) pp 35–45
Grady D E 1998a Shock wave compression of brittle solids *Mech. Mater.* **29** 181–203
Grady D E 1998b Scattering as a mechanism for structured shock waves in metals *J. Mech. Phys. Solids* **46** 2017–32
Grady D E 2000 Shock profile studies on selected silicon carbide ceramics with application to dynamic yield mechanism *Shock Compression of Condensed Matter* (American Institute of Physics) pp 629–32
Grady D E 2003 The physics of ceramics from shock-wave experiments *Proc. 27th Int. Cocoa Beach Conf. on Advanced Ceramics and Composites* ed W M Kriven WM and H T Lin (American Ceramic Society) pp 389–99
Grady D E 2006 Effects of meso-scale stress equilibration on the delayed failure of brittle solids, in anisotropy, texture, dislocations and multiscale modeling in finite plasticity and viscoplasticity and metal forming *Proc. 12th Int. Symp. on Plasticity, (Halifax, Canada, 17–22 July)* ed A S Khan and R Kazmi pp 130–2
Grady D E 2007 The shock wave profile *Shock Compression of Condensed Matter* (American Institute of Physics) pp 3–11
Grady D E 2010a Structured shock waves and the fourth-power law *J. Appl. Phys.* **107** 013506
Grady D E 2010b Dynamic equation of state and strength of boron carbide *34th Int. Conf. on Advanced Ceramics and Composites (ICACC) (Daytona Beach, FL, 25–29 January)*
Grady D E 2015 Hugoniot equation of state and dynamic strength of boron carbide *J. Appl. Phys.* **117** 165904
Grady D E, Hollenbach R E, Schuler K W and Callender J F 1976 Compression Wave Studies in Blair dolomite, *Sandia National Laboratories Technical Report* SAND76-0005, February.
Grady D E and Hollenbach R E 1977 Rate controlling processes in the brittle failure of rock *Sandia National Laboratories Technical Report* SAND76-0659, February.
Grady D E, Hollenbach R E, Schuler K W and Callender J F 1977 Strain rate dependence in dolomite inferred from impact and static compression studies *J. Geophys. Res.* **82** 1325–33

Grady D E, Hollenbach R E and Schuler K W 1978 Compression wave studies in calcite rock *J. Geophys. Res.* **83** 2839–49

Grady D E and Lipkin J 1980 Criteria for impulsive rock fracture *Geophys. Res. Lett.* **7** 255–8

Grady D E and Kipp M E 1989 Fragmentation of solids under dynamic loading *Structural Failure* ed J Wierzbicki (New York: Wiley) pp 1–40

Grady D E and Mashimo T 1992 Shock and release properties of yttria-doped tetragonal and cubic zirconia *J. Appl. Phys.* **71** 4868–74

Grady D E and Wise J L 1993 Dynamic properties of ceramic materials, *Sandia National Laboratories Technical Report* SAND93-0610, September.

Grady D E and Zhugin Y N 1994 Critical transition stress in the shock compression of SiO_2 *Bull. Am. Phys. Soc.* **39** 410–1

Grady D E 1995 *Sandia National Laboratories Technical Report* SAND94-3266, February.

Grady D E and Moody R L 1996 Shock compression profiles in ceramics, *Sandia National Laboratories Technical Report* SAND96-0551, March.

Grady D E and Chhabildas L C 1997 Shock-wave properties of soda-lime glass *Proc. 4th US Symp. on Solid Mechanics (Myrtle Beach, SC, 16–18 October)* (Battelle) pp 29–38

Green S J and Perkins R D 1968 Uniaxial compression tests at varying strain rates on three geological materials *Proc. 10th Symp. on Rock Mechanics* ed K E Gray pp 35–54

Griffith A A 1924 Theory of rupture. *Proc. First Int. Congress on Applied Mechanics (Delft)* pp 55–63

Gupta Y M 1977 Effect of crystal orientation on the dynamic strength of LiF *J. Appl. Phys.* **48** 5067–77

Gust W H and Royce E B 1971 Dynamic yield strengths of B_4C, BeO, and Al_2O_3 ceramics *J. Appl. Phys.* **42** 276–94

Hadley K 1976 Comparison of calculated and observed crack densities and seismic velocities in Westerley granite *J. Geophys. Res.* **81** 3484–94

Heard H C 1963 Effect of large changes in experimental deformation of yule marble *J. Geol.* **71** 163

Heard H C, Duba A, Abey A E and Schock R N 1973 Mechanical properties of Blair dolomite, *Lawrence Livermore Laboratory Report* UCRL-51465.

Hemley R J, Jephcoat A P, Mao H K, Ming M H and Manghnani M H 1988 Pressure-induced amorphization of crystalline silica *Nature* **334** 254–8

Herrmann W and Nunziato J W 1973 Nonlinear constitutive equations *Dynamic Response of Materials to Intense Impulsive Loading* Air Force Materials Laboratory, Wright-Paterson AFB) pp 123–281

Hirth J P and Lothe J 1968 *Theory of Dislocations* (New York: McGraw Hill)

Isenberg J 1972 Nuclear geoplosics part two: mechanical properties of earth materials, *Defense Nuclear Agency Report* 1285H2, November.

Hongliang H, Sekine T, Kobayashi T, Hirosaki H and Suzuki T 2000 Shock-induced phase transition of β-Si_3N_4 to *c*-Si_3N_4 *Phys. Rev.* B **62** 17 11412 (6)

Jamieson J C 1953 Phase equilibrium in the system calcite–aragonite *J. Chem. Phys.* **21** 1385–90

Janach W 1976 The role of bulking in brittle failure of rocks under rapid compression *Int. J. Rock Mech. Min. Sci.* **113** 1–10

Janach W 1977 Failure of granite under compression *Int. J. Rock Mech. Min. Sci.* **14** 209–15

Jaeger J C and Cook N G 1969 *Fundamentals of Rock Mechanics* (London: Methuen)

Jeanloz R 1989 Shock wave equation of state and finite strain theory *J. Geophys. Res.* **94** 5873–86

Johnson J N 1992 Calculation of path-dependent shock-wave hardening *J. Appl. Phys.* **72** 797–9

Jones O E and Graham R A 1968 Shear strength effects on phase transition pressures determined from shock-compression experiments, *Proc. Symp. on Accurate Characterization of the High-Pressure Environment, (Gaithersburg, MD, 14–18 October)*

Kalthoff J F and Shockey D A 1977 Instability of cracks under impulse loads *J. Appl. Phys.* **48** 986–93

Kanel G I, Rasorenov S V and Fortov V E 1992 The failure waves and spallation in homogeneous brittle materials *Shock Waves in Condensed Matter* (Amsterdam: Elsevier) 451–4

Kerley G I 1989 Equations of state for calcite minerals: theoretical modeling for dry calcium carbonate *High Pressure Res.* **2** 29–47

Kerley G I 1990 Theory of calcite equation of state *Shock Waves in Condensed Matter* (New York: AIP) pp 613–6

Kipp M E and Grady D E 1989 Shock compression and release in high-strength ceramics, *Sandia National Laboratories Technical Report* SAND89-1461, July.

Kipp M E and Grady D E 1990 Shock compression and release in high-strength ceramics *Shock Waves in Condensed Matter* (New York: AIP) pp 377–80

Kipp M E and Grady D E 1994 Shock phase transformation and release properties of aluminum nitride *J. Physique* 249–56

Kipp M E, Grady D E and Chen E P 1980 Strain-rate dependent fracture initiation *Int. J. Fracture* **16** 471–8

Kimberleya J, Ramesha K T and Daphalapurkara N P 2012 A scaling law for the dynamic strength of brittle solids *J. Mech. Phys. Solids.*

Kolmogorov A N 1941 *Doklady Akad. Nauk., SSSR*; **30** 9; **31** 99; **32** 16

Kondo K A, Sawaoka A, Sato K and Ando M 1982 Shock compression and phase transformation of AlN and BP *Shock Waves in Condensed Matter* (New York: AIP Press) pp 325–3295

Kumar A 1968 The effect of stress rate and temperature on the strength of basalt and granite *Geophysics* **33** 501–10

Lankford J 1976 Dynamic strength of oil shale *Soc. Petrol. Eng. J.* **17** 17–22

Lankford J 1981 Mechanisms responsible for strain rate dependent compressive strength of ceramic materials *J. Amer. Ceramic. Soc* **64** C33-C34

Larsen D B, Rodean H C, Spataro S J and Hearst J R 1971 Seismic evasion research program progress report, *Lawrence Livermore Laboratory Report* UCID-15948

Lawn B R and Wilshaw T R 1975 *Fracture of Brittle Solids* (Cambridge: Cambridge University Press)

Lindholm U S, Leakley L M and Nagy A 1974 The dynamic strength and fracture properties of Dresser basalt *Int. J. Rock. Mech. Min. Sci.* **11** 181–91

Longy F and Cagnoux J 1989 Plasticity and microcracking in shock loaded alumina *J. Amer. Ceramic. Soc.* **72** 971–9

McQueen R G, Marsh S P, Tayler J W, JFritz J N and Carter W J 1970 The equation of state of solids from shock wave studies *High Velocity Impact Phenomena* ed R Kinslow (New York: Academic)

Mescheryakov Y I, Atroshenko S A, Vasilkov V B and Chernyshenko A I 1992 Criteria of transition from translational to rotational motion of media under shock loading *Shock Compression of Condensed Matter* (Amsterdam: Elsevier) pp 407–10

Mott N F 1948 Brittle fracture in mild steel plates *Engineering* **165** 16

Merlini M, Hanfland M and Crichton W A 2012 $CaCO_3$-III and $CaCo_3$-VI high-pressure polymorphs of calcite: possible host structures for carbon in the Earth's crust *Earth Planetary Sci. Lett.* **333–334** 265–71

Munson D E and Lawrence R J 1979 Dynamic deformation of polycrystalline alumina *J. Appl. Phys.* **50** 6272–82

Murrel S A F 1963 A criterion for brittle fracture of rocks and concrete under triaxial stress and the effect of pore pressure on the criterion *Rock Mechanics* ed C Fairhurst (Oxford: Pergamon) pp 563–77

Nakamura A and Mashimo T 1994 Shock-induced phase transition in AlN *High-Pressure Science and Technology* (New York: AIP Press) 981–4

Nickolaevskii V N 1981 Limit velocity of fracture front and dynamic strength of brittle solids *Int. J. Eng. Sci.* **19** 41–56

Nunziato J C, Walsh E K, Schuler K W and Barker L M 1974 *Wave Propagation in Nonlinear Viscoelastic Solids (Handbuch der Physik vol 4)* (Springer: Berlin)

Paris V, Frage N, Dariel M P and Zaretsky 2011 Divergent impact study of the compressive failure threshold in SiC and B_4C *Int. J. Impact Eng.* **38** 228–37

Passman S L, Grady D E and Rundle J B 1980 The role of inertia in the fracture of rock *J. Appl. Phys.* **51** 4070–5

Petersen C F 1969 Shock wave studies of selected rocks *PhD thesis* Stanford University

Podurets A M and Trunin R F 1971 A property of the shock compressibility of quartzite *Sov. Phys. Dokl.* **15** 1117–8

Raiser G V and Clifton R J 1993 Failure waves in uniaxial compression of an aluminosilicate glass *High Pressure Science and Technology* ed S C Schmidt *et al* (New York: AIP Press) pp 1039–42

Raiser G V, Wise J L, Clifton R J, Grady D E and Cox D E 1994 Plate impact response of ceramics and glasses *J. Appl. Phys.* **75** 3862–72

Ravichandran G and Subhash G 1995 A micromechanical model for high strain rate behavior of ceramics *Int. J. Solids Struct.* **32** 2627–46

Robertson E C 1955 Experimental study of the strength of rocks *Geol. Soc. Amer. Bull.* **66** 1275

Rosenberg Z 1995 On the correlation between dynamic compressive strengths of strong ceramics and their indentation hardness *Shock Waves in Condensed Matter* (New York: AIP) pp 543–6

Rosenberg Z, Brar N S and Bless S J 1988 Elastic precursor decay in ceramics as determined with Manganin stress gauges *J. Physique* 707–11

Rubin M B 1990 Analysis of weak shocks in 6061-T6 aluminum *Shock Waves in Condensed Matter* (Amsterdam: Elsevier) 321–8

Sato T and Akimoto S 1979 Hydrostatic compression of four corundum-type compounds *J. Appl. Phys.* **50** 5285–91

Schock R N, Heard H C and Stephens D R 1973 Stress strain behavior of a granodiorite and two graywackes on compression to 20 kilobars *J. Geophys. Res.* **78** 5922–41

Schock R N and Heard H C 1974 Static mechanical properties and shock loading response of granite *J. Geophys. Res.* **79** 1662–6

Schotz C H and Kranz R 1974 Notes on dilatancy recovery *J. Geophys. Res.* **79** 2132–5

Schuler K W, Lysne P C and Stevens 1976 Dynamic mechanical properties of two grades of oil shale *Int. J. Rock Mech. Min. Sci.* **13** 91–5

Serdengecti S and Boozer G D 1961 The effects of strain rate and temperature on the behavior of rocks subject to triaxial compression *Fourth Symp. Rock Mechanics* Pennsylvania State University) p 83

Senf H, Strausburger E and Rothenhausier H 1994 Stress wave damage and fracture in impacted glasses *Proc. EuroDYMAT 94, Int. Conf. of Mechanical and Physical Behavior of Materials under Dynamic Loading (Oxford, 26–30 September)* pp 741–6
Shih C J, Meyers M A and Nesterenko V F 1998 High strain rate deformation and comminution of silicon carbide *J. Appl. Phys.* **83** 4660–71
Shockey D A, Curran D R, Seamann L, Rosenberg J T and Petersen C F 1974 Fragmentation of rock under dynamic loads *Int. J. Rock Mech. Min. Sci.* **11** 303–17
Shono Y 1988 Shock-induced phase transitions in solids: their understanding at atomistic level *Shock Waves in Condensed Matter* (Amsterdam: Elsevier) pp 19–26
Sikka S K 1992 Pressure induced glasses *Recent Trends in High Pressure Physics* (Oxford and IBH) pp 254–8
Steverding B and Lehnigk S H 1976 The fracture penetration depth of stress pulses *Int. J. Rock. Mech. Min. Sci.* **13** 75–80
Swegle J W and Grady D E 1985 Shock viscosity and the prediction of shock wave rise times *J. Appl. Phys.* **58** 692–701
Swift R P 1971 The dynamic response of westerly granite to spherical stress waves, *Physics International Co. Technical Report* DASA 2712, San Leandro, CA.
Tse J S and Klug D D 1992 High pressure phase transitions in silica and berlinite by molecular dynamics simulations *Recent Trends in High Pressure Physics* (Oxford and IBH) pp 274–9
Tuler F R and Butcher B M 1968 A criterion for the time dependence of dynamic fracture *Int. J. Frac. Mech.* **4** 431–7
Turner F J, Griggs D T, Clark R H and Dixon R H 1956 Deformation of yule marble *Geol. Soc. Amer. Bull.* **67** 1259
Vizgirda J and Ahrens T J 1982 Shock compression and implications for the equation of state of carbonates *J. Geophys. Res.* **87** 4747–58
Vogler T J, Reinhart W D and Chhabildas L C 2004 Dynamic behavior of boron carbide *J. Appl. Phys.* **95** 4173
Vogler T J, Reinhart W D, Chhabildas L C and Dandekar D P 2006 Hugoniot and strength behavior of silicon carbide *J. Appl. Phys.* **99** 023512
Vollstadt H E, Ito E, Akaishi M, Akimoto S and Fukunaga O 1990 High pressure synthesis of rocksalt type AlN *Proc. Japan Acad.* **66B** 7–9
Wackerle J 1962 Shock wave compression of quartz *J. Appl. Phys.* **33** 922–37
Walsh J B and Brace W F 1972 Elasticity of rock in uniaxial strain *Int. J. Rock Mech. Min. Sci.* **9** 7–15
Wise J L and Chhabildas L C 1986 Laser interferometer measurements of refractive index in shock-compressed materials *Shock Waves in Condensed Matter* (New York: Plenum) pp 441–5
Wise J L and Grady D E 1993 Dynamic multiaxial impact response of confined and unconfined ceramic rods, *AIRAPT/APS High Pressure Science and Technology Conf. (Colorado Springs, CO, 28 June–2 July)* pp 777–81
Xia Q, Xia H and Ruoff A L 1994 New high pressure phases of the III–IV compounds AlN, GaN, In N *Technology* (New York: AIP Press) pp 307–10
Zhugin Y N and Krupnikov K K 1987 Structural peculiarities of plastic front in quartz in the mixed phase region *Trans. 11th Int. IUPAP-IUTAM Symp. on Non-Linear Acoustics, Part II, Novosibirsk, Russia* pp 196–200
Zlatin N A, Kozhushko A A and Rykova I I 1988 A high strength state of glass at high loading velocities *Sov. Tech. Phys. Lett.* **14** 852

IOP Publishing

Physics of Shock and Impact, Volume 2
Materials and shock response
Dennis Grady

Chapter 10

The shock wave equation of state

The present chapter addresses equation-of-state (EOS) concerns of condensed matter when subjected to intense shock compression. Both lattice compression and thermal energy combine in determining the state of material in responding to the pressure and temperature extremes brought about by shock compression. The chapter starts with discussion of EOSs for gases under shock compression. Emphasis is on the parallels to EOS practices common to the shock compression of condensed matter. This introduction is followed by a description and applications of the Mie–Gruneisen EOS. A comparison with the analogous Rice–Walsh EOS is provided, demonstrating that this EOS has analytic conveniences for certain shock compression applications. Universal compression relations for solids are discussed that are required as reference compression equations necessary for the description of the complete thermodynamic EOS. Calculation of temperature under shock compression is addressed. The chapter closes with EOS experiments and analysis of material mixtures under shock compression.

10.1 Introduction

Strength, viscosity, compaction and fracture are key to the structure and evolution of large-amplitude wave propagation in solids in the shock and impact environment. The equation of state (EOS) of condensed matter provides the remaining component of the constitutive description. The EOS of a material provides the framework about which strength and failure models are constructed.

The thermo-elastic EOS of matter is fundamental to the phenomena of shock compression. A number of shock EOS issues are explored in earlier chapters. The present chapter focuses principally on the EOS properties of condensed matter as they relate to the analysis and underlying physics of shock compression. Some introductory description is provided of the shock compression EOSs of gases.

The Hugoniot EOS, namely the compression path in pressure, volume and energy space achieved through successively higher shock compression events, is basic to

doi:10.1088/978-0-7503-1257-8ch5

shock physics. More complete EOSs are necessary for the prediction of thermodynamic states removed from the principle Hugoniot. Such EOSs are commonly referenced to the principal Hugoniot. Alternatively, EOSs constructed from *ab initio* methods can predict Hugoniot properties of matter. Decomposition of the lattice and the thermal energies, and relationships among isothermal, isentropic and Hugoniot compressibility, are fundamental to the EOS and EOS analysis.

The Mie–Gruneisen EOS, based on an additive mixing of lattice compression and thermal pressure, has enjoyed a prominent role in shock physics analysis. The Gruneisen thermal properties of matter and the Gruneisen thermodynamic parameter are intrinsic to the development of shock wave EOSs and the application of EOSs to shock analysis. The molecular resistance of matter to compression within the description of shock wave EOSs spans the breadth of experimental observation through first-principles theories. The semi-empirical relations of Tait, Murnaghan, Birch and others continue to provide expedient functional forms. Early intermolecular potential models such as the Morse, Rydberg and Lennard-Jones relations also find application. A universal character to the compressibility of crystalline materials has been explored and provides the EOS relations identified with the names Rose–Vinet, Zharkov–Kalinin and Alchagirov, among others.

The Mie–Gruneisen and related isochoric EOS constructions are occasionally problematic when the shock response of highly distended solids and certain material mixtures is of concern. The Rice–Walsh and the Anderson–Gruneisen forms of EOSs, based on an alternative enthalpic iso-pressure development, are found to be appropriate in selected cases for shock EOS descriptions of such materials.

When required, temperature states achieved under shock wave compression and isentropic decompression of matter must result from the shock EOS analysis. Specific heat descriptions of the media are joined with the shock wave EOS. Laws of material melting, such as that due to Lindemann, are necessary for a multi-phase description of shock compression.

Calculations of the shock compression response of complex mixtures of materials benefit from theories that provide for the joining of EOSs of the mixture elemental components. Additive mixture models have found practical application in shock EOS analysis. Significant among such applications is the experimental testing of a mixture and the application of mixture models to assess EOS properties of an entrained component.

10.2 Equation-of-state compressibility of a solid

The problem posed is this: given a material at an initial pressure P_o, temperature T_o and energy E_o, what is the resulting pressure $P(v)$, temperature $T(v)$ and energy $E(v)$ when the material is subjected to shock compression from an initial specific volume v_o to a specific volume $v < v_o$? Solutions of this problem require an appropriate thermodynamic EOS description of the material. Here an EOS refers to any state relation between the thermodynamic properties of a material, such as pressure as a function of specific volume $P(v)$ or specific volume as a function of temperature $v(T)$. A complete EOS would be the energy as a function of specific volume and entropy $E(v,S)$, or the other free energies, from which any required EOS can be calculated.

The theoretical and experimental approaches to this issue have been many. Here the focus is on EOS descriptions developed for solids subjected to shock compression. The range of intensity of shock compression should be clarified. Some aspects of extreme shock compression are considered, however discussions mainly relate to relatively modest shock compression of condensed matter to increases in density of perhaps a factor of two or three. For many materials, this compression range achieves shock pressures up to hundreds of GPa. Shock compressions to higher levels approach the regime of extreme energy density (e.g. Drake 2010) and issues of material physics that are not addressed here in detail.

Over the range of compressions of principal interest, material strength can, to varying degrees, play a role in the shock stress levels that are achieved. Material strength is largely ignored in the present EOS discussions of shock compression. Usually this is a reasonable simplification, although the possible effects of strength at the shock state should never be fully discounted.

A further issue of interest is the shock compression of distended or porous solids. The combination of large distention and intense shock compression can achieve quite large thermal energy and corresponding thermal pressure through the shock dissipation process. The EOS description of matter over the range of temperatures and pressures achieved through shock compression of distended material is of practical interest and is addressed at several points through the chapter. Additional discussions of EOSs and distended materials is provided in chapter 7.

10.2.1 Some elementary equations of state

Before addressing the issue of the EOS of solids, some discussion of more familiar EOSs of simple gases in the shock environment lends insights to the problems of interest. EOSs for ideal gases and the slightly more complex van der Waals gases can be framed within the same functional forms that have evolved as the most convenient for the description of solids in the shock compression environment.

10.2.1.1 Ideal gas equation of state

Perhaps the most fundamental EOS is that for an ideal gas. When written in the form

$$P(v,E) = \frac{\gamma - 1}{v}E, \tag{10.1}$$

for the pressure in the gas as a function of the specific energy and specific volume, a perspective on the EOS framework common to solids begins to emerge. The constant ratio of the specific heats is provided by $\gamma = c_p/c_v$. For air $\gamma \sim 1.4$ is appropriate. For an ideal gas, the pressure depends entirely on the magnitude of thermal energy in the material. A further EOS is needed to fully characterize the material. Temperature and energy are related through the specific heat at constant volume c_v through the relation

$$\left.\frac{\partial E}{\partial T}\right|_{v} = c_v(T). \tag{10.2}$$

When any gas is adequately characterized by the ideal gas EOS relations provided in equations (10.1) and (10.2), pressure, energy and temperature states through any compression path can be calculated.

Compression of an ideal gas at constant entropy requires

$$dE = -P\,dv. \tag{10.3}$$

Joining equations (10.1) and (10.3) provides the ideal-gas isentropic compression relation

$$Pv^{\gamma} = P_o v_o^{\gamma}. \tag{10.4}$$

Compression through a shock wave is governed by the Rankine–Hugoniot energy relation

$$E = E_o + \frac{1}{2}(P + P_o)(v_o - v). \tag{10.5}$$

Eliminating energy between the differentials of equations (10.1) and (10.5) leads to the differential relation

$$\frac{dv}{\left(v - \frac{\gamma - 1}{\gamma + 1}v_o\right)} + \frac{dP}{\left(P + \frac{\gamma - 1}{\gamma + 1}P_o\right)} = 0. \tag{10.6}$$

Integrating provides the pressure versus volume Hugoniot for shock compression of an ideal gas,

$$\left(P + \frac{\gamma - 1}{\gamma + 1}P_o\right)\left(v - \frac{\gamma - 1}{\gamma + 1}v_o\right) = \frac{4\gamma}{(\gamma + 1)^2}P_o v_o. \tag{10.7}$$

Inspection of equation (10.7) reveals that a limiting compression ratio of

$$\frac{\rho_{\infty}}{\rho_o} = \frac{\gamma + 1}{\gamma - 1} \tag{10.8}$$

is approached at sufficiently high pressures. For air at ambient conditions ($\gamma = 7/5$) a limiting compression of six is predicted. Experiment and theoretical calculation predict a limiting compression ratio closer to ten for ambient air (Zeldovich and Raiser 1966). A Hugoniot for air with standard temperature and pressure conditions ahead of the shock wave, taken from a table from Zeldovich and Raiser, is shown in figure 10.1. The Hugoniot is based on detailed calculations including excited vibrational modes and dissociation, and is provided from several sources referenced therein.

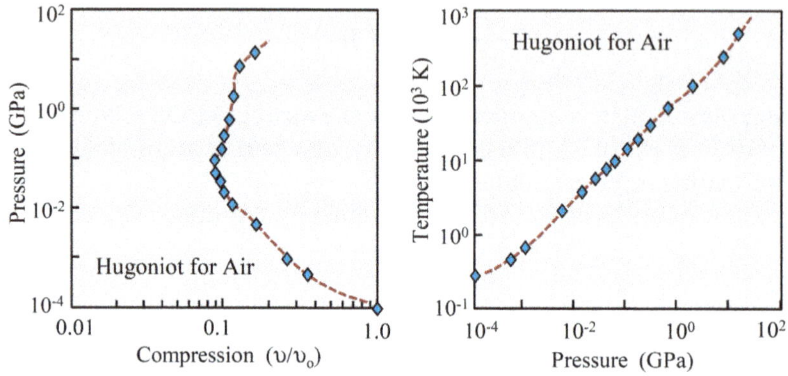

Figure 10.1. Pressure (left) and temperature (right) Hugoniot for air from STP initial state (Zeldovich and Raiser 1966). Based on calculations referenced therein.

10.2.1.2 Van der Waals gas equation of state

Perhaps more illustrative of EOS issues encountered in solids is that of a van der Waals gas. The van der Waals EOS is commonly written in the form

$$\left(P + \frac{a}{v^2}\right)(v - b) = \frac{R}{\mu}T, \tag{10.9}$$

where R is the gas constant, μ the molecular weight, and v the specific volume. Two terms in the van der Waals EOS allow improved description of real gases over that of the idea gas EOS. First, the term a/v^2, called the internal pressure, accounts for the forces of intermolecular attraction between gas molecules. The second term b is dubbed the covolume, and accounts for the volume occupied by the gas molecules.

The temperature and thermal energy when related through a constant specific heat yield

$$T = \frac{1}{c_v}(E - E_i(v)). \tag{10.10}$$

$E_i(v)$ corresponds to the internal pressure through

$$E_i(v) = E_o - \int P_i(v)dv = E_o - \frac{a}{v}. \tag{10.11}$$

Joining equations (10.9), (10.10) and (10.11) yield

$$\left(P + \frac{a}{v^2}\right)(v - b) = \frac{R}{\mu c_v}\left(E - \left(E_o - \frac{a}{v}\right)\right) = (\gamma - 1)\left(E - \left(E_o - \frac{a}{v}\right)\right) \tag{10.12}$$

or

$$P(v,E) = -\frac{a}{v^2} + \frac{\gamma - 1}{v - b}\left(E - \left(E_o - \frac{a}{v}\right)\right). \tag{10.13}$$

The van der Waals EOS, as written in equation (10.13), is in a form common to the shock EOS description of solids. The pressure is partitioned into a non-thermal pressure that is dependent only on the specific volume, and a thermal pressure that depends on both volume and energy. The definition of the thermodynamic Gruneisen parameter provides for a van der Waals gas

$$\Gamma(v) = v\frac{\partial P}{\partial E}\bigg|_v = \frac{\gamma - 1}{1 - b/v}, \tag{10.14}$$

which approaches that of an ideal gas as the specific volume becomes large. Later it is shown that the van der Waals EOS follows from a degenerated form of the Mie–Gruneisen potential joined with this specific relation for the Gruneisen parameter.

As in the case for an ideal gas, equation (10.13) for a van der Waals gas can be joined with equation (10.3), or the Rankine–Hugoniot relation in equation (10.5), to determine the isentropic or the Hugoniot compression properties of the material. In contrast to an ideal gas, however, the van der Waals EOS can exhibit a multivalued pressure–volume behavior replicating a gas-to-liquid phase transformation on the principal compression curves.

A special case of the van der Waals EOS that has useful application is achieved by setting the constant $a = 0$. The EOS in the form

$$P(v,E) = \frac{\gamma - 1}{v - b}E \tag{10.15}$$

is, depending on the author, referred to as the Nobel–Abel, the Abel–Nobel, or just the Abel EOS. This EOS is appropriate where strictly compression of a gaseous medium is of interest and the detailed EOS response is not essential. The Nobel–Abel EOS is found appropriate, for example, in the description of propellant gases within the interior ballistics of guns (Johnson 2005). The covolume parameter b for propellants is reported to be of order 0.001 $m^3\ kg^{-1}$.

A modest modification of the Nobel–Abel EOS that is referenced to zero (negligible) pressure at specified specific volume is found useful in the description of the shock compression of selected condensed and dispersed materials. The isentropic or cold compression curve is provided by

$$P(\mu) = K_o\frac{1 - \mu_b}{\gamma}\left[\frac{(1 - \mu_b)^\gamma}{(\mu - \mu_b)^\gamma} - 1\right], \tag{10.16}$$

where $\mu = v/v_o$ is the compression referenced to the initial specific volume and $\mu_b = b/v_o$ is the limiting compression at high pressure. The constant K_o provides the material bulk modulus at the reference density. Equation (10.16) is readily inverted.

10.2.2 Mie–Gruneisen equation of state

There are a number of approaches that lead to comparable shock EOS representations of condensed matter. Here, the approach starts with the thermodynamic definition of the Gruneisen property of the material,

$$v\frac{\partial P}{\partial E}\bigg|_V = \Gamma. \tag{10.17}$$

The Gruneisen parameter is a dimensionless measure of the rise in pressure when heat (thermal energy) is added to a body while maintaining the volume of the material constant. At any specific volume this differential is written in the integral form

$$\int_{P_r(v)}^{P} dP = \int_{E_r(v)}^{E} \frac{\Gamma}{v} dE, \tag{10.18}$$

where integration is from some reference pressure and energy $P_r(v)$ and $E_r(v)$ to any P and E state of the material. Theory and experience suggests that for many materials the Gruneisen Γ is only weakly dependent on energy. The assumption is commonly made that the Gruneisen Γ is a function of volume only. This assumption is found to be consistent with a large body of shock compression studies (e.g. McQueen *et al* 1970). The simplification appears to be a sensible approximation for many materials over a reasonable range of shock compression. Assumption of a volume-dependent-only Gruneisen Γ simplifies the integral in equation (10.18) leading to the linear relation between pressure and energy

$$P(v,E) = P_r(v) + \frac{\Gamma(v)}{v}(E - E_r(v)). \tag{10.19}$$

The relation in equation (10.19) is the Mie–Gruneisen form of the shock EOS for condensed matter. It is widely used in the EOS description of shock compression. Born (1939) noted that when the thermal vibrations in a solid are harmonic, the lattice compressional and vibrational energies are additive and pressure decomposes into the additive form of equation (10.19). (See also Bridgman (1952).) An informative tutorial is put forth by Johnson (1968) and is expanded on in the following section. EOSs as written for an ideal gas in equation (10.1) and a van der Waals gas in equation (10.13) are of the form of the Mie–Gruneisen EOS in equation (10.19).

The volume-dependent reference EOS relations for the pressure $P_r(v)$ and energy $E_r(v)$ are a matter of choice. The choice is usually governed by the intent of the solution and the availability of properties for the material. The cold curve, or the compression curve at absolute zero temperature, is commonly used in theoretical studies. The cold curve conveniently separates energy (and pressure) into contributions due strictly to compression of the cold lattice (ignoring zero-point energy) and that due to lattice and electronic thermal contributions. The principal isentrope (referenced from ambient temperature, pressure and energy) is also appropriate as the Mie–Gruneisen EOS reference curve. The principal isentrope of a solid is a convenient reference relation when ambient ultrasonic properties of the material are available. The principal Hugoniot of the solid also provides a suitable reference curve for the Mie–Gruneisen EOS. Hugoniot data are available for many materials and provide a reliable measure of compression at large shock pressures. The principal Hugoniot for the solid is a common EOS reference relation used when assessing shock compression of porous bodies of the same material.

10.2.2.1 Gruneisen parameter

To proceed further with the EOS analysis, functional forms must be specified for both the reference compression curve and the volume-dependent Gruneisen parameter. Various functional forms are assumed for volume dependence of the Gruneisen parameter. A functional form proposed by Altshuler and Sharipdzhanov (1971) is appropriate and captures most of the simpler assumptions

$$\Gamma(v) = \Gamma_o \mu^n + \Gamma_\infty (1 - \mu^n), \tag{10.20}$$

with $\mu = v/v_o$. In equation (10.20), in terms of more familiar thermodynamic properties, the Gruneisen parameter at ambient conditions is provided by

$$\Gamma_o = \frac{\alpha K_S}{\rho_o c_P} = \frac{\alpha K_T}{\rho_o c_V}, \tag{10.21}$$

while $\Gamma_\infty \simeq 0.5$ is appropriate at extreme energy states of the material. A functional form commonly assumed and analytically convenient is extracted from equation (10.20) by ignoring the second term and setting $n = 1$. Namely, the assumption is that $\rho\Gamma = \rho_o\Gamma_o$ is a constant. A constant $\rho\Gamma$ is assumed for some of the following EOS analysis and further simplifies equation (10.19) for the Mie–Gruneisen EOS to the relation

$$P(v,E) = P_r(v) + \rho_o\Gamma_o(E - E_r(v)). \tag{10.22}$$

The ambient ultrasonic bulk modulus K_o of a solid provides a measure of the harmonic tension of the molecular lattice. The change in the bulk modulus as pressure is applied to the solid is denoted by K_o'. This pressure derivative quantifies the increase of the bulk modulus with increasing pressure and captures the anharmonic character of the lattice. Theoretical efforts have found that the ambient Gruneisen parameter Γ_o is related to K_o'. Theories of Slater (1939), Dugdale and MacDonald (1953) and Vaschenko and Zubarev (1963) arrive at Γ_o equal to $(K_o' - 1/3)/2$, $(K_o' - 1)/2$ and $(K_o' - 5/3)/2$, respectively.

The Gruneisen parameter resides at the heart of shock wave EOS analysis of solid matter. Johnson (1968) provides an instructive tutorial that reveals the underlying physics of the Gruneisen parameter. The discussions here follows the approach of Johnson, although his ideas are extended to examine the Gruneisen consequences of several familiar reference EOS compression relations, namely the Murnaghan, the linear inertial modulus (linear-Z) and the Mie–Born EOSs. More is said of these reference EOSs in later sections of the chapter.

Consider the one-dimensional model of a unit cell of a solid as illustrated in figure 10.2. When the central mass m is at rest at $x = 0$, each spring is compressed a distance Δ in response to a pressure $P_o(\Delta)$. Unit areas normal to the axis of compression are assumed so that pressure and force, as well as displacement and volume, are synonymous. The unit cell size at zero pressure is v_o, and under

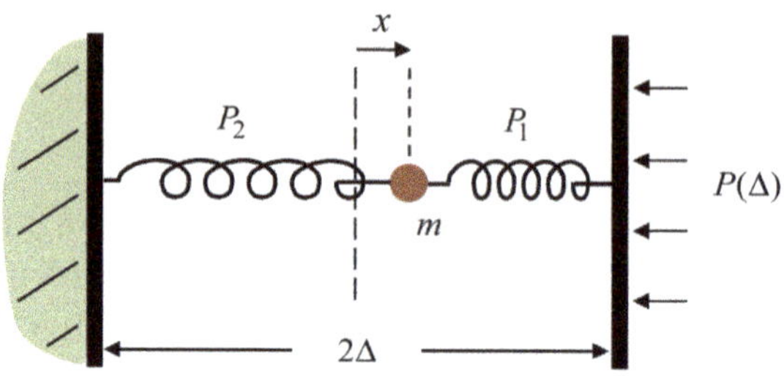

Figure 10.2. One-dimensional single-particle model of cold compression and thermal pressure in a solid.

compression $2\Delta = v_o - v$. The parameter v can be considered as a length scale, a volume scale, or a specific volume if normalized by the mass m in the cell. The pressure $P_o(\Delta)$, or $P_o(v)$, is identified with the cold compression of the unit cell.

When the unit cell is subjected to a compressing of 2Δ, and then the mass m is moved a distance x from the central point, the forces of the springs, P_1 and P_2, are unequal. They are, to second-order in the displacement x,

$$P_1 = P_o(\Delta + x) = P_o(\Delta) + P_o'(\Delta)x + \frac{1}{2}P''_o(\Delta)x^2 \qquad (10.23)$$

and

$$P_2 = P_o(\Delta - x) = P_o(\Delta) - P_o'(\Delta)x + \frac{1}{2}P''_o(\Delta)x^2. \qquad (10.24)$$

Equation of motion of the mass m is governed by the expression

$$m\frac{d^2x}{dt^2} = -(P_1 - P_2) = -2P_o'(\Delta)x \qquad (10.25)$$

and has the solution

$$x(t) = x_o \sin \omega t, \quad \omega = \sqrt{2P_o'(\Delta)/m}. \qquad (10.26)$$

When the cell is compressed to a volume 2Δ, and the mass m subject to the motion prescribed by equation (10.25), then the appropriate pressure is provided by $P_1(\Delta,x)$ from equation (10.23). This pressure is time varying through the solution for $x(t)$. But, in practice this time dependence refers to molecular vibrational frequencies, and the applicable pressure is that averaged over time, or

$$P(\Delta) = \langle P_1(\Delta,x)\rangle = \frac{1}{\tau}\int_0^\tau P_1(\Delta,x)dt, \qquad (10.27)$$

with τ equal to one vibrational period. Carrying out the averaging of equation (10.23) yields

$$P(\Delta) = P_o(\Delta) + \frac{1}{4}P_o''x_o^2. \qquad (10.28)$$

The energy of vibration of the mass m is equal to the kinetic energy as the mass passes through the central point, or

$$E_T = E - E_o(\Delta) = \frac{1}{2} m\omega^2 x_o^2. \tag{10.29}$$

The energy E_T is identified with the thermal energy of the unit cell. It is also the difference between the total energy E and the cold energy

$$E_o(\Delta) = \int_0^{\Delta} P_o(\underline{\Delta})d\underline{\Delta}. \tag{10.30}$$

Joining equations (10.28) and (10.29) results in

$$P(\Delta) = P_o(\Delta) + \frac{1}{2}\frac{P_o''}{m\omega^2}E_T = P_o(\Delta) + \frac{1}{4}\frac{P_o''}{P_o'}E_T. \tag{10.31}$$

The latter step incorporates the expression for the vibrational frequency from equation (10.26).

Reverting to volume dependence of the pressure through $2\Delta = v_o - v$, equation (10.31) becomes

$$P(v) = P_o(v) + \frac{1}{v}\left(-\frac{1}{2}\frac{d\ln dP/dv}{d\ln v}\right)E_T. \tag{10.32}$$

The dimensionless expression in the thermal term is identified with the Gruneisen parameter for the material

$$\Gamma(v) = -\frac{1}{2}\frac{d\ln dP/dv}{d\ln v} = \frac{d\ln\omega}{d\ln v}. \tag{10.33}$$

The second step again makes use of the frequency of vibration expression from equation (10.26). In this form it is recognized as equivalent to the Gruneisen parameter resulting from crystal quantum mechanical analysis with ω a measure of the average normal mode vibrational frequency of the lattice.

It is instructive to examine the Gruneisen parameter provided by equation (10.33) for several common reference EOS relations. Gruneisen parameter expressions that result from a Murnaghan, a linear-Z and a Mie–Born equation for the cold compression curve are calculated. Again, these reference EOS relations are discussed in further detail later in the chapter. The same method would apply to any analytic relation for the pressure versus specific volume cold compression curve.

The compression relation for the Murnaghan equation is

$$P(v) = \frac{K_o}{K_o'}\left(\left(\frac{v}{v_o}\right)^{-K_o'} - 1\right). \tag{10.34}$$

Carrying out the derivatives required from equation (10.33) yields a Gruneisen parameter of

$$\Gamma = \frac{K_o' + 1}{2}, \tag{10.35}$$

that is a constant independent of the magnitude of volume compression.

A linear representation of the inertial modulus $Z = \rho K$ in the pressure

$$Z(P) = Z_o + Z_o' P \tag{10.36}$$

leads to the pressure versus specific volume relation

$$P(v) = \frac{Z_o}{Z_o'}\left(e^{Z_o'(v_o - v)} - 1\right), \tag{10.37}$$

where $Z_o = \rho_o K_o$ and $Z_o' = \rho_o(K_o' + 1)$. Again, completing the appropriate derivatives in equation (10.33) yields the volume-dependent Gruneisen parameter

$$\Gamma(v) = \frac{K_o' + 1}{2}\frac{v}{v_o}. \tag{10.38}$$

The present model applied to the linear-Z EOS leads to the commonly used assumption of $\rho\Gamma$ equal to a constant. At zero pressure the Murnaghan and the linear-Z equations yield the same value for the ambient Gruneisen parameter.

The procedure is readily applied to the more general Mie–Born EOS. Using the pressure versus specific volume expression provided from the later equation (10.132), a Gruneisen function is obtained of the form

$$\Gamma(v) = \frac{1}{6}\frac{(m+3)(m+6)(v_o/v)^{(m+6)/3} - (n+3)(n+6)(v_o/v)^{(n+6)/3}}{(m+6)(v_o/v)^{(m+6)/3} - (n+6)(v_o/v)^{(n+6)/3}}. \tag{10.39}$$

Equation (10.39) is readily checked. For the special case of $n = -3$ and $m = 3(K_o' - 1)$, the Mie–Born equation degenerates to the Murnaghan equation, and equation (10.39) reproduces equation (10.35). A further special case is provided by $n = 2$ and $m = 4$, where the Mie–Born equation yields the second-order Birch equation provided later in equation (10.147), and the pressure derivative of the bulk modulus is precisely $K_o' = 4$. The predicted volume dependence of the Gruneisen parameter for the several cold-compression EOSs are compared in figure 10.3. An ambient $K_o' = 4$ is assumed for each case. An inverse first-power dependence on the specific volume is exhibited by the linear-Z compression equation. The initial volume dependence of the Birch equation is comparable to the linear-Z equation, but lessens with increasing compression. Similar behavior is exhibited for other reasonable values of the Mie–Born constants.

At zero pressure the Mie–Born relation and equation (10.39) yields an ambient Gruneisen parameter of

$$\Gamma_o = \frac{1}{6}(m + n + 9) = \frac{1}{2}(K_o' + 1), \tag{10.40}$$

where the relation of $K_o' = (m + n + 6)/3$ for the Mie–Born equation is applied in the second step. The same ambient Gruneisen parameter results from the Murnaghan, the linear-Z and the Mie–Born reference EOSs.

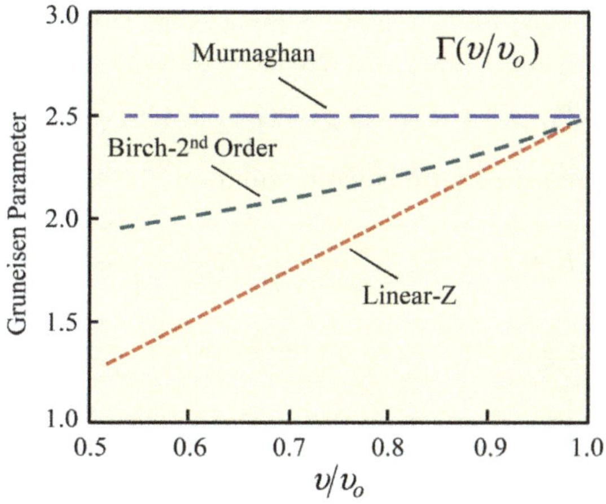

Figure 10.3. Volume functional dependence of the Gruneisen parameter for selected cold compression EOSs based on the model of Johnson (1968).

The predicted ambient Gruneisen parameter of equation (10.40) from the present elemental one-dimension model is not out of line with observed values. Values of K_o' for thirty-six metals reported by Steinberg (1982) provide values of Γ_o from equation (10.40) that range from 1.7–3.7. More detailed models have been proposed that lead to the succinct form, noted earlier in this subsection, of $\Gamma_o = (K_o' - s - 1/3)/2$, with $s = 0$, 2/3 and 4/3 (see Jeanloz 1989). Dimensional arguments would suggest that the volume dependence of the Gruneisen parameter calculated from the present one-dimensional model and provided in equation (10.33) would be generally appropriate within a constant multiplier probably modestly less than unity.

10.2.2.2 Reference equations of state

The reference EOS, identified by the reference pressure $P_r(\upsilon)$ and reference specific energy $E_r(\upsilon)$ for the Mie–Gruneisen EOS in equation (10.19), can be any convenient compression path. The principal isotherm, the principal isentrope, the principal Hugoniot or the absolute zero temperature cold compression curve are all appropriate reference EOS relations for the Mie–Gruneisen EOS.

As an example, for a gamma law gas, as written in equation (10.1),

$$P(\upsilon,E) = \frac{\gamma - 1}{\upsilon}E \tag{10.41}$$

is in an appropriate Mie–Gruneisen form, referenced to the cold compression curve. Namely, there is no cold compression curve as pressure in the gas is entirely thermal. Alternatively, the isentropic compression of a gamma law gas from initial state, P_o, ρ_o and E_o, is provided by the pressure and energy relations

$$P_r(\sigma) = P_o\sigma^\gamma, \quad E_r(\sigma) = E_o + \frac{P_o}{\rho_o}\frac{1}{\gamma - 1}(\sigma^{\gamma-1} - 1), \tag{10.42}$$

where $\sigma = \rho/\rho_o$ is a measure of the relative compression. The Mie–Gruneisen EOS from equation (10.19) for a gamma law gas referenced from the compression isentrope is

$$P(\sigma,E) = P_r(\sigma) + \rho_o\sigma\Gamma(E - E_r(\sigma)). \tag{10.43}$$

Inserting for the reference equations from equation (10.42) and rearranging,

$$\frac{P}{\sigma} - \rho_o\Gamma E + P_o\sigma^{\gamma-1}\left(1 - \frac{\Gamma}{\gamma - 1}\right) = P_o\frac{\Gamma}{\gamma - 1} - \rho_o\Gamma E_o. \tag{10.44}$$

Setting the Gruneisen gamma to the gamma law gas exponent $\Gamma = \gamma - 1$ reduces equation (10.44) to

$$\frac{P}{\sigma} - \rho_o(\gamma - 1)E = P_o - \rho_o(\gamma - 1)E_o. \tag{10.45}$$

The reference energy E_o is arbitrary within a constant. The right-hand side can be zeroed and the standard form for the EOS of a gamma law gas, namely, equation (10.41), is recovered.

In a similar vein, a common representation of a Mie–Gruneisen EOS for solids is constructed with the Murnaghan equation description of the ambient isentropic compression as the reference EOS. Again, in the functional form of equation (10.43) the Murnaghan isentropic compression expressions for the pressure and energy are

$$P_r(\sigma) = \frac{\rho_o C_o^2}{n}(\sigma^n - 1), \quad E_r(\sigma) = E_o + \frac{C_o^2}{n}\left(\frac{\sigma^{n-1}}{n - 1} + \frac{1}{\sigma}\right), \tag{10.46}$$

where initial state conditions are here selected to be density ρ_o, and with $E_o = 0$ and $P_o = 0$. The Murnaghan compression relation follows from the commonly observed pressure dependence of the bulk modulus $K = K_o + K_o'P$ with $\rho_o C_o^2 = K_o$ and $n = K_o'$ in the equation (10.46) relations. Introducing the reference equations from equation (10.46) into equation (10.43) yields after some reorganizing

$$P(\sigma,E) = P_r(\sigma)\left(1 - \frac{\Gamma}{n - 1}\right) - \rho_o C_o^2(1 - \sigma)\frac{\Gamma}{n - 1} + \rho_o\sigma\Gamma(E - E_o). \tag{10.47}$$

As with the expression for the gamma-law gas, equation (10.47) simplifies considerably for the special case of $\Gamma = \Gamma_o = n - 1$,

$$\frac{P + \rho_o C_o^2}{\sigma} - \rho_o\Gamma_o E = \rho_o C_o^2 - \rho_o\Gamma_o E_o. \tag{10.48}$$

Selecting the reference energy E_o to zero on the right-hand side the EOS relation yields

$$P + \rho_o C_o^2 = \rho_o\Gamma_o\sigma E. \tag{10.49}$$

Equation (10.49) is recognized as the Tait EOS (Tait 1888) found useful for the description of solids and liquids (see also Davis (1997)). The Tait equation is shown to provide a reasonable representation of water and sea water over modest temperatures and pressures (Li 1967). A Gruneisen gamma of $\Gamma_o = n - 1$ where $n = K_o'$, although not unreasonable for solids, is generally found to be somewhat high.

For much of the analysis performed in this chapter a convenient cold compression or isentropic reference EOS of the functional form

$$P_r(\varepsilon) = P_s(\varepsilon) = \frac{K_o}{\alpha}(e^{\alpha\varepsilon} - 1), \tag{10.50}$$

where a volume compression strain measure is provided by $\varepsilon = 1 - v/v_o$, will be used with the Mie–Gruneisen EOS. Further motivation for this particular reference relation is provided later in the chapter. Equation (10.50) is referred to interchangeably as the linear-Z or the modified Murnaghan EOS. Briefly, equation (10.50) follows from the assumption that the inertial modulus Z, rather than the bulk modulus K, exhibits a linear dependence on the pressure. Namely, $Z(P) = \rho K = Z_o + Z_o'P$. From initial state of density ρ_o, and with $P_o = 0$ and $E_o = 0$, the corresponding reference specific energy follows from the integral of the pressure,

$$E_r(\varepsilon) = E_s(\varepsilon) = \frac{1}{\rho_o}\int_0^{\varepsilon} P_s(\varepsilon)d\varepsilon = \frac{1}{\alpha}\frac{K_o}{\rho_o}\left(\frac{1}{\alpha}e^{\alpha\varepsilon} - \frac{1}{\alpha} - \varepsilon\right). \tag{10.51}$$

In a later section this particular functional form for the reference isentropic compression will be contrasted with other more common EOSs. Both the Murnaghan (1944) and the Birch (1947) (also referred to as the Birch–Murnaghan) equations have been long-standing analytic descriptions of the high-pressure compression of solids. The relation in equation (10.50) is shown to generally provide a better description of the compression of many solids than does either the Murnaghan or the Birch equation (Grady 2012).

There have, of course, been a number of efforts to generate more universal EOS relations for the compression of solids to quite high compressions (e.g. Rose *et al* 1984, Vinet *et al* 1986, Holzapfel 1996, Alchagirov *et al* 2001). Discussion of such EOSs is provided in a later section. Equation (10.50) is shown to quite adequately replicate these universal relations to compressions approaching about a factor of two or $\varepsilon = 0.5$. The analytic convenience of such universal EOSs are lost, however, when further physics of the compression process, notably phase transformations, complicate the EOS surface. Tabular EOSs then become more convenient in the practical applications to shock EOS studies (e.g. Lyon and Johnson 1992). Alternatively, analytic forms can be used to represent independent phases, and appropriate transformation functions used to carry material from one phase to the next. Examples of the latter approach are provided in chapter 6 to characterize shock transformation in ceramic compounds.

The present equation for the reference isentrope has the analytic convenience of being invertible to dependence of the compression strain and energy on the pressure,

$$\varepsilon_s(P) = \frac{1}{\alpha}\ln\left(1 + \alpha\frac{P}{K_o}\right) \tag{10.52}$$

and

$$E_s(P) = \frac{1}{\alpha}\frac{K_o}{\rho_o}\left(\frac{P}{K_o} - \frac{1}{\alpha}\ln\left(1 + \alpha\frac{P}{K_o}\right)\right). \tag{10.53}$$

This isentropic relation has the conceptual difficulty of approaching a finite pressure at the limiting strain of one (infinite compression). This problem is not a serious issue as the relation is intended for application in the range of compression not to exceed about a factor of two (or less than approximately $\varepsilon = 0.5$).

The isentropic bulk modulus is provided by the derivative

$$K_s = -v\frac{dP_s}{dv} = (1 - \varepsilon)\frac{dP_s}{d\varepsilon} = K_o(1 - \varepsilon)e^{\alpha\varepsilon}. \tag{10.54}$$

The bulk modulus is commonly expressed as a function of pressure in the power series

$$K_s(P) = K_o + K_o'P + \frac{1}{2}K_o''P^2 + \cdots. \tag{10.55}$$

Measurements of elastic wave speeds as a function of modest pressure are available for many materials and provide the constants K_o, K_o' and (much less commonly) K_o''. With some analytic effort equations (10.50) or (10.52) provide K_o as the ambient isentropic bulk modulus whereas

$$K_o' = \alpha - 1 \text{ and } K_o'' = -\alpha/K_o. \tag{10.56}$$

10.2.2.3 Hugoniot equation of state

Joining equations (10.50) and (10.51) for the reference pressure and energy curves with the equation (10.19) Mie–Gruneisen relation provides an analytic EOS for pressure as a function of specific volume and energy. In particular, the principal Hugoniot originating from the same ambient state can be calculated as can the Hugoniot for any degree of distention (porosity) for the initial state of the same solid. Entropy increases under shock compression to the Hugoniot state. The principal Hugoniot will reside at a higher pressure and energy than the principal isentrope at a specified volume. Compression curves that reside on the EOS surface for the material are identified in the pressure–volume plane and on the EOS surface in figure 10.4.

The principal Hugoniot is calculated by introducing the Rankine–Hugoniot energy relation for a shock Hugoniot transition from the ambient state $v = v_o$, $P = P_o = 0$ and $E = E_o = 0$, to the shock Hugoniot state v, $P_h(v)$ and $E_h(v)$,

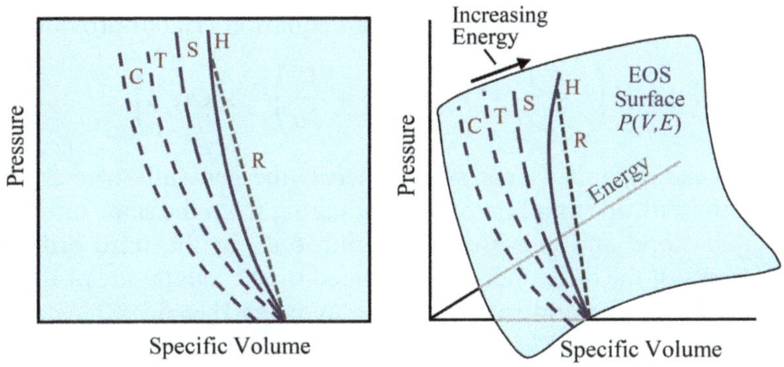

Figure 10.4. EOS compression curves in the pressure versus specific volume plane (left) and on the pressure–volume–energy surface (right). The respective compression curves shown are the cold curve (C), ambient temperature isotherm (T), isentrope (S), Hugoniot (H) and Rayleigh line (R).

$$E_h(v) = \frac{1}{2}P_h(v_o - v) = \frac{1}{2\rho_o}P_h\varepsilon. \tag{10.57}$$

Eliminating energy between equations (10.22) and (10.57) yields

$$P_h(\varepsilon) = P_r(\varepsilon) + \rho_o\Gamma_o\left(\frac{1}{2\rho_o}P_h(\varepsilon)\varepsilon - E_r(\varepsilon)\right), \tag{10.58}$$

where $P_h(\varepsilon)$ identifies the Hugoniot pressure versus volume compression relation. Note that the simplifying assumption of a constant $\rho\Gamma = \rho_o\Gamma_o$ is also employed. Constant $\rho\Gamma$ is consistent with the harmonic model and the linear-Z compression relation (see figure 10.3).

Equation (10.58) is explicitly solved for dependence of the Hugoniot pressure on the compression strain through the reference pressure and energy relations

$$P_h(\varepsilon) = P_r(\varepsilon) + \rho_o\Gamma_o\left(1 - \frac{\Gamma_o}{2}\varepsilon\right)^{-1}\left(\frac{1}{2\rho_o}P_r(\varepsilon)\varepsilon - E_r(\varepsilon)\right). \tag{10.59}$$

Note that the principal Hugoniot as the reference compression path zeros the second expression on the right and recovers the expected identity. Incorporating the functional relations for the isentropic reference compression from equations (10.50) and (10.51), and solving for the Hugoniot pressure yields

$$P_h(\varepsilon) = \frac{K_o}{\alpha}\left(1 - \frac{\Gamma_o}{2}\varepsilon\right)^{-1}\left[\left(1 - \frac{\Gamma_o}{\alpha}\right)(e^{\alpha\varepsilon} - 1) + \Gamma_o\varepsilon\right]. \tag{10.60}$$

A power series expansion of the principal isentrope relation from equation (10.50) yields

$$P_s(\varepsilon) = K_o\left(\varepsilon + \frac{1}{2}\alpha\varepsilon^2 + \frac{1}{6}\alpha^2\varepsilon^3 + O(\varepsilon^4)\right). \tag{10.61}$$

An expansion of the principal Hugoniot from equation (10.60) provides

$$P_h(\varepsilon) = K_o\left(\varepsilon + \frac{1}{2}\alpha\varepsilon^2 + \frac{1}{6}\alpha^2\left(1 + \frac{\Gamma_o}{2\alpha}\right)\varepsilon^3 + O(\varepsilon^4)\right). \tag{10.62}$$

Comparison of the expanded expressions reveals the generally true condition that the compression isentrope and Hugoniot originating from the same initial state have the same initial slope and curvature, differing only in the third-order term and higher. The extent of the difference is determined by the magnitude of the Gruneisen parameter. The Hugoniot and isentrope overlay when $\Gamma_o = 0$.

Many materials are found to exhibit a linear relation in an experimental plot of the Hugoniot shock velocity versus particle velocity. Such behavior is not observed for all materials and changes of state under shock compression, such as melting or solid state phase transformation, usually affect a break in linear behavior. Continued linearity is also not consistent with the limiting compression of materials at extreme pressure. Nonetheless, experimentally observed linearity is sufficiently pervasive that considerable shock wave analysis has been constructed around the observed linearity. Shock-velocity versus particle-velocity linearity is expressed through the relation $U = C_o + Su$, where the Hugoniot constants C_o and S have been measured for a large body of materials. (A condensed compilation of the Los Alamos National Laboratory (LANL) GMX-6 Hugoniot data is provided in the text edited by Asay and Shahinpoor (1993).) When joined with the Hugoniot conservation relations for mass and momentum, an analytic expression for the Hugoniot compression relation is obtained,

$$P_h(\varepsilon) = \frac{\rho_o C_o^2 \varepsilon}{(1 - S\varepsilon)^2}. \tag{10.63}$$

A Taylor expansion of equation (10.63) yields

$$P_h(\varepsilon) = \rho_o C_o^2\left(\varepsilon + 2S\varepsilon^2 + 3S^2\varepsilon^3 + 4S^3\varepsilon^4 + O(\varepsilon^5)\right). \tag{10.64}$$

Equation (10.62) for the Hugoniot relation based on the modified Murnaghan EOS of choice can be compared term by term with equation (10.64) from the experimentally observed linear shock versus particle-velocity behavior. The first-order term yields $K_o = \rho_o C_o^2$ while the second-order term requires $\alpha = 4S = K_o' + 1$. The latter equality follows from equation (10.56). The coefficient of the third-order term in equation (10.62) can be written as

$$3S^2\left(1 + \frac{\Gamma_o - S}{9S}\right) \tag{10.65}$$

and is equal to the third-order term in equation (10.64) only for the special case of $S = \Gamma_o$. The functional form for the reference isentrope from equation (10.50) and the Mie–Gruneisen EOS leads to a Hugoniot relation in equation (10.60) that is a reasonable approximation to the experimentally observed linear behavior between

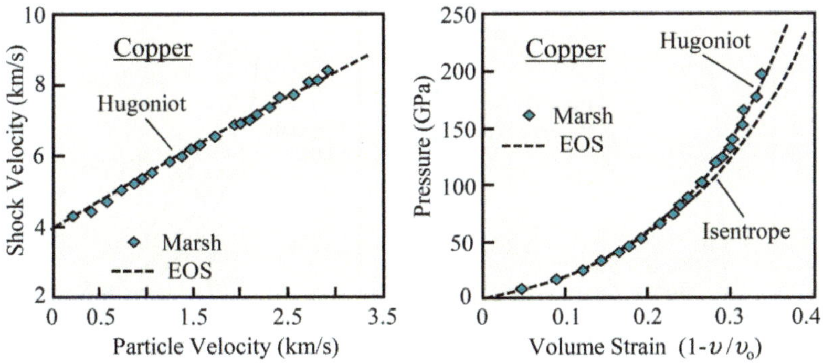

Figure 10.5. Experimental Hugoniot data for copper (Marsh *et al* 1980). Comparison with the Mie–Gruneisen EOS isentrope and Hugoniot based on ambient conditions and ultrasonic and thermodynamic properties (Sikka 1989).

shock and particle velocity. Differences between the two Hugoniot relations occur only in the third and higher order terms.

Application of the present Mie–Gruneisen EOS with the reference isentrope provided by equation (10.50) based on ambient properties for copper is illustrated in figure 10.5. Values for the ultrasonic properties of K_o = 142 GPa and K_o' = 5.25 are from Sikka (1989). The ambient Gruneisen parameter for copper is $\Gamma_o = 1.99$. Experimental linear shock-velocity versus particle-velocity properties for copper are $C_o = 3.94$ km s^{-1} and $S = 1.49$. The EOS isentrope and Hugoniot data are compared with the Hugoniot data of for copper in the figure. Visual inspection shows that the ultrasonic K_o leads to a slightly higher C_o than indicated by the data, and the ultrasonic K_o' provides and overall slope that is somewhat smaller than the S extracted from a linear fit to the data. Nonetheless, over the range of the experimental data the EOS Hugoniot based on ambient ultrasonic and thermodynamic properties provides a reasonable description of the data.

Relationships between the pressure derivative of the isentropic bulk modulus K_o' and the slope S of the shock-velocity versus particle-velocity data initially explored by Steinberg (1982), are discussed further in section 10.3.2. Issues of linearity between Hugoniot shock velocity and particle velocity, and universal EOS descriptions of solids, are explored in section 10.3.5.4.

10.2.2.4 Isentropic release from Hugoniot state

The common shock wave event is that of an intense but brief pressure loading followed by a more gradual release of pressure back to the initial pressure state. This process is usually modeled as shock compression of the material to a Hugoniot state followed by isentropic decompression to the reduced pressure level. This picture of the shock compression event is illustrated by the sketches provided in figure 10.6. On the left is shown the time history of compression and release at a point in the material that is subjected to intense shock wave loading. Shock compression occurs through a steady structured shock wave. Shock compression carries the material to a

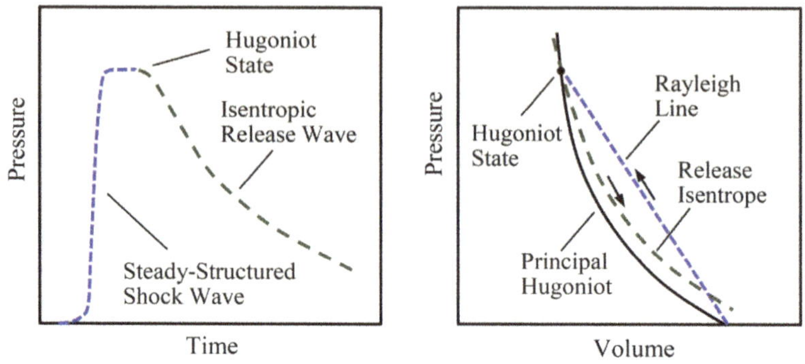

Figure 10.6. The time history of a material point subjected to shock compression and isentropic decompression (left). The corresponding Rayleigh-line shock loading to a Hugoniot state followed by pressure release along an isentropic path (right).

pressure–volume state on the principal Hugoniot along a straight pressure versus volume path identified as the Rayleigh line. Decompression of the material occurs through an isentropic release wave also shown on the left in figure 10.6. This release wave is not steady, but gradually spreads with propagation distance as dictated by the decreasing wave speed at different pressure levels on the release isentrope. The release isentrope governing decompression is that specific isentrope originating at the shock Hugoniot state as illustrated on the right in figure 10.6. A unique release isentrope is associated with each Hugoniot state on the Hugoniot curve.

This simple model of Hugoniot shock compression and isentropic decompression is necessarily approximate. Shock compression need not be through a steady structured shock, and a Hugoniot state on the principal Hugoniot may not be strictly achieved. For example, shock loading through sudden impact or from explosive loading through a gradation of materials does not immediately lead to a steady structured shock wave necessary to achieving a principal Hugoniot state. Some level of wave propagation and wave evolution is needed before steady wave conditions are achieved. Correspondingly, decompression from a shock state is not isentropic. Numerous non-isentropic processes can occur in the pressure release of the material that leads to a degree of entropy production. Decompression deformation is not isotropic and shear flow will lead to dissipation through viscosity or plasticity processes. Also, some level of heat conduction can occur during the time frame of the release wave.

Nonetheless, in the preponderance of intense shock wave applications, the assumption of Hugoniot compression and isentropic release is a very good approximation. Analysis of the isentropic release response of a material from a Hugoniot state can be accomplished through application of principles of continuum shock wave EOS thermodynamics. What follows illustrates one approach for analyzing isentropic decompression from a Hugoniot state achieve through shock wave compression.

In exceptional cases, usually generated experimentally, isentropic recompression from a shock Hugoniot state is achieved. Isentropic decompression or recompression is determined by the unique isentrope that intercepts the shock Hugoniot at the Hugoniot state achieved.

10.2.2.5 Hugoniot release or recompression isentrope

An isentropic path on the EOS surface of a material follows from the differential relation

$$dP = \left.\frac{\partial P}{\partial E}\right|_{v} dE + \left.\frac{\partial P}{\partial v}\right|_{E} dv \tag{10.66}$$

and the thermodynamic law constrained to the isentrope

$$dE = T\ dS - P\ dv = -P\ dv. \tag{10.67}$$

Joining the two relations provides the ordinary differential equation for the isentrope

$$\left.\frac{dP}{dv}\right|_{S} + g(v,E)P = \partial P/\partial v|_{E}, \tag{10.68}$$

where

$$g(v,E) = \Gamma(v,E)/v \tag{10.69}$$

is the Gruneisen function for any EOS of the form $P(v,E)$. Energy is eliminated from the isentrope differential equation in (10.68) through the $P(v,E)$ EOS relation. Equation (10.66) and equation (10.67) can equally well be joined to provide a differential equation assessing the volume dependence of the energy on the same isentrope.

The additivity of the pressure $P(v,E) = P_r(v) + P_t(v,E)$ is a common theoretical and modeling EOS assumption, where $P_r(v)$ is a reference compression pressure and $P_t(v,E)$ is the additional thermal pressure. Again commonly, the reference EOS $P_r(v)$ is the cold pressure $P_c(v)$ accounting for the molecular resistance to compression at absolute zero temperature. For the present purpose of calculating the isentrope centered at a material Hugoniot state, other reference EOSs may be convenient, such as the ambient isentrope or the cold Hugoniot. The cold Hugoniot is identified as that of a porous material in the limit of zero porosity, namely the ambient theoretical-density solid Hugoniot. The cold Hugoniot is frequently used as the reference EOS when Hugoniot and isentrope properties of a material initially subjected to shock from a porous or powdered state are of interest.

The introduction of the additive pressure $P(v,E) = P_r(v) + P_t(v,E)$ into equation (10.68) leads to the differential equation for the thermal pressure

$$\left.\frac{dP_t}{dv}\right|_{S} + g(v,E)P_t = -g(v,E)P_r + \partial P_t/\partial v|_{E}, \tag{10.70}$$

where the right-hand side reduces to a function $f(v,P_t)$. The terms that drop out follow from $P_r(v)$ being a function of volume only.

A further simplification of the differential relation results from the assumption of energy independence of the Gruneisen function $g(v) = \Gamma(v)/v$ from which it follows that the thermal pressure is of the form

$$P_t(v,E) = g(v)(E - E_r(v)). \quad (10.71)$$

Equation (10.70) reduces to the inhomogeneous linear differential equation

$$\left.\frac{dP_t}{dv}\right|_S + (g - g'/g)P_t = -g(P_r + dE_r/dv), \quad (10.72)$$

where $g' = dg(v)/dv$ and $E_r(v)$ is the internal energy on the reference EOS $P_r(v)$. For example, if $P_r(v)$ is the cold Hugoniot $P_{ho}(v)$ then $E_r(v) = E_{ho}(v) = (P_{ho}(v)/2)(v_o - v)$ when centered from the initial state $P_o = E_o = 0$ and $v = v_o$. Equation (10.72) is readily solved.

The solutions to equation (10.72) are simplified further if any isentrope (including the zero-temperature cold curve) is selected as the reference EOS. On a reference isentrope $dE_r/dv = -P_r$ and equation (10.72) reduces to the homogeneous differential equation

$$\left.\frac{dP_t}{dv}\right|_S + (g - g'/g)P_t = 0. \quad (10.73)$$

Equation (10.73) is integrated, providing the thermal pressure for any volume-dependent function of the Gruneisen parameter,

$$P_t(v) = P_{th}\frac{g}{g_h}e^{-\int_{v_h}^{v} g\,dv}, \quad (10.74)$$

where P_{th} and g_h are the thermal pressure and Gruneisen function evaluated at the Hugoniot state. The isentrope centered at the Hugoniot state (P_h, v_h) is then

$$P_s(v) = P_r(v) + (P_h - P_r(v_h))\frac{g(v)}{g_h}e^{-\int_{v_h}^{v} g\,dv}, \quad (10.75)$$

with $P_r(v)$ the reference isentrope. Simple solutions are provided from the commonly applied assumption of a constant $\rho\Gamma = g_o$,

$$P_t(v) = P_{th}e^{-g_o(v-v_h)}, \quad (10.76)$$

or a constant Γ_o with $g(v) = \Gamma_o/v$,

$$P_t(v) = P_{th}(v_h/v)^{\Gamma_o - 1}. \quad (10.77)$$

Hugoniot data for tungsten carbide (WC) ceramic provides an application of the isentropic decompression calculation. The Hugoniot properties of full-density tungsten carbide shown in figure 10.7 are from the Hugoniot shock compendia of Marsh (1980) and Trunin (2001). A linear fit to the shock-velocity versus particle-velocity data shown on the left yields an intercept $C_o = 5.14$ km s^{-1} and a slope $S = 1.15$. A reference EOS, the principal isentrope shown on the right in figure 10.7,

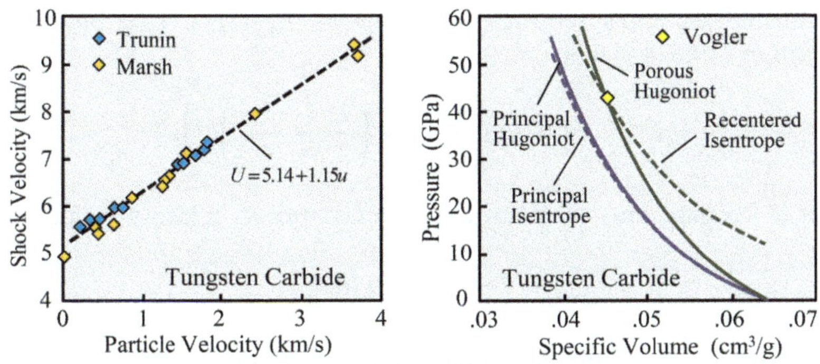

Figure 10.7. Hugoniot and isentropic expansion of solid and distended tungsten carbide. Left: Hugoniot data for full-density WC (15.6 g cm^{-3}) (Marsh 1980, Trunin *et al* 2001). Right: A single Hugoniot measurement for powdered WC (8.57 g cm^{-3}) (Vogler *et al* 2011), along with calculated Hugoniot and isentrope properties.

of the linear-Z form with $K_o = \rho_o C_o^2$ and $\alpha = 4S$ initiates the calculation. (See also discussions of the linear-Z or modified Murnaghan EOS later in this chapter.)

A single Hugoniot point obtained by Vogler *et al* (2011) on WC powder with initial density of 8.67 g cm^{-3} (distention of 1.82) constrains the thermal EOS of the material. A single Hugoniot point requires an assumption of the functional form for the Gruneisen parameter. A constant $\rho\Gamma = \rho_o\Gamma_o$ is used. A value of $\Gamma_o = 0.7$ adjusts the porous Hugoniot and is shown on the right in figure 10.7. The Gruneisen EOS constrained by this Hugoniot point and using equations (10.75) and (10.76) provides the isentrope shown on the right in figure 10.7 recentered at the experimental Hugoniot state.

10.2.2.6 Abridged isentrope equation of state

Commonly the range of compression of concern is such that the reference Hugoniot is remarkably well described by a cubic polynomial in the compression strain. Representative would be the EOS description of the unreacting Hugoniot of various energetic materials. This representation of the principal Hugoniot is written as

$$p_h(\varepsilon) = a\varepsilon + b\varepsilon^2 + c\varepsilon^3. \tag{10.78}$$

The principal isentrope over the same compression range is similarly captured through a cubic relation and differs only in the coefficient of the cubic term. Namely

$$p_s(\varepsilon) = a\varepsilon + b\varepsilon^2 + c'\varepsilon^3. \tag{10.79}$$

Specific energy on the Hugoniot follows from

$$\rho_o E_h(\varepsilon) = \frac{1}{2}p_h(\varepsilon)\varepsilon = \frac{1}{2}a\varepsilon^2 + \frac{1}{2}b\varepsilon^3 + \frac{1}{2}c\varepsilon^4 \tag{10.80}$$

and on the isentrope from

$$\rho_o E_s(\varepsilon) = \int p_s(\varepsilon)d\varepsilon = \frac{1}{2}a\varepsilon^2 + \frac{1}{3}b\varepsilon^3 + \frac{1}{4}c'\varepsilon^4. \tag{10.81}$$

The Mie–Gruneisen EOS requires $p_h - p_s = \rho\Gamma(E_h - E_s)$. Combining equations (10.78) through (10.81) yields

$$(c - c')\varepsilon^3 = (\rho\Gamma/\rho_o)\left(\frac{1}{6}b\varepsilon^3 + \left(\frac{1}{2}c - \frac{1}{4}c'\right)\varepsilon^4\right). \tag{10.82}$$

Solving for c' keeping only terms to cubic in compression strain yields

$$c' = c - \frac{1}{6}(\rho\Gamma/\rho_o)b. \tag{10.83}$$

The linear shock-velocity versus particle-velocity relation for the Hugoniot pressure,

$$p_h(\varepsilon) = \rho_o C_o^2 \frac{\varepsilon}{(1 - S\varepsilon)^2}, \tag{10.84}$$

expanded to order cubic in the compression strain is

$$p_h(\varepsilon) = \rho_o C_o^2(\varepsilon + 2S\varepsilon^2 + 3S^2\varepsilon^3), \tag{10.85}$$

and for the Hugoniot energy assuming zero energy at the origin

$$E_h(\varepsilon) = \frac{1}{2}C_o^2(\varepsilon^2 + 2S\varepsilon^3 + 3S^2\varepsilon^4). \tag{10.86}$$

Identifying corresponding terms for a, b and c in equation (10.78), the relation for the isentrope in equation (10.79) becomes

$$p_s(\varepsilon) = \rho_o C_o^2\left(\varepsilon + 2S\varepsilon^2 + 3S^2\left(1 - \frac{\Gamma_o}{9S}\right)\varepsilon^3\right). \tag{10.87}$$

The corresponding energy on the isentrope is

$$E_s(\varepsilon) = \frac{1}{2}C_o^2\left(\varepsilon^2 + \frac{4}{3}S\varepsilon^3 + \frac{3}{2}S^2\left(1 - \frac{\Gamma_o}{9S}\right)\varepsilon^4\right). \tag{10.88}$$

Replacing $\rho\Gamma/\rho_o = \Gamma_o$ a constant in equations (10.87) and (10.88) is appropriate to the level of approximation. Pressures and energies on the Hugoniot and isentrope diverge in the third-order term. Equation (10.87) as the reference isentrope is joined with equation (10.76) to provide the isentropic pressure path positioned from any elevated pressure on the Hugoniot.

10.2.3 Enthalpy-based equations of state

The Mie–Gruneisen EOS has found practical application over the years for addressing problems in which the thermo-elastic response of materials plays a key role in applications of impact and explosive loading. When only wave propagation and deformation history are of interest, the pressure as a function of volume and energy $P(v,E)$ provided by the EOS is adequate for the purposes. When temperature excursions are of interest, the Gruneisen EOS is supplemented with a relation for the specific heat. The assumption of a temperature-independent Gruneisen parameter $\Gamma(v)$ leads to particularly convenient analytic forms for the Gruneisen EOS. Even when temperature independence is not justified, the Gruneisen EOS provides for reasonable estimates of the EOS surface for modest extrapolations from the reference EOS (the principal Hugoniot for example).

There are applications, however, in which the use of a Gruneisen EOS becomes awkward. Well known are the intense shock compression of highly distended solids such as foams and powders. The extreme shock heating can lead to volume expansions that render the Hugoniot backward leaning (volume increasing with increasing pressure). There are also cases for which the Hugoniot is multivalued (exhibiting several pressure states corresponding to the same specific volume). For such applications the Gruneisen EOS can be inconvenient as analyses are carried out along paths of constant volume and cumbersome analytic methods must be used to account for the non-unique solution. A further difficulty occurs for extreme expanded states (Hugoniot specific volumes in excess of the ambient crystal specific volume). The reference EOS is necessarily extrapolated into tension to accommodate the Gruneisen EOS framework.

10.2.3.1 Rice–Walsh equation of state

An EOS of comparable stature to the Gruneisen EOS that has analytic conveniences for the EOS analysis of the shock compression of distended solids was proposed by Rice and Walsh (1957). The Rice–Walsh EOS starts with the thermodynamic differential

$$P\frac{\partial v}{\partial H}\bigg|_P = \zeta(P), \tag{10.89}$$

where the parameter $\zeta(P)$ is assumed to depend on pressure only. Implications of this differential parameter were previously discussed in the earlier chapter on crushable media. The Rice–Walsh parameter and the Gruneisen parameter are related through

$$\zeta(P) = \frac{\Gamma}{K_s}P, \tag{10.90}$$

where K_s is the isentropic bulk modulus. Integration of equation (10.89) leads to the Rice–Walsh form of the EOS

$$v(P,H) = v_r(P) + \eta(P)(H - H_r(P)). \tag{10.91}$$

For subsequent analytic convenience the pressure-dependent property

$$\eta(P) = \frac{\partial v}{\partial H}\bigg|_P = \frac{\zeta(P)}{P} = \frac{\Gamma}{K_s} = \frac{\alpha}{\rho c_P} \tag{10.92}$$

will be identified as the Rice–Walsh factor. The last steps in equation (10.92) relates the Rice–Walsh factor to either the ratio of the Gruneisen gamma and the isotropic bulk modulus, or the coefficient of thermal expansion α and the specific heat capacity at constant pressure c_p. On paths of constant pressure the Rice–Walsh EOS provides a linear relation between specific volume and enthalpy. Equation (10.91) is also recognized as a mathematical expansion of the thermal volume in powers of the excess enthalpy where the model assumes that dependence on higher powers of the enthalpy excess can be ignored. In contrast to the Gruneisen EOS, the Rice–Walsh EOS circumvents the analytic difficulties introduced by backward leaning and multivalued Hugoniot relations. Nor does the Rice–Walsh EOS delve into the tensile domain of the reference EOS curve.

It is recognized that this is a phenomenological model intended for engineering applications. The model is benchmarked to available data and intended for interpolation within the range of the data. The range of data leading to the observed thermal Hugoniot anomalies span thermodynamic states where thermal and lattice energies are anticipated to be comparable. Some level of ionization is expected. The present model approximates thermal states within a complex span of material behavior where even detailed physics-based theories are challenged.

The reference compression path is provided by the pressure-dependent volume and enthalpy relations, $v_r(P)$ and $H_r(P)$. The isentrope, isotherm or Hugoniot (the cold Hugoniot) of the fully dense medium are convenient reference relations. The Hugoniot is more commonly experimentally available. The isentrope is usually more convenient when both shock and decompression analysis is required. Joining the Hugoniot enthalpy relation

$$H_h(P) = \frac{1}{2}P(v_{oo} + v_h) \tag{10.93}$$

with the Rice–Walsh EOS in equation (10.91) provides a relation for the Hugoniot centered at any distended specific volume $v_{oo} \geqslant v_o$,

$$v_h(P) = v_r(P) + \frac{\eta}{1 - \eta P/2}\left(\frac{1}{2}P(v_{oo} + v_r(P)) - H_r(P)\right). \tag{10.94}$$

If the reference compression relation is the Hugoniot of the solid material $v_r(P) = v_{ho}(P)$ centered from the initial full-density specific volume v_o, then the Hugoniot centered from any distended state v_{oo} further simplifies to

$$v_h(P) = v_{ho}(P) + \frac{\eta}{1 - \eta P/2}\left(\frac{1}{2}P(v_{oo} - v_o)\right). \tag{10.95}$$

To complete the EOS model the thermal Rice–Walsh factor $\eta(P)$ must be specified. Guidance in the construction is provided from the more common Mie–Gruneisen formulation. The common observation of $\rho\Gamma$ constant on the reference isentrope can be written

$$\Gamma(P) = \Gamma_o\frac{v}{v_o} \cong \Gamma_o e^{-P/K_o}. \tag{10.96}$$

The second approximate step closely replicates a constant $\rho\Gamma$ over a reasonable range of compression, and has the analytic convenience of extending to a relation for gamma that extrapolates to a theoretical value of $\Gamma_\infty = 0.5$ at extreme compressions (e.g. Zeldovich and Raiser 1966), namely

$$\Gamma(P) = \Gamma_o e^{-P/K_o} + \Gamma_\infty(1 - e^{-P/K_o}). \tag{10.97}$$

This relation mirrors, in the isobaric EOS framework, that of Altshuler and Sharipdzhanov (1971) provided in equation (10.20). A useful form of the Rice–Walsh factor is then

$$\eta(P) = \frac{\Gamma(P)}{K_s(P)} = \frac{\Gamma_o e^{-P/K_o}}{K_o + K_o' P}, \tag{10.98}$$

where a bulk modulus of the Murnaghan form is assumed for the denominator.

Figure 10.8 illustrates an application in which the Rice–Walsh enthalpy-based EOS demonstrates its practical utility (Fenton *et al* 2015). The material is tantalum oxide (Ta_2O_5) with varying levels of initial distention. The higher initial density (1.2, 3.0 and 7.42 gm cm^{-3}) Hugoniot measurements are from Vogler *et al* (2011). Hugoniot data on tantalum oxide aerogel are from Miller *et al* (2007). The curves are EOS model representations of the shock Hugoniots for the respective initial distention of the material. The tantalum oxide cold Hugoniot (8.20 gm cm^{-3}) provides the Rice–Walsh reference EOS. Appropriate selection of the Rice–Walsh factor determines the Hugoniot-temperature-induced fan of Hugoniot curves displayed in the plot.

A further, more elemental application of the enthalpy-based EOS description of the shock compression of highly distended solids is illustrated in figure 10.9. The material is the compound cerium dioxide. Diamond anvil cell (DAC) compression of cerium (IV) oxide has been established to 70 GPa (Duclos *et al* 1988). CeO_2 is a

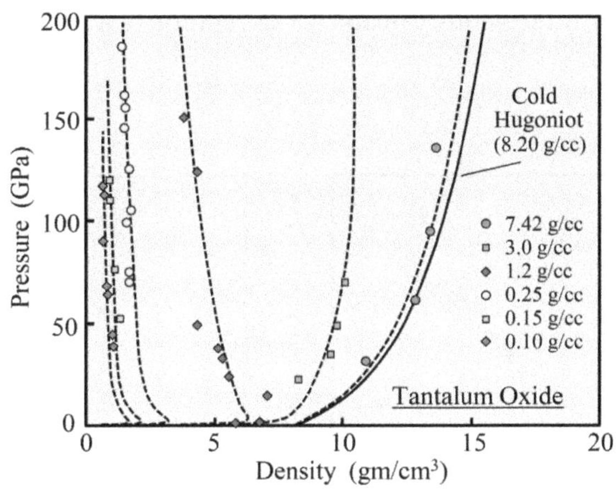

Figure 10.8. Extreme pressure Hugoniot properties of distended tantalum oxide (Ta_2O_5). The initial density of starting materials is identified in the plot. The higher density material of 1.2, 3.0 and 7.42 gm cm^{-3} are from Vogler *et al* (2011). The Hugoniot data on reduced density aerogel are from Miller *et al* (2007). Enthalpy-based EOS from Fenton *et al* (2015). The very low-pressure compression structure is a consequence of the compaction model.

reasonably stiff oxide with the fluorite crystal structure. Single-phase compression to approximately 40 GPa and 10% volume strain is achieved before phase transformation to an orthorhombic crystal structure. DAC compression to 40 GPa is adequately described by an ambient bulk modulus of $K_o = 230$ GPa and pressure derivative $K_o' = 4$ through any one of several analytic reference equations. The crystal density of cerium oxide is $\rho_o = 7.215$ g cm^{-3}. For the present purposes this two-parameter description of the crystal compression adequately describes a reference isentrope for the enthalpy-based EOS.

Shock Hugoniot data adequate to constraining the dynamic EOS of cerium oxide is limited, but nonetheless sufficient for present purposes. Again, the EOS analysis is restricted to pressures less than about 40 GPa, avoiding description of phase transformation above this pressure level. Early Hugoniot data for CeO_2 are reported in the LANL Compendium (Marsh 1980). No additional references are cited. The material is identified as powdered and unpressed. The average density is 1.133 g cm^{-3}. Further, the seven data points partition into experimentally discernable sets of three initial densities of approximately 0.84 g cm^{-3}, 1.13 g cm^{-3} and 1.34 g cm^{-3}. These density differences are noted in the plot of the data in figure 10.9. Additional shock Hugoniot data within the pressure range of interest on cerium oxide were reported more recently (Fredenburg and Chisolm 2014). Initial state of the material is also distended with a nominal density of 4.03 g cm^{-3}. Combined with the DAC compression study, these data constrain the dynamic EOS of CeO_2 within the range of interest.

To compression strains of 10% any number of analytic relations based on the initial bulk modulus and pressure derivative can be used for the reference relation. Here, a model is used that is based on the experimental observation that the isentropic inertial modulus is linear in pressure, $Z = Z_o + Z_o'P$, where $Z_o = \rho_o K_o$ and $Z_o' = \rho_o(K_o' + 1)$, adequately describes the compression of many materials. This constraint on the compressibility leads to a relation for the reference isentropic volume compression

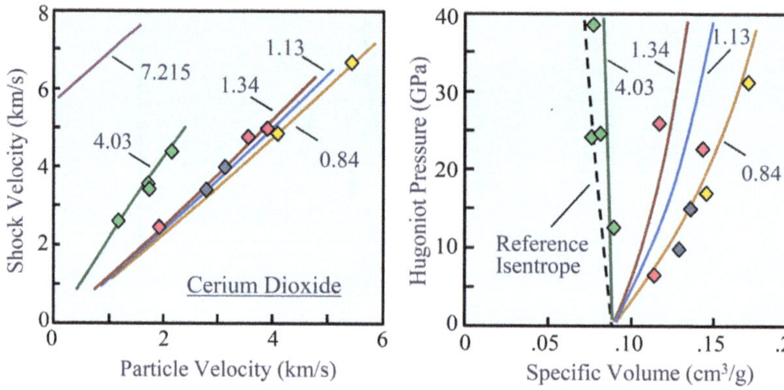

Figure 10.9. Hugoniot data for cerium oxide (Marsh 1980, Fredenburg and Chisolm 2014) and enthalpy-based EOS representation. The initial density is identified with the respective EOS Hugoniot curves.

$$v_s(P) = v_o\left[1 - \frac{1}{K_o' + 1}\ln\left(1 + \frac{K_o' + 1}{K_o}P\right)\right]. \tag{10.99}$$

The corresponding enthalpy on the reference isentrope is provided by the integral

$$H_s(P) = \int_0^P v_s(P)dP. \tag{10.100}$$

Writing the enthalpy-based EOS in the Rice–Walsh form from equation (10.91),

$$v(P,H) = v_s(P) + \eta(P)(H - H_s(P)), \tag{10.101}$$

and with

$$\eta(P) = \frac{\Gamma}{K_s} = \frac{\rho\Gamma}{\rho K_s} = \frac{\Gamma_o}{K_o + (K_o' + 1)P}, \tag{10.102}$$

yields the EOS

$$v(P,H) = v_s(P) + \frac{\Gamma_o}{K_o + (K_o' + 1)P}(H - H_s(P)). \tag{10.103}$$

Equation (10.102) assumes constant $\rho\Gamma = \rho_o\Gamma_o$ and makes use of the linear pressure dependence of the isentropic inertial modulus. Thermo-elastic compression of cerium oxide based on equation (10.103) depends only on the three EOS parameters, K_o, K_o' and Γ_o. The first two are constrained by the compression on the reference isentrope. Only Γ_o is adjustable to characterize thermal expansion properties of the material.

The experimental DAC and Hugoniot compression data in figure 10.9 are used to constrain the enthalpy-based EOS for cerium oxide of the form provided by equation (10.103). Reference isentropic compression is provided by the DAC compression for CeO_2. Reference compression and enthalpy is described through equations (10.99) and (10.100) with an ambient bulk modulus of $K_o = 230$ GPa and pressure derivative $K_o' = 4$. Only the Gruneisen constant remains to be constrained through the available Hugoniot data. Within the functional constraints of the selected EOS model, an optimum fit to the Hugoniot data is achieved with a Gruneisen gamma of $\Gamma_o = 2$. The EOS description in both the shock-velocity–particle-velocity plane and pressure–volume plane is shown with the data in figure 10.9. The EOS representation of the Hugoniot data is sensible, although the limited data suggest that thermal expansion at the pressure provided by the model is perhaps over-predicted at lower temperatures and under-predicted at the higher temperatures. An alternative enthalpy-based EOS formulation is described in the next subsection.

10.2.3.2 Anderson–Gruneisen equation of state

A further EOS based on an isobaric formulation was put forth by Anderson (1967) and has been actively pursued by workers in the geosciences community. The EOS of Anderson has subsequently been dubbed the Anderson–Gruneisen EOS, presumably because of the reliance on principles of the thermal Gruneisen properties of solids in its development. The Anderson–Gruneisen EOS appears founded on somewhat sounder physical evidence than is the Rice–Walsh EOS.

Details of the Anderson–Gruneisen EOS development can be found in the Anderson paper and the many subsequent references. Here a straightforward plausibility analysis reveals the essence of the EOS. The EOS construction initiates from the experimental observation that the product of the thermal expansion and the bulk modulus is reasonably constant over the experimental range of interest. This product can be written in terms of the density, Gruneisen parameter and specific heat,

$$\alpha K_S = \rho \Gamma c_p. \tag{10.104}$$

Since c_p is sensibly constant at or near its Dulong–Petit limit, this constancy would appear sensible, as constancy of the product $\rho\Gamma$ is commonly observed and applied in shock physics applications.

An experimental observation made popular through the EOS development of Murnaghan (1944) concerns the linearity or near-linearity of the bulk modulus with applied pressure. A similar observation was reported by Tait (1888) while investigating the compressibility of sea water in preparation for the Challenger scientific expedition. Namely

$$K_S(P) = K_o + K_o' P. \tag{10.105}$$

Anderson discusses in detail and applies the Murnaghan relation in development of the model. Where careful ultrasonic measurements have been attempted, the second pressure derivative K_o'' is small and commonly negative. Note in equation (10.56) that for the linear-Z, or modified Murnaghan, equation that a small negative K_o'' results.

In terms of a volume strain measure $\mu = v_o/v - 1$, the pressure can be written

$$P = K_o \mu + \cdots, \tag{10.106}$$

and joining with equation (10.105),

$$K_S = K_o(1 + K_o' \mu + \cdots). \tag{10.107}$$

To the order displayed the expansion is equivalent to

$$K_S = K_o(1 + \mu)^{K_o'} = K_o \left(\frac{v}{v_o} \right)^{-K_o'}. \tag{10.108}$$

Accepting the assumption of Anderson that αK_S is constant leads to

$$\alpha = \alpha_o \left(\frac{v}{v_o} \right)^{\delta}, \tag{10.109}$$

for the thermal expansion where $\delta = K_o'$ is identified in the literature as the Anderson–Gruneisen parameter. The Anderson–Gruneisen parameter is also provided in the differential form

$$\delta = \left.\frac{\partial \ln K_s}{\partial \ln v}\right|_P . \tag{10.110}$$

Equation (10.109) provides the differential form of the Anderson–Gruneisen EOS. Namely, forming the differential equation

$$\left.\frac{1}{v}\frac{\partial v}{\partial T}\right|_P = \left.\frac{c_P}{v}\frac{\partial v}{\partial H}\right|_P = \alpha_o\left(\frac{v}{v_o}\right)^{\delta}, \tag{10.111}$$

and integration provides the Anderson–Gruneisen EOS

$$v(P,H) = v_o\left[\left(\frac{v_r(P)}{v_o}\right)^{-\delta} - \delta\frac{\rho_o\Gamma_o}{K_o}(H - H_r(P))\right]^{-1/\delta} . \tag{10.112}$$

Equation (10.112) can be compared with the Rice–Walsh form of the isobaric EOS in equation (10.115). Equation (10.112) reproduces the Rice–Walsh equation only for the special case of $\delta = -1$ and $\eta(P) = \Gamma_o/K_o$. For most materials $\delta = K_o'$ ranges from about 3 to 5. It is noteworthy that the integration carries through if the Anderson–Gruneisen parameter δ is a function of pressure, as would be the case if higher order terms for the expansion of the bulk modulus in equation (10.105) are retained.

Equation (10.112) readily represents the modest pressure and thermal expansion range of the EOS interests of the planetary-interior geophysics community. Whether the Anderson–Gruneisen EOS offers advantages over that of the Rice–Walsh EOS over the temperature extremes spanned in the shock compaction of highly distended solids requires further study.

10.2.3.3 Isentropic release in the enthalpy-based framework

Calculation of isentropic decompression (or recompression) from states of Hugoniot compression follows from an analysis comparable to that previously described for the Mie–Gruneisen EOS. In an enthalpy description of the EOS, the differential

$$dv = \left.\frac{\partial v}{\partial H}\right|_P dH + \left.\frac{\partial v}{\partial P}\right|_H dP, \tag{10.113}$$

joined with the isotropy condition $dH = v\, dP$ leads to the differential equation for the isentrope

$$\left.\frac{dv}{dP}\right|_S - \eta(P)v = \left.\frac{\partial v}{\partial P}\right|_H, \tag{10.114}$$

where $\eta(P)$ is the pressure-dependent Rice–Walsh factor identified in equation (10.92).

When either the ambient isentrope or the cold compression curve (absolute zero-temperature isentrope) for the solid is used for the reference compression path $v_r(P) = v_{so}(P)$, the Rice–Walsh EOS from equation (10.91) is written

$$v_s(P) - v_{so}(P) = \eta(P)(H_s(P) - H_{so}(P)) \tag{10.115}$$

or

$$v_t(P) = \eta(P)H_t(P). \tag{10.116}$$

In equation (10.116) notation has been simplified in relating the thermal expansion volume $v_t(P)$ to the excess enthalpy $H_t(P)$ through the Rice–Walsh property $\eta(P)$. The isentropic pressure derivative yields

$$\left.\frac{\partial v_t}{\partial P}\right|_S = \eta' H_t + \eta \left.\frac{\partial H_t}{\partial P}\right|_S. \tag{10.117}$$

Substituting from equation (10.116) and replacing $\partial H_t/\partial P|_S = v_t$ provides the homogeneous differential equation

$$\frac{dv_t}{dP} - \left(\frac{\eta'}{\eta} + \eta\right)v_t = 0. \tag{10.118}$$

The solution of equation (10.118) provides the isentrope centered at the Hugoniot state (P_h, v_h)

$$v_s(P) = v_{so}(P) + v_{th}\frac{\eta}{\eta_h}e^{\int_{P_h}^{P} \eta\, dP}, \tag{10.119}$$

where v_{th} and η_h are the thermal volume expansion and the Rice–Walsh factor evaluated at the Hugoniot pressure P_h. Namely, from equation (10.94),

$$v_{th}(P_h) = \frac{\eta_h}{1 - \eta_h P_h/2}\left(\frac{1}{2}P_h(v_{oo} + v_{so}(P_h)) - \int_0^{P_h} v_{so}(P)dP\right). \tag{10.120}$$

A fully analytic solution for the Hugoniot state recentered isentrope can be calculated for a special case. Writing the Rice–Walsh factor

$$\eta(P) = \frac{\rho\Gamma}{\rho K_s} = \frac{\rho_o\Gamma_o}{Z_s}. \tag{10.121}$$

In the second step the commonly use constancy of the product $\rho\Gamma$ is used while the inertial modulus $Z_s = \rho K_s$ is assumed to be linear in the pressure $Z_s = Z_o + Z_o'P$. The assumed simplifications lead to a pressure dependence of the Rice–Walsh factor

$$\eta(P) = \frac{\Gamma_o}{K_o + \alpha P}, \tag{10.122}$$

with $\alpha = K_o' + 1$, and where K_o and K_o' are the initial and first pressure derivative of the isentropic bulk modulus. Equation (10.119) has the solution for the Hugoniot state recentered isentrope,

$$v_s(P) = v_{so}(P) + v_{th}\left(\frac{K_o + \alpha P}{K_o + \alpha P_h}\right)^{\frac{\Gamma_o}{\alpha}-1}. \tag{10.123}$$

Again, v_{th} is provided by equation (10.120) for the specified reference isentrope and Hugoniot state.

10.3 Universal equations of state

The response of matter when subjected to extremes of pressure in excess of ambient at the surface of planet Earth is of interest to many areas of science. The seminal work of Bridgman (1952) is perhaps foremost in defining this topical area of science. Fundamental to this subject is the question of the extent of densification as matter is exposed to states of increasing pressure. The question resides at the heart of numerous planetary, astrophysical and cosmological issues. Densification under extremes of pressure is a concern in applications involving intense shock compression from explosion or impact. Issues relating to the densification of matter subject to pressure can begin with a journey to the depths of Earth's seas. At the depths of the Marianas Trench (almost 7 miles) pressures of about 100 MPa are achieved and the compression of water approaches 5%. The pressure at the Earth's core is of the order of 300 GPa, and that of the center of the neighboring Giant Planets markedly more. Extremes of pressure and densification are reached within our Sun and other stellar interiors. Even higher extremes of densification of matter are conceived within the theories of the genesis of this universe. Certainly many engineering and scientific concerns regarding the physics and chemistry of pressure-induced densification reside within the range of these natural phenomena.

Such questions have encouraged many efforts in exploring what has been identified as a universal character to the compressibility of matter when subject to intense pressure. Intriguing systematics observed in both static-high-pressure and shock-compression studies have provided stimulus for these efforts. Geophysical scientific issues have provided impetus for finding unifying EOS principles. A number of analytic relations that depend on several fundamental parameters have been put forth to describe the compression of matter under pressure. The literature on these efforts extends back at least to the beginning of the last century. The results of these studies have shown varying degrees of success that depend primarily on the range of compression over which the application is required. Recognize also that some authors reject the notion of universal EOSs. So the success of such ventures is to a certain degree dependent on the objectives of the application.

Such universal EOSs are more attractive if parameters for specific materials are made available from properties determined at or near ambient conditions. The nature of the intermolecular forces responsible for resistance to compression are manifest by a number of ambient properties including initial density, elastic properties, binding energy, thermal expansion and Gruneisen pressure. The density, bulk modulus and pressure derivative of the bulk modulus have been the properties of choice in the majority of universal compressibility relations proposed.

Of course thermodynamic, structural and electronic phase transformations are recognized as impediments to a universal compression behavior. In the smaller view universal compression is suggested in the limited range excluding phase transformation where deformation of the unit molecular cell remains nominally self-similar. In the larger view phase transitions—like the effects of war and famine on human population growth—can be viewed as modest perturbations on the inevitable outcome.

Further, attempts to uncover a universal character of compressibility of matter have generally restricted thinking to a thermodynamic region in which atomic and electronic elements maintain a sensibly ordered and correlated structure. High-energy-density states of matter exist in which resistance to compression is dominated by kinetic energy of the molecular and atomic constituents (e.g. Drake 2010). Such extreme temperature states of matter are not generally addressed by such universal compression relations. Although a topic rich in scientific and engineering potential, high-energy-density physics is not addressed in the discussions to follow.

10.3.1 Murnaghan equation of state

Although not necessarily chronological, perhaps the most widely recognized early attempt at a universal description of the compression of solids is that of Murnaghan (1944). It should be noted that the efforts of Tait (1888) skirt close to the observations and ideas that motivated Murnaghan. Observing that the bulk modulus for many materials plots linearly as a function of pressure within a modest range near ambient, the now-well-known Murnaghan power-law compression relation emerges from extrapolation of the observed trend. Although recognized to have application only over a limited compression range, the Murnaghan equation continues to find application because of simplicity and analytic convenience. Thus, Murnaghan made the assumption that to a reasonable approximation the solid bulk modulus is described by the linear relation

$$K(P) = K_o + K_o'P. \tag{10.124}$$

The values for K_o and K_o' can apply to the measured isothermal modulus or adjusted through the appropriate thermodynamic relations to the isentropic bulk modulus. When joined with a differential definition of the bulk modulus, the Murnaghan assumption leads to the differential equation

$$-v\frac{dP}{dv} = K_o + K_o'P. \tag{10.125}$$

Equation (10.125) integrates to the Murnaghan EOS

$$P(v) = \frac{K_o}{K_o'}\left(\left(\frac{v}{v_o}\right)^{-K_o'} - 1\right). \tag{10.126}$$

Like the previous equation (10.50), the Murnaghan equation can be used as a reference isothermal or isentropic compression relation with the Gruneisen EOS.

Equation (10.126) is easily inverted and also provides a convenient reference EOS in the Rice–Walsh isobaric EOS formulation. When ambient elastic properties are used for K_o and K_o', the Murnaghan equation has been found to become unreliable above about 15%–20% compression. When these elastic properties are established from other higher pressure data (DAC for example), equation (10.126) can, of course, be used as a convenient analytic EOS over a broader pressure range. As one example, this was effectively done by Kalashnikov *et al* (1973) to describe the shock compression properties of porous calcium carbide to pressures approaching 100 GPa. The study of Kalashnikov is discussed in a later section of this chapter.

10.3.2 Linear shock-velocity–particle-velocity equation of state

Application of the Murnaghan equation in the EOS description of a solid requires a measure of the initial pressure derivative of the bulk modulus K_o'. This critical property is most readily obtained from ultrasonic measurements over a modest range of pressures in excess of ambient. This is a rather specialized test and such data are not always available. Further, when the data are available for a material from several laboratories, the experimental differences can be disturbing. An alternative measure of K_o' is provided from shock Hugoniot data. Earlier in this chapter and elsewhere in the book it is shown that the compression isentrope and Hugoniot centered from the same ambient state are identical to second-order in a Taylor expansion in volume strain about the ambient state. The two compression paths diverge in the third and higher order terms. This identity requires that $K_o = \rho_o C_o^2$ and $K_o' = 4S - 1$, where C_o and S are the intercept and slope of the shock Hugoniot data in a linear shock-velocity versus particle-velocity, or U–u, plot of the Hugoniot data. Or, the same constants are the same intercept and first-order coefficient for a higher order polynomial fit to the Hugoniot U–u data. The Gruneisen thermal properties of the material lead to divergence of the isentrope and Hugoniot in the third and higher order terms.

With K_o and K_o' provided by $\rho_o C_o^2$ and $4S - 1$, respectively, from the shock Hugoniot data, the Murnaghan equation is uniquely constrained and provides an appropriate reference EOS for the cold curve or any convenient reference compression path. Alternatively, the functional form provided by the linear U–u relation can serve equally well for the reference compression curve. The functional form, obtained through joining the linear U–u relation with the Hugoniot conservation of mass and momentum relations, namely

$$P(v) = \rho_o C_o^2 \frac{(1 - v/v_o)}{(1 - S(1 - v/v_o))^2}, \tag{10.127}$$

is comparably analytically convenient. The inverse relation is

$$v(P) = v_o\left[1 - \frac{a(P)}{S}\left(1 - \sqrt{1 - a(P)^{-2}}\right)\right], \quad a(P) = 1 - \frac{\rho_o C_o^2}{2SP}. \tag{10.128}$$

If a linear U–u equation is used for the reference cold curve or ambient isentrope, then the resulting principal Hugoniot will not be linear in a U–u plot. It is near-linear, however. To volume compressions of a factor of two the Murnaghan and the linear *U–u* equations plot quite closely, with the latter slightly softer. For the Murnaghan equation, by definition $K_o'' = 0$. For the linear *U–u* equation K_o'' is slightly negative. Specifically,

$$K_o'' = \frac{2S}{K_o}(S - 2), \qquad (10.129)$$

where for the preponderance of materials the Hugoniot parameter S is less than two.

A reasonable question concerns the reliability of K_o' (and K_o for that matter) evaluated from shock Hugoniot data. This concern was pursued by Steinberg (1982). The ultrasonic and shock data yielding K_o' for 36 metals that were collected and compared by Steinberg are reproduced in figure 10.10. In the left-hand graph $4S - 1$ from shock Hugoniot data on the ordinate is plotted against the ultrasonic K_o' on the abscissa. The same data are repeated on the right where the ratio $(4S - 1)/K_o'$ is plotted against the ultrasonic K_o'.

In general the comparison is not very encouraging. A fit to the data shows the pressure derivative inferred from shock data is on average about 13% lower. The alkali metals are the exception, providing comparable or slightly higher values. Tantalum, palladium and niobium also compare reasonably well, but for most of the materials $4S - 1$ provides a markedly reduced estimate of the pressure derivative.

A perusal of the LANL shock Hugoniot compendium (Marsh 1980) reveals several concerns with estimates of K_o' from shock Hugoniot data. First, K_o' from ultrasonic methods is usually acquired from measurements over a pressure range of a few tenths of a GPa above ambient. Only in recent years have techniques emerged to provide ultrasonic properties to somewhat higher pressures. Shock Hugoniot data from which the slope of the *U–u* plot is determined typically start at Hugoniot pressures in excess of 10 GPa. Further, for a number of the metals, the *U–u* plots are not linear because of shock-induced phase transformations on the Hugoniot. For these metals a linear lower pressure segment of the plot must be deduced that presumably represents shock compressibility of the ambient phase of the material.

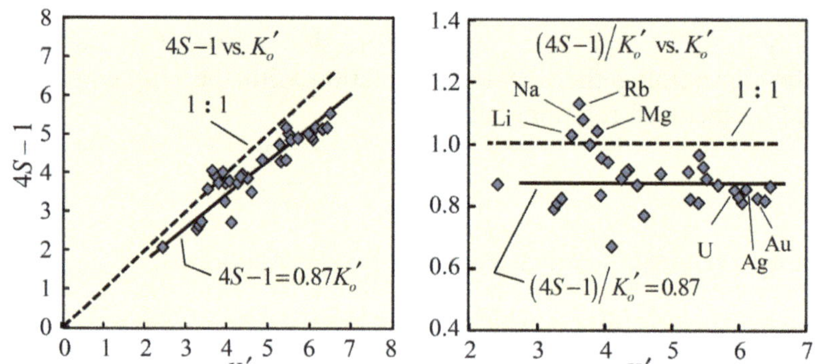

Figure 10.10. Comparison of the pressure derivative of the bulk modulus from ultrasonic and shock Hugoniot data. From Steinberg (1982).

The extrapolated intercept of the U–u plot from which C_o is determined is also a concern in this correlation of shock and ultrasonic measurements. For most of the materials C_o is not equal to the ultrasonic bulk velocity $C_b = \sqrt{K_o/\rho_o}$. An ultrasonic C_b markedly larger than C_o is usually indicative of a phase transformation on the Hugoniot. For other metals the opposite is observed in the Steinberg study. Namely, C_o exceeds the bulk velocity C_b. In particular, this is the case for the metals molybdenum, uranium, silver and gold identified in the figure 10.10 plot.

If the ultrasonic K_o and the shock Hugoniot C_o are required to coincide through $K_o = \rho_o C_o^2$, which seems reasonable, then the implication is that the U–u Hugoniot in the lower pressure range below that of available Hugoniot data is not linear. This explanation is difficult to test because measurement of the required hydro Hugoniot properties within this pressure range is complicated by material strength. A nonlinearity of the U–u EOS Hugoniot in the lower pressure region is, however, dictated by the requirement that $K_o' = 4S - 1$ as the Hugoniot and the isentrope merge at the origin. Materials for which the lower pressure Hugoniot is more readily assessed, water for example, exhibit curvature in the U–u Hugoniot at lower pressures and transition to linear behavior at the higher pressures. The Woolfolk and modified Woolfolk equation discussed later in the present section are appropriate for analytic representation of this character.

10.3.3 Intermolecular potential equations of state

Prior to the studies of Murnaghan, models of the intermolecular potential of solids were explored that were subsequently extended to analytic EOS relations for purposes of describing the high-pressure compressibility of materials. Generally, these models provide physically reasonable functional expressions for both the attractive and repulsive forces between molecular elements as a function of intermolecular spacing.

Perhaps the earliest is the Mie, or Mie–Gruneisen, potential in which opposing power-law functions were used to separately describe the attractive and repulsive forces (Mie 1903). The same potential form was also used by Gruneisen (1912). The Mie energy potential is written as a function of the intermolecular spacing r,

$$U(r) = \frac{\varepsilon}{m-n}\left(n\left(\frac{r_o}{r}\right)^m - m\left(\frac{r_o}{r}\right)^n\right). \tag{10.130}$$

The equilibrium intermolecular spacing, or equilibrium molecular cell size, is r_o when at ambient pressure conditions in the solid. The power constants m and n, with $m > n$, determine the relative strengths of the repulsive and attractive forces, respectively. The constant $\varepsilon > 0$ is the intermolecular binding or cohesive energy in units of energy per unit mass. The binding energy per cell is provided be $\rho_o r_o^3 \varepsilon$ where the ambient density of the material is ρ_o.

If r_o is the principal length scale governing hydrostatic compression or expansion of the material, then replacement of the size r_o with the specific volume through $r/r_o = (v/v_o)^{1/3}$ leads to the energy versus specific volume relation

$$U(v) = \frac{\varepsilon}{m-n}\left(n\left(\frac{v_o}{v}\right)^{m/3} - m\left(\frac{v_o}{v}\right)^{n/3}\right). \tag{10.131}$$

The pressure–volume EOS follows from equation (10.131) through $P(v) = -dU/dv$ and the bulk modulus from $K(v) = -v\, dP/dv$. The pressure versus volume Mie EOS becomes

$$P(v) = \frac{3K_o}{m-n}\left(\left(\frac{v_o}{v}\right)^{\frac{m}{3}+1} - \left(\frac{v_o}{v}\right)^{\frac{n}{3}+1}\right) \tag{10.132}$$

and the bulk modulus

$$K(v) = \frac{K_o}{m-n}\left((m+3)\left(\frac{v_o}{v}\right)^{\frac{m}{3}+1} - (n+3)\left(\frac{v_o}{v}\right)^{\frac{n}{3}+1}\right). \tag{10.133}$$

Expanding the latter in a polynomial in pressure results in expressions for the first- and second-order terms

$$K_o = \frac{mn}{9}\frac{\varepsilon}{v_o}, \quad K_o' = \frac{1}{3}(m+n+6). \tag{10.134}$$

The present energy and pressure–volume EOS is usually applied to the cold compression curve of the material. Namely, compression at absolute zero temperature in which only the lattice potential energy is accounted for (neglecting the quantum mechanical zero point thermal energy). With modestly less rigor the Mie potential, and other similar potential models, can be applied to the ambient isothermal or isentropic compression of a material. The Mie potential and the pressure versus compression EOS are characterized by the four parameters r_o, ε, m and n. The four Mie constants are not fully constrained by the ambient solid density ρ_o, and the bulk modulus properties K_o and K_o'. Additional information is needed such as K_o'', or knowledge of the binding energy.

Restricted forms of the Mie potential have found application in the EOS description of solids. Perhaps most familiar is the Lennard-Jones 6-12 potential (Lennard-Jones 1924). The power $m = 12$ was proposed to account for the repulsive force due to exchange interactions from overlapping electron orbitals while the $n = 6$ power replicates molecular attraction through van der Waals or dispersive forces. A value of $K_o' = 8$ is obtained from equation (10.134) for the Lennard-Jones EOS, somewhat higher than the range for K_o' of 3–7 observed for most solids. Values of $m = 4$ and $n = 2$ lead to a form of the Birch EOS to be discussed later in this section. Ionic solids have been shown to be reasonably described through Mie parameters of $m = 9$ and $n = 1$ (Walton 1983). The Mie EOS from equation (10.132) for $m = 3K_o' - 3$ and $n = -3$ is found to yield the Murnaghan equation provided in equation (10.126).

The Mie EOS potential is constrained by four material parameters. Selected potential functions and tension EOS curves are illustrated for the Mie EOS in figure 10.11. The initial specific volume υ_o and the binding energy ε provide scale parameters for the energy potential curves. Similarly, the initial specific volume and the ambient bulk modulus K_o scale the negative pressure curves. The power constants m and n provide the shape parameters for both sets of curves. As expected from equation (10.134), increasing the power constants increases the stiffness of the material response. Less obvious is the reduction in the maximum tension the material can support with increasing values of the power constants.

The tension EOS for the Mie energy model illustrated on the right in figure 10.11 is related to the compression EOS of interest in the present section through the same set of material constants that constrain the energy potential. The tension EOS is also of interest to the dynamic tensile strength or spall strength of solids addressed in earlier chapters. The Mie potential for the parameters used shows that the maximum or theoretical tensile strength is about 10%–20% of the bulk modulus of the material. This result is found to hold generally true for other functional representations of the intermolecular potential. Tensile strengths from the Mie potential are compared in figure 10.11 with one Griffith estimate of the theoretical strength of crystalline solids. In addition to the theoretical strength, the potential model also provides the theoretical expansion strain to spall. For the Mie energy potential the theoretical spall strain (and strength) is reasonable straightforward to calculate. By equating the bulk modulus to zero in equation (10.133) a spall strain $\mu_{\text{spall}} = \upsilon_{\text{spall}}/\upsilon_o$ is calculated,

$$\mu_{\text{spall}} = \left(\frac{m+3}{n+3}\right)^{3/(m-n)} \quad \text{or } \mu_{\text{spall}}(m=2n) = \left(\frac{2K_o' - 1}{K_o' + 1}\right)^{3/(K_o'-2)}. \tag{10.135}$$

The second relation provides the spall strain for the restricted case of $m = 2n$ in the Mie EOS. This is the case, for example for the Lennard-Jones EOS, $m = 12$, $n = 6$ and $K_o' = 8$, or the Birch EOS, $m = 4$, $n = 2$ and $K_o' = 4$. The theoretical spall strength is calculated by solving equation (10.132) at the spall expansion strain. The spall strength relations provide, for example, $\mu_{\text{spall}} = 1.657$ and $P_{\text{spall}}/K_o = -0.185$ for the case of $m = 4$ and $n = 2$.

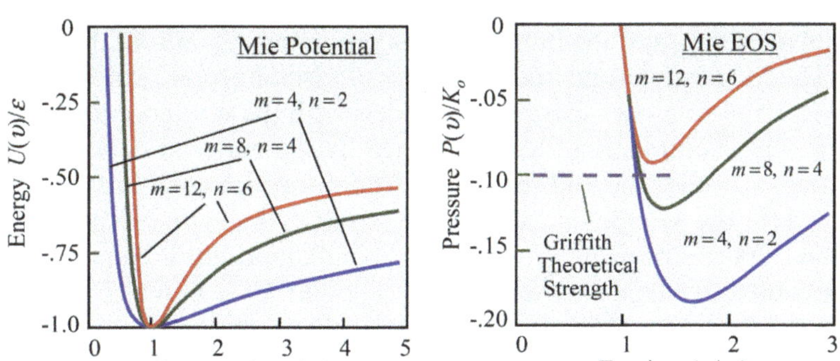

Figure 10.11. Mie potential (left) and the Mie tension (negative pressure) EOS (right) for selected values of the potential shape parameters. Values of the pressure derivative of the bulk modulus are $K_o' = 4$, 6 and 8 for the identified increasing values of the power constants.

An alternative intermolecular potential model that has received considerable attention over the years was proposed by Morse (1929). The Morse potential, in contrast to the Mie potential, uses dueling exponential functions rather than power functions to model the opposing intermolecular forces. The Morse potential relation corresponding to equation (10.130) for the Mie potential is written

$$U(r) = \varepsilon(\mathrm{e}^{2\alpha(1-(r/r_o))} - 2\mathrm{e}^{\alpha(1-(r/r_o))}). \tag{10.136}$$

Again, relating the cell size r_o to the material specific volume through $r/r_o = (v/v_o)^{1/3}$,

$$U(v) = \varepsilon(\mathrm{e}^{2\alpha(1-(v/v_o)^{1/3})} - 2\mathrm{e}^{\alpha(1-(v/v_o)^{1/3})}). \tag{10.137}$$

The pressure versus specific volume for the Morse potential follows from $P(v) = -dU/dv$,

$$P(v) = 3\frac{K_o}{\alpha}\left(\frac{v}{v_o}\right)^{-2/3}(\mathrm{e}^{2\alpha(1-(v/v_o)^{1/3})} - \mathrm{e}^{\alpha(1-(v/v_o)^{1/3})}), \tag{10.138}$$

and bulk modulus from $K(v) = -v\, dP/dv$,

$$K(v) = K_o\left(\left(\frac{v}{v_o}\right)^{-1/3}(2\mathrm{e}^{2\alpha(1-(v/v_o)^{1/3})} - \mathrm{e}^{\alpha(1-(v/v_o)^{1/3})}) + \frac{2}{\alpha}\left(\frac{v}{v_o}\right)^{-2/3}(\mathrm{e}^{2\alpha(1-(v/v_o)^{1/3})} - \mathrm{e}^{\alpha(1-(v/v_o)^{1/3})})\right). \tag{10.139}$$

The initial bulk modulus and pressure derivative for the Morse EOS are shown to be

$$K_o = \frac{2}{9}\frac{\alpha^2}{v_o}\varepsilon, \quad K_o' = \alpha + 1. \tag{10.140}$$

The Morse potential, in contrast to the Mie potential, is a three-parameter model characterized by the parameters, r_o, ε and α. The Morse EOS is uniquely determined by the initial density ρ_o and the bulk modulus properties K_o and K_o' through equation (10.140). (The Morse potential and a number of similar atomic potential models are frequently referred to as two-parameter models with the reference atomic radius, or the initial specific volume, readily constrained by the material density.)

The character of the Mie and the Morse EOSs are compared in figure 10.12. $m = 2n$ is assumed for the Mie potential. The tension EOS is shown on the right while the compression EOS to a pressure of ten times the bulk modulus is displayed on the left. The comparisons are made for values of $K_o' = 4$ and $K_o' = 7$, which spans the range of K_o' observed for most solids. Both models predict a theoretical tensile strength in the neighborhood of 10%–20% of the bulk modulus. In compression the Morse potential generally predicts more compliant behavior than the Mie potential. This observation remains true for reasonable variations of the powers m and n for a fixed K_o' for the Mie EOS.

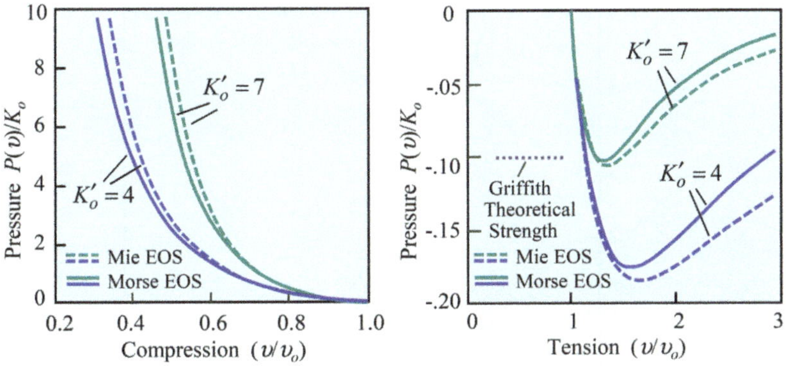

Figure 10.12. Compression (left) and tension (right) EOSs based on the Mie and Morse potential models.

Other phenomenological descriptions of the intermolecular binding energy that can lead to EOS descriptions of solids are possible. For example, Zharkov and Kalinin have proposed the potential functional form

$$U(v) = \frac{\varepsilon}{\alpha - n}\left(n e^{\alpha(1-(v/v_o)^{1/3})} - \alpha\left(\frac{v}{v_o}\right)^{-n/3}\right), \tag{10.141}$$

with pressure relation

$$P(v) = \frac{1}{3v_o}\frac{\alpha n \varepsilon}{\alpha - n}\left(\frac{v}{v_o}\right)^{-2/3}\left(\mathrm{e}^{\alpha(1-(v/v_o)^{1/3})} - \left(\frac{v}{v_0}\right)^{-(n+1)/3}\right). \tag{10.142}$$

This EOS is shown to reasonably describe the shock compression properties of silicate and oxide minerals (Zharkov and Kalinin 1971, Kalinin 1972). The Zharkov–Kalinin potential combines an exponential representation of the repulsive intermolecular forces with a power-law description of the attractive bonding forces.

For the present concerns, the objective of developing an analytic description of the intermolecular bonding potential is for the purpose of predicting the resistance of solids to compression when subjected to high pressures. Constants for the functional forms for the participating intermolecular forces are quantified through the measurement of properties near ambient pressure conditions. Specifically, the initial solid density ρ_o, the elastic bulk modulus K_o and the pressure derivative K_o' determine the constants in the EOS model and allow for the prediction of pressure versus compression to levels desired in the scientific application. There is, of course, no guarantee that the selected functional forms will adequately described the intermolecular forces over the range of interest, or that the selected ambient material properties will fully constrain the model. Ultimately, such phenomenological EOS models will survive or not through their ability to predict experimental observation.

A comparison of intermolecular potential EOS models with experimental data is shown in figure 10.13. Appropriate for this purpose are the experimental data of Nellis *et al* (1988) where a combination of absolute shock data and theory provides a reference 300 K isothermal compression curve for aluminum to pressures approaching

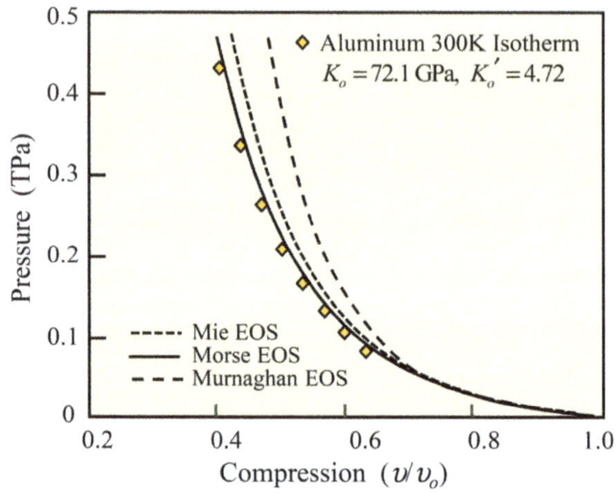

Figure 10.13. Comparison of the Mie, the Morse and the Murnaghan EOS with the 300 K isothermal compression for aluminum (Nellis *et al* 1988). Ultrasonic properties from Sikka (1989).

0.5 TPa and compressions in excess of a factor of two. The authors also provide reference isothermal data for copper and lead. Values of the bulk modulus and the pressure derivative for aluminum from different experimenters show some modest differences. Values of $K_o = 72.1$ GPa and $K_o' = 4.72$ from the study of Sikka (1989) are representative, however, and modest experimental variations do not change the results of the comparison. These values of K_o and K_o' for aluminum in the Mie, the Morse and the Murnaghan EOSs yield the pressure–compression curves that are compared with the experimental data of Nellis *et al* in figure 10.13. For the Mie EOS $m = 2n$ is assumed for the power constants, although reasonable variations in m and n for the same $K_o' = (m + n + 6)/3$ do not significantly modify the compression curve. Overall the Morse EOS provides a reasonable prediction of the isothermal compression of aluminum to 0.5 TPa, capturing the compression within several percent over the full range of the data. Later authors have generally accepted the exponential nature of the repulsive intermolecular forces exploited in the Morse potential. The markedly stiffer compression predicted by the Murnaghan equation indicates that a higher order description of the bulk modulus would require $K_o'' < 0$.

There are a number of other EOSs models based on principles of interatomic pair potential. Any attempt to more exhaustively survey these alternative EOSs would not be productive here. Most have marked similarities, combining exponential and power-law functions in attempting to replication of the repulsive and attractive forces responsible for the stability and stiffness of solid matter. Representative for example are the Rydberg (1931) and the Born–Mayer (1932) EOSs which will be referred to later in the chapter. Expressed as a pressure versus volume compression relation the Rydberg equation is

$$P(v) = 3K_o\left(\frac{v}{v_o}\right)^{-2/3}\left(1 - \left(\frac{v}{v_o}\right)^{1/3}\right)e^{\alpha(1-(v/v_o)^{1/3})}, \tag{10.143}$$

while the Born–Mayer equation is frequently written as

$$P(v) = 3K_o\left(\frac{v}{v_o}\right)^{-2/3}\left(e^{\alpha(1-(v/v_o)^{1/3})} - \left(\frac{v}{v_o}\right)^{-2/3}\right). \tag{10.144}$$

The similarity of the Born–Mayer EOS and that proposed by Zarkov and Kalinin (1971) in equation (10.142) can be noted. The Rydberg EOS is identical to the Rose–Vinet EOS with $\alpha = K_o'$ discussed later in this section (Rose *et al* 1984, Vinet *et al* 1986).

10.3.4 Finite-strain equations of state

Following the seminal efforts of Murnaghan, Birch (1947, 1978) pursued a finite-strain theory of elasticity for the purposes of describing the compressibility of solids at high pressures. The principal impetus for the study of Birch was to provide a description of the elastic properties of the Earth's interior. Namely, to describe the response of solids subjected to small non-hydrostatic strains (such as sonic waves) superimposed on a finite hydrostatic compression. A major consequence of the work of Birch was to provide analytic EOS relations for solids that have since been widely applied by the high-pressure community.

Let $\eta(v)$ be a measure of the finite volumetric strain of a solid material. The energy expanded in a power series in finite strain is written (Jeanloz 1988, 1989)

$$E(v) = 9v_oK_o\left(\frac{1}{2}\eta^2 + \frac{1}{3}a_1\eta^3 + \cdots\right). \tag{10.145}$$

The energy is the Helmholtz free energy if an isothermal EOS is pursued, or the internal energy for an isentropic EOS. Accordingly K_o is the corresponding isothermal or isentropic bulk modulus.

In principle any formulation of finite strain is appropriate for the theory. Ambiguity arises from non-uniqueness in defining an initial frame of reference. Birch presented the theory for both a Lagrangian and an Eulerian finite-strain representation. A generalized definition of the volumetric finite strain can be written (e.g. Jeanloz 1989)

$$\eta(v) = \frac{1}{n}\left(\left(\frac{v_o}{v}\right)^{n/3} - 1\right), \tag{10.146}$$

where the Lagrangian and an Eulerian finite strain definition correspond to $n = -2$ and $n = 2$, respectively. Calculating the pressure from equation (10.145) and joining with equation (10.146), Birch provided from Eulerian finite elastic strain theory to second-order (quadratic term in equation (10.145)) the EOS relation

$$P(v) = \frac{3}{2}K_o\left(\left(\frac{v_o}{v}\right)^{7/3} - \left(\frac{v_o}{v}\right)^{5/3}\right). \tag{10.147}$$

Equation (10.147) yields a pressure-dependent bulk modulus with a first pressure derivative of $K_o' = 4$. The relation provides a reasonable first-order EOS for many solids. It is worth noting that equation (10.147) is identical to the Mie EOS from equation (10.132) with power constants of $m = 4$ and $n = 2$.

The Birch EOS from Eulerian finite strain when extended to third-order (cubic term in equation (10.145)) is

$$P(v) = \frac{3}{2}K_o\left(\left(\frac{v_o}{v}\right)^{7/3} - \left(\frac{v_o}{v}\right)^{5/3}\right)\left(1 + \frac{3}{4}(K_o' - 4)\left(\left(\frac{v_o}{v}\right)^{2/3} - 1\right)\right). \tag{10.148}$$

The constant $a_1 = (3/2)(K_o' - 4)$ for an Eulerian finite-strain representation in equation (10.145). Birch provided arguments that an Eulerian strain representation of the theory would more rapidly converge with the least use of higher order terms. Jeanloz (1989) reached the same conclusion following an extensive study of high-pressure shock wave data and finite-strain theory. Equation (10.148) is the most commonly used form of the Birch finite-strain EOS.

10.3.5 Further universal equations of state

More recent efforts have contributed additional EOSs that propose to be relations that have wide-ranging application to the high-pressure compressibility of solids. Covalent, ionic and metallic materials and perhaps others are demonstrated to be encompassed by such EOSs. These universal EOSs necessarily build on aspects of earlier work, but offer new perspectives on the underlying physics governing the resistance of solid material to high-pressure compression.

10.3.5.1 Rose–Vinet binding energy equation of state

In a series of papers (e.g. Rose *et al* 1984, Vinet *et al* 1986) a study was undertaken to develop and test a universal EOS that has broad application to the high-pressure compressibility of solids. The authors demonstrated that when the dimensionless length scale

$$a = \frac{r_o - r}{l} = \alpha\left(1 - \left(\frac{v}{v_o}\right)^{1/3}\right) \tag{10.149}$$

is used to describe the functional dependence of the volumetric compression and dilation of the cohesive energy, the available data for a number of materials collapse to a universal curve. In equation (10.149) r_o is the dimension of the molecular cell while $l = r_o/\alpha$ is a characteristic measure of the width of the binding energy curve. Again, as was discussed when considering the previous intermolecular-potential EOSs, the observed scaling with two length scales would be appropriate only for materials where volumetric compression is sufficiently insensitive to internal intermolecular degrees of freedom. The authors also caution that when applied to the broader range of solids (i.e. covalent, metallic, ionic), universality of the energy function is restricted to compression and only modest dilation near equilibrium. Attractive forces governing larger dilation differ markedly for the different materials. Note in the second equality that the relation is expressed as the product of a dimensionless parameter and a Lagrangian finite-strain measure of the volumetric compressibility.

An analytic relation is then proposed for the cohesive energy of the form

$$U(a) = (f_0 + f_1 a + f_2 a^2 + \cdots)e^a. \tag{10.150}$$

An exponential function is judged to be the most appropriate for describing the energy in compression. The polynomial is a slowly varying function of compression that is truncated at the linear term. Constants are determined from the binding energy and compliance properties at the equilibrium zero pressure state of the material. Joining equations (10.149) and (10.150) provides the binding energy relation

$$U(v) = \varepsilon\left(\frac{3}{2}\alpha\left(1 - \left(\frac{v}{v_o}\right)^{1/3}\right) - 1\right)e^{3/2\alpha(1-(v/v_o)^{1/3})}. \tag{10.151}$$

Solving for the pressure $P(v) = -dU/dv$ and bulk modulus $K(v) = -v\, dP/dv$ yields

$$K_o = \frac{1}{9}\frac{\alpha^2}{v_o}\varepsilon, \qquad K_o' = \alpha + 1. \tag{10.152}$$

The Rose–Vinet binding energy EOS is then

$$P(v) = 3K_o\left(1 - \left(\frac{v}{v_o}\right)^{1/3}\right)\left(\frac{v}{v_o}\right)^{-2/3} e^{3/2(K_o'-1)(1-(v/v_o)^{1/3})}. \tag{10.153}$$

The Rose–Vinet binding energy EOS has demonstrated reasonable success in describing the high-pressure EOS properties of diverse solid material. The authors ascribe such universal behavior to a convergence of repulsive energy effects in the compression region of materials where overlap of atomic electron densities account for much of the compression resistance. In figure 10.14 the reference isothermal data for aluminum and copper from the study of Nellis *et al* (1988) are shown and compared with the predictions of the Rose–Vinet EOS based on measured ultrasonic values for K_o and K_o'.

Commonalities should be noted between the Rose–Vinet EOS and the earlier EOSs of Morse (1929) and of Zharkov and Kalinin (1971) provided in equations (10.137) and (10.141), respectively. The Rose–Vinet EOS is identical to the much earlier Rydberg EOS provided in equation (10.143). Both the Morse and the Zharkov–Kalinin, EOSs offer an exponential functional form in the description of compression resistance due to electron orbit overlap. Further, both of the earlier EOSs propose functional dependence on the same two-length-scale volume strain parameter used in the Rose–Vinet EOS (10.149). The earlier EOSs choose to describe the competing forces of molecular repulsion and attraction through the sum of opposing functional forms. Two exponential functions are used in the case of the Morse potential. The same competition is achieved within the Rose–Vinet EOS

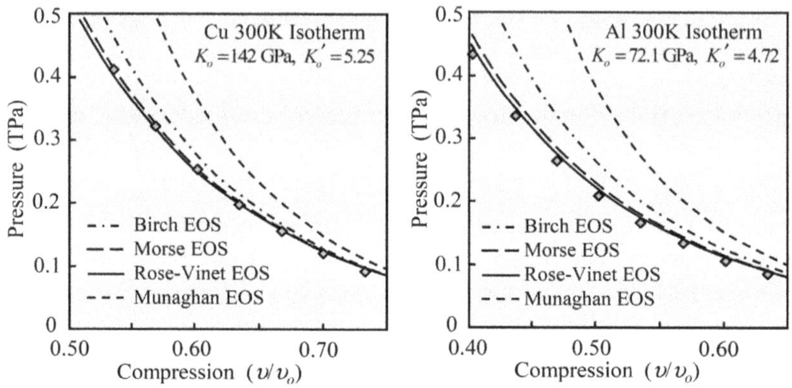

Figure 10.14. Predictions of the Murnaghan, Birch, Morse and Rose–Vinet EOSs with the 300 K isothermal compression of aluminum (left) and copper (right) (Nellis *et al* 1988). Ultrasonic properties are from Sikka (1989).

through the product of a repulsive exponential function and a truncated polynomial. Predictions of the Morse EOS from equation (10.138) based on the same set of elastic properties for aluminum and copper are also shown in figure 10.14. Both the Morse and the Rose–Vinet EOSs are equally adequate in predicting the compression data. Predictions based on the Murnaghan and the Birch third-order EOSs are less satisfactory.

10.3.5.2 Uniform electron gas equation of state

A further EOS development effort with the potential of universal application is that of Alchagirov *et al* (2001). This EOS model is perhaps the most solidly grounded in the underlying physics of interatomic forces governing crystal bonding in solids. Although specifically developed for metals, the dominance of electron core overlap in determining resistance to extreme compression suggests possible application to other solids. The EOS is based on the uniform electron gas (or jellium) model of metallic cohesion (e.g. Hughes 2006). This model is shown to predict a number of physical features associated with metallic bonding. For the present application an additional theoretical component is required. A pseudo-potential is introduced to account for the Coulombic energy of the ionic lattice that stabilizes the material at the correct initial density. It is hence also referred to as the stabilized pseudo-potential model.

The energy EOS is written as a sum of powers of the specific volume

$$U(v) = a\frac{v_o}{v} + b\left(\frac{v_o}{v}\right)^{2/3} + c\left(\frac{v_o}{v}\right)^{1/3} + d, \tag{10.154}$$

where each term relates to specific interatomic energies and forces responsible for cohesive properties of the material. The first two terms are the repulsive component of the electronic–ionic pseudo-potential and the electron gas kinetic energy, respectively. The latter terms account for the Coulombic electrostatic energy and the exchange-correlation energy. The EOS model as developed has similarities to the virial EOS for gases where theoretical models are used to calculate the virial coefficients from the intermolecular forces.

A volume derivative of the energy leads to the pressure EOS

$$P(v) = \frac{1}{3v_o}\left(3a\left(\frac{v_o}{v}\right)^2 + 2b\left(\frac{v_o}{v}\right)^{5/3} + c\left(\frac{v_o}{v}\right)^{4/3}\right). \tag{10.155}$$

The EOS constants are determined by the constraints of the binding energy, the bulk modulus, and the pressure derivative at the pressure of the equilibrium state. The model constants are

$$\begin{aligned} a &= \frac{9}{2}K_o v_o\left(K_o' - 3\right), \quad b = \frac{9}{2}K_o v_o\left(10 - 3K_o'\right), \\ c &= \frac{9}{2}K_o v_o\left(3K_o' - 11\right), \quad d = -\varepsilon + \frac{9}{2}K_o v_o\left(4 - K_o'\right). \end{aligned} \tag{10.156}$$

The EOS model is not as analytically convenient as some of the others discussed previously, and does not at present appear to be broadly used. Where tested, however, the uniform electron gas EOS is shown to provide reliable compressibility predictions from ambient compliance properties of the material.

10.3.5.3 Extreme compression equations of state

EOSs that have been discussed to this point are recognized, at least tacitly, to have limits to the range of compression for which they are applicable. In practical shock and static high-pressure compaction of solids this range is substantially less than one order of magnitude in volume compression. Experimental methods continue to evolve, however, such that this limit may in the near future be exceeded with increasing regularity.

The EOS efforts of Holzapfel (1996) are unique in attempting to provide EOS relations that extrapolate in extreme compression limits to states predicted from fundamental physical principles. In the extreme pressure limit as $v/v_o \rightarrow 0$ with the corresponding extreme overlap of the ion cores all electrons ionize into the valence band. Under such compressions matter should approach the Thomas–Fermi limit where compressibility is governed by the Fermi gas pressure relation $P_{fg} = a_{fg}(Z/v)^{5/3}$. The electron number is Z and a_{fg} is a fundamental electronic constant. The Fermi gas bulk modulus $K_{fg} = (5/3)P_{fg}$ is readily calculated and yields a limiting pressure derivative of $K_\infty' = 5/3$.

Holzapfel demonstrated that several of the pair potential models are readily modified to provide the correct limiting compression of $K_\infty' = 5/3$. In particular, working with the Rydberg equation (see (10.143)) Holzapfel proposed the extreme compression EOS relation

$$P(v) = 3K_o\left(\frac{v}{v_o}\right)^{-5/3}\left(1 - \left(\frac{v}{v_o}\right)^{1/3}\right)e^{\alpha(1-(v/v_o)^{1/3})}. \tag{10.157}$$

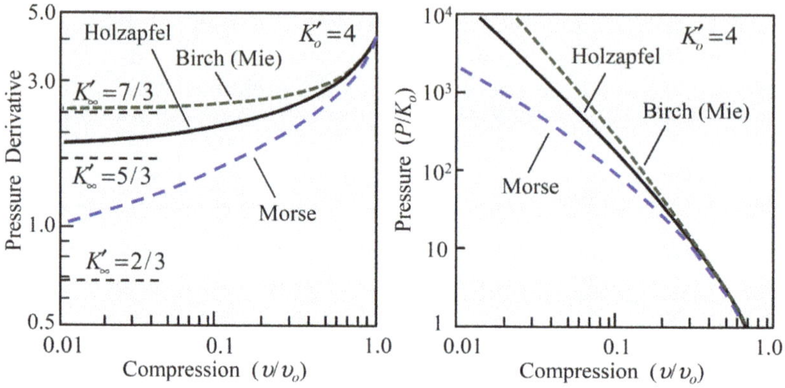

Figure 10.15. Comparison of the Holzapfel extreme pressure EOS with the Birch and the Morse equations. The initial pressure derivative of the bulk modulus is 4. Left: Limiting values of the pressure derivative for the Birch (7/3), Holzapfel (5/3) and the Morse (2/3) equations. Right: Pressure versus compression in units of the initial bulk modulus.

Equation (10.157) quite adequately replicates the desired limiting physics. However, through some added terms introduced into the equation, Holzapfel provided a further relation that more exactly achieves the Fermi gas pressure in the limit.

The Holzapfel EOS from equation (10.157) is illustrated in figure 10.15 for a volume compression of two orders of magnitude and contrasted with both the Morse and the Birch finite-strain EOS. An initial pressure derivative of $K_o' = 4$ is assumed in each case. The pressure derivative of the bulk modulus as a function of pressure is also compared in the figure. The Morse equation diverges from below, reflecting a limiting $K_\infty' = 2/3$ that is markedly lower than the Fermi limit. The not unreasonable agreement between the Birch equation and the Holzapfel equation reflects the unique choice of $K_o' = 4$ for which the Birch equation is second-order (or equivalent to the Mie equation with $m = 4$ and $n = 2$ as discussed in the paragraph following equation (10.147)) and exhibits a limiting $K'_\infty = 7/3$, fairly close to the Fermi $K_\infty' = 5/3$ limit. For any $K_o' \neq 4$ the Birch equation is third-order with a limiting value of $K_\infty' = 3$ and will exhibit increasing divergence from the Fermi limit with increasing compression. The plot of K' as a function of compression on the left in figure 10.15 has interest in illustrating the extent of compression required in achieving the limiting rate of stiffening of the bulk modulus.

10.3.5.4 Shock velocity versus particle velocity and the universal equations of state

At this point a revisit with the inconsistency between ultrasonic and shock Hugoniot measurements pointed out by Steinberg (1982) and discussed in section 10.3.2 is of interest. Those that have developed universal EOSs relations for purposes of describing the high-pressure compression of matter seldom go the extra mile and plot the relations in the shock-velocity versus particle-velocity space so revered by the shock physicist.

The comparisons in figure 10.14 show that both the Morse potential and the Rose–Vinet EOSs reasonably describe the compression of aluminum and copper over the range of compression achieved in the preponderance of shock Hugoniot data. The Birch equation less so and the Murnaghan equation provides poor agreement at compressions exceeding about 20%–25% compression strain.

Here the Morse EOS is used to examine some of the implications of a universal EOS in description of the shock-velocity versus particle-velocity characteristics of a solid. A Mie–Gruneisen EOS joined with the commonly assumed Gruneisen relation of $\rho\Gamma$ equal to a constant and with the Morse equation energy and pressure from equations (10.137) and (10.138) as the reference EOS relations, leads to a principal Hugoniot $P_h(v)$ provided by equation (10.58).

The corresponding shock velocity and particle velocity are provided by

$$U(v) = \sqrt{\frac{P_h(v)}{\rho_o(1 - \rho_o v)}}, \qquad u(v) = \sqrt{\frac{P_h(v)(1 - \rho_o v)}{\rho_o}}. \tag{10.158}$$

Shock-velocity versus particle-velocity plots, based on a Morse reference EOS, are shown in figure 10.16 for metallic copper. EOS properties of $K_o = 142$ GPa and $K_o' = 5.25$ are from Sikka (1989). Shock velocity is normalized by the calculated intercept $C_o = \sqrt{K_o/\rho_o}$. Also shown in the plot is the derivative of the shock-velocity relations identified as $S(u) = dU/du$; namely the local slope of the shock-velocity versus particle-velocity curve. Three cases are shown. First, the initial second-order tangency of the reference isentrope (the Morse equation) and the Hugoniot require $S = (K_o' + 1)/4 = 1.563$. A constant slope S provides the dashed curves in the figure. The second two sets of curves are calculated from equation (10.158) based on the Morse equation with a constant $\rho\Gamma$ Mie–Gruneisen EOS for both $\Gamma_o = 0$ and $\Gamma_o = 2.02$. The latter is the appropriate value for copper (Steinberg 1996).

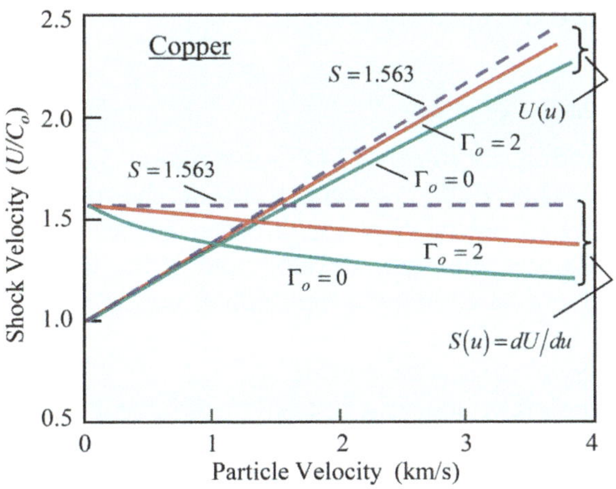

Figure 10.16. Shock-velocity versus particle-velocity plots and first derivative for copper based on Morse reference equation and Mie–Gruneisen EOS.

The first observation is that shock velocity versus particle velocity based on a Morse universal EOS is not linear. Rather, the relation has a modestly negative curvature, most clearly revealed by the derivative plotted in the figure. The second observation is the influence of the Gruneisen parameter. The appropriate curve for copper is that of $\Gamma_o = 2$. The predicted shock-velocity versus particle-velocity relation actually matches the linear relation reasonably closely when Γ_o is about three.

The experimental shock-velocity versus particle-velocity Hugoniot data for copper in the LANL compendium (Marsh 1980) span the particle-velocity range from a few tenths of 1 km s^{-1} to a few tenths over 3 km s^{-1} (the middle range of the plot in the figure). Over this range the slope is on average about 4%–5% below the constant value of 1.563 corresponding to the linear *U–u* dashed curve in the figure. A straight-line fit to the Hugoniot data over this range would consequently yield a value of *S* approximately 4%–5% less than 1.563. The LANL compendium reports *S* = 1.51 for copper. Steinberg (1982) reports a value of *S* for copper that is about 9% lower than that calculated from the ultrasonic pressure derivative of the bulk modulus (Steinberg uses a pressure derivative of 5.48 which is higher than the 5.25 value reported by Sikka (1989)). Discussions of Jeanloz (1988) in comparisons of the Rose–Vinet universal EOS and the Birch third-order EOS also relate to this issue.

The present observations provide a possible explanation for the discrepancies between the values of *S* determined from experimental Hugoniot data and the calculated value from $S = (K_o' + 1)/4$ as reported and discussed by Steinberg (1982). Namely, shock-velocity data plotted against particle-velocity data are not rigorously a straight line, but rather have very modest negative curvature as predicted by the present Morse EOS. Experimental scatter over the range of the Hugoniot data negates determination of this very modest curvature.

Extrapolation of the linear fit to zero particle velocity should correspondingly result in a value of the shock-velocity–particle-velocity intercept C_o that exceeds the ultrasonic $C_o = \sqrt{K_o/\rho_o}$. Although this offset is not observed for copper, perusal of the LANL compendium reveals that this expectation is the case for a number of the other metals examined by Steinberg (1982).

10.3.6 Linear-Z equation of state

The linear inertial modulus or linear-*Z* EOS warrants special attention in the present text (Grady 2012). This EOS is a construct of the present author and will consequently be allowed prerogative not afforded earlier EOSs. A small level of bias is noted. Before proceeding with development of the EOS, some fundamentals of the elastic compliance of condensed materials is considered.

10.3.6.1 Compliance of condensed matter

The following discussion is concerned strictly with the bulk or volumetric compliance of material. Similar considerations can be addressed to the elastic shear properties of a material, but that is not the present purpose. A common property of matter is the compliance of a body of material when the pressure is elevated by an increment ΔP and the body responds with a corresponding contraction in the volume ΔV. Here consideration is a body of volume V containing a homogeneous mass M of the material of interest. A measure of the compliance is the ratio of the change in pressure to the change in volume $\Delta P/\Delta V$. This ratio is not an appropriate property of the material, however. If the initial volume V of the body were increased for example by a factor of two, the volume change ΔV would correspondingly change by a factor of two and $\Delta P/\Delta V$ would accordingly change and consequently not be a constant of the material. The obvious solution, and what is done, is to normalize the ratio by the current volume V. The compliance property then becomes

$$K = -V\frac{dP}{dV} = -v\frac{dP}{dv}. \tag{10.159}$$

A minus sign is introduced to make K a positive quantity (with units of pressure) and K is commonly referred to as the bulk modulus of the material. The bulk modulus is an appropriate property of the material and is independent of any reference configuration. The second step in equation (10.159) is a recognition that the relation can also be written in terms of the specific volume $v = V/M$. Details of the modulus of course, depends on the nature of the pressure increment—whether the process is isothermal or isentropic for example. When related to the cold curve of the material this of course is not an issue. This level of detail is not of concern here.

Note that modulus K does not depend on the mass M of the material within the volume V. It depends only on the strength of the intermolecular bond and is independent of the molecular mass. If the magnitudes of the molecular bonds are similar then the values of the bulk moduli will be similar. Whether the material has the mass of beryllium or of uranium is does not influence the bulk modulus.

An alternative normalization of the compliance is possible, however, and also leads to an appropriate property of the material. Namely, the compliance can be normalized by the mass M of the material within the volume V,

$$Z = -M\frac{dP}{dV} = -\frac{dP}{dv}. \tag{10.160}$$

This compliance is referred to as the inertial modulus. In contrast to the bulk modulus, the inertial modulus is dependent on both the molecular mass and the intermolecular potential. The bulk modulus and inertial modulus are independent properties of the material and are related through $Z = \rho K$. The inertial modulus Z is also the square of the impedance of the material $z = \rho c = \sqrt{\rho K}$, where c is the Eulerian bulk wave speed.

A third bulk elastic modulus (not independent of K and Z) can be introduce which adds a certain symmetry to the relations. Referred to here as the acoustic modulus C, it is provided by

$$C = \frac{dP}{d\rho}. \tag{10.161}$$

Where $c = \sqrt{C}$ provides the sonic velocity of the material in units of distance per unit time, $z = \sqrt{Z}$ is the inertial velocity of the material in units of mass per unit time (actually mass per unit area per unit time). The bulk modulus is provided by $K = \sqrt{ZC}$ and recognized as the geometric average of the inertial modulus and the acoustic modulus.

Early in chapter 1 a simple linear mass–spring model of the molecular nature of a solid with spring constant k, mass m and spacing a resulted in an Eulerian sonic velocity and material density of

$$c = a\sqrt{\frac{k}{m}}, \quad \rho = \frac{m}{a}. \tag{10.162}$$

Using the relation $K = \rho c^2$, and expressions from above, yields

$$K = ak, \quad Z = mk, \quad C = \frac{a^2 k}{m}, \tag{10.163}$$

for the bulk modulus, the inertial modulus and the acoustic modulus, respectively. Note that the bulk modulus does not depend on the molecular mass while the inertial modulus does not depend on the molecular spacing. The acoustic modulus depends on all defining properties of the unit cell.

10.3.6.2 The linear-Z equation of state

The bulk modulus of a material determined through ultrasonic measurements is commonly observed to exhibit a linear dependence on the applied pressure. This observation has given rise to the commonly used Murnaghan EOS and is illustrated in figure 10.17 with experimental data for tungsten carbide (Amulele *et al* 2008). The inertial modulus for tungsten carbide is similarly plotted and is also observed to be linear with pressure with a comparable goodness of fit (Grady 2012). Both, of course, cannot be precisely linear. The ultrasonic data, however, are not adequate for determining between the two.

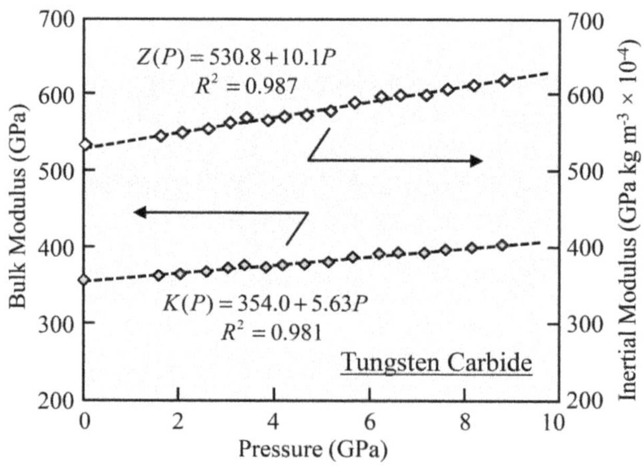

Figure 10.17. Comparison of the bulk modulus and inertial modulus for tungsten carbide ceramic (Grady 2012). A measure of the goodness of fit is provided in the figure. Ultrasonic data to 8 GPa are from Amulele *et al* (2008).

This chapter does not advocate for, or against, any of the various universal EOSs that have been discussed. Recognized, however, is the continuing need for analytically convenient EOSs that sensibly reproduce the shock and static compression of solids over the span of applications encountered by the shock wave and impact community. This need is attested to by continued application of the Murnaghan EOS in spite of recognized limitations. An alternative reference EOS is proposed here that is based on a variation of Murnaghan's equation that leads to an equally convenient analytic EOS form. Further, this EOS is observed to reproduce experimental high-pressure data to levels of compression markedly higher than the Murnaghan EOS. Comparisons with compression data also suggest somewhat better agreement than with the Birch third-order EOS, and comparable agreement with available potential models to compressions of about a factor of two.

The inertial modulus expanded to first-order about the ambient pressure state is

$$Z(P) = Z_o + Z_o'P. \tag{10.164}$$

Like the bulk modulus, $Z(P)$ is determined from ultrasonic data and when plotted over the pressure range of most ultrasonic data (several GPa) will appear linear in P if $K(P)$ appears linear in P. The first- and second-order expansion parameters for the bulk and inertial modulus are related through

$$Z_o = \rho_o K_o, \text{ and } Z_o' = \rho_o(K_o' + 1). \tag{10.165}$$

Neither modulus is precisely linear in pressure when extended to higher levels of compression. The variation from linearity of either modulus can be examined through inspection of higher fidelity EOS models. A linear dependence of the bulk modulus on pressure results in the Murnaghan EOS

$$P(v) = \frac{K_o}{K_o'}\left(\left(\frac{v}{v_o}\right)^{-K_o'} - 1\right). \tag{10.166}$$

The corresponding EOS assuming a linear dependence on pressure of the inertial modulus (referred to as the linear-Z EOS) is derived by equating equations (10.160) and (10.164), and integrating,

$$P(v) = \frac{Z_o}{Z_o'}\left(e^{Z_o'(v_o - v)} - 1\right). \tag{10.167}$$

EOS compression of aluminum and copper for the Murnaghan, the Birch third-order, and the linear-Z EOS are compared with experimental compression data of Nellis *et al* (1988) in figure 10.18. Ambient bulk modulus and pressure derivative for the two metals are from Sikka (1989).

Compression data for aluminum and copper to nearly a factor of two favor extrapolation of the ultrasonic determined compressibility with the linear-Z EOS over that of both Murnaghan and Birch (Grady 2012). The linear-Z EOS diverges below the compression data when further extrapolated to the full factor of three

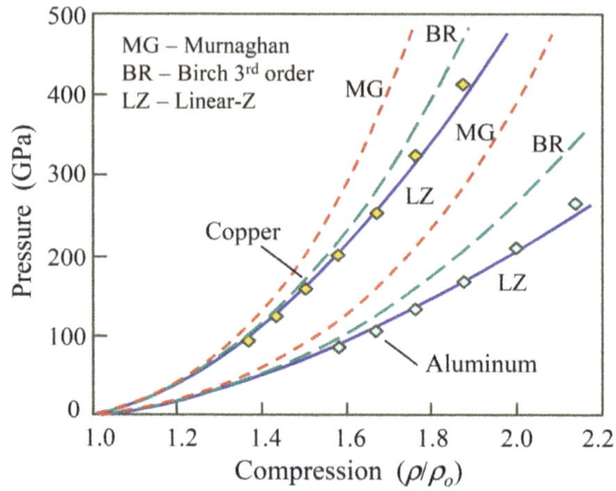

Figure 10.18. Reference isothermal compression of aluminum and copper from shock Hugoniot data of Nellis *et al* (1988). Murnaghan, Birch and linear-*Z* compression curves based on ultrasonic elastic moduli from Sikka (1989): aluminum $K_o = 72.1$ GPa, $K_o' = 4.72$; copper $K_o = 142$ GPa, $K_o' = 5.25$.

compression range of the experimental data. Of the higher level models discussed earlier in the section, the Morse EOS best replicates the compression of aluminum and copper over the full range of the data.

Limited comparisons with nonmetals suggest similar improved prediction with the linear-*Z* EOS. The linear-*Z* EOS is also found to provide an appropriate analytic Hugoniot representation for compressions to approximately a factor of two. A plot in the shock-velocity–particle-velocity plane is nearly linear with a slight negative curvature.

10.3.6.3 Application of linear-Z equation of state

The linear inertial modulus reference equation within the broader framework of the enthalpy-based Rice–Walsh EOS is here applied to magnesium oxide. Extensive shock Hugoniot data for magnesium oxide are reported in the LANL shock Hugoniot compendium (Marsh 1980). Magnesium oxide is one of the more thoroughly characterized compounds regarding shock wave properties and has been identified as a reference oxide material for high-pressure studies (Carter *et al* 1971). The LANL compendium provides experimental Hugoniot data for crystalline magnesium oxide as well as Hugoniot data for porous magnesium oxide from three starting densities over a Hugoniot pressure range from 10 GPa to in excess of 100 GPa.

Magnesium oxide (MgO) is the naturally occurring mineral periclase. It occurs in the cubic halite crystal structure with ambient density of 3.58 g cm^{-3}. The cubic structure is stable in MgO to in excess of 200 GPa, well into the pressure range of Earth's lower mantle. As an end member of alleged major lower mantle mineralogy, considerable elasticity, high-pressure static compression and shock wave data are available for this ionic compound. The crystal elasticity of MgO at pressure is investigated in the study of Karki *et al* (1997). Duffy *et al* (1995) provide background on DAC compression and strength-at-pressure to in excess of 100 GPa in MgO. Exemplary shock Hugoniot data are available for MgO (Marsh 1980) and provide the data subjected to the present EOS assessment.

The intent is to use the linear inertial modulus (linear-Z) reference EOS in conjunction with the Rice–Walsh representation of high-pressure thermo-elasticity of MgO to capture the span of available Hugoniot data. Linearity of the inertial modulus in conjunction with constant $\rho\Gamma = \rho_o\Gamma_o$ leads to the a Rice–Walsh EOS of the form

$$v(P,H) = v_s(P) + \frac{\rho_o\Gamma_o}{Z_o + Z_o'P}(H - H_s(P)), \tag{10.168}$$

with the reference isentrope of the linear-Z form, the inverse of equation (10.167),

$$v_s(P) = v_o - \frac{1}{Z_o'}\ln\left(1 + \frac{Z_o'}{Z_o}P\right). \tag{10.169}$$

Properties of the bulk modulus and the inertial modulus are related through $Z_o = \rho_o K_o$ and $Z_o' = \rho_o(K_o' + 1)$.

All shock Hugoniot data from the LANL compendium (Marsh 1980) for MgO are plotted in both a shock-velocity versus particle-velocity and Hugoniot-pressure versus specific-volume representation in figure 10.19. A best fit of the EOS provided by equations (168) and (169) to the collection of shock Hugoniot data is shown with the data in figure 10.19. The initial bulk modulus K_o and pressure derivative K_o' of the reference isentrope are assessed through a best fit to crystal density Hugoniot data. Gruneisen parameter Γ_o is determined from optimum fit to the porous Hugoniot data. An analytic EOS representation of the Hugoniot data provided by equations (10.168) and (10.169) is provided by the dashed curves in both plots in the figure.

The principal isentrope compression relation resulting from the EOS fit is identified in the right-hand plot. The optimum EOS parameters of K_o =153 GPa,

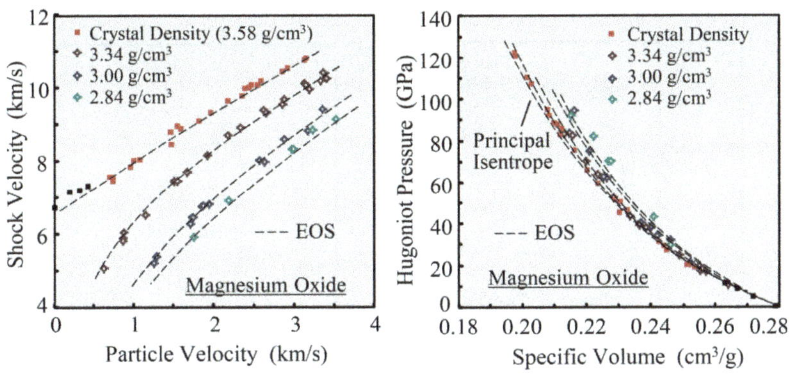

Figure 10.19. Shock Hugoniot data for magnesium oxide (Marsh 1980). Dashed curves identify the best EOS fit (K_o = 153 GPa, K_o' = 4.8, Γ_o = 1.2). The principal isentrope assessed from the EOS fit is identified in the right-hand plot. Three data points in the 0.2–0.5 km s^{-1} particle-velocity range are from Grady (1977).

$K_o' = 4.8$ and $\Gamma_o = 1.2$ are not markedly altered when an alternative reference relation, such as the Morse equation, is used. Compare with ultrasonic measurements of K_o =157–163 GPa, K_o'= 4.1–4.3 (Karki *et al* 1997).

The modest reduction in K_o assessed from the Hugoniot data below that from ultrasonic data was also noted in the EOS investigation of Carter *et al* (1971). A phase transformation on the Hugoniot at pressure below 20 GPa was suggested in that work. Later studies have not supported this explanation. More likely, a two-parameter representation of the compression isentrope over the range of the data is not adequate.

The EOS analysis does not address possible strength effects at states of Hugoniot compression. Duffy and Ahrens (1993) provide four Hugoniot measurements on near-full-density (~3.57 g cm^{-3}) polycrystalline MgO over the pressure rang of 15–130 GPa. Within the range of 15–50 GPa, through comparisons of Hugoniot states with the static isotherm assessed from ultrasonic data (Jackson and Niesler 1982), they report a strength offset of 1.5–2.5 GPa. Strength measurements from DAC studies on MgO report strengths of order 10 GPa over the range of 50–100 GPa (Duffy *et al* 1995).

Measurements of time-resolved shock wave profiles in [100] oriented single-crystal MgO at pressures over the range of 5–11 GPa provide some indications of the strength characteristics of this mineral (Grady 1977). The experiments entail planar-impact shock loading and measurement of the transmitted particle velocity with velocity interferometry (VISAR) at the interface between the test sample and a sapphire window material (Barker and Hollenbach 1970). Measured wave profiles are shown on the left in figure 10.20. The precursor wave is indicative of initial strength and subsequent stress relaxation in the failure characteristics. The stress amplitudes of the precursor features are displayed in the right-hand plot. Analyses of the measured profiles yield the Hugoniot states shown on the right. These several Hugoniot measurements constitute the lowest pressure single-crystal Hugoniot points plotted in figure 10.19.

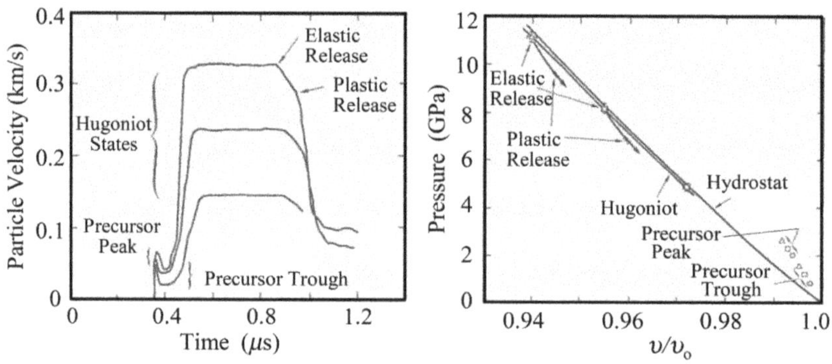

Figure 10.20. Particle-velocity profiles in [100] oriented crystal magnesium oxide at the sample and sapphire window interface following 3.3 mm propagation from the planar impact surface (Grady 1977). Precursor wave stress amplitudes and Hugoniot states are identified on the right.

Comparison with hydrostatic compression of MgO determined from ultrasonic data shows that the Hugoniot states and the hydrostatic compression curve overlay within an uncertainty of about 0.2 GPa. The subsequent release wave structure reveals, however, that the material retains strength at the Hugoniot state. Shock behavior where transient loss of strength in the initial shock wave leads to Hugoniot states with no or minimal stress offset and strength recovery when subjected to subsequent stress wave stimulus has been observed in other material. Notable is that of boron carbide (e.g. Vogler *et al* 2004). Note that the principal Hugoniot data for MgO in the LANL shock compendium is on single-crystal material. Observation of differences in the shock strength characteristics of single-crystal and polycrystalline samples of the same compound is not uncommon.

10.3.7 Woolfolk Hugoniot equation of state and modifications

Woolfolk *et al* (1973) examined the shock Hugoniot properties of selected liquids and noted the nonlinear downward-curvature character of the shock-velocity versus particle-velocity (U–u) data before transitioning at higher shock pressures to the more common linear U–u behavior observed in the preponderance of condensed matter Hugoniot response. Based on these observations and efforts of earlier workers, these authors proposed the Hugoniot EOS relation

$$U(u) = C_o + Su - \delta e^{-u/u*}. \tag{10.170}$$

Similar Hugoniot anomalies have been observed in organic polymers. Explosive binders are notable. The underlying physics of the accelerated stiffening early in the Hugoniot compression appears to be associated with the molecular free volume present in liquids and polymer solids. The initial reduced stiffness of the defected materials is accompanied with an initial high-pressure derivative K' that transitions to more common behavior at higher Hugoniot compressions.

Steinberg (1996) noted the computational convenience of couching the description of nonlinear U–u behavior in the functional form of shock-velocity dependence on the shock Hugoniot strain $U(\varepsilon)$. The pressure versus compression strain Hugoniot relation then takes the immediate form

$$P(\varepsilon) = \rho_o U^2(\varepsilon)\varepsilon. \tag{10.171}$$

The Woolfolk Hugoniot EOS can accordingly be modified to the form

$$U(\varepsilon) = \frac{C_o}{1 - S\varepsilon} - \delta e^{-\varepsilon/\varepsilon*}, \tag{10.172}$$

where approximately $\varepsilon* = u*/C_o$.

The initial bulk wave speed in either the Woolfolk or the modified Woolfolk Hugoniot relation of $C_o^* = C_o - \delta$ is reduced by the decrement δ from the linear intercept C_o. At compression strains in excess of $\varepsilon*$, the U–u Hugoniot transitions to a straight line of slope S and intercept C_o. Evidence of markedly reduced but similar anomalous behavior is suggested from a more detailed examination of shock

Hugoniot data for metals. See, for example, the paper by Steinberg (1982) and the discussion of this study in subsection 10.3.5.4.

At no added expense, equation (10.172) can be further modified by adding a shape parameter, yielding

$$U(\varepsilon) = \frac{C_o}{1 - S\varepsilon} - \delta e^{-(\varepsilon/\varepsilon^*)^n}, \tag{10.173}$$

allowing better characterization of the anomalous region of the Hugoniot when justified by the data. Equation (10.173) has found application, for example, in the Hugoniot EOS description of the explosive material PBXW-128 (Lindfors 1996, Grady 1997, Grady *et al* 2000). This explosive is approximately 60% HMX and 40% binder by volume. Experimental Hugoniot data and the EOS representation are shown in figure 10.21.

The Woolfolk relation and modifications suggested by the EOS development efforts of Steinberg (1996) are used here for purposes of analytic representation of the Hugoniot response of certain materials—notably liquids and organic polymers. The same functional relations are also appropriately used as reference EOSs to describe the ambient isentrope or cold curve of the material and joined with a Mie–Gruneisen or comparable thermo-elastic EOS for a fuller description of the material response. It is noteworthy that for selected values of the EOS parameters, computational applications of equation (10.173) can lead to the formation of rarefaction shock waves.

A convenient extension and application of the modified Woolfolk Hugoniot EOS as displayed in equation (10.173) is for materials that undergo shock-induced phase transformation at relatively modest compression on the Hugoniot and then transition to linear *U–u* behavior at higher Hugoniot compression. There are a number of materials that exhibit such shock transformation behavior that could serve as illustrative examples. Not widely known is the study of Sandstrom *et al* (1993) in which the unreacting compression properties of ammonium nitrate and ammonium perchlorate are assessed through both shock Hugoniot and static high-

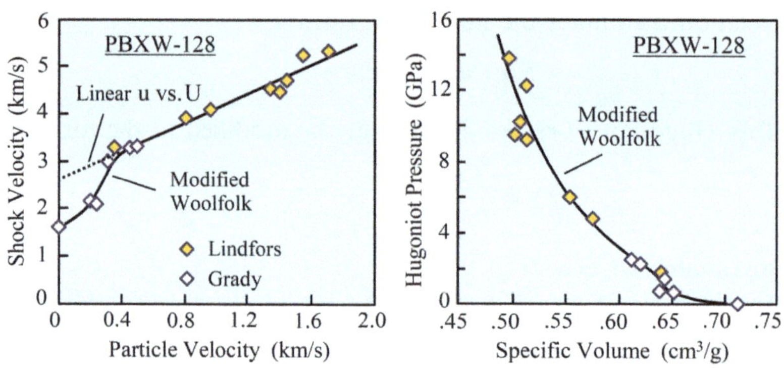

Figure 10.21. Unreacting Hugoniot data for PBXW-128 explosive (left; Lindfors 1996, Grady 1997) and representation with the modified Woolfolk Hugoniot EOS (right; Grady *et al* 2000). Parameters for the representation are $C_o = 2.62$ km s^{-1}, $S = 1.5$, $\delta = 1.02$ km s^{-1}, $\varepsilon^* = 0.11$ and $n = 10$.

pressure measurements. The study reveals a shock-induced phase change in ammonium nitrate within the pressure range of about 5–8 GPa. No phase change was observed in ammonium perchlorate to shock pressures in excess of 20 GPa. Crystal structure data from static compression, on the other hand, suggested a phase transition at about 4.7 GPa in ammonium perchlorate.

The modified Woolfolk EOS is used here to provide an analytic description of the Hugoniot response of ammonium nitrate and is compared with the data in figure 10.22. In this application the initial bulk sound speed $C_o^* = C_o - \delta$ is greater than C_o with a negative value of the decrement parameter of $\delta = -0.84$. Limited parameters in the EOS do not allow for independent specification of the slope of the U–u curve below the phase change. Rather $S^* = S + \delta/(C_o - \delta) = 1.91$ is modestly less than that of the slope $S = 2.18$ above the transformation.

The Hugoniot description of ammonium nitrate in the figure is constructed from both the isothermal static compression data and shock Hugoniot data of Sandstrom *et al* (1993). The shock-velocity versus particle-velocity measurements of Sandstrom are for a material that is about 5%–6% porous. Consequently, the data in the panel on the left are corrected to the theoretical density Hugoniot. This synchronization of both static compression and shock Hugoniot data to the crystal density is achieved by calculating corrected particle velocity and shock velocity from the shock relations,

$$u' = \sqrt{\frac{P(1 - \rho_o v)}{\rho_o}}, \qquad U' = \sqrt{\frac{P}{\rho_o(1 - \rho_o v)}}. \tag{10.174}$$

The reported Hugoniot pressure and specific volume data are P and v, while ρ_o in equation (10.174) is the crystal density of ammonium nitrate. The shock phase transition in ammonium nitrate is not well constrained by the data. The transition in not observed under static compression and apparently requires inelastic deformation intrinsic to the shock compression event to initiate and carry through phase transformation. The analytic modified Woolfolk model provided by equation

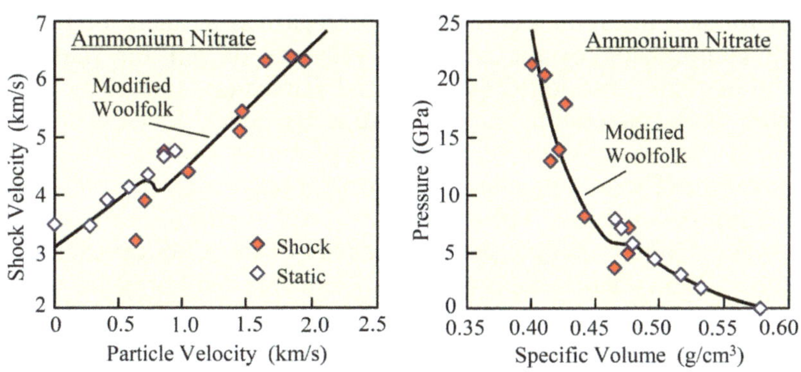

Figure 10.22. Unreacting Hugoniot data for ammonium nitrate (Sandstrom *et al* 1993). The modified Woolfolk EOS provides an analytic description including shock phase change. The parameters for the representation are $C_o = 2.23$ km s^{-1}, $S = 2.18$, $\delta = -0.84$ km s^{-1}, $\varepsilon^* = 0.19$ and $n = 15$.

(10.173), however, readily captures the observed Hugoniot behavior including phase transformation.

10.3.8 Assessment of universal equations of state

Considerable theoretical effort has been expended in developing analytic EOSs for describing the high-pressure compressibility of solids. A number of these potentially universal EOSs have been described in the previous subsections. Such EOSs have several purposes. As analytic descriptions they can be readily integrated or differentiated for purposes of providing state properties of the solid (shock and acoustic wave speeds for example) in EOS calculations of the shock process. This application can be appropriate even for materials that exhibit marked phase transformation properties with concomitant anomalies in the material compressibility within the range of interest. For such materials analytic EOSs can be applied piecewise over the range of normal analytic compressibility of the lattice. Appropriate expressions for regions of phase transformation are then joined to provide analytic continuity. One approach is described in chapter 6 where EOSs of the linear shock-velocity versus particle-velocity form are piecewise joined to describe the compressibility of boron carbide through two phase transformation regions. Alternatively, the modified Woolfolk EOS offers a convenient analytic description of shock wave phase transformation appropriate in certain cases.

Of perhaps more importance, and certainly of more significance from a theoretical perspective, is the ability of such EOSs to predict the high-pressure compressibility of matter from a limited number of readily measurable properties. These might include for example ambient measurements of elastic compliance, thermal expansion, specific heats and heats of sublimation. Some successes in this regard have been demonstrated. In no case, however, are such extrapolations made with full confidence.

In this perspective the extreme EOSs proposed by Holzapfel (1996) are of interest. Few materials achieved such extremes without incurring some form of structural transformation. As discussed in the introduction to this section, phase transformations to the Holzapfel EOSs would be regarded as mere perturbations on the overall compression of the material. Whether or not the material at the compression extremes envisioned by Holzapfel retains any memory of the K_o and K_o' properties of the material at zero pressure is arguable.

The pressure versus compression properties of matter cannot be fully decoupled from temperature. Thermal pressure (and energy) plays an important EOS role in the shock compression event. The present universal compressibility discussions do not address the extreme thermal energy portions of the EOS domain (e.g. Drake 2010). From ambient conditions, isothermal, isentropic and Hugoniot compressions are restricted to ranges where thermal pressures and thermal expansions are reasonably addressed through conventional models of the Gruneisen properties of the material. The exception noted here is the application of the enthalpy-based Rice–Walsh EOS to shock compression of highly distended solids. Thermal states of matter are achieved that suggest the use of such EOSs be pursued with some caution.

10.4 Temperatures under shock compression

The calculation of temperature is not necessary to the analysis of shock wave and large-amplitude wave propagation in solids. It is an additional calculation that can be performed if warranted by the application and if the necessary thermal properties of the material are available.

Shock-induced heating and the accompanying temperature rise in the shock wave is a complex nonequilibrium event. Non-isentropic compression and deformation dissipation contribute to the additional temperature increase. On the continuum scale, temperature rise in a structured shock wave is near-adiabatic with thermal conduction occurring on time scales far larger than that of wave propagation. Shock heating on the local scale, however, is far from adiabatic. Heterogeneous dissipation and thermal conduction are competing processes on the microscale that are often first-order influences on the physical, chemical and metallurgical processes occurring in the shock. Despite such complexities, sensible calculations of temperature based on continuum thermodynamic principles are tractable and provide useful bounds on the shock-heating event. Further perspective from a phononic description of thermal equilibration within the shock is considered in chapter 8.

10.4.1 Thermodynamics and temperature

The rise in the continuum temperature of a solid subjected to shock compression is calculated through fundamental thermodynamic principles. Temperature increase on the compression isentrope or the Hugoniot is determined through appropriate manipulation of the equilibrium differential form of the first and second law of thermodynamics

$$dE = TdS - Pdv, \tag{10.175}$$

where the symbols are, respectively, the system energy, temperature, entropy, pressure and volume. The temperature–volume differential of the entropy provides

$$dS = \left.\frac{\partial S}{\partial T}\right|_v dT + \left.\frac{\partial S}{\partial v}\right|_T dv. \tag{10.176}$$

Extensive parameters are here considered to be specific values (per unit mass) although this is not required. Through the definition of the specific heat and the appropriate Maxwell relation equation (10.176) is reduced to

$$dS = \frac{c_v}{T}dT + \left.\frac{\partial P}{\partial T}\right|_v dv. \tag{10.177}$$

Replacing $(\partial P/\partial T)_v = c_v\Gamma/v$ yields

$$dS = \frac{c_v}{T}dT + \frac{c_v\Gamma}{v}dv. \tag{10.178}$$

Equations (10.175) and (10.178) join to eliminate entropy and provide a differential expression for the dependence of the temperature differential on the specific volume and energy,

$$c_v dT = -c_v \Gamma T \frac{dv}{v} + P\,dv + dE. \tag{10.179}$$

10.4.2 Temperature on the isentrope

Under isentropic compression with $dS = 0$, equation (10.175) provides for a differential increase in energy on the isentrope from the increment in compressive work, $dE = -P\,dv$, and equation (10.179) yields the differential equation for the isentropic temperature rise,

$$\frac{dT}{dv} + \frac{\Gamma}{v} T = 0. \tag{10.180}$$

The Gruneisen parameter Γ is commonly assumed to be independent of temperature, and power-law dependent on the volume,

$$v \left.\frac{\partial P}{\partial E}\right|_v = \Gamma(v) = \Gamma_o \left(\frac{v}{v_o}\right)^n. \tag{10.181}$$

For $n = 0$ the Gruneisen $\Gamma = \Gamma_o$ is a constant independent of volume. Equation (10.180) yields for the isentropic temperature rise the power-law relation

$$T_S(v) = T_o \left(\frac{v}{v_o}\right)^{-\Gamma_o}. \tag{10.182}$$

For $n = 1$ the commonly assumed functional dependence for the Gruneisen gamma of $\rho\Gamma = \rho_o\Gamma_o$ is obtained. An exponential increase in isentropic temperature results from equation (10.180),

$$T_S(v) = T_o e^{\Gamma_o(1 - v/v_o)}. \tag{10.183}$$

In each of the above relations Γ_o is provided by the Gruneisen parameter at ambient conditions, and is related to the more common thermal expansion, incompressibility and specific heat properties through

$$\Gamma_o = \frac{\alpha K_S}{\rho_o c_p} = \frac{\alpha K_T}{\rho_o c_v}. \tag{10.184}$$

The temperature at initiation of compression is T_o. Isentropic temperature solutions provided by either equations (10.182) or (10.183) reveal that the absolute zero temperature isotherm and isentrope are coincident, and that the temperature on any isentrope depends only on the specific volume, independent of the functional form of the isentropic EOS.

10.4.3 Temperature on the Hugoniot

The specific energy for single-shock compression on the principal Hugoniot from a reference state of volume v_o, pressure $P_o = 0$ and energy $E_o = 0$ is given by the Rankine–Hugoniot equation

$$E = \frac{1}{2}P(v_o - v), \tag{10.185}$$

and the differential form is

$$dE = \frac{1}{2}(v_o - v)dP - \frac{1}{2}P\,dv. \tag{10.186}$$

Joining the energy expression in equation (10.186) with the temperature relation of equation (10.179) yields

$$c_v\,dT = -c_v\Gamma T\frac{dv}{v} + \frac{1}{2}(v_o - v)dP + \frac{1}{2}P\,dv, \tag{10.187}$$

and the more commonly presented differential equation for temperature on the Hugoniot,

$$c_v\frac{dT}{dv} + c_v\frac{\Gamma(v)}{v}T = \frac{1}{2}(v_o - v)P_h'(v) + \frac{1}{2}P_h(v). \tag{10.188}$$

This differential equation for temperature on the Hugoniot was derived in the early work of Walsh and Christian (1955), and applied to shock-induced melting of metals by Carter (1973). See also Horie and Sawaoka (1993) and Forest (1996). At this point no restrictions have been placed on the volume dependence of the Gruneisen gamma or the temperature dependence of the specific heat.

Before further exploring temperature states on the Hugoniot, equation (10.188) is specialized to a constant $\rho\Gamma = \rho_o\Gamma_o$ relation for the Gruneisen parameter and transformed to Hugoniot strain $\varepsilon = 1 - v/v_o$ as the independent variable, with v_o the ambient, or reference, specific volume,

$$c_v\frac{dT}{d\varepsilon} - c_v\Gamma_o T = \frac{1}{2}v_o\big(\varepsilon P_h'(\varepsilon) - P_h(\varepsilon)\big). \tag{10.189}$$

Assuming that the specific heat is constant, equation (10.189) is readily solved for the Hugoniot temperature yielding

$$c_v T_h(\varepsilon) = c_v T_o e^{\Gamma_o\varepsilon} + \frac{1}{2}V_o\int_0^{\varepsilon}\big(\xi P_h'(\xi) - P_h(\xi)\big)e^{\Gamma_o(\varepsilon-\xi)}d\xi. \tag{10.190}$$

The first term on the right provides the isentropic temperature rise to strain ε while the second term is the excess temperature resulting from shock dissipation in the compression event. Using the Hugoniot expression for a material that is reasonably described by a linear shock-velocity versus particle-velocity relation

$$P_h(\varepsilon) = \frac{\rho_o C_o^2 \varepsilon}{(1 - S\varepsilon)^2}, \tag{10.191}$$

equation (10.190) simplifies to

$$c_v T_h(\varepsilon) = c_v T_o e^{\Gamma_o \varepsilon} + C_o^2 S \int_0^{\varepsilon} \frac{\xi^2 e^{\Gamma_o(\varepsilon - \xi)}}{(1 - S\xi)^3} d\xi. \tag{10.192}$$

Familiarization with the shock dissipation term in the temperature expression provides insight into the underlying mechanisms. Ignore for the moment the exponential term in the integrand of equation (10.190), which is of order unity. (Equivalently, expand $e^{\Gamma_o(\varepsilon-\xi)} = 1 + \Gamma_o(\varepsilon - \xi) + \cdots$, and consider only the first term.) The integral is then written

$$\frac{1}{2} v_o \int_0^{\varepsilon} (\xi P_h'(\xi) - P_h(\xi)) d\xi = v_o \int_0^{\varepsilon} d\left(\frac{1}{2}\xi P_h(\xi)\right) - v_o \int_0^{\varepsilon} P_h(\xi) d\xi. \tag{10.193}$$

In this form the integral is recognized as the area under the Rayleigh line minus the area under the principal Hugoniot. This region in a plot of the Hugoniot and isentropic compression curves is identified in figure 10.23. The first term on the right is the total area enclosed by the triangular region whereas the second term is the area under the Hugoniot curve (includes the area under the isentrope identified in the figure plus the unshaded area between the isentrope and the Hugoniot).

Consider a power series expansion of the principal Hugoniot. Without loss of generality, constants in the expansion descriptive of the linear shock-velocity versus particle-velocity representation from equation (10.191) can be used,

$$P_h(\xi) = \rho_o C_o^2 (\varepsilon + 2S\varepsilon^2 + \cdots). \tag{10.194}$$

Applying the integral expression in equation (10.190), completing expansion of the integrand and carrying out the integration yields

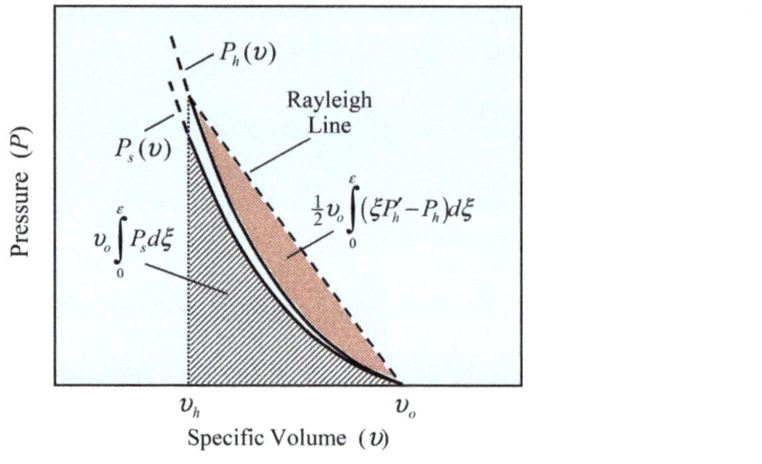

Figure 10.23. Graphical energy components to Hugoniot temperature rise in the shock compression of a solid.

$$c_v T_h(\varepsilon) = c_v T_o e^{\Gamma_o \varepsilon} + \frac{1}{3} C_o^2 S\left(\varepsilon^3 + \frac{1}{4}\Gamma_o \varepsilon^4 + \cdots\right). \tag{10.195}$$

Equation (10.195) provides the thermal energy in the shock compression event to a Hugoniot volume strain ε, and the contribution to temperature rise resulting from shock dissipation. In powers of the compression strain the first term (third power in the strain) gives the energy encompassed by the Rayleigh line and the Hugoniot compression as provided by equation (10.193). The second term (fourth power in strain) accounts for the additional area (or dissipation energy) residing between the isentrope and the Hugoniot identified by the unshaded area in figure 10.23 between the isentrope and the Hugoniot.

When a constant specific heat does not adequately describe the thermal properties of the material over the range of shock compression of interest, the calculation of temperature from equation (10.188) has to be performed numerically. The Debye temperature for many ceramic materials is, for example, reasonably high. The application of a temperature-dependent Debye specific heat provides improved temperature estimates.

10.4.4 Hugoniot temperature calculations

For illustrative purposes calculations of temperature rise on the principal shock Hugoniot are made for aluminum, copper and iron with the EOSs and thermal properties provided in table 10.1. The Hugoniot and Gruneisen gamma properties are from Marsh (1980). Equation (10.192) is appropriate for the Hugoniot temperature calculations for these and many other solid metals:

The shock Hugoniot curves are adequately described by a linear shock-velocity versus particle-velocity $U = C_o + Su$ Hugoniot relation. Representation of the iron Hugoniot in the 10–15 GPa range is approximate because of a solid–solid phase transformation. Also, a quadratic term in the LANL shock-velocity versus particle-velocity Hugoniot for iron (Marsh 1980) is ignored that probably reflects the ε–γ phase transformation.

The Debye temperatures are reasonably close to room temperature (430 K, 340 K and 470 K for aluminum, copper and iron, respectively). The limiting high-temperature Dulong–Petit constant specific heat is appropriate for the present calculations. Room temperature specific heats provided in table 10.1 for the three metals are sensibly close to the Dulong–Petit limit. A constant $\rho\Gamma = \rho_o\Gamma_o$ description

Table 10.1. Metal properties.

Metal	Density(kg m^{-3})	C_o(km s^{-1})	S	Γ_o	c_v(J kg^{-1} K)	T_m(K)
Aluminum	2707	5386	1.339	1.97	876	933
Copper	8930	3930	1.489	2.02	384	1358
Iron	7850	3574	1.920	1.69	453	1811
Tantalum	16690	3410	1.200	1.67	140	3290

for the volume dependence of the Gruneisen parameter is a reasonable approximation for the three metals. When more complex functional dependence of the specific heat or the Gruneisen parameter is required, Hugoniot temperature calculations are perform through numerical solution of equation (10.188).

Temperature calculations for aluminum, copper and iron to Hugoniot strains approaching $\varepsilon = 0.3$ are shown in figure 10.24. Significant divergence of the isentropic and Hugoniot temperature rise starts at a strain of about $\varepsilon = 0.1$.

10.4.5 Melting on the Hugoniot

One objective of temperature calculations in the shock compression event is to determine the shock wave intensity sufficient to cause melting of the material. The onset of melting on the principal Hugoniot is intrinsic to shock temperature analysis. Calculations of melting on the Hugoniot require knowledge of the pressure dependence of the solid melt curve. Fundamental thermodynamics provides the Clausius–Clapeyron equation for the pressure dependence of the melt temperature

$$\frac{d \ln T_m}{dP} = \frac{\Delta v_m}{L}, \tag{10.196}$$

where Δv_m and L are the transformation volume increment and the latent heat of melting (specific volume and specific energy are used).

The Gruneisen $P(v,E)$ EOS of the solid material is intimately related to the melt curve (Stacy and Irvine 1977). Consider a solid just at the melt boundary at a pressure P and add to the body an amount of thermal energy equal to the latent heat L maintaining the specific volume constant. The Gruneisen EOS of the solid requires that pressure increase by an amount

$$\Delta P = \rho \Gamma L. \tag{10.197}$$

Now allow the body to relax under adiabatic conditions back to the pressure P on the melt boundary. The material will expand by an amount

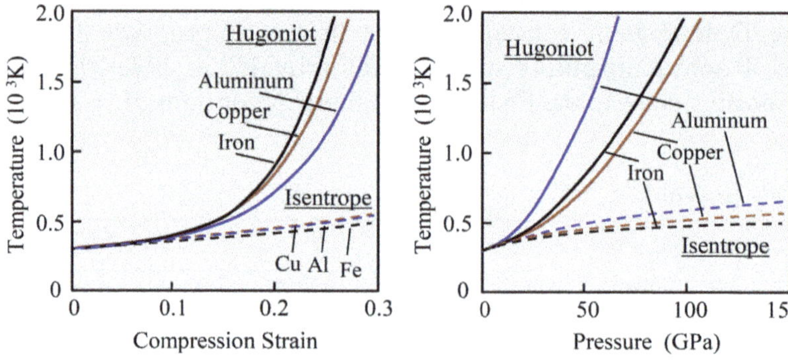

Figure 10.24. Hugoniot and isentropic temperature dependence on compression strain (left) and pressure (right) of selected metals.

$$\Delta v_m \cong \frac{\Delta P}{\rho K}, \tag{10.198}$$

where K is the bulk modulus. (The bulk modulus of the liquid and the solid phase are comparable.) Equations (10.197) and (10.198) yield approximately $\Delta v_m/L = \Gamma/K$, and the differential relation for the melt boundary becomes

$$K\frac{d\ \ln T_m}{dP} = -\frac{d\ \ln T_m}{d\ \ln v} \cong \Gamma(v). \tag{10.199}$$

Lindemann (1910) was inspired by the observation that for many solids the product of the volume coefficient of expansion and the melt temperature is constant and given approximately by

$$\alpha T_m \cong 0.07. \tag{10.200}$$

Based on the hypothesis that all solids melt when the amplitude of thermal vibrations are large enough to cause atomic collisions, Lindemann arrived at the relation

$$-\frac{d\ \ln\ T_m}{d\ \ln\ v} = 2(\Gamma(v) - 1/3). \tag{10.201}$$

The Lindemann relation does not differ markedly from the approximate derivation of Stacy and Irvine resulting in equation (10.199). Equation (10.201) is readily integrated and, when the Gruneisen gamma is a constant $\Gamma = \Gamma_o$, provides

$$T_m/T_{mo} = (v/v_{mo})^{-2(\Gamma_o-1/3)}. \tag{10.202}$$

Alternatively, when the Gruneisen gamma satisfies $\rho\Gamma = \rho_o\Gamma_o$

$$T_m/T_{mo} = (v/v_{mo})^{2/3} e^{2\rho_o\Gamma_o(v_{mo}-v)}. \tag{10.203}$$

The zero pressure melt temperature and specific volume of the solid at onset of melt are T_{mo} and v_{mo}, respectively.

Calculation of the melt boundary in temperature–pressure space requires joining equation (10.203) for the melt boundary with the Gruneisen EOS. Written in terms of the temperature and volume strain referenced from the ambient state

$$P(\varepsilon,T) = P_h(\varepsilon) + \rho_o\Gamma_o c_v(T - T_h(\varepsilon)), \tag{10.204}$$

the volume strain can be eliminated between equations (10.203) and (10.204) to provide the requisite P–T melt boundary. Again, $\rho\Gamma$ and c_v are assumed constant. The Gruneisen pressure is referenced from the principal Hugoniot with $P_h(\varepsilon)$ and $T_h(\varepsilon)$ provided by equations (10.191) and (10.192) for a linear shock-velocity versus particle-velocity material. When equation (10.203) is expressed in volume strain, care must be taken in accounting for volume expansion to the melt temperature. In terms of volume strain referenced from ambient density ρ_o and temperature T_o, equation (10.203) is written

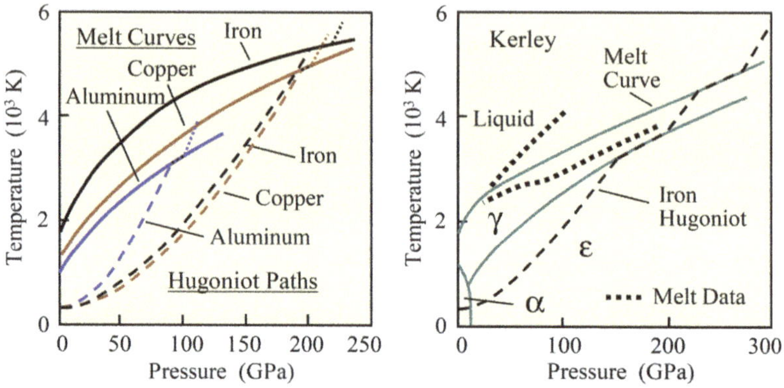

Figure 10.25. Left: Shock Hugoniot and Lindemann melt curves for selected metals based on analysis in the text. Right: The Hugoniot and melt curve for iron using a physics-based tabular EOSs (Kerley 1994). Dotted curves are experimental measurements of iron melt (Williams *et al* 1987, Boehler 1993).

$$T_m = T_{mo}\left(\frac{1}{\beta}(1 - \varepsilon)\right)^{2/3} e^{2\Gamma_o(\beta-1+\varepsilon)}, \tag{10.205}$$

where $\beta = \exp[\alpha(T_{mo} - T_o)]$, and where α is the thermal expansion coefficient.

Melt curves based on the Lindemann model are calculated and shown for solid aluminum, copper and iron in figure 10.25. Hugoniot curves for the three metals, based on a linear U–u representation of the shock Hugoniot data, are compared with the melt boundaries. It is seen that shock pressures in the neighborhood of 100–200 GPa are necessary to achieve temperatures sufficient for melt in the selected metals. Hugoniot curves through and above the melt boundary are shown by the dotted extension, and are for illustrative purposes, but are qualitatively correct.

A more detailed phase diagram for iron provided by Kerley (1994) is shown on the right in figure 10.25. The three solid state phases in addition to the liquid phase are shown that are accessed through Hugoniot shock compression of solid iron. Comparison of the two plots shows that the Lindemann model, ignoring subsequent crystallographic transformations, leads to somewhat higher melt temperatures for iron at pressure. Additionally, the high-pressure ε–γ phase transformation softens the Hugoniot, leading to lower temperature and higher pressure onset of melt on the Hugoniot than is predicted by the Lindemann model. The predictions of the Kerley melt phase boundary are in good agreement with experimental indications of melt on the Hugoniot reported by Brown and McQueen (1986). Dashed curves are experimental measurements of the pressure–temperature melt boundary (Williams *et al* 1987, Boehler 1993). The wide disparity between the two sets of data illustrates the difficulties of experimental identification of the melt transition under high-pressure conditions.

10.4.6 Plastic dissipation contribution to the shock temperature

Calculations of Hugoniot temperatures have assumed to this point that temperature rise is dominated by the shock pressure versus volume compression of the material.

Contributions to temperature rise brought about by deviator stresses and inelastic shear dissipation are ignored. At sufficiently high shock pressure, the contributions to dissipation from deviator strength effects become small. An assumption of a hydrodynamic response of the solid is sensible. At more modest shock pressures the contributions from strength-induced dissipation should be considered.

Dissipation in the plastic shear of a solid does not necessarily reside in thermal energy. The generation of defects, and the concomitant configurational energy and entropy, in the deformation process provides a sink for athermal dissipation. Common lore reports that 80%–90% of deformation dissipation typically contributes to the thermal energy. Here, for purposes of estimating the relative magnitude of shear dissipation in the shock temperature rise, all dissipation is assumed to contribute to the thermal energy. If required, the contribution to athermal dissipation can be accounted for.

In an elastic-plastic solid the axial stress resides above the mean stress under the constraint of uniaxial strain, or shock wave, loading by an amount of $2Y/3$. The unconfined yield strength of the solid is Y and is considered here a material constant. To account for plastic dissipation in the temperature rise a work term of $dE = \frac{2}{3}(Y/\rho_o)d\varepsilon$ is included in equation (10.179) in addition to the shock dissipation term provided by equation (10.186). Since shock pressures are modest, equation (10.195) to lowest order in shock strain, supplemented by the plastic dissipation term, is appropriate for a linear *U–u* material,

$$c_v T_h(\varepsilon) = c_v T_o e^{\Gamma_o \varepsilon} + \frac{1}{3} C_o^2 S \varepsilon^3 + \frac{2}{3}\frac{Y}{\rho_o}\varepsilon. \qquad (10.206)$$

Temperature calculations based on equation (10.206) are illustrated for tantalum metal in figure 10.26. Properties for tantalum are provided in table 10.1. In the shock environment, tantalum exhibits a Hugoniot elastic limit of about 2.8 GPa and a yield stress of about $Y = 1.5$ GPa (Furnish *et al* 2006). Temperature contributions

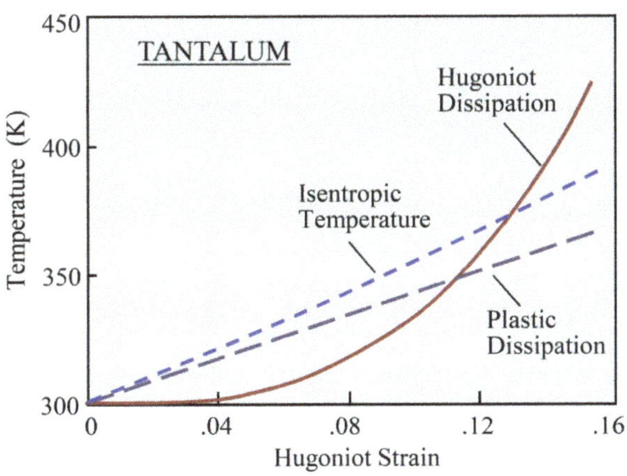

Figure 10.26. Contributions of isentropic compression, shock dissipation and plastic work to the Hugoniot temperature rise in an elastic-plastic solid (tantalum). A strain of 0.11 corresponds to a Hugoniot pressure of about 20 GPa.

from each of the three terms in equation (10.206) are illustrated separately in the figure. Temperature rise from plastic work initially exceeds shock dissipation to a strain of order $\varepsilon = (2Y/C_o^2 S)^{1/2} = 0.11$ (about 20 GPa) at which point the two terms are comparable. Isentropic temperature rise is of the same order at strains of 0.10–0.15, but is recoverable on decompression from the shock state.

10.4.7 The porous Hugoniot

Temperature calculations for shock compression of a porous solid require knowledge of the shock Hugoniot unique to the specified distention. Experimentally acquired Hugoniot compression curves for the porous solid are appropriate. Alternatively, a Mie–Gruneisen EOS can be assumed for the material.

The Mie–Gruneisen EOS for a volume-dependent Gruneisen gamma $\Gamma(v)$ is written

$$P(v,E) = P_{ho}(v) + \rho\Gamma(E - E_{ho}(v)), \tag{10.207}$$

with $P_{ho}(v)$ and $E_{ho}(v)$ the pressure and energy versus volume relations for the principal solid (undistended or cold) Hugoniot. Density and specific volume are interchangeable through $v = 1/\rho$. The Hugoniot compression for any initial distention of the same material specified by an initial specific volume $v_{oo} = 1/\rho_{oo}$ is then

$$P_h(v) = P_{ho}(v) + \rho\Gamma(E_h(v) - E_{ho}(v)). \tag{10.208}$$

The Hugoniot energy for the distended solid is provided through the Rankine–Hugoniot relation

$$E_h(v) = E_{oo} + \frac{1}{2}P_h(v)(v_{oo} - v). \tag{10.209}$$

Ignoring differences in the reference energy between the solid and distended material equation (10.208) provides

$$P_h(v)\left(1 - \frac{1}{2}\rho\Gamma(v_{oo} - v)\right) = P_{ho}(v)\left(1 - \frac{1}{2}\rho\Gamma(v_o - v)\right) \tag{10.210}$$

or

$$P_h(v) = P_{ho}(v)\frac{\left(1 - \frac{1}{2}\rho\Gamma(v_o - v)\right)}{\left(1 - \frac{1}{2}\rho\Gamma(v_{oo} - v)\right)}. \tag{10.211}$$

The limiting compression at extreme pressure is determined when the denominator on the right is zero, yielding

$$\rho = \rho_{oo}\left(1 + \frac{2}{\Gamma}\right) \tag{10.212}$$

or, for $\rho\Gamma = \rho_o\Gamma_o$ a constant

$$\rho = \rho_{oo}\left(1 - \frac{2}{\Gamma_o}\frac{\rho_{oo}}{\rho_o}\right)^{-1}. \tag{10.213}$$

Equations (10.212) or (10.213) are applicable provided the cold Hugoniot $P_{ho}(v)$ does not achieve an asymptotic limit at a lower density; difficulties that are not encountered with a Rice–Walsh or isobaric formulation of the EOS. Equation (10.212) has an important implication. Namely, if the Gruneisen parameter is truly temperature-independent, then the porous Hugoniot will achieve a vertical asymptote in the limit of high pressure. This experimental result is not commonly observed.

An alternative representation of the porous Hugoniot is frequently more convenient. Starting with the Mie–Gruneisen EOS constrained to any reference EOS $P_r(v)$,

$$P(v,E) = P_r(v) + \rho\Gamma(E - E_r(v)), \tag{10.214}$$

and introducing the porous Hugoniot energy from equation (10.209), again ignoring differences in the reference energy between the solid and distended material, yields

$$P_h(v) = P_r(v) + \rho\Gamma\left(\frac{1}{2}P_h(v)(v_{oo} - v) - E_r(v)\right). \tag{10.215}$$

Solve for the Hugoniot pressure,

$$P_h(v) = P_r(v) + \frac{\rho\Gamma}{1 - (\rho\Gamma/2)(v_{oo} - v)}\left(\frac{1}{2}P_r(v)(v_{oo} - v) - E_r(v)\right). \tag{10.216}$$

When referenced to the cold Hugoniot $P_{ho}(v)$, equation (10.216) simplifies to

$$P_h(v) = P_{ho}(v) + \frac{\rho\Gamma/2}{1 - (\rho\Gamma/2)(v_{oo} - v)}P_{ho}(v)(v_{oo} - v_o). \tag{10.217}$$

Comparable relations for the porous Hugoniot in the Rice–Walsh EOS framework are provided by equations (10.94) and (10.95).

10.4.8 Post-shock residual temperatures

Continuous calculation of the temperature during decompression from the shock Hugoniot state entails additional complications. The shock Hugoniot state resides on a different isentrope than that identified with the ambient state of the material prior to shock compression. Assuming that decompression is isentropic, temperature recovery during decompression is associated with the shock state isentrope.

In applications, concern is usually with the post-shock temperature of the body following completed decompression. This temperature is readily approximated from any of the previous temperature relations by excluding the isentropic temperature rise. In equations (10.190) or (10.192), for example, the second term on the right provides a sensible calculation of the post-shock temperature. For many cases an estimate of post-shock temperature based on the area (energy) between the shock

Hugoniot and the Rayleigh line is adequate. Issues on post-shock temperatures are belabored further in chapter 7.

10.4.9 Enthalpy equation of state and temperature on the Hugoniot

The analysis in section 10.4.3 leads to a differential equation for temperature on the Hugoniot. Solutions of the differential equation in (10.188) provide Hugoniot temperatures $T_h(v)$ as a function of the specific volume. It is at times convenient to work with shock EOSs in an enthalpy formulation. An enthalpy representation with the Rice–Walsh EOS, for example, has been shown convenient for highly distended solids (e.g. Fenton *et al* 2012, Grady *et al* 2013). It is necessary to develop and alternative differential relation for temperature on the Hugoniot when using an enthalpic EOS description of the media. The differential equation for temperature with an enthalpy-based EOS is outlined and Hugoniot temperature calculations are presented for solid tantalum oxide (Ta_2O_5).

The differential expression for specific enthalpy is

$$dH = TdS + v\, dP. \tag{10.218}$$

The temperature–pressure differential for the entropy is

$$dS = \left.\frac{\partial S}{\partial T}\right|_P dT + \left.\frac{\partial S}{\partial P}\right|_T dP. \tag{10.219}$$

The temperature derivative is related to the specific heat, while the pressure derivative is transformed through the appropriate Maxwell relation

$$\left.\frac{\partial S}{\partial P}\right|_T = -\left.\frac{\partial v}{\partial T}\right|_P = -c_p\left.\frac{\partial v}{\partial H}\right|_P = -c_p\frac{\zeta}{P}, \tag{10.220}$$

where ζ is the Rice–Walsh parameter first introduced in section 7.8 (see also section 10.2.3). Equation (10.219) for the entropy differential is then

$$dS = \frac{c_p}{T}dT - c_p\frac{\zeta}{P}dP. \tag{10.221}$$

A differential expression for the temperature on the isentrope follows directly from equation (10.221),

$$\frac{dT}{dP} - \frac{\zeta(P)}{P}T = 0. \tag{10.222}$$

In pursuing the calculation of temperature on the Hugoniot, equations (10.218) and (10.221) are joined to eliminate entropy,

$$dH = c_p dT + \left(v - c_p\frac{\zeta}{P}\right)dP. \tag{10.223}$$

The enthalpy differential on the Hugoniot is

$$dH = \frac{1}{2}(v + v_o)dP + \frac{1}{2}Pdv. \tag{10.224}$$

Enthalpy is eliminated yielding the differential equation

$$c_p\frac{dT}{dP} - c_p\frac{\zeta(P)}{P}T = \frac{1}{2}Pv'_h(P) + \frac{1}{2}(v_o - v_h(P)). \tag{10.225}$$

Equation (10.225) for temperature on the Hugoniot can be compared with the previous temperature relation in equation (10.188) developed in the energy formulation. The homogeneous solution of equation (10.225) provides the isentropic temperature rise to Hugoniot pressure P. The inhomogeneous solution accounts for the additional temperature on the Hugoniot.

To carry out a temperature calculation on the Hugoniot an analytic expression of the linear-Z form is used for a representation of the Hugoniot compression. Namely

$$P_h(v) = \frac{K_o}{\alpha}(e^{\alpha(1-v/v_o)} - 1), \tag{10.226}$$

or, for an isobaric representation,

$$v_h(P) = v_o\left(1 - \frac{1}{\alpha}\ln\left(1 + \frac{\alpha}{K_o}P\right)\right). \tag{10.227}$$

The Hugoniot properties are $K_o = \rho_o C_o^2$ and $\alpha = 4S$, where C_o and S are the linear shock-velocity versus particle-velocity properties of the material. Equations (10.226) or (10.227) adequately replicates experimental Hugoniot data, although their representation in U–u space is modestly nonlinear.

Temperature calculations are undertaken for solid tantalum oxide (Ta_2O_5). The material properties $\rho_o = 8200$ kg m^{-3}, $C_o = 3.25$ km s^{-1}, $S = 1.2$, $\Gamma_o = 1.0$ and c_v, or c_p, of 324 J kg^{-1} K adequately characterize the Hugoniot compression and thermoelastic character of tantalum oxide. No attempt is made to account for temperature dependence of the specific heat or latent heat of transformation.

The differential equations for temperature on the Hugoniot, applying the Hugoniot relations from equations (10.226) or (10.227), are either

$$\frac{dT}{dv} + \frac{\Gamma(v)}{v}T = \frac{K_o}{2c_v}\left[\frac{1}{\alpha}(e^{\alpha(1-v/v_o)} - 1) - \left(1 - \frac{v}{v_o}\right)e^{\alpha(1-v/v_o)}\right] \tag{10.228}$$

or

$$\frac{dT}{dP} - \frac{\zeta(P)}{P}T = \frac{1}{2c_p}\left[\frac{v_o}{\alpha}\ln\left(1 + \frac{\alpha}{K_o}P\right) - \frac{v_o}{K_o}\frac{P}{1 + \alpha P/K_o}\right]. \tag{10.229}$$

Either differential equation is readily solved with sensible expressions for the Gruneisen parameter $\Gamma(v)$ or the Rice–Walsh parameter $\zeta(P)$.

The solutions do not, in general, lead to the same temperature on the Hugoniot. A volume-dependent Gruneisen parameter $\Gamma(v)$ or a pressure-dependent Rice–Walsh parameter $\zeta(P)$ are independent assumptions. Each leads to a tractable analytic EOS for the thermo-elastic description of the solid. The two parameters are related through the thermodynamic identity

$$\frac{\zeta}{P} = \frac{\Gamma}{K_s}. \tag{10.230}$$

Equations (10.228) and (10.229) yield identical solutions for the temperature on the Hugoniot if this identity is enforced on the Hugoniot curve.

One assumption leads to tractable expressions for either parameter. Writing

$$\frac{\zeta}{P} = \frac{\Gamma}{K_s} = \frac{\rho\Gamma}{\rho K_s} = \frac{\rho_o\Gamma_o}{Z_s}, \tag{10.231}$$

in the last step the assumption of a constant $\rho\Gamma$ is applied. The inertial modulus, written as

$$\rho K_s = Z_s = \rho_o(K_o + \alpha P), \tag{10.232}$$

leads to the linear-Z form of the EOS. Modest differences between the isentropic and Hugoniot modulus are ignored in equation (10.232). The argument of the integrating factor in equation (10.229) becomes $\zeta(P)/P = \Gamma_o/(K_o + \alpha P)$; that of equation (10.228) becomes $\Gamma(v)/v = \rho_o\Gamma_o$.

The solutions of either equation (10.228) or equation (10.229) lead to the Hugoniot temperature versus pressure curve plotted in figure 10.27. The solutions differ by only a few degrees kelvin at a pressure of 200 GPa. The approximation of the Hugoniot modulus, rather than the isentropic modulus, used in the argument of the integrating factor in equation (10.229) introduces a small error in the solution. The temperature rise on the isentrope is plotted for comparison. The Lindemann melt curve for tantalum oxide is also shown. The Hugoniot and isentropic pressure versus specific volume are provided in the left-hand plot.

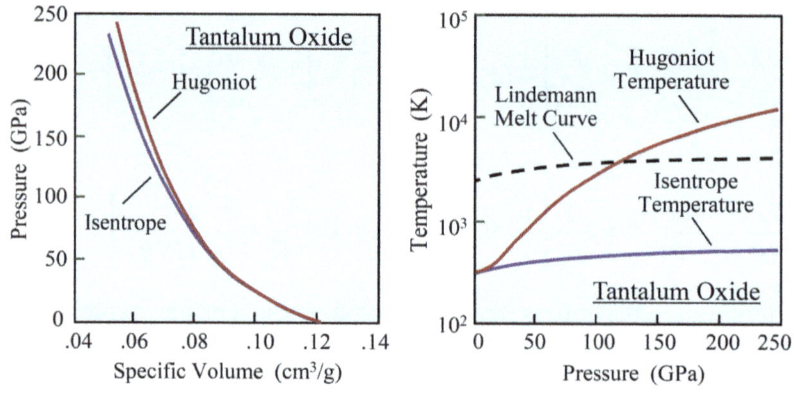

Figure 10.27. Temperatures for tantalum oxide on the principal Hugoniot and isentrope.

10.5 Additive mixture theory and experiment under intense shock compression

Prediction of the mechanical and EOS properties of mixtures of materials from the mechanical and EOS properties of the mixture constituents received considerable attention through the latter few decades of the past century. Estimates of elastic properties based on the methods of Voigt and Ruess, and later improvements of such elastic mixture methods such as the theories of Budiansky (1965), are well known. Theories of mixtures based on solid rational mechanics and thermodynamics principles have been pursued in some detail. The review article by Bedford and Drumheller (1983) provides an excellent entry to this approach.

Here the focus is on experiment and theory directed towards the more extreme shock pressure EOS of mixtures. Efforts towards these ends appear to have been initiated in the Soviet Union in the early 1960s with the study of Dremin and Karpukhin (1960). Extensive experimental activity in that country continued with the works of Altschuler and Sharipdzhanov (1971) and of Kalashnikov *et al* (1973) on geologic mixtures of silicate and carbonate rocks.

Comparable studies in the US within the same era were more diverse in motivation. Extensive high-pressure Hugoniot data on rock and mineral materials were produced at LANL. These earlier data are most readily accessed through the LANL shock Hugoniot compendium (Marsh 1980). Additive mixture theories were pursued (e.g. Anderson *et al* 1966, Riney *et al* 1970). Perhaps unique are the studies of Hugoniot mixtures using the Anderson seismic parameter in assessing mineral composition of the Earth's interior (e.g. Anderson 1967, Davies and Gaffney 1973).

Additive mixture methods to assess the mechanical crush and EOS response of powders are central to the p-lambda computational compaction model described in chapters 6 and 7. The range of applicability and the accuracy of additive constitutive models must be established through experimental testing. Earlier shock Hugoniot data are available that can be used to investigate additive mixture theories. The present section examines some of these earlier shock data.

The application of additive models to extreme shock compaction of material mixtures requires a level of understanding of the physics and the physical processes experienced by material in the shock compression environment. Issues of nonequilibrium pressure and temperature, and the possible transmutation of chemical species, in the intense shock compaction event are of concern to additive mixture theories. Changes in chemical state relate to the intimacy of initial mixing and available paths for transmutation on the time scale of the shock compaction event. These physical issues and the applicability of additive mixture models are examined here within the context of these earlier seminal shock compression data.

The first several subsections address several levels of mixture EOS formulations for application in the shock event, along with experimental studies undertaken for purposes of testing the mixture theories. Application of mixture theories are examined on both the granular and the molecular scale. In the latter subsections techniques are explored, whereby shock test data on mixtures are used to extract dynamic compression properties of one component through subtraction of the known compression properties of the companion component.

10.5.1 Shock equation of state through additive mixture models

The constitutive response of a mixture of two or more solid materials with disparate thermo-mechanical properties is dependent on the strength of the mixture constituents. Material strength will affect the stress state both on the continuum scale and at the granular scale of the mixture. The strengths of all materials are finite, however, and in the intense shock compression of a material mixture, strength becomes an increasingly less important feature of the constitutive response. In the extreme pressure shock compression regime of solid mixtures, to a reasonable approximation, only the fluid, or non-strength, thermodynamic properties of the mixture and mixture constituents need be considered. Of concern to the EOS response are the changes through the shock wave of the intensive properties of pressure and temperature and the extensive properties of specific volume and entropy.

Dissipation and entropy production in the mixture material during passage of a strong structured shock wave is a turbulent and heterogeneous process. Pressure and temperature equilibration following passage of the shock wave are not immediately achieved. Pressure equilibration is delayed in most applications by a few stress wave transits of the mixture granular structure. Equilibration of temperature within the granular microstructure is more problematic, however, and is achieved somewhat later. Length scales over which thermal transport necessary for thermal equilibration proceeds are determined by both initial mixture granularity and by the evolving deformation microstructure responsible for the requisite dissipation and entropy production. The assumption of a shock state of the mixture other than equilibrated temperature is difficult to justify. For example, independent adiabatic compression of each mixture component has been suggested as an alternative bounding response. Following pressure equilibration at the shock state, subsequent temperature equilibration has limited influence on wave propagation. Thus, an assumption of pressure and temperature equilibrium in the post-shock state is the most expedient, and probably the most sensible.

Additive mixture modeling assumes that mixture constituents remain chemically inert. That is, components of the composite do not mix and undergo a chemical change of state within the shock wave. The issues concerning whether shock EOS properties of the components of a chemical species can be additively joined to provide shock properties of the chemical have previously been addressed. One example described here is whether the shock Hugoniot properties of calcium oxide (CaO) and carbon dioxide (CO_2) will yield through additive mixture theory the Hugoniot EOS properties of calcium carbonate ($CaCO_3$). Although exceptions are found, such EOS mixing on a molecular scale is generally not appropriate.

10.5.2 Additive mixtures modeling through Gibbs energy

For a solid mixture for which material strength can be sensibly ignored and that undergoes shock compression to a uniform state of pressure and temperature through a steady structured shock wave, additive EOS mixing is expressed through the thermodynamic Gibbs energy (Duval and Taylor 1971)

$$G(P,T) = \lambda_1 G_1(P,T) + \lambda_2 G_2(P,T). \tag{10.233}$$

Mass fractions of the mixture constituents are λ_1 and λ_2 with $\lambda_1 + \lambda_2 = 1$. Extension to three or more mixture components is apparent. Specific volume and entropy are correspondingly additive,

$$\left.\frac{\partial G}{\partial P}\right|_T = v(P,T) = \lambda_1 v_1(P,T) + \lambda_2 v_2(P,T), \tag{10.234}$$

$$-\left.\frac{\partial G}{\partial T}\right|_P = S(P,T) = \lambda_1 S_1(P,T) + \lambda_2 S_2(P,T). \tag{10.235}$$

Second derivatives of the Gibbs energy provide additive relations for the three fundamental thermodynamic properties—compressibility, specific heat and thermal expansion. They are for the isothermal compressibility

$$\left.\frac{\partial^2 G}{\partial P^2}\right|_T = \beta v = \lambda_1 \beta_1 v_1 + \lambda_2 \beta_2 v_2, \tag{10.236}$$

the specific heat at constant pressure

$$-T\left.\frac{\partial^2 G}{\partial T^2}\right|_P = c_P = \lambda_1 c_{P1} + \lambda_2 c_{P2} \tag{10.237}$$

and the coefficient of thermal expansion

$$\frac{\partial^2 G}{\partial T \partial P} = \alpha v = \lambda_1 \alpha_1 v_1 + \lambda_2 \alpha_2 v_2. \tag{10.238}$$

Hugoniot properties of the mixture are obtained by joining the Rankine–Hugoniot energy relation,

$$E = E_o + \frac{1}{2}(P + P_o)(v_o - v), \tag{10.239}$$

with the first and second law of thermodynamics,

$$dE = \frac{1}{2}(v_o - v)dP - \frac{1}{2}(P + P_o)dv = TdS - Pdv. \tag{10.240}$$

The differentials dS and dv are eliminated through derivatives with respect to P and T. The resulting differential equation provides the mixture Hugoniot in the pressure–temperature plane.

10.5.3 Mixture modeling through additive Hugoniot relations

The determination of shock EOS properties through the Gibbs energy formalism is tractable in principle. In practice, a less rigorous approach through direct additivity of the principal Hugoniot curves of the mixture constituents has shown reasonable success (e.g. Dremin and Karpukhin 1960). In this additive mixture method, the relation corresponding to equation (10.234) directly adds the component Hugoniot specific volumes,

$$v_h(P) = \lambda_1 v_{h1}(P) + \lambda_2 v_{h2}(P), \tag{10.241}$$

to provide the Hugoniot for the two-component mixture. A similar relation can be written for the Hugoniot energy. Temperature differences associated with the component Hugoniot curves are ignored at this level of additive mixture modeling. Because of the lack of thermodynamic rigor, experimental testing is warranted to establish the applicability of equation (10.241).

Alternatively, one can establish reference isentropes for the components and through additive mixing with a relation comparable to equation (10.241) determine an isentrope for the mixture. Then, through appropriate averaging of the component Gruneisen parameters, determine a Gruneisen parameter for the mixture. This approach is more appropriate when a degree of porosity must be accounted for in calculating the Hugoniot response of the mixture.

Experimental testing of additive mixture theories to the shock compression environment has not been extensive—especially in the extreme shock environment where such testing is particularly demanding. Experimental Hugoniot studies performed by several institutes in the Soviet Union in the 1960s and 1970s provide some of the best data available for testing additive mixture theories.

These early data are not generally well known. The shock data are largely for geologic-related rocks and minerals and reflect the emphasis on nuclear ground shock effects that were of concern during that era. The authors of the studies tested additive mixture theories with the shock Hugoniot data acquired by them. The same data are presented here and subjected to further analysis regarding the applicability of additive mixture theories. Comparable experimental data acquired in the US are similarly examined.

10.5.3.1 Mixtures of calcium carbonate (calcite) and paraffin

Possibly the seminal experimental investigation of additive mixture modeling of shock Hugoniot EOS data was performed by Dremin and Karpukhin (1960). The authors of the study provide shock Hugoniot data for both calcium carbonate (calcite in the form of metamorphic marble) and paraffin. Additive mixture theory was then tested through shock Hugoniot experiments on a mixture of 25%–75% by weight paraffin and calcite. In the same study, the authors also investigated mixtures of quartz and Wood's metal, and mixtures of quartz and paraffin. The present re-examination of the data focuses on the investigation of the shock properties of calcite and paraffin, and mixtures of the two. Both shock-velocity versus particle-velocity data, and Hugoniot pressure versus relative compression data from the study of Dremin and Karpukhin (1960) are shown in figure 10.28.

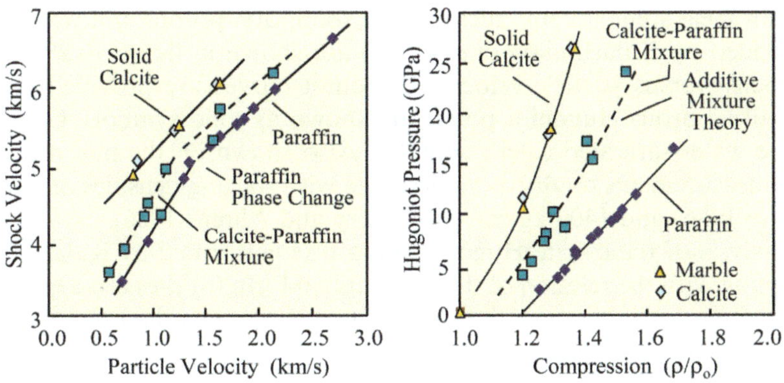

Figure 10.28. Hugoniot data for calcite, paraffin and a 25%–75% by weight mixture of calcite and paraffin (Dremin and Karpukhin 1960). Initial density for paraffin is $\rho_o = 910$ kg m^{-3} and for calcite is $\rho_o = 2710$ kg m^{-3}. The comparable Hugoniot points identified by open diamonds are for single-crystal calcite (Grady 1986). The dashed line is the predicted Hugoniot response based on additive mixture theory.

The shock data of Dremin and Karpukhin (1960) examine Hugoniot compression of the materials to several tens of GPa. Shown on the left are shock-velocity versus particle-velocity measurements for calcite, paraffin and the calcite–paraffin mixture. Hugoniot measurements for single-crystal calcite from a later study (Grady 1986) are compared with the earlier data of Dremin and Karpukhin for marble. The three measurements closely match the earlier data.

Hugoniot data for paraffin reveals a marked phase transformation on the Hugoniot at about 6 GPa. This same transformation is evident in the experimental Hugoniot for the calcite–paraffin mixture. The same data are plotted in the Hugoniot pressure versus relative compression plane on the right in figure 10.28. Comparison of the additive mixture model from equation (10.241) is observed to be in satisfactory agreement with the measured Hugoniot for the calcite–paraffin mixture. The Hugoniot data for paraffin can be compared with that of Marsh (1980) discussed later in the present section.

10.5.3.2 Porous calcium carbonate (calcite) and water

One of the more extensive investigation of the extreme shock compression of solids undertaken to test additive mixture theory is that of Kalashnikov *et al* (1973). This study is focused on carbonate minerals, both in the dry–porous and water-saturated states. Hugoniot measurements are provided for calcite ($CaCO_3$) in the solid state (limestone with initial density of 2665 kg m^{-3}, slightly less than the ambient calcite crystal density of 2710 kg m^{-3}) and porous calcite (chalk) prepared as compacts of calcite powder with porous initial densities of 2020 kg m^{-3} and 1700 kg m^{-3}. In one further series, porous calcite fully saturated with water to an initial density of 2200 kg m^{-3} was tested. The water content in this material was 13.6% by weight. The authors also performed an extensive series of Hugoniot measurements on the calcite–dolomite–magnesite carbonate solid solution series. The Hugoniot properties of the various carbonate minerals are provided to pressures in excess of 100 GPa.

Hugoniot measurements for calcite in the solid, dry–porous and water-saturated states provided by Kalashnikov *et al* (1973) are shown in figure 10.29. The fan of shock-velocity versus particle-velocity Hugoniot curves are provided on the left. Solid and dry–porous Hugoniot points are shown as solid symbols. Open symbols identify the water-saturated calcite samples. Also shown on the plot are additional Hugoniot measurements on dry–porous calcite with starting densities of 1370 kg m^{-3} (Grady *et al* 1986) and 1400 kg m^{-3} (Tyburczy and Ahrens 1986).

EOS analysis of the calcite Hugoniot data was performed by Kalashnikov *et al* (1973) starting with the relation of the Murnaghan form for the zero degree isotherm (an adequate approximation for the ambient temperature isotherm over the range of Hugoniot temperatures achieved),

$$P_c(\sigma) = \frac{\rho_o C_o^2}{n}(\sigma^n - 1), \tag{10.242}$$

where $\sigma = \rho/\rho_o$ is the compression.

Equation (10.242) follows from the expression for the cold lattice energy,

$$E_c(\sigma) = E_o + \frac{C_o^2}{n}\left(\frac{\sigma^{n-1}}{n-1} + \frac{1}{\sigma} - \frac{n}{n-1}\right). \tag{10.243}$$

From equation (10.243), the initial bulk modulus K_o and pressure derivative K_o' are

$$\rho\frac{dP_c}{d\rho}\bigg|_{\sigma=1} = K_o = \rho_o C_o^2, \quad \frac{dK}{dP_c}\bigg|_{\sigma=1} = K_o' = n. \tag{10.244}$$

The principal Hugoniot is obtained through the Mie–Gruneisen EOS

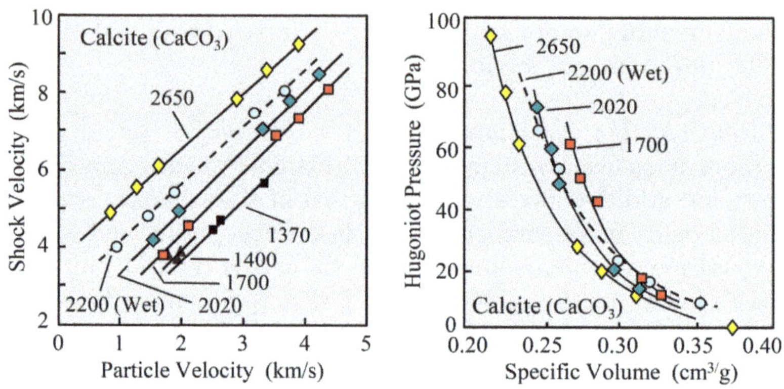

Figure 10.29. Principal shock Hugoniot data for solid, dry–porous and water-saturated calcium carbonate (calcite) (Kalashnikov *et al* 1973). The numbers identify the initial density of the test material. Ambient crystal density of calcite in the calcite I phase is 2710 kg m^{-3}. Water-saturated porous calcite (13.6% by weight) is identified by open symbols and a dashed curve. Square and triangle symbols are shock data of Grady *et al* (1986) and Tyburczy and Ahrens (1986), respectively.

$$P_h(\sigma) = P_c(\sigma) + \rho\Gamma(E_h(\sigma) - E_c(\sigma)), \tag{10.245}$$

and the Rankine–Hugoniot energy relation

$$E_h(\sigma) = E_o + \frac{1}{2}P_h(\sigma)(v_{oo} - v) = E_o + \frac{1}{2}\frac{P_h}{\rho_o}\left(\alpha - \frac{1}{\sigma}\right), \tag{10.246}$$

where $\alpha = \rho_o/\rho_{oo}$ is the initial distention of the porous solid. For the present porous calcite $\rho_o = 2710\ \mathrm{kg\ m^{-3}}$ is the crystal density while $\rho_{oo} = 1700\ \mathrm{kg\ m^{-3}}$ and $\rho_{oo} = 2020\ \mathrm{kg\ m^{-3}}$ are the densities of the calcite powder compacts.

Joining equations (10.242), (10.243), (10.245) and (10.246) yields the principal Hugoniot relation,

$$\begin{aligned} P_h(\sigma) = {} & \frac{\rho_o C_o^2}{n}(\sigma^n - 1) \\ & + \rho\Gamma\left[\frac{1}{2}\frac{P_h(\sigma)}{\rho_o}\left(\alpha - \frac{1}{\sigma}\right) - \frac{C_o^2}{n}\left(\frac{\sigma^{n-1}}{n-1} + \frac{1}{\sigma} - \frac{n}{n-1}\right)\right]. \end{aligned} \tag{10.247}$$

The assumption that the Gruneisen $\Gamma = \Gamma_o$ is constant, and solving for the Hugoniot pressure, yields

$$P_h(\sigma) = \frac{\rho_o C_o^2}{n(h_\infty - \alpha\sigma)}\left[\left(h_\infty - \frac{n+1}{n-1}\right)\sigma^n + \frac{2n}{n-1}\sigma - (h_\infty + 1)\right], \tag{10.248}$$

where $h_\infty = \rho_\infty/\rho_o = (\Gamma_o + 2)/\Gamma_o$ is the limiting compression on the Hugoniot for the fully dense solid. Limiting compression of the porous solid is reduced by the distention factor $\sigma_\infty = h_\infty/\alpha$. A comparable analytic expression for the Hugoniot pressure is obtained assuming that $\rho\Gamma = \rho_o\Gamma_o$ is constant in the Gruneisen EOS. The calculation also assumes the same reference energy E_o in equations (10.243) and (10.246). This energy is, for example, the lattice binding energy in equation (10.243). The authors show that equation (10.248) provides a quite satisfactory representation of both the solid and dry–porous calcite shock Hugoniot data.

Hugoniot data for the calcite–water mixture are re-plotted in figure 10.30, and compared with the Hugoniot for solid calcite (limestone with initial density of $2665\ \mathrm{kg\ m^{-3}}$) and water (density of $998\ \mathrm{kg\ m^{-3}}$). The additive Hugoniot model from equation (10.241) is tested through additive mixing of the solid calcite and water Hugoniot curves. Additive mixing although not unreasonable, modestly under predicts Hugoniot densification of the mixture. Within the lower pressure range (10–20 GPa), one explanation rests with differences in the known shock-induced phase transformation in this pressure range between the solid and the water-saturated calcite (e.g. Grady 1986). Similar disparities are not, however, observed for the previous calcite–paraffin mixture of Dremin and Karpukhin (1960) over the same pressure range. Disparity with the high-pressure data (50–70 GPa) is address in more detail in the next several subsections.

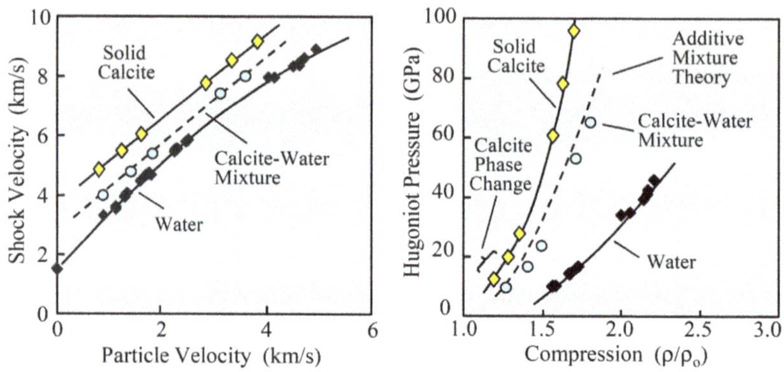

Figure 10.30. Shock Hugoniot data for solid calcite (limestone 2665 kg m^{-3}), water-saturated calcite (2200 kg m^{-3}) and water (998 kg m^{-3}). Water-saturated porous calcite is 13.6% by weight. The dashed curves provide predictions of water-saturated calcite Hugoniot based on additive mixture theory. The calcite Hugoniot data is that of Kalashnikov *et al* (1973). The water Hugoniot is from Marsh (1980).

10.5.3.3 Rock-simulating grout and water

A rock-simulating grout was prepared for application in a nuclear test dubbed Mini-Jade (MJ) conducted at the Nevada Test Site (NTS), Nevada. MJ grout is a high quartz-content material, carefully prepared and characterized for purposes of post-test analysis of shock coupling and shock wave propagation. The EOS and dynamic-strength characterization of MJ grout includes shock Hugoniot measurements over the pressure range of 5–70 GPa (Grady and Furnish 1988, 1990). Hugoniot measurements on MJ grout are appropriate for assessment of the application of additive mixture modeling in the development of shock EOSs.

The constitution of MJ grout is by weight 63% SiO_2 (α-quartz), 22% H_2O, 8% CaO, 3% Al_2O_3 and less than 1% contributions of several additional oxides. The initial test density was 2000 kg m^{-3}. The test samples were fully water saturated.

Constituent compression properties appropriate to application of additive mixing analysis are shown in figure 10.31. Component compressibility is shown through dependence of the Hugoniot pressure on the relative density ρ/ρ_0. Compressibility of quartz is displayed in two parts. That from 30–80 GPa is an optimum fit of the high-pressure Hugoniot data (above the α-quartz to stishovite phase transformation) from both Marsh (1980) and Trunin *et al* (2001) on full-density quartz. Compression for quartz to 20 GPa (below the (α-quartz to stishovite phase transformation) is from static hydro-compression of Angel *et al* (1997). Hugoniot compression relations of aluminum oxide and calcium oxide are from Marsh (1980). Hugoniot compression for water is principally from Marsh (1980) and Trunin *et al* (2001). Sensitivity studies demonstrate that compression of the mixture is dominated by that of quartz and water.

Shock Hugoniot data for MJ grout over the pressure range of 5–70 GPa are compared with the component compression relations in figure 10.31. The Hugoniot data in the lower pressure range were acquired with a single-stage powder gun, whereas higher pressure data relied on two-stage, light gas-gun techniques. Increased

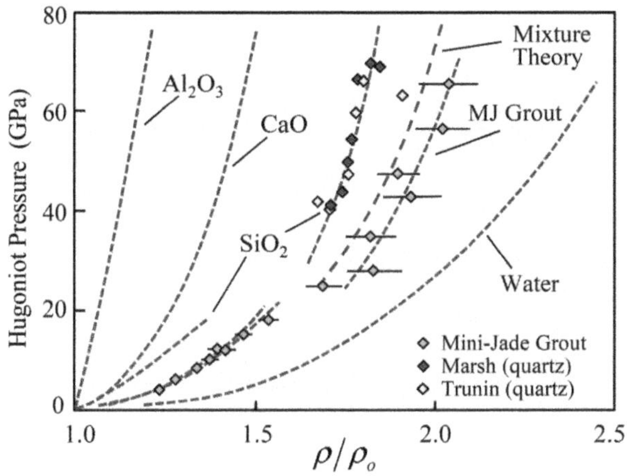

Figure 10.31. Hugoniot measurements on rock-matching MJ grout (Grady and Furnish 1990). Hugoniot compression are shown for component materials. High-pressure quartz data from Marsh (1980) and Trunin *et al* (2001).

scatter in the higher pressure data are a consequence of material variability and the smaller sample size necessitated by two-stage gas-gun technology.

Additive Hugoniot mixing based on equation (10.241) extended to a four component system provides the Hugoniot mixture compression relation shown with the Hugoniot data in the figure. Although Hugoniot data uncertainties negate definitive conclusions, systematic differences with the measured Hugoniot response, at least in the higher pressure range, suggest under prediction of the mixture compressibility by five to ten percent. Although Al_2O_3 and CaO are accounted for in the mixture analysis, their contribution is negligible and the test material is adequately described as a quartz and water mixture.

Various explanations can be offered for the discrepancy between the test data and the predictions of mixture theory based on additive component Hugoniot compression, including systematic uncertainties in material constitution and Hugoniot test methods. One physics-based explanation stands out and warrants further exploration. Application of the principal Hugoniot for water in the additive mixing may be increasingly inappropriate at the higher pressure range of the data. The intense heating of water in single-shock compression and the concomitant Hugoniot compressibility of water may differ from the compression of water to pressure in shock compression of the mixture. Further consideration requires a closer look at the shock EOS of water.

10.5.3.4 Equation of state of water

Shock Hugoniot data to in excess of 100 GPa is available for water. Hugoniot data are in both the LLNL (van Thiel *et al* 1977) and LANL (Marsh 1980) shock wave compendia. Hugoniot data for water from Soviet studies are provided in Trunin *et al* (2001). These data are supplemented with measurements of Skidmore and Morris (1962), and Mitchell and Nellis (1982). Double shock measurements for water were

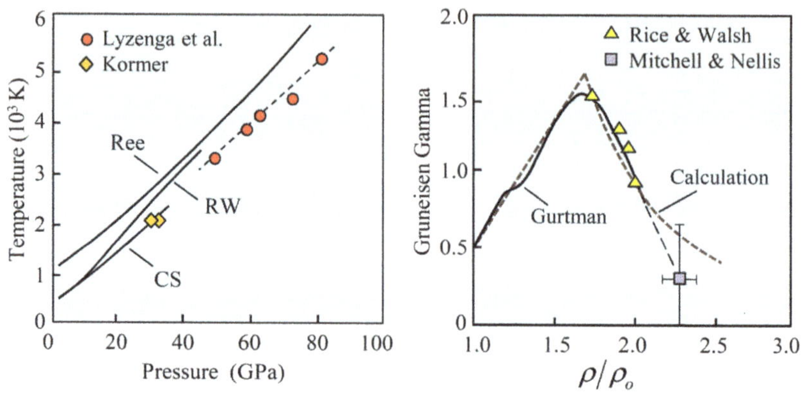

Figure 10.32. Temperature (left) and Gruneisen parameter (right) on the Hugoniot of water. Temperature measurements are from Lyzenga *et al* (1982) and Kormer (1968). Temperature calculations are those of Rice and Walsh (1957), Cowperthwaite and Shaw (1970) and Ree (1982). Gruneisen parameter measurements from double shock experiments of Rice and Walsh (1957) and Mitchell and Nellis (1982). The theoretical curve is from Gurtman *et al* (1971).

obtained by Walsh and Rice (1957). These Hugoniot data are shown later in figure 10.33 and are reasonably consistent below about 50 GPa with increasing scatter at higher pressures.

The unusual thermal properties of water through the higher Hugoniot pressure range of interest are illustrated in figure 10.32. Direct measurements of temperature through optical pyrometry are reported by Lyzenga *et al* (1982). Temperatures spanning Hugoniot pressures from 50 to 80 GPa range over 3300–5300 K. Temperatures are consistent with lower pressure measurements of Kormer (1968). Experimental measurements are compared with temperature calculations of Rice and Walsh (1957), Cowperthwaite and Shaw (1970) and Ree (1982).

The Gruneisen parameter for water increases with density from about 0.5 at ambient to about 1.5 at a compression near $\rho/\rho_o = 1.7$ (Hugoniot pressure of about 10 GPa). The physics underlying the Gruneisen properties of water is explored by Gurtman *et al* (1971). Double shock Hugoniot experiments (Walsh and Rice 1957) reveal a subsequent reduction of the Gruneisen parameter with density. One further double shock experiment at $\rho/\rho_o = 2.3$ (Hugoniot pressure of about 50 GPa) supports further reduction with increasing density (Mitchell and Nellis 1982). Steinberg (1987) develops and applies a Mie–Gruneisen EOS for water for purposes of explosive shock loading, based on the earlier work of Gurtman *et al* (1971).

Description of the Gruneisen parameter dependence on density allows for estimating the ambient isentrope of water through the Hugoniot pressure ranged achieved in the MJ grout shock test series. In particular the dashed curve approximation of the Gruneisen parameter on the right in figure 10.32 provided by the straight line from ambient density to a compression of 1.7 and a power relation of $\Gamma(\rho) = 10/(\rho/\rho_o)^{-3.5}$ for $\rho/\rho_o > 1.7$ is used for the calculation. The resulting ambient isentrope is shown with the Hugoniot data for water in figure 10.33. Additive mixing through equation (10.241) using the ambient isentrope for water

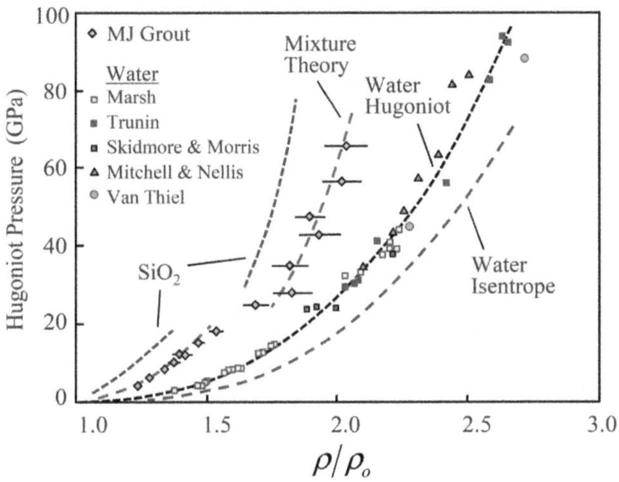

Figure 10.33. Hugoniot and isentrope EOS of water. The latter is based on the Mie–Gruneisen EOS and density-dependent Gruneisen parameter. The mixture calculation of the MJ grout Hugoniot is based on the isentropic compression of water. The references for the water Hugoniot data are provided in the text.

rather than the Hugoniot produces remarkably good agreement with the MJ grout data.

The calculation and comparison suggests that dissipation and temperature rise in the water component of the MJ grout mixture subject to steady shock compression is appreciably less than that of water subject to single-shock compression to the same Hugoniot pressure. It is perhaps more realistic to expect the water in the mixture to achieve pressure through several shock steps resulting in markedly lower temperatures. Nonetheless the mixture is subject to steady shock compression with concomitant Hugoniot dissipation. Partitioning of dissipation among the mixture components is complex, but intuition suggests that much of the dissipation is still accommodated by the water. Reassessment of the several calcite-water high-pressure Hugoniot points of Kalashnikov *et al* (1973) in figure 10.29 based on isentropic compression of water achieves similar improved agreement.

Comparisons with available shock data of additive mixing of component Hugoniot compressibility at extreme pressures when there are marked differences in component compressibility suggest difficulties. The data on mixtures of rock media and water are illustrative. Other simple additive mixture models are possible. For example, additive mixing of component cold or ambient isentrope curves, combined with an appropriate averaging of component Gruneisen properties, is a sensible alternative. Exploring such alternatives warrants more definitive mixture shock Hugoniot data.

10.5.3.5 Nevada Test Site (NTS) alluvium

Shock Hugoniot data for NTS alluvium provided by McQueen and Marsh (1962) illuminate several issues critical to extreme-pressure EOS modeling of granular mixtures. These Hugoniot data are plotted in figure 10.34. First, the data

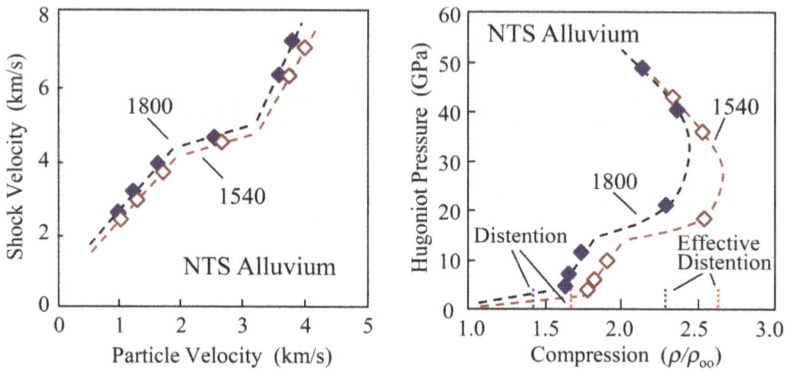

Figure 10.34. Shock Hugoniot data for NTS alluvium (McQueen and Marsh 1962). Hugoniot data are for materials with initial bulk densities of 1540 kg m^{-3} and 1800 kg m^{-3}. Nominal initial density of the fully compacted alluvium is assumed that of alpha-quartz (~2600 kg m^{-3}). Vertical dashed lines on the abscissa identify the magnitude of the initial sample distension of $\alpha \sim 1.4$ and $\alpha \sim 1.7$, and the effective distension, including phase transformation volume, of $\alpha \sim 2.3$ and $\alpha \sim 2.7$.

demonstrate that Hugoniot curves that are multi-valued functions of the compression can be achieved at extreme shock compression when the effects of intense shock heating overwhelm lattice compressibility. Second, phase transformation on the Hugoniot contributes substantially to shock dissipation, and adds to the complexity of the Hugoniot EOS analysis.

Alluvium is a silt, sand and gravel composite. At the NTS, alluvium is composed principally of oxide minerals. Approximately 70%–75% SiO_2 and 10%–15% aluminum oxide with percentage levels of other oxides is representative of the NTS alluvium (Bass *et al* 1963). Two initial sample densities of 1540 kg m^{-3} and 1800 kg m^{-3} were tested in the study of McQueen and Marsh (1962). The data are also provided in the LLNL compendium (Marsh 1980). Assuming an average crystal density of about 2600 kg m^{-3}, comparable to SiO_2 alpha-quartz, suggests initial bulk distension of about $\alpha \sim 1.7$ and $\alpha \sim 1.4$ with $\alpha = \rho_o/\rho_{oo}$ for the two alluvium materials.

Silicon dioxide, the principal component of the present alluvium, undergoes a phase transformation with large volume change on the Hugoniot in the pressure range of 15–30 GPa (Wackerle 1962). This shock-induced phase change is evident in both the shock-velocity versus particle-velocity and pressure versus compression plots in figure 10.34. (Note the abscissa on the right is densification relative to initial distended density.) SiO_2, with oxygen in four-coordination with silicon, transforms to a collapsed lattice structure with oxygen in six-coordination with silicon (stishovite). The ambient density of the high-pressure phase is about 4100 kg m^{-3} (Stishov and Popova 1961). Under extreme shock compression, shock compaction associated with both transformation volume (phase change) and void volume (porosity) are sources of comparable dissipation and entropy production in the wake of a strong shock wave. Assuming that 4100 kg m^{-3} is representative of the nominal alluvium crystal density accounting for the transformation volume, the effective distension of the two materials is increased to about $\alpha \sim 2.7$ and $\alpha \sim 2.3$, respectively. The actual distension, and the effective distension including the transformation volume, are identified for purposes of comparison with the shock compression in the plot of pressure versus compression in figure 10.34. The decompression trending (double-valued) Hugoniot for alluvium with increasing Hugoniot pressure over the range of about 30–50 GPa is a consequence of the extensive shock dissipation brought about by both void collapse and lattice collapse (phase transformation).

10.5.4 Additive Hugoniot mixing at the molecular level

The expansion of shock EOSs to a wider range of materials through EOS mixing theories has practical applications. Efforts towards that goal, however, have been mixed (no pun intended) when such theories have been applied to material mixtures at the molecular level. The chemistry necessary to form particular molecular species from its molecular components may not occur within the time durations, deformation history and thermodynamic conditions imposed by a strong shock wave.

Several examples of the application of additive mixture theory to solids on the molecular level are considered. Again, the shock Hugoniot studies of Kalashnikov *et al* (1973) provides the experimental data for the discussion. Calcite, dolomite, and magnesite are unique molecular structures of the (Mg,Ca)CO_3 carbonate solid-solution series. The Hugoniot data from any two, in principle, can be used to predict the Hugoniot response of the third through additive mixing. The same authors also provide Hugoniot data for calcium oxide (CaO) and CO_2 (dry ice). Such data, in principle, can be applied to predict the Hugoniot response of calcite ($CaCO_3$) through the additive mixture model.

10.5.4.1 Calcite–dolomite–magnesite carbonate minerals

Calcite ($CaCO_3$) and magnesite ($MgCO_3$) are end members of the (Mg,Ca)CO_3 solid solution series. Near 50–50 atomic percent calcium–magnesium, the mineral dolomite forms with Ca and Mg atoms at alternate lattice sites. The Hugoniot mixing rule can be applied to determine if the shock compression of dolomite is adequately described by the appropriate weighted volume average of the Hugoniot curves for calcite and magnesite.

The shock Hugoniot data for calcite, dolomite and magnesite from Kalashnikov *et al* (1973) are shown in figure 10.35. A calculation of the dolomite Hugoniot compression based on the additive Hugoniot rule from equation (10.241) is plotted as the dashed curve on the right in the figure. The prediction is reasonable although some separation of calculation and experiment is noted towards the higher pressure range. The mixture calculation may not capture high-pressure phase transition complexities of the dolomite mineral. Phase transformation in dolomite is observed through more detailed shock wave profile measurements (Grady *et al* 1976), and more recent static high-pressure studies (Santillán *et al* 2003) reveal complex phase structures at pressure in this mineral. Disturbing differences in EOS properties of this mineral as assessed through shock wave and static DAC studies are reported in Grady (2015a). The same EOS issues of dolomite are explored further in chapter 11.

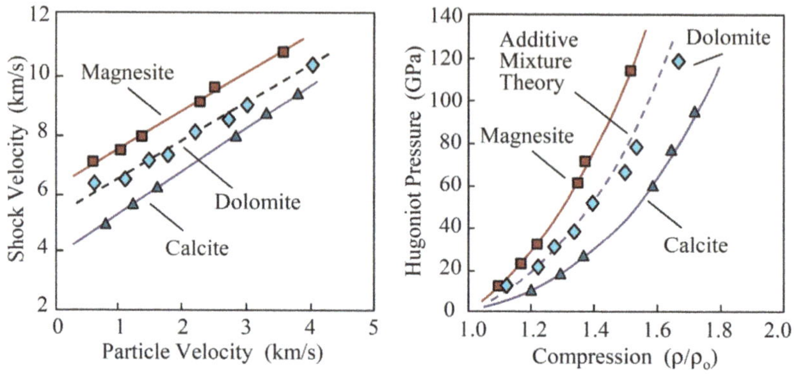

Figure 10.35. Shock Hugoniot data for calcite, dolomite and magnesite carbonate minerals (Kalashnikov *et al* 1973). Initial densities are for calcite (2665 kg m^{-3}), for dolomite (2840 kg m^{-3}) and for magnesite (2975 kg m^{-3}). Prediction of additive Hugoniot mixing is shown as the dashed curve.

10.5.4.2 Calcite, calcium oxide and dry ice

A further application of additive mixing by Kalashnikov *et al* (1973) is that using the measured Hugoniot compression curves of calcium oxide (CaO) and carbon dioxide (CO_2) to predict the Hugoniot compression of calcite ($CaCO_3$). This calculation, using the additive Hugoniot rule, is shown in figure 10.36 and is shown to provide a quite poor result.

The chemical bonding, associated with formation of the calcite crystal structure, results in a starkly smaller specific volume. The Hugoniot results suggest that little or no chemical transformation occurs within the shock wave. Hugoniot curves for calcium oxide and calcite actually overlie at the higher pressures.

To some extent the present example suffers from the same issues revealed in the quartz–water (MJ–grout) mixture calculation described in an earlier subsection. Namely, temperatures and corresponding Hugoniot specific volumes achieve on the dry ice Hugoniot are probably excessive compared to that achieved in shock compression of a mixture of CO_2 and CaO. Although mixing instead with the dry ice principal isentrope would improve the prediction it would not account for the discrepancy at the highest pressures.

10.5.5 Dynamic isentrope from shock compression of mixtures

A test method only infrequently explored over past decades is the immersed-mixture shock wave experiment for determining dynamic isentropic properties of selected solids. The method uses the same mixture concepts introduced in the previous subsections with an alternative perspective. The immersed-mixture dynamic isentrope experimental method is, in particular, relevant to shock EOS studies of high-strength solids where data on monolithic solids are complicated by material strength issues.

When materials with strength are tested under static high-pressure compression, particulate samples are imbedded in a material identified as the pressure transmitting medium (PTM). A PTM material is selected that exhibits hydrostatic (or nearly

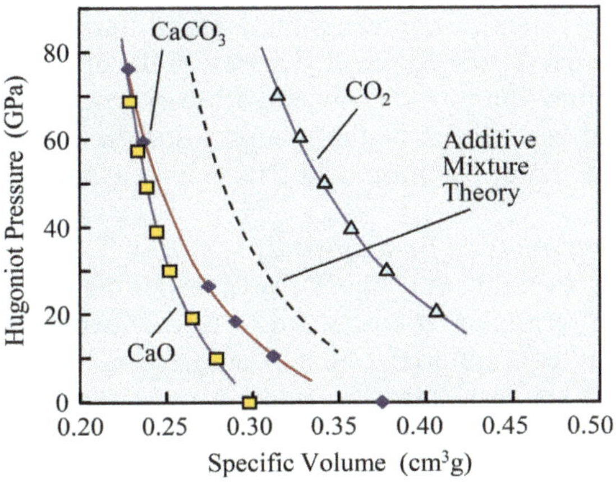

Figure 10.36. Shock Hugoniot data for calcium oxide (CaO), carbon dioxide (CO_2) and calcite ($CaCO_3$) (Kalashnikov *et al* 1973). Additive Hugoniot mixing to predict the calcite Hugoniot illustrates difficulties with the additive model when applied Hugoniot mixing on the molecular level.

hydrostatic) response to compression such that uniform (or nearly uniform) pressure is applied to the test material. The samples under pressure are interrogated by the diagnostics common to static high-pressure physics.

A similar approach intended for the assessment of the dynamic isentropic compression of high-strength materials has been explored (Dremin and Karpukhin 1960, Kanel and Pityulin 1984, Tang and Gupta 1988, Grady and Wise 1993, Grady 1994), although extensive application of the technique is admittedly sparse. The concept, as in the static experiment, is to imbed the material of interest in a PTM that responds as nearly as possible hydrodynamic to the dynamic application of pressure. Thus, the experiment is one of producing a granular mixture of the high-strength material with a ductile low-strength material, such as an appropriate metal or polymer, and subjecting the mixture to a Hugoniot experiment with the appropriate shock wave diagnostics. Dynamic compression of the material of interest is then extracted from experimental shock wave measurements on the mixture through the known thermo-elastic properties of the PTM and appropriate additive mixing analysis of the data.

Several sets of shock Hugoniot measurements performed on granular mixtures of selected compounds with metals and polymers are reported in the LANL shock compendium (Marsh 1980). The data are appropriate for assessing the immersed mixture dynamic isentrope method. The shock data examined here are several. Particularly illuminating are shock data on a tungsten carbide and copper mixture and, to a lesser extent, tungsten carbide mixed with metallic silver and cobalt. Hugoniot data on a granular mixture of boron carbide and copper are also instructive. These unique data provide the opportunity to test the immersed mixture dynamic isentrope shock wave experiment and apply the method to EOS development of high-strength compounds of interest.

Paraffin is a natural choice for use as a PTM as this material offers considerable convenience in the construction and performance of shock experiments.

Accordingly, the LANL shock compendium contains ample test data on materials mixed with paraffin. Here, the required dynamic EOS of paraffin is assessed and shock data are examined on paraffin mixed with periclase and quartz. The section closes with a critical assessment of the complications and the merits of additive mixing EOS analysis and application of the technique as an experimental tool.

10.5.5.1 Concept and analysis fundamentals

The premise of the experimental shock wave experiment is that steady shock compression of the granular mixture leads to isentropic (or near-isentropic) compression of the test material. First, compression of the test component must be adiabatic. This issue must be address on a material-by-material basis. But, for many compounds of interest, thermal diffusivity is appropriately low. As an example thermal diffusivity for boron carbide is approximately $\chi = 25 \times 10^{-6}\ \mathrm{m^2\ s^{-1}}$. A characteristic thermal diffusion time is of order $a^2/2\chi$ whereas risetime of the shock wave in the mixture is of order $2a/U$. For a mixture grain size of say $a = 20\ \mu\mathrm{m}$ a thermal diffusion time of approximately 10 μs and shock risetime of 10 ns differ by three orders of magnitude, rendering the pressure rise of the test compound near adiabatic within the experiment measurement time. Second, the peak compression of the high-strength test material in the mixture is expected to proceed through several reverberations, thus minimizing dissipation and entropy production. The dissipation necessitated by steady shock compression of the mixture is assumed to be concentrated in the lower strength PTM. Although intuitively sensible, the mesoscopic details of this extreme thermo-mechanical dynamics is far from clear and warrants closer investigation. Relative impedances, compressibilities and strengths will play determining roles.

At the pressure-equilibrated Hugoniot state of the mixture, additivity of the specific volume and specific enthalpy provides two relations for the corresponding component properties,

$$v(P) = \mu v_m(P) + (1 - \mu)v_c(P), \qquad h(P) = \mu h_m(P) + (1 - \mu)h_c(P). \quad (10.249)$$

The test PTM mass fraction is μ, while subscripts c and m refer to specific volumes and specific enthalpies of the component test material and PTM, respectively. A third relation is provided by the known EOS $v_m(h_m, P)$ of the PTM. Closure is achieved through the assumption of isentropic compression $dh_c = v_c\, dP$ of the test material.

10.5.5.2 Application to tungsten carbide

Shock Hugoniot data on material mixtures of tungsten carbide combined, respectively, with metallic copper, silver and cobalt are available in the LANL shock compendium (Marsh 1980). In examining the tungsten carbide mixture data for purposes of testing the immersed-mixture dynamic isentrope measurement, several issues are explored. Is the dynamic isentrope for tungsten carbide inferred from the mixture data, when compared with independent Hugoniot data for monolithic tungsten carbide, sensibly consistent? The dynamic isentrope determined from mixture Hugoniot data should be independent of the selected PTM. Is this the case for the data for tungsten carbide combined with copper, with silver and with cobalt? Tungsten carbide and the several metals in the mixture composites have different compliance moduli over the range of the data. Does the microstructure strength and relative crush of the disparate compliance components in the mixture compromise the isentrope analysis?

Consider first the shock Hugoniot data on a mixture of tungsten carbide and copper (Marsh 1980). This Hugoniot set consists of 11 data points spanning a shock pressure range of 10 to 195 GPa. The mixture is a commercially available ceramic–metal composite identified as Elkonite TC10. The mass fraction of copper is 0.44 (volume fraction of 0.57). The reported density of the mixture is 11.65 g cm^{-3}, close to the calculated theoretical density of 11.77 g cm^{-3}. Porosity or impurities account for less than one percent of the material.

The Hugoniot shock-velocity versus particle-velocity data for the Elkonite TC10 spanning the pressure range from 10 GPa to nearly 200 GPa are plotted on the left in figure 10.37. The data are generally consistent, although modest scatter is noted in the lowest pressure Hugoniot points. A linear best fit to the shock-velocity versus particle-velocity data yields intercept $C_o = 3.93$ km s^{-1} and slope $S = 1.55$, although a statistically measurable nonlinear trend is noted in the data.

The immersed mixture analysis outlined in the earlier subsection is applied to the copper and tungsten carbide Hugoniot data. The results of this analysis of the data are shown on the right in the figure. The experimental mixture Hugoniot states provide the central curve in the figure. Using the known EOS for copper the principal dynamic isentrope states for tungsten carbide are calculated from the mixture data.

The dynamic isentrope for tungsten carbide extracted from the Hugoniot data on the copper and tungsten carbide metal–ceramic composite is replotted as a function of the compression strain $1 - \rho_0/\rho$ in the left-hand plot in figure 10.38. The isentrope data are subjected to an optimum nonlinear fit to a reference EOS of the linear inertial modulus form (also referred to as the linear-Z or the modified Murnaghan EOS) introduced earlier in this chapter.

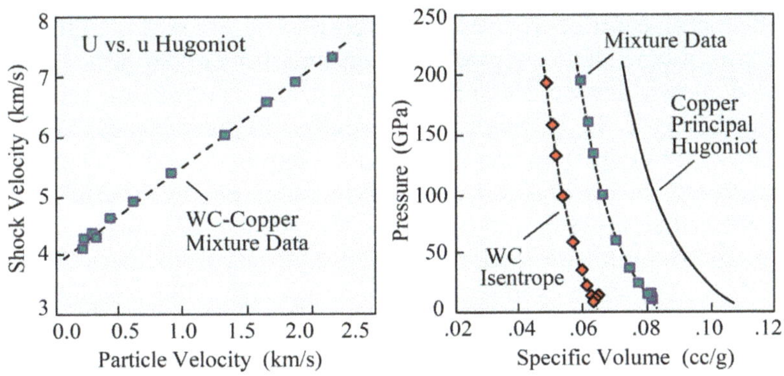

Figure 10.37. Hugoniot data for Elkonite TC10, a tungsten carbide and copper metal–ceramic composite (Marsh 1980). Principal Hugoniot for copper metal, and dynamic isentrope for tungsten carbide from immersed mixture analysis.

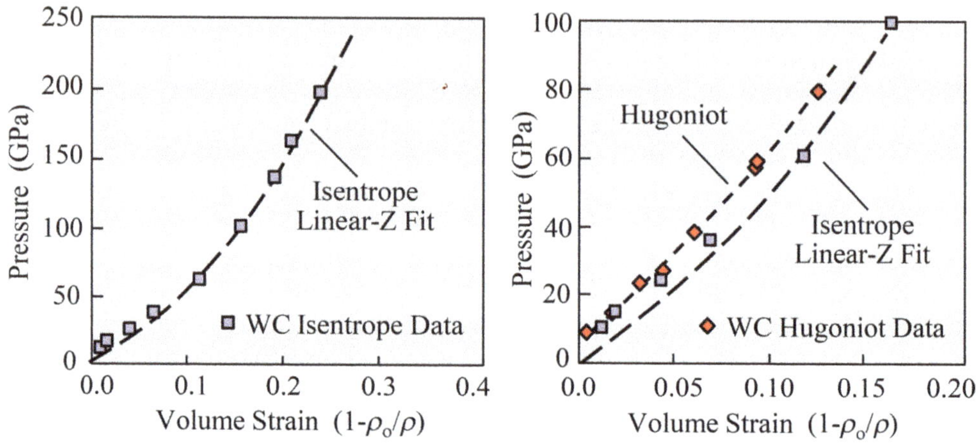

Figure 10.38. Pressure versus volume strain representation of the dynamic isentrope for tungsten carbide. Comparison with principal Hugoniot measurements on tungsten carbide ceramic (Dandekar and Grady 2002, Grady 2009).

Hugoniot data have been obtained on a near-full-density hot-pressed tungsten carbide ceramic manufactured by Cercom, Inc., to approximately 80 GPa (Dandekar and Grady 2002, Grady 2009). These Hugoniot data are plotted on the right in figure 10.38 and compared with the same linear-Z fit to the dynamic isentrope shown in the left-hand plot. The difference in the two curves is identified as a non-hydrodynamic stress offset at the Hugoniot state and provides a measure of strength of the tungsten carbide at pressure. Assessment of strength on the Hugoniot through comparisons with the present dynamic isentrope measurement is in agreement with earlier analysis of Hugoniot strength through alternative methods (Dandekar and Grady 2002).

The pressure states on the dynamic isentrope shown in figure 10.38 in the lower range of 10–40 GPa are observed to lie somewhat above the linear-Z analytic fit to the isentrope. These points have negligible influence on the statistical fit to the data. The data in this range are nonetheless of interest for other reasons. The data in this range for the metal–ceramic mixture from figure 10.37 are replotted in figure 10.39. Shown are the independent compression relations for copper metal and tungsten carbide. Also shown are iso-pressure and iso-strain compression relations appropriate to the Elkonite TC10 mixture of copper and tungsten carbide within the context of a p-lambda crush model described in chapters 6 and 7. There is scatter in the Hugoniot data in this pressure range and assignment of experimental errors is uncertain. Nevertheless, Hugoniot states in the 10–15 GPa range suggest microstructure crush strength effects due to the disparate compliances of copper and tungsten carbide. Evidence of microstructure crush effects is negligible at pressures above about 20 GPa. It is notable that the compliances of copper and tungsten carbide become comparable in the 100–200 GPa shock pressure range.

The dynamic isentrope measurement through the immersed mixture technique should of course be independent of the material selected for the PTM. An observed dependence on the PTM selected can be a consequence of several problems. The

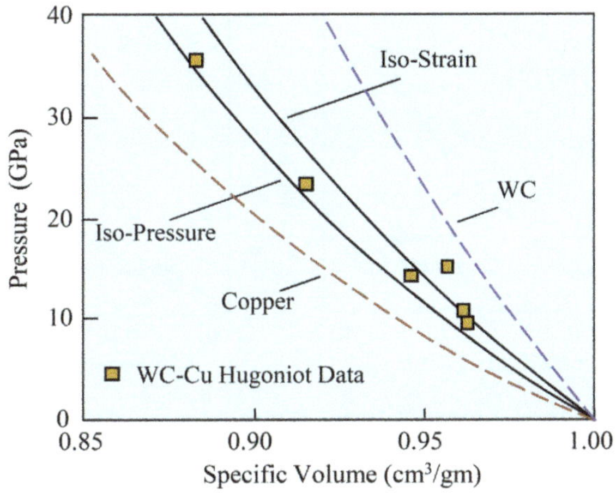

Figure 10.39. P-lambda characterization of the dynamic compression of copper and tungsten carbide composite. Dashed curves identify independent compression of copper and tungsten carbide, whereas solid curves are iso-pressure and iso-strain compression relations specific to the LANL data for Elkonite TC10.

EOS of the PTM may not be adequately known for the application. The structure of the test mixture may not be appropriate in terms of the relative volume fractions of the test material and PTM components. The granular composition of the mixture may not adequately create the near-hydrodynamic compression necessary to the experimental technique.

Hugoniot data for a composite of tungsten carbide and silver metal identified as the commercial product Elkonite G-12 is also provided in the LANL compendium. This metal–ceramic composite is 60 wt% silver (volume fraction of silver is 0.68). Four Hugoniot measurements over the range of 30–220 GPa are provided. The data are subjected to the same dynamic isentrope analysis and are compared with that for the tungsten carbide and copper composite on the left in figure 10.40. The isentrope assessed from the WC–silver composite are in reasonable agreement with that from the WC–copper composite although the four data points are sparse and data scatter is amplified by the large volume fraction of silver.

On the right in figure 10.40 is shown the results for the same dynamic isentrope analysis applied to a cobalt and tungsten carbide composite from the LANL shock compendium. The cobalt metal content in this composite is 5% (volume fraction of cobalt is about 0.08). There is disparity in the comparison of results for the cobalt PTM and that of the copper PTM. A likely explanation is that the fraction of cobalt metal is insufficient to achieve the hydrodynamic compression of the tungsten carbide essential to the dynamic isentrope experiment. This tungsten carbide composite exhibits strength on the Hugoniot much like that of the monolithic tungsten carbide shown in figure 10.38. In addition, the compliance difference (the bulk modulus of tungsten carbide is about twice that of cobalt) would lead to microstructure crush (p-lambda) effects over much of the range of the data. A reasonably uniform pressure state in the tungsten carbide would be unlikely in this range.

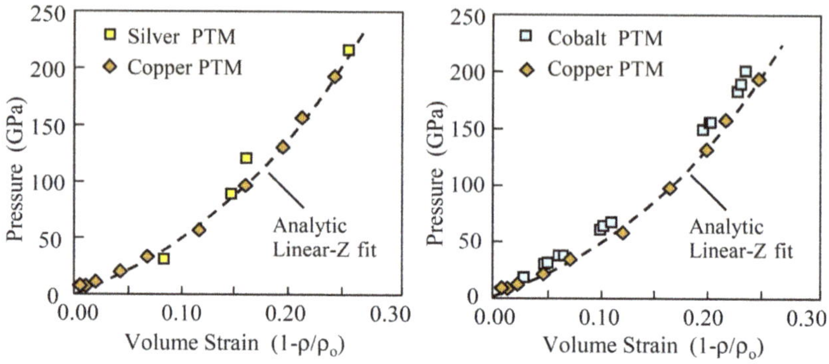

Figure 10.40. Dynamic isentrope assessment from tungsten carbide and metal composites with different (silver, cobalt) metal PTM.

The lesson learned from examination of the several tests is that the relative fraction of the metal PTM must be tailored to optimize the experiment. Too little PTM leads to unacceptable non-hydrodynamic loading of the immersed test compound. Too much PTM, although assuring near-hydrodynamic loading, leads to inadequate accuracy of the EOS test.

Nonetheless, assessment of the dynamic isentrope measurement into the extreme pressure regime through the immersed mixture method can be achieved. Application to Hugoniot data on tungsten carbide and copper mixture demonstrate efficacy of the dynamic isentrope measurement for tungsten carbide to approximately 200 GPa pressure. Further, shock studies on monolithic tungsten carbide demonstrate complications of strength at pressure in assessing EOS strictly from Hugoniot measurements. Thus, pursuit of the dynamic isentrope technique is warranted in the development of extreme pressure EOSs for high-strength materials of interest.

10.5.5.3 Application to boron carbide

Shock Hugoniot measurements performed on a granular mixture of boron carbide and copper are reported in the LANL shock compendium (Marsh 1980). These data were collected in the same time frame as the LANL test data on monolithic boron carbide and the test material is likely from the same supplier. This Hugoniot set consists of 13 data points spanning 15–125 GPa shock pressure. Phase transformation under shock compression is suspected in boron carbide. Further, unexplained discrepancies are noted among the various shock Hugoniot studies on boron carbide (Grady 2010, 2015b).

The immersed mixture assessment of dynamic isentropic compression of boron carbide is applied to the Hugoniot data for the boron carbide and copper mixture. The reported mass fraction of boron carbide and the density of the mixture are 0.272 and 4.84 g cm^{-3}, respectively. The calculated theoretical density of the mixture is 5.29 g cm^{-3}. Specific volumes corresponding to the starting (S.D.) and theoretical densities (T.D.) are identified on the abscissa on the left in figure 10.41. The reported Hugoniot data for the mixture and the known Hugoniot for copper are shown in the plot. The dynamic isentrope is calculated by the methods described. Excess

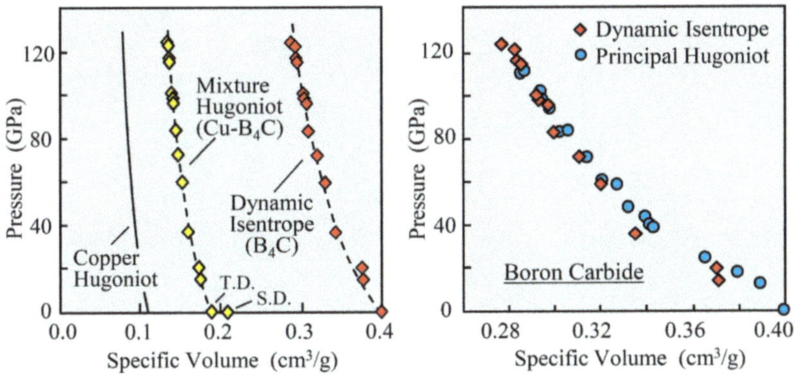

Figure 10.41. Left: Dynamic isentrope of boron carbide from shock compression data on copper–boron carbide mixture material (Marsh 1980). Right: Comparison with the Hugoniot of boron carbide ceramic (3%–7% initial porosity).

dissipation is assigned to the copper in assessing the specific volume of the boron carbide at the Hugoniot pressure state.

The dynamic isentrope for boron carbide determined from the mixture Hugoniot data is compared with the LANL Hugoniot for monolithic boron carbide (Marsh 1980) on the right in figure 10.41. The reported LANL Hugoniot data are for two sets of porous materials with initial starting porosities of 3%–7% and 22%–25%, respectively. Only the Hugoniot data for the higher density ceramic (3%–7% porosity) is shown in the plot. Hugoniot data for the higher porosity ceramic (22%–25% porosity) overlay that of the lower porosity material where the data overlap (see chapter 6).

The dynamic isentrope measurements reside statistically at modestly higher densities than the Hugoniot although convergence of the two is noted at the highest pressures. Uncertainties in the inferred dynamic isentrope are not readily assessed. The principal errors result from inaccuracies in the mixture Hugoniot measurements, and uncertainties in the mixture mass ratio and the EOS of copper. The first accounts for the observed scatter whereas the latter two would entail a measure of systematic error.

The dynamic isentrope assessed from the mixture Hugoniot data is replotted in figure 10.42 as a function of the volume compression $\varepsilon = 1 - \upsilon/\upsilon_o$ with $\rho_o = 2.52 \text{ g cm}^{-3}$ the ambient crystal density of boron carbide. Maximum volume compressions of about 30% is achieved or about a 10% reduction of the lattice dimension. The data are examined with a two-parameter semi-empirical reference EOS of the linear-Z form. An optimum fit (not shown in figure 10.42) to the entire thirteen data points yields $K_o = 182$ GPa and $K_o' = 4.1$. This value of K_o is markedly less than the ambient ultrasonic bulk modulus of 235 GPa for boron carbide (Manghnani *et al* 2000), indicative of marked softening on the dynamic isentrope, most likely phase transformation.

Alternatively, application of a three parameter fit of the reference equation to the higher pressure span of data in figure 10.42 (pressures greater than 35 GPa), omitting the two lower pressure points, and allowing for adjustment of the ambient

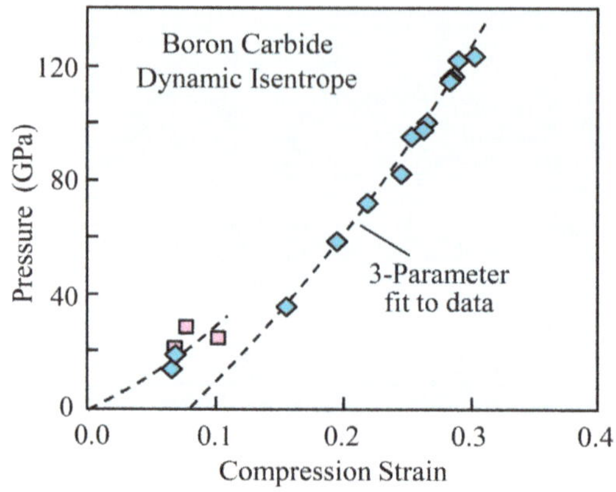

Figure 10.42. Dynamic isentrope for boron carbide and empirical EOS fit to low- and high-pressure data. Diamonds assessed from LANL data (Marsh 1980). Squares from Grady (1994).

density, yields $K_o = 395$ GPa, $K_o' = 1.3$ and an intercept density of 2.72 g cm^{-3}. The fit to the data is shown in figure 10.42. This decomposition of the data is warranted as assessment of available shock and static DAC compression data on boron carbide suggest first-order phase transformation in the 20–30 GPa pressure range (Grady 2010, 2015b). See also further discussions in chapter 6.

The two lowest pressure dynamic isentrope points from the LANL data at 15.0 and 20.8 GPa yield $K_o = 225$ GPa, assuming a reasonable value of $K_o' = 4$, in sensible agreement with the ambient ultrasonic bulk modulus. The data of Grady (1994) in this pressure range, also shown in the plot, were performed for the purpose of assessing dynamic isentrope properties of boron carbide. Although exhibiting more experimental scatter, they are consistent with the trend of the LANL two lower pressure data points.

Again, as in the earlier assessment of tungsten carbide, the calculations for boron carbide lend encouraging support for the dynamic isentrope experiment.

10.5.5.4 Paraffin wax as a dynamic pressure transmitting medium (PTM)

The LANL shock compendium provides shock Hugoniot data for mixtures of paraffin wax with various compounds including magnesium oxide, silicon dioxide and aluminum oxide among others. Paraffin would appear a likely candidate for a known PTM in the immersed-mixture shock wave experiment. The early shock Hugoniot measurements on calcite and paraffin mixtures by Dremin and Karpukhin (1960) is considered in section 10.5.3. The LANL data offer further opportunity to assess the dynamic isentrope measurement on selected compounds through the immersed mixture method. Assessment will here focus on the paraffin mixture with magnesium oxide (periclase) and with silicon dioxide (quartz). Results provide further encouragement for the test method. The exploration, however, brings to light further complications of the analysis method that are necessary to achieving an optimum assessment of the dynamic compression of the test compound.

Reasonable knowledge of the EOS of paraffin is necessary to the analysis extracting dynamic compression of the test material. Hugoniot compression of a commercially available paraffin to approximately 50 GPa is available in the LANL compendium. Hugoniot shock compression of the compound and paraffin mixture involves appreciable shock dissipation. A sensible assumption is to assign dissipation in the shock event to the paraffin. Consequently, knowledge of the thermal expansion, or alternatively, the Gruneisen properties of the paraffin is necessary to the analysis. Thermal properties of commercial paraffin waxes vary widely. A study of thermal expansion and specific heat properties of commercial waxes by Ukrainczyk *et al* (2010) reveals a reasonable correlation with initial density. Comparison with this correlation indicates an ambient Gruneisen gamma of 3.4 is appropriate for the LANL paraffin Hugoniot data. Correction for thermal expansion of the paraffin is further amplified because of several percent porosities in the paraffin–oxide mixtures.

Hugoniot data for paraffin from the LANL compendium are shown in figure 10.43. The LANL Hugoniot data are adequately represented by a linear shock-velocity versus particle-velocity relation. Alternatively, analytic representation of the principle isentrope and Hugoniot in the figure is provided through a Mie–Gruneisen EOS with a linear-Z description of the reference isentrope. The appropriate analytic relations are provided in section 10.2.2. Initial density of 0.917 g cm^{-3} is near the upper end of commercial paraffin densities tested by Ukrainczyk *et al* (2010). Comparison is illustrated with the ambient isentrope based on an initial Gruneisen parameter of $\Gamma_o = 3.4$ and volume dependence assuming $\rho\Gamma$ is equal to a constant. This value of the Gruneisen parameter is in accord with the thermo-physical data for paraffin of Ukrainczyk *et al.* Thermal expansion on the Hugoniot near the upper range of the data is extensive. Assumption of a constant $\rho\Gamma$ volume dependence can certainly be questioned. Applicability of the EOS rests on comparative assessments of the immersed-mixture analysis. Note that the data of Marsh (1980) span the range above the approximately 7 GPa phase transformation observed in the paraffin Hugoniot data of Dremin and Karpukhin (1960) shown in figure 10.28.

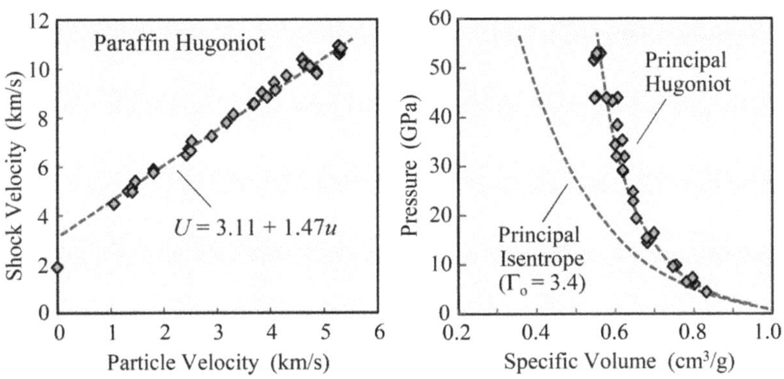

Figure 10.43. Hugoniot data for paraffin (left; Marsh 1980) and comparison with the principal isentrope (right).

10.5.5.5 Application to periclase and quartz

The immersed mixture analysis method is applied to Hugoniot data for the magnesium oxide and paraffin mixture. The set of Hugoniot data consists of seven experimental points spanning 5–50 GPa shock pressure. The reported mass fraction of MgO is 0.842. The initial densities of the mixture test samples were 2.33–2.37 g cm^{-3} with average of 2.353 g cm^{-3}. Based on an initial density for paraffin of 0.917 g cm^{-3} (same density as material in the LANL paraffin Hugoniot data), the theoretical density of the mixture would be 2.454 g cm^{-3}. The discrepancy in measured density and theoretical density implies a mixture distention of $\alpha = 1.043$ (a void volume fraction of 0.043).

The dynamic isentrope for magnesium oxide is calculated by the methods described previously. Excess dissipation is assigned to the paraffin in assessing the specific volume of the magnesium oxide at the mixture Hugoniot pressure. The left-hand plot in figure 10.44 shows the experimental Hugoniot for the MgO and paraffin mixture, the Hugoniot for paraffin and the calculated dynamic isentrope. The plot on the right shows the dynamic isentrope states for magnesium oxide determined from the mixture data based on an initial Gruneisen parameter for paraffin of $\Gamma_o = 3.4$, in line with the Gruneisen properties of paraffin determined from the study of Ukrainczyk *et al* (2010). For perspective, analysis is also performed with $\Gamma_o = 1.0$ to illustrate the contribution of paraffin thermal expansion. A constant $\rho\Gamma$ is assumed in the EOS description of paraffin. These results for the dynamic isentrope assessed from mixture analysis are compared with the isentrope determined from the independent EOS analysis of the Hugoniot data on full-density and porous magnesium oxide provided previously in figure 10.19.

For a $\Gamma_o = 3.4$ the dynamic isentrope determined from the mixture analysis and the isentrope calculated from the full-density and porous MgO (periclase) Hugoniot data are nearly coincident, at least in the 30–50 GPa pressure range. Agreement is less convincing in the lower pressure range, although scatter in the data and potential microstructure crush complications compromise the data in this range.

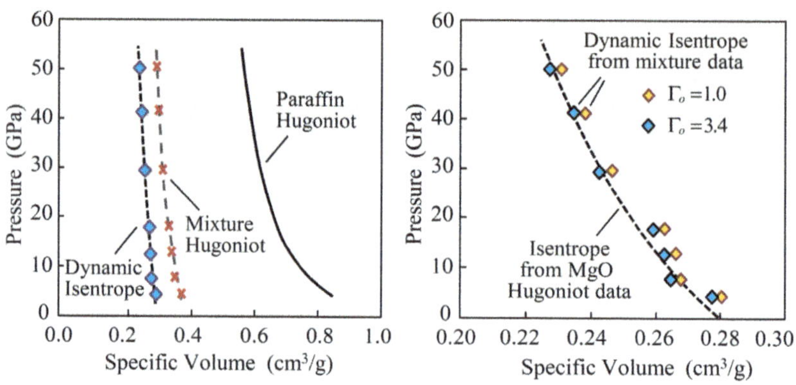

Figure 10.44. Dynamic isentrope for magnesium oxide determined from the immersed mixture method and comparison with the isentrope extracted from full-density and porous Hugoniot data on monolithic MgO. Hugoniot data for the mixture are from Marsh (1980).

Compressibility of MgO and paraffin differ by more than an order of magnitude at low pressure but are comparable at higher pressure as is observed in the left-hand plot of figure 10.44. Comparison of the immersed mixture isentropic compression for MgO and that from analysis of the solid and porous MgO data is encouraging. Overlay of the two requires a relatively large value for the Gruneisen property of paraffin, and as noted previously is in accord with the thermo-physical data for paraffin of Ukrainczyk *et al.*

The present study offers an alternative perspective. The independent analysis of the MgO data, and the separate analysis of the MgO–paraffin mixture, can be regarded as a calibration of the response of the paraffin within the mixture in the shock compression event. The Gruneisen parameter is regarded as an adjustable parameter that is assigned a value that brings the two isentropic compression relations for MgO into coincidence. Once calibrated, the mixture of paraffin with other compounds of interest can be used to assess isentropic compression properties. This intention is in line with the objectives of MgO as a high-pressure standard as proposed in Carter *et al* (1971).

Crystalline quartz (SiO_2) is the dominant mineral in the Earth's crust. The common form is α-quartz with trigonal crystal symmetry and oxygen in tetragonal coordination with silicon. The crystal density of α-quartz is 2.65 g cm^{-3}. The definitive high-pressure phases of quartz are coesite and stishovite. The α-quartz to stishovite phase transformation is dominant under shock compression. With an ambient crystal density for stishovite of 4.29 g cm^{-3}, transformation volume strain is massive and is a determining influence on the high-pressure quartz Hugoniot. Phase transformation on the Hugoniot is reported to span a broad pressure range of about 15–35 GPa (Wackerle 1962). A more definitive shock transition pressure of about 23 GPa is identified in a later study (Zhugin and Krupnikov 1987).

The LANL compendium (Marsh 1980) contains shock Hugoniot data for a mixture of α-quartz and paraffin, as well as independent Hugoniot data for α-quartz. The mixture data consist of seven data points ranging over 4–40 GPa. The reported mass fraction of the paraffin is 0.187. The average density of the mixture test samples is 1.920 g cm^{-3}. Based on crystal quartz density and a density for paraffin of 0.917 g cm^{-3}, a volume fraction of paraffin of 0.423 is calculated as well as a modest porosity distention of 1.019.

The dynamic isentrope for α-quartz is calculated by the immersed-mixture analysis. The left-hand plot in figure 10.45 shows the experimental Hugoniot for the quartz and paraffin mixture, the Hugoniot for paraffin, and the calculated dynamic isentrope states. On the right the dynamic isentrope states for quartz from the immersed mixture method are compared with hydrostatic compression of α-quartz (Angel *et al* 1997) and stishovite (Ross *et al* 1990) and with shock Hugoniot data for quartz (Marsh 1980).

Dynamic isentrope data in the range of 5–15 GPa are in reasonable agreement with the static compressibility of α-quartz. Hugoniot data in the same range show evidence of incomplete crush. (Hugoniot measurements are for material with starting porosity of about 20%.) The isentrope datum at 24 GPa is consistent with partial transformation of the quartz to the stishovite phase. The highest two

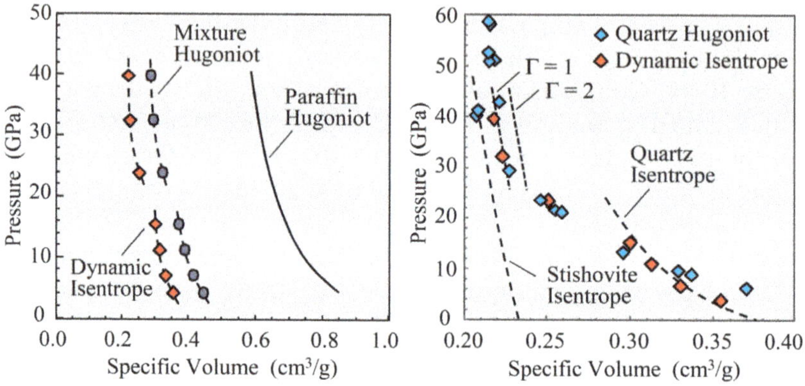

Figure 10.45. Left: Dynamic isentrope for alpha quartz determined from the immersed mixture method with paraffin immersion medium. Right: Comparison with quartz Hugoniot and with α-quartz and stishovite isentropic compression from static and ultrasonic studies. Hugoniot data for the mixture are from Marsh (1980).

dynamic isentrope data points at 33 and 40 GPa appear fully transformed to the stishovite phase. These data are in sensible agreement with the Hugoniot data, although the latter exhibits considerable scatter. Comparison is shown with the static isentrope for stishovite (Ross *et al* 1990). The dynamic isentrope resides on a higher enthalpy isentrope as a consequence of shock compression from the initial α-quartz phase. Although the transformation path is not well captured by the data, displacement of the dynamic isentrope from the ambient stishovite isentrope is reasonably estimated through

$$\Delta v = \frac{\Gamma}{K_s} p_T \Delta v_T. \tag{10.250}$$

Transformation pressure and volume change are p_T and Δv_T, respectively. Using a $p_T = 23$ GPa from Zhugin and Krupnikov (1987) (not out of line with the intermediate dynamic isentrope datum at 24 GPa), a transformation volume calculated from the ambient densities of α-quartz and stishovite (2.65 g cm^{-3} and 4.29 g cm^{-3}, respectively), and a bulk modulus of stishovite of $K_s = 306$ GPa, elevated enthalpy isentrope relations are calculated. Assuming a Gruneisen gamma of 1 and of 2 for stishovite provides the two offset curves that are compared with the highest pressure dynamic isentrope states on the right in figure 10.45.

Within the limited data available, a dynamic isentrope assessment of quartz from the immersed mixture method appears reasonably successful. Also, phase transformation proceeds in quartz under the quasi-isentropic compression of the mixture method, although the detailed compression path requires further study.

10.5.5.6 Assessments of additive mixing

Additive mixture theories are an expedient method for generating shock EOS relations for composite mixtures of materials when shock compression data for the component materials are available. At lower shock pressures, material strength complicates material response. As shock pressures become increasingly larger than the strength of the materials, however, a fluid approximation of the thermo-elastic compressibility becomes increasingly applicable.

Material mixtures can range from the granular to the molecular. At one end of the range are mixtures of powders or pressed granular composites. At the other end are mixtures of chemical species that form more complex molecular solids through chemical bonding. The availability of mixing theories for purposes of providing shock EOS relations would be useful in all such applications.

The achievement of pressure and temperature equilibrium following passage of a strong shock wave through the composite material is a concern in developing EOS mixing rules. Temperature equilibrium through thermal conduction over length scales determined by both granularity and shock deformation microstructure is particularly problematic. Nonetheless, pressure and temperature equilibrium at the shock state is the most expedient, and probably the most sensible, assumption in developing an additive EOS mixture theory. Conversely, near-adiabatic compression of components is assumed in the immersed mixture dynamic isentrope experiment.

Additive mixing through the thermodynamic Gibbs energy is theoretically the most appropriate when fluid-like response with equilibrium pressure and temperature is assumed at the shock state. Unfortunately, the thermo-elastic material properties necessary for this more thorough EOS development are seldom available.

Shock Hugoniot data are available for many materials, however, and can be reasonably estimated in certain cases from ambient elastic properties of materials. Shock EOS mixing through additive Hugoniot relations has been shown to be an appropriate approximation for certain material mixtures. Some seminal experimental studies of additive mixing of shock Hugoniot data are assessed in earlier sections. Generally, it is found that additive Hugoniot mixing on a granular scale is reasonably effective. This result is found even when stark differences in material compressibility are tested. The test shock data described for a crystalline solid mixed with paraffin and with water, both reveal applicability of additive mixing of shock Hugoniot relations although some concerns were raised by tests conducted on a calcite–water mix and on a water-saturated grout mix where application of the water isentrope achieves better agreement with available data.

Additive mixing on a molecular scale is found to be problematic. For the two examples described, one (dolomite) provides reasonable agreement, whereas the second (CaO–CO_2) is seriously in error. Probably additive mixing on a molecular scale is not generally appropriate, although further study is needed.

Alternatively, shock experiments on well-prepared mixture can be exploited to assess the dynamic isentropic (or near-isentropic) compression of selected materials. The material under test is mixed with a material with known dynamic compression properties. Dynamic compression of the material of interest is then extracted from experimental shock wave measurements on the mixture through appropriate mixture analysis of the data.

It is reasonable to ask as to the purpose of this additional test method. The shock Hugoniot experiment has been refined over nearly seventy years of testing at many laboratories. Hugoniot compressibility of many materials have been established. Quite mature methods have been developed for constraining high-pressure EOSs through the material Hugoniot. The reasons are several, as follows.

Material strength. Material strength can complicate the Hugoniot measurement (e.g. Dandekar and Grady 2002, Vogler *et al* 2004, Paris *et al* 2011). The Hugoniot test does not necessarily provide an EOS measurement in the tested material. Deviator stress states can persist to high levels of Hugoniot compression. This is particularly true for high-strength compounds of interest. Corrections to strength offset on the Hugoniot cannot be made with confidence. Intensities of strength offset are material-dependent. The underlying causes are not sufficiently understood to account for material strength effects.

Shock-induced defects. EOSs based on shock Hugoniot measurements necessarily assume dissipation in the shock wave transitions to thermodynamically equilibrated temperature through the lattice and electronic structure of the material at the Hugoniot state. There is evidence of defect generation within the shock wave providing an alternative athermal sink for the shock Hugoniot energy (Ruoff 1980, Graham 1981). Shock studies on Al_2O_3 provide evidence of large increase in configurational entropy with only modest temperature increase to nearly 400 GPa Hugoniot pressure (Nellis *et al* 2009). Shock defect structure can support a non-equilibrium pressure component, further complicating equilibrium EOS assessment from Hugoniot data (Ruoff 1980).

Shock compaction. Discounting measurements on single-crystal materials, near-full-density compounds prepared for purposes of cold Hugoniot testing will necessarily encompass some levels of porosity. Uncertainties in the shock compaction of high-strength compounds contribute to the complications of equilibrium EOS analysis of such compounds.

Shear-induced phase transformation. Thermodynamic solid–solid phase transformation is a necessary facet of the baseline EOS for many compounds of interest. Shock experiments provide evidence that phase transformation is sensitive to inelastic shear deformation in the shock compression event leading to the Hugoniot state (Grady *et al* 2013). Shear-induced phase transformation will necessarily be part of the constitutive model yielding a description of the dynamic response of the materials to intense shock loading. Extracting baseline EOSs from Hugoniot measurements are, however, further complication by this shock phase transformation effect.

References

Alchagirov A B, Perdew J P, Boettger J C, Albers R C and Fiolhais C 2001 Energy and pressure versus volume: equations of state motivated by the stabilized jellium model *Phys. Rev.* **B 63** 22415 1–16

Altschuler L V and Sharipdzhanov I I 1971 Additive equation of state of silicates at high pressure *Phys. Solid Earth* **3** 167–77

Amulele G M, Manghnani M H, Marriappan S, Hong L, Li F, Qin X and Liermann H P 2008 Compression behavior of WC and WC-6%Co up to 50 GPa determined by synchrotron x-ray diffraction and ultrasonic technique *J. Appl. Phys.* 103 2008

Anderson D L 1967 A seismic equation of state *Geophys. J. R. Astr. Soc.* **13** 9–30

Anderson D L 1967 Equation for thermal expansivity in planetary interiors *J. Geophys. Res.* **72** 3661–8

Anderson G D, Duvall G E, Erkman H O, Fowles G R and Peltzer C P 1966 Investigation of equation of state of porous earth materials AFWL-TR-65-146 *Air Force Weapons Laboratory Report* (New Mexico: Kirtland Air Force Base)

Angel R J, Allen D R, Miletich R and Finger L W 1997 The use of quartz as an internal pressure standard in high-pressure crystallography *J. Appl. Cryst.* **30** 461–6

Asay J R and Shahinpoor M (ed) 1993 *High-Pressure Shock Compression of Solids* (New York: Springer)

Barker L M and Hollenbach R E 1970 Shock wave studies of PMMA, fused silica and sapphire *J. Appl. Phys.* **41** 4208–26

Bass R C, Hawk H L and Chabai A J 1963 Hugoniot data for some geologic materials, *Sandia National Laboratories Report* June

Bedford A and Drumheller D S 1983 Theories of immiscible and structured mixtures *Int. J. Eng. Sci.* **21** 863–960

Birch F 1947 Finite elastic strain in cubic crystals *Phys. Rev.* **71** 809–24

Birch F 1978 Finite strain isotherm and velocities in single-crystal and polycrystalline NaCl at high pressures and 300 K *J. Geophys. Res.* **83** 1257–66

Boehler R 1993 Temperatures in the Earth's core from melting point measurements of iron at static high pressures *Nature* **363** 534–6

Born M 1939 *J. Chem. Phys.* **7** 591

Born M and Mayer J 1932 *Z. Phys* **75** 1

Bridgman P W 1952 *The Physics of High Pressure* (London: Bell and Sons)

Brown J M and McQueen R G 1986 Phase transitions, Gruneisen parameter, and elasticity for shocked iron between 77 GPa and 400 GPa *J. Geophys. Res.* **91** 7485–94

Budiansky B 1965 *Mech. Phys. Solids* **13** 223

Carter W J 1973 Relation between dynamic and static phase transformation studies *Metallurgical Effects at High Strain Rates* ed R W Rohde, B M Butcher, J R Holland and C H (KarnesPlenum) pp 171–84

Carter W J, Marsh S P, Fritz J N and McQueen R G 1971 The equation of state of selected materials for high-pressure references *Symp. Accurate Characterization of the High-Pressure Environment, (National Bureau of Standards, Gaithersburg, MD, 14–18 October 1968)* (Washington, DC: NBS) Special publication 326

Cowperthwaite M and Shaw R 1970 Equation of state for liquids. calculation of the shock temperature of carbon tetrachloride, nitromethane, and water in the 100 kbar region *J. Chem. Phys.* **53** 555

Dandekar D P and Grady D E 2002 Shock equation of state and dynamic strength of tungsten carbide *Shock Compression of Condensed Matter–2001* ed M Furnish, N Thadhani and Y (HorieAmerican Institute of Physics) pp 783–6

Davis W C 1997 Shock waves; rarefaction waves; equation of state: *Explosive Effects and Applications* ed J A Zukas and W P Walters (Springer) pp 47–113

Davies G F and Gaffney E S 1973 Identification of high-pressure phases of rocks and minerals from Hugoniot Data *J. R. Astr. Soc* **33** 165–83

Drake R P 2010 High energy density physics *Phys. Today* **6** 28–33

Dremin A N and Karpukhin I A 1960 Methods of determining the shock adiabats of dispersed substances *Zh. Prikl. Mekh. Tek. Fiz* **3** 184–8

Duffy T S and Ahrens T J 1993 Shock compression and release of polycrystalline magnesium oxide *High-Pressure Science and Technology* ed S C Schmidt, J W Shaner, G A Samara and M Ross (New York: American Institute of Physics) pp 1107–10

Duffy T S, Hemley R J and Mao H K 1995 Equation of state and shear strength at multi-megabar pressures: magnesium oxide to 227 GPa *Phys. Rev. Letts.* **74** 1371–4

Duclos S J, Vohra Y K, Ruoff A L, Jayaraman A and Espinosa G P 1988 High-pressure x-ray diffraction study of CeO_2 to 70 GPa and pressure-induced phase transformation from the fluorite structure *Phys. Rev.* B **38** 7755–8

Dugdale J S and MacDonald D K 1953 The thermal expansion of solids *Phys. Rev.* **85** 832–4

Duvall G E and Taylor S M 1971 Shock parameters in a two component mixture *J. Compos. Mater.* **5** 130–9

Fenton G, Grady D and Vogler T 2012 Intense shock compression of porous solids: application to WC and Ta_2O_5 *Shock Waves in Condensed Matter—2011* ed M L Elert, W T Buttler, J P Borg, J L Jordan and T J Vogler (New York: AIP) pp 1463–6

Fenton G, Grady D E and Vogler T J 2015 Shock compression modeling of distended mixtures *J. Dynamic Behav. Mater* **1** 103–13

Forest C A 1996 Isentrope energy and Hugoniot temperature and the Mie–Grüneisen equation of state *Shock Compression of Condensed Matter—1995* ed S C Schmidt and W C Tao (New York: American Institute of Physics) pp 31–4

Fredenburg D A and Chisolm E D 2014 equation of state and compaction modeling for CeO_2 *Los Alamos National Laboratory Report* LA-UR-14-28164 October

Furnish M D, Reinhart W D, Trott W M, Chhabildas L C and Vogler T J 2006 Variability in dynamic properties of tantalum: spall, Hugoniot elastic limit and attenuation *Shock Waves in Condensed Matter* (Plenum) pp 615–8

Grady D E 1977 Processes occurring in shock wave compression of rocks and minerals *High Pressure Research: Applications in Geophysics* (New York: Academic) pp 389–438

Grady D E 1986 High-pressure release-wave measurements and phase transformation in $CaCO_3$ *Shock Waves* ed Condensed Matter *Y M Gupta* (New York: Plenum) pp 589–93

Grady D E 1994 Shock compression properties of metal–ceramic mixtures *DYMAT* J **1** 1–13

Grady D E 1997 *PBXW-128 energetic material modeling and testing Applied Research Associates Final Report for Naval Surface Warfare Center, Dahlgren, VA* DO 0046, AMSAA Contract DAAA15-94D-005 1997

Grady D E 2009 Analysis of shock and high-rate data for ceramics: equation of state properties and fragmentation in the ballistic environment, prepared for US Army TARDEC Contract No. W56HZV-08-P-L594 *Applied Research Associates Technical Report* May

Grady D E 2010 Dynamic equation of state and strength of boron carbide *Advances in Ceramic Armor vol 6 (Ceramic Engineering and Science Proceedings)* ed J Swab (Hoboken, NJ: Wiley) pp 115–42

Grady D E 2012 *Equation of state for solids Shock Compression of Condensed Matter—2011* ed M D Elert, W T Buttler, J P Borg, J L Jordan and T J Vogler (New York: American Institute of Physics) pp 800–3

Grady D E 2015a Equation of state of dolomite from shock Hugoniot and static compression studies *Proc. APS Topical Group on Shock Compression of Condensed Matter* (Tampa, FL, 15–19 June)

Grady D E 2015b Hugoniot equation of state and dynamic strength of boron carbide *J. Appl. Phys.* **117** 165904

Grady D E, Murri W J and Mahrer K D 1976 Shock compression of dolomite *J. Geophys. Res.* **81** 889–93

Grady D E, Moody R L and Drumheller D S 1986 Release equation of state of dry and water-saturated porous calcite *Sandia National Laboratories Report SAND86* 2110 November

Grady D E and Furnish M D 1988 Shock- and release-wave properties of MJ-2 grout *Sandia National Laboratories Technical Report SAND88* **1642** December

Grady D E and Furnish M D 1990 Hugoniot and release properties of a water-saturated high-silica-content grout *Shock Compression of Condensed Matter—1989* ed S C Schmidt, J N Johnson and L W Davison (Elsevier Science) pp 621–4

Grady D E and Wise J L 1993 Dynamic properties of ceramic materials *Sandia National Laboratories Technical Report* SAND93-0610 September

Grady D E, Chhabildas L C, Reinhart W D and Wilson L T 2000 Dynamic equation-of-state and strength properties of unreacted PBXW-128 explosive *Proc. 11th Int. Detonation Symposium (Snowmass, CO 31 August–4 September 1998)* Office Naval Research 33300-5 1049–57

Grady D, Fenton G and Vogler T 2013 Equation of state and evidence of enhanced phase transformation for shock compression of distended compounds *Int. J. Impact Eng.* **56** 19–26

Graham R A 1981 Active measurements of defect processes in shock-compressed metals and other solids *Shock Waves and High-Strain Rate Phenomena in Metals* ed M A Meyers and L E Murr (Plenum) pp 375–86

Gruneisen E 1912 *Ann. Phys. (Leipzig)* **39** 257–306

Gurtman G A, Kirsch J W and Hastings J R 1971 Analytical equation of state of water compressed to 300 kbar *J. Appl. Phys.* **42** 851

Holzapfel W B 1996 Physics of solids under strong compression *Rep. Prog. Phys.* **59** 29–90

Horie Y and Sawaoka A B 1993 *Shock Compression Chemistry of Materials* (Tokyo: KTK Scientific)

RIG Hughes 2006 Theoretical practice: the Bohm–Pines quartet *Perspect. Sci* 14 457–524

Jackson I and Niesler H 1982 The elasticity of periclase to 3 GPa and some geophysical implications *High Press. Res. Geophys* 93–133

Jeanloz R 1988 Universal equation of state *Phys. Rev.* B **38** 805–7

Jeanloz R 1989 Shock wave equation of state and finite strain theory *J. Geophys. Res.* **94** 5873–86

Johnson I A 2005 The Nobel–Abel Equation of State: Thermodynamic Derivations for Ballistic Modeling (Edinburgh, SA: Defense Science Technology Organization)

Johnson J N 1968 Single particle model of a solid: the Mie–Gruneisen equation *Am. J. Phys.* **36** 917–9

Kalashnikov N G, Pavlovskiy M N, Simakov G V and Trunin R F 1973 Dynamic compressibility of calcite group minerals *Phys. Solid Earth* **2** 80–4

Kalinin V A 1972 A universal equation of state for solids *Phys. Solid Earth* **4** 16–23

Kanel G I and Pityulin A N 1984 *Fiz. Goren. Vzryva* **28** 85

Kerley G I 1994 A new multiphase equation of state for iron *High Pressure Science and Technology—1993* (New York: AIP) pp 903–6

Kormer S B 1968 Optical study of the characteristics of shock-compressed condensed dielectrics *Sov. Phys. Usp* **11** 229–54

Lennard-Jones J E 1924 On the determination of molecular fields. I. from the variation of the viscosity of a gas with temperature *Proc. R. Soc.* A **106** 441–62

Lennard-Jones J E 1924 On the determination of molecular fields. II. From the equation of state of a gas *Proc. R. Soc.* A **106** 463–77

Li Y H 1967 Equation of state of water and sea water *J. Geophys. Res.* **72** 2665–78

Lindemann F A 1910 The calculation of molecular vibration frequencies *Z. Phys* **11** 609–15

Lindfors A J 1996 Unpublished data *Naval Air Warfare Center*

Lyon S P and Johnson J D 1992 SESAME: the Los Alamos National Laboratory equation of state database *LANL Technical Report LA-UR* 92–3407

Lyzenga G A, Ahrens T J, Nellis W J and Mitchell A C 1982 The temperature of shock compressed water *J. Chem. Phys.* **76** 6282

Manghnani M H, Wang Y, Zinin F and Rafaniello W 2000 Elastic and vibrational properties of B4C to 21 GPa *Science and Technology of High Pressure vol 2* ed M H Manghnani, W J Nellis and M F Nicol (Hyderabad: Hyderabad University Press) pp 945–8

Marsh S P 1980 *LASL Shock Hugoniot Data* (University of California Press)

McQueen R G and Marsh S P 1962 Equation of state of Nevada alluvium *Los Alamos Scientific Laboratory Report* LAMS-2760.

McQueen R G, Marsh S P, Taylor J W, Fritz J N and Carter W J 1970 The equation of state of solids from shock wave studies *High Velocity Impact Phenomena* ed R Kinslow (New York: Academic) pp 294–419

Mie G 1903 Zur Kinetischen Theorie der Einatomigen Körper *Ann. Phys* **11** 657–97

Miller J, Boehly T, Meyerhofer D and Eggert J 2007 Equation-of state measurements in Ta2O5 aerogel *Shock Compression of Condensed Matter—2007* (New York: American Institute of Physics) pp 71–4

Mitchell A C and Nellis W J 1982 Equation of state and electrical conductivity of water and ammonia shock to the 100 GPa pressure range *J. Chem. Phys* **76** 6273

Morse P M 1929 Diatomic molecules according to the wave mechanics. II. Vibrational levels *Phys. Rev.* **34** 57–64

Murnaghan F D 1944 The compressibility of media under extreme pressures *Proc. Natl Acad. Sci* **30** 244–7

Nellis W J, Moriarty J A, Mitchell A C, Ross M, Dandrea R G, Ashcroft N W, Holmes N C and Gathers R G 1988 Metal physics at ultrahigh pressure: aluminum, copper and lead as prototypes *Phys. Rev. Lett.* **60** 1414–7

Nellis W J, Kanel G I, Razorenov S V, Savinykh A S and Rajendran A M 2009 *Entropy-dominated dissipation in sapphire shock-compressed up to 400 GPa (4 Mbar), arXiv* 09054730

Paris V, Frage N, Dariel M P and Zaretsky 2011 Divergent impact study of the compressive failure threshold in SiC and B_4C *Int. J. Impact Eng.* **38** 228–37

Ree F H 1982 Molecular interaction of dense water at high temperature *J. Chem. Phys.* **76** 6287

Rice M H and Walsh J M 1957 Equation of state of water to 250 kilobars *J. Chem. Phys.* **26** 824–30

Riney T D, Gary S K, Kirsh J W, Morland L W and Hastings C R 1970 Stress wave effects in inhomogeneous and porous earth materials *Systems, Science and Software Report* 3SR-267, La Jolla, California.

Rose J H, Smith J R, Guinea F and Ferrante J 1984 Universal features of the equation of state for metals *Phys. Rev.* B **29** 2963–9

Ross N L, Shu J F, Hazen R M and Gasparik T 1990 High Pressure Crystal Chemistry of Stishovite *Amer. Mineral.* **75** 739–47

Ruoff A L 1980 Estimates of uncertainty in isotherms deduced from Hugoniots resulting from shock wave generated defects *J. Appl. Phys.* **51** 6221
Rydberg R 1931 *Z. Phys.* **73** 376–85
Sandstrom F W, Persson P A and Olinger B 1993 Isothermal and shock compression of high density ammonium nitrate and ammonium perchlorate *Proc. 10th Int. Detonation Symp. (Boston, MA,* 12–16 *July)* pp 766–74
Santillán J, Williams Q and Knittle E 2003 Dolomite-II: a high-pressure polymorph of CaMg $(CO_3)_2$ *Geophys. Res. Lett.* **30** 1054
Sikka S K 1989 Empirical equation of state theories at ultra-high pressures *Phys. Lett.* A **135** 129–31
Skidmore I C and Morris L 1962 *Thermodynamics of Nuclear Materials* (Vienna: International Atomic Energy Agency/Aldermaston: Atomic Weapons Research Establishment) 173
Slater J C 1939 *Introduction to Chemical Physics* (New York: McGraw-Hill)
Steinberg D J 1982 Some observations regarding the pressure dependence of the bulk modulus *J. Phys. Chem. Solids* **43** 173–5
Steinberg D J 1987 Spherical explosions and the equation of state of water *LLNL Report* UCID 20974 February
Steinberg D J 1996 Equation of state and strength properties of selected materials *Lawrence Livermore National Laboratory Report* UCRL-MA-106439, February
Stacy F D and Irvine R D 1977 Theory of melting: thermodynamic basis of Lindemann's law *Austral. J. Phys.* **30** 631–40
Stishov S M and Popova S V 1961 A new dense modification of silica *Geokhimiya* **16** 923–6
Tait P G 1888 Reports on some of the physical properties of fresh water and sea water *Rept. Sci. Results Voy. HMS Challenger, Phys. Chem.* **2** 1–76
Tang Z P and Gupta Y M 1988 Shock-induced phase transformation in cadmium sulfide dispersed in an elastomer *J. Appl. Phys.* **64** 1827–37
Trunin R F 2001 *Experimental Data on Shock Compression and Adiabatic Expansion of Condensed Matter (Sarov) RFNC-VNIIF*
Trunin R F, Gudarenko L F, Zhernokletov M V and Simakov G V 2001 Experimental Data on Shock Compression and Adiabatic Expansion of Condensed Matter Sarov: RFNC-VNIIF; ISBN 5-85165-624-7
Tyburczy J A and Ahrens T J 1986 Dynamic compression and volatile release of carbonates *J. Geophys. Res.* **91** 4730–44
Ukrainczyk N, Kurajica S and Šipušiæ J 2010 Thermophysical comparison of five commercial paraffin waxes as latent heat storage materials *Chem. Biochem. Eng.* Q **24** 129–37
Van Thiel M, Shaner J and Salinas E 1977 Compendium of shock wave data *Lawrence Livermore Laboratory Rept.* UCRL-50108 **vol 1, 2 and 3** June
Vaschenko V and Zubarev V N 1963 Concerning the Gruneisen constant *Sov. Phys. Solid State* **5** 643–55
Vinet P, Ferrante J, Smith J R and Rose J H 1986 A universal equation of state for solids *J. Phys.* C **19** 467–73
Vogler T J, Reinhart W D and Chhabildas L C 2004 Dynamic behavior of boron carbide *J. Appl. Phys.* **95** 4173
Vogler T J, Root S, Knudson M D and Reinhart W D 2011 High pressure shock behavior of WC and Ta_2O_5 powders *Sandia National Laboratories Technical Report* SAND2011 **6770** October

Wackerle J 1962 Shock-wave compression of quartz *J. Appl. Phys.* **33** 922–37

Walsh J M and Christian R H 1955 Equation of state of metals from shock wave measurements *Phys. Rev* **97** 1544–56

Walsh J M and Rice M H 1957 Dynamic compression of liquids from measurements of strong shock waves *J. Chem. Phys.* **26** 815

Walton A J 1983 *Three Phases of Matter* (Oxford University Press)

Williams Q, Jeanloz R, Bass J, Svendson B and Ahrens T J 1987 The melting curve of iron to 250 GPa: a constraint of the temperature of the Earth's core *Science* **236** 181–2

Woolfolk R W, Cowperthwaite M and Shaw R 1973 A universal Hugoniot for liquids *Thermochim. Acta* **5** 409–14

Zeldovich Y B and Raiser Y P 1966 *Physics of Shock Waves and High-Temperature Hydrodynamic Phenomena* vol 1 (New York: Academic)

Zharkov V N and Kalinin V A 1971 Equations of State for Solids at High Pressures and Temperatures (New York: Consultants Bureau)

Zhugin Y N and Krupnikov K K 1987 Structural peculiarities of plastic front in quartz in the mixed phase region *Trans. 11th Int. IUPAP-IUTAM Symp. on Non-Linear Acoustics part II* (Novosibirsk, Russia) pp 196–200

IOP Publishing

Physics of Shock and Impact, Volume 2
Materials and shock response
Dennis Grady

Chapter 11

Shock wave analysis and applications

The present chapter pursues methods used in the analysis and interpretation of shock wave data. Pressure–particle-velocity and distance–time plots are fundamental to the design of experiments and the interpretation of experimental data. The methods of Lagrangian analysis are explored as applied to the measurements of time-resolved stress and particle-velocity waves. The basic analysis methods are introduced. Heuristic examples provide understandings into the use of the shock analysis tools. Further insights are provided through applications to relevant experimental shock wave data.

11.1 Introduction

The shock wave physics of condensed materials encompasses unique facets of large-amplitude nonlinear wave propagation. Accordingly, specialized analysis tools have evolved to dissect and interpret the experimental and theoretical outcomes of shock physics studies.

Plots of both distance versus time and pressure versus particle velocity are intrinsic to the analysis of shock data and the design of shock wave experiments. Distance–time plots allow for the tracking of large-amplitude waves through multiple materials and complex wave interactions. Correspondingly, such wave complications carry material regions through pressure and particle-velocity states that can be rendered understandable through the application of pressure–particle-velocity plots.

With increasing sophistication and resolution in experimental techniques for measuring transient stress and particle velocity in shock wave studies, assumptions of steady state wave propagation and the Hugoniot jump conditions are often unacceptable approximations. Lagrangian analysis emerged as a technique to extract dynamic material response constitutive properties from structured and unsteady time-resolved measurements of stress and particle velocity.

doi:10.1088/978-0-7503-1257-8ch11

These analysis tools that are basic to the interpretation of shock wave phenomena are described. Examples are provided that illustrate the use of the tools. Experimental shock physics results are used to further illustrate the purposes of the distance–time and the pressure–particle-velocity analysis methods. The techniques of Lagrangian analysis are described and demonstrated on selected shock physics experimental studies.

Exploration of the analysis of shock wave measurements draws on experimental data that range widely over material types and shock properties. Geological materials provide a rich array of phenomena under shock compression. Silicate and carbonate rocks including quartzite and dolomite are selected for representative shock analysis studies. Further perspective is provided from shock wave measurements on oil shale. Shock data on concrete is also examined. Assessment of the shock response of metals is explored with shock data obtained on several alloys of uranium. Shock measurements on these materials and others are used to demonstrate applications of the shock wave analysis tools that have evolved in parallel with ever improving shock measurement techniques.

11.2 Analysis in shock wave applications

As in all areas of science, the field of shock physics has specialized analysis tools to design, interpret and assess specialized experiments exploring the effects of shock waves. These include the distance–time plot allowing for the convenient tracking of interacting shock and release waves, and the pressure–particle-velocity plot essential to the preparation and subsequent interpretation of experiments in shock physics. These shock wave methods, and other tools central to the analysis and interpretation of shock wave phenomena, are addressed in this section.

11.2.1 Wave interaction analysis

The analysis of interacting shock waves is one of the principal tools used in assessing the consequences of a shock wave event. Such analyses can, of course, be carried out with an appropriate wave propagation code. The intuitive understanding afforded by the graphical solution of such problems, however, is usually beneficial in the design of shock wave experiments, in the assessment of a particular shock wave event and in understanding the results of a computational simulation. Several simple examples of shock wave interactions are illustrated in figure 11.1 and provide cases for an initial description of the formalism of shock wave interaction analysis.

In the example on the left a shock wave is normal incident on an interface between two dissimilar materials. The consequence of the interaction is the transmission of a shock wave into the opposing material and an accompanying wave reflection into the initial material. The reflected wave can be either a shock wave or a release wave, depending on the relative impedances of the two materials. The second example on the right is that of the wave interaction of two normal colliding shock waves within one material. Two reflected shock waves of differing amplitudes is the consequence of this wave interaction. Similar thinking applies, of course, to the interactions of more elementary acoustic waves.

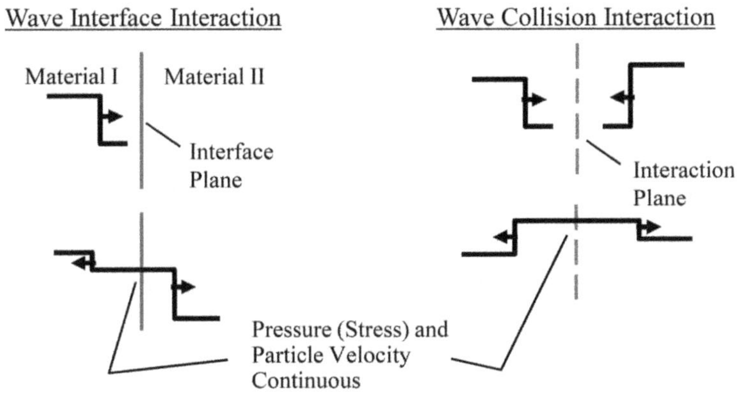

Figure 11.1. Wave interactions at an interface (left) and colliding shock waves (right).

11.2.1.1 Pressure versus particle-velocity Hugoniot and release curves

The consequence to note subsequent to the wave interaction in either of the examples posed in figure 11.1 is that the pressure (stress on planes normal to the direction of shock propagation) and the particle velocity are continuous in the new state created by the shock wave interaction (assuming that the interface does not separate). This condition is identified in the figure. There are wave interactions in which material separation can occur at the point of interaction. Spall fracture is a notable example. Such special cases can also be usefully analyzed through the methods described here. Pressure will be referred to throughout the discussion. The techniques, however, are equally applicable to materials with strength where the stress normal to the plane of wave interaction serves the same role as the pressure.

Continuity of pressure (normal stress) and particle velocity subsequent to the wave interaction suggests that analysis will be performed in a plot of pressure versus particle velocity. Consequently, essential to the analysis will be a pressure–particle-velocity Hugoniot for the materials participating in the wave interaction event. If, through either experimental measurement or through theoretical equation-of-state (EOS) analysis, a pressure–volume Hugoniot $P_h(v)$ is known for the material, then a pressure–particle-velocity Hugoniot $P_h(u)$ is readily determined through appropriate application of the Hugoniot conservation relations. The overwhelming preponderance of Hugoniot data available, however, is provided in the shock-velocity versus particle-velocity $U(u)$ representation. This is the case, for example, in the Los Alamos National Laboratory (LANL) compendium of shock data (Marsh 1980) where the fundamental data are measured states of shock velocity and particle velocity. The particle-velocity Hugoniot is then immediately provided through the Hugoniot momentum relation

$$P_h(u) = \rho_o U(u)u. \tag{11.1}$$

Hugoniot data in the LANL compendium are provided, or can usually be fit, in the form

$$U(u) = C_o + Su + Qu^2, \tag{11.2}$$

in which case the pressure–particle-velocity Hugoniot is the polynomial form

$$P_h(u) = \rho_o C_o u + \rho_o S u^2 + \rho_o Q u^3. \tag{11.3}$$

For many materials the coefficient Q is negligible, and a quadratic pressure versus particle-velocity Hugoniot is appropriate. For other materials, more complex $U(u)$ relations may be necessary. Polymethyl methacrylate (PMMA), commonly used as an interferometer window material, requires a fourth-order polynomial $U(u)$ relation for an adequate description of the $P_h(u)$ Hugoniot (Barker and Hollenbach 1972).

On the left in figure 11.2 is shown a plot of the primitive pressure versus particle-velocity Hugoniot. Any Hugoniot that represents a translation of the primitive Hugoniot, or a reflection and translation of the primitive Hugoniot, is an equivalent Hugoniot. Each represents the shock response of a material subjected to either a backward facing or forward facing shock wave propagating into media at initial particle velocity u^* in front of the shock wave. Such equivalent Hugoniot curves are used in the wave interaction analysis.

Commonly, the analysis requires the construction of a translated and/or reflected equivalent Hugoniot through some point (P_r, u_r) in the pressure–particle-velocity plane. Consideration of the necessary translation and/or reflection leads to the relation

$$P_h(\kappa(u - u_r) + u_{ro}), \tag{11.4}$$

where u_{ro} is the particle velocity on the principal Hugoniot at pressure P_r. The number κ takes the value of $+1$ or -1 for translation or for translation and reflection, respectively.

11.2.1.2 Basic pressure versus particle-velocity analysis

The technique to assess wave interactions through the analysis of pressure–particle-velocity plots can be applied to the elementary wave interactions illustrated in figure 11.1. The first example of a shock wave incident on an interface between two materials with different shock impedance is illustrated in figure 11.3. Prior to wave interaction at the interface between material 1 and material 2, a forward facing

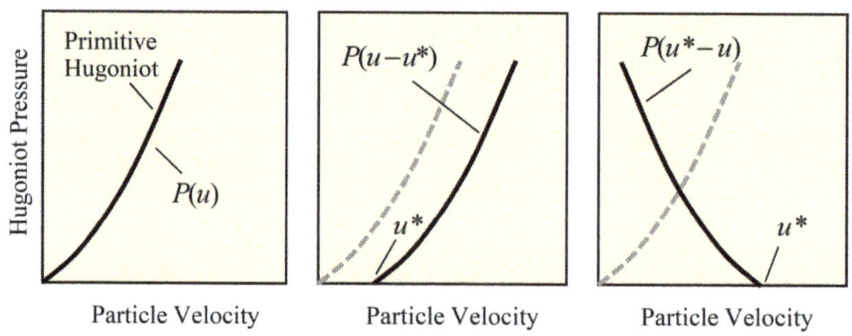

Figure 11.2. Equivalent pressure versus particle-velocity Hugoniot curves.

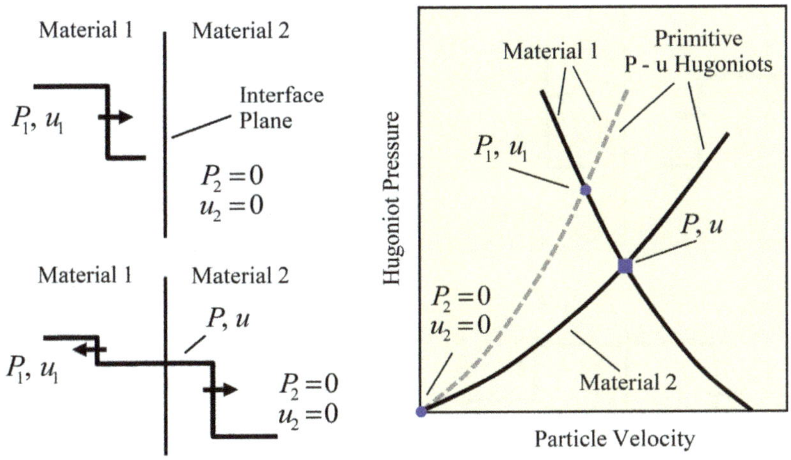

Figure 11.3. Pressure–particle-velocity analysis of shock wave interaction at an interface between dissimilar materials.

shock wave propagates into a material at rest and at zero pressure. The shock wave carries material 1 to Hugoniot pressure and particle-velocity state (P_1, u_1). Subsequent to the wave interaction a forward facing shock wave propagates into material 2 at initial state ($P_2 = 0$, $u_2 = 0$). A corresponding backward facing wave propagates in material 1 into the state (P_1, u_1). The state (P, u) behind the waves is shared by both. The values of P and u, and whether the backward facing wave in material 1 is a shock wave or a release wave, is determined by the analysis.

For the forward facing shock wave in material 2 a forward-directed Hugoniot that passes through the state ($P_2 = 0$, $u_2 = 0$) is provided by the primitive Hugoniot for material 2. For the backward facing wave in material 1 a reflected and translated copy of the material 1 primitive Hugoniot is constructed as in equation (11.4) through the state (P_1, u_1). The intersection of the two constructed Hugoniot curves provides the wave interaction state (P, u). That the backward facing wave in material 1 is a release wave, is indicated from the pressure reduction $P < P_1$.

The second example from figure 11.1 is illustrated in figure 11.4. In this interaction event, a forward facing shock wave carrying the material to Hugoniot state (P_1, u_1) and a backward facing shock wave achieving Hugoniot state (P_2, u_2) collide at an interface in the material identified by the dashed line on the left in the figure. The state of the material ahead of both shock waves identified as $P_0 = 0$ and u_0 in the figure is necessarily the same. Construct an equivalent forward-directed Hugoniot through the state (P_1, u_1) through a translation of the principal Hugoniot a requisite extent, and an equivalent backward-directed Hugoniot through the state (P_2, u_2) through a translation and reflection of the principal Hugoniot. The intersection of the two Hugoniot curves determines the Hugoniot state (P, u) behind the shock waves emerging from the plane of shock interaction.

A practical example is provided by a shock wave experiment leading to spall failure in a boron carbide ceramic. The experiment is that of a PMMA plate (2.0 mm thickness) undergoing planar impact on a plate of boron carbide ceramic (10.5 mm

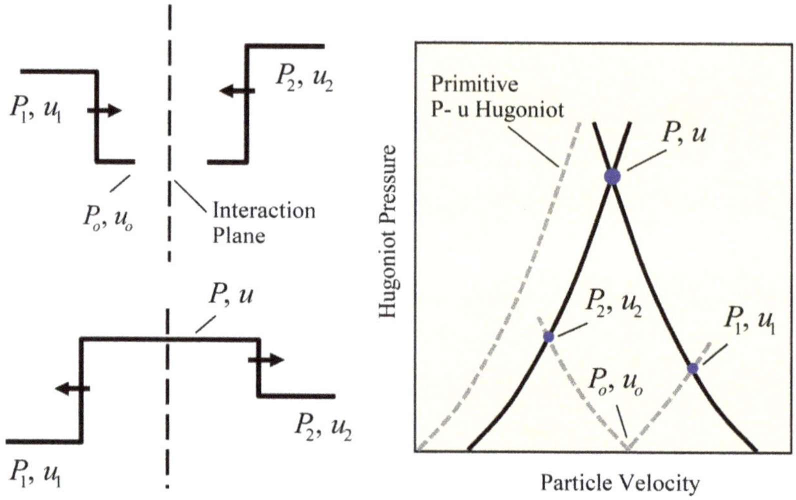

Figure 11.4. Pressure–particle-velocity analysis of shock wave interaction at the interface of collision within one material.

thickness) at a velocity of 0.91 km s^{-1}. Measurement of the velocity history resulting from the stress wave interaction event is provided by VISAR recording at the interface between the back of the ceramic and a PMMA window. Shock wave interaction within the boron carbide sample leads to spall fracture. The spall signal is observed in the velocity recording shown in figure 11.5. Analysis of the impact using the known pressure–particle-velocity Hugoniot for PMMA and elastic Hugoniot for boron carbide provides a shock state in the ceramic of $u_h = 0.096$ km s^{-1} and $P_h = 3.41$ GPa. Starting from this Hugoniot state, appropriate translated and reflected Hugoniot relations are necessary to complete the spall analysis.

Arrival of the shock wave at the recording interface results in the reflection of a returning release wave in the boron carbide and transmission of a shock wave in the PMMA window. A translated and reflected Hugoniot for boron carbide is constructed centered at the Hugoniot state (P_h, u_h). Equation (11.4) is applied with $u_r = u_h$, with u_{ro} also equal to u_h, and with $\kappa = -1$. The release state in the window PMMA is provided by the intersection of this constructed release Hugoniot curve and the principal Hugoniot for PMMA as is illustrated on the right in figure 11.5. Identify this release state as (P_{rl}, u_{rl}).

The release path to the spall state is constructed from this point again using equation (11.4) with $u_r = u_{rl}$, with u_{ro} provided by the particle velocity at pressure P_{rl} on the boron carbide principal Hugoniot, and with $\kappa = +1$. The analysis is completed by constructing a further path from the spall state to that imaged at the PMMA recording interface again using equation (11.4). A spall stress of $P_s = -0.54$ GPa is determined when the intersection of this path and the principal PMMA Hugoniot is equal to the spall pullback particle velocity of the recorded VISAR measurement (~120 m s^{-1}). A second path is shown in figure 11.5 centered at the intersection of the spall release path pressure equal to zero. Intersection of this path with the principal PMMA Hugoniot determines the post-spall velocity

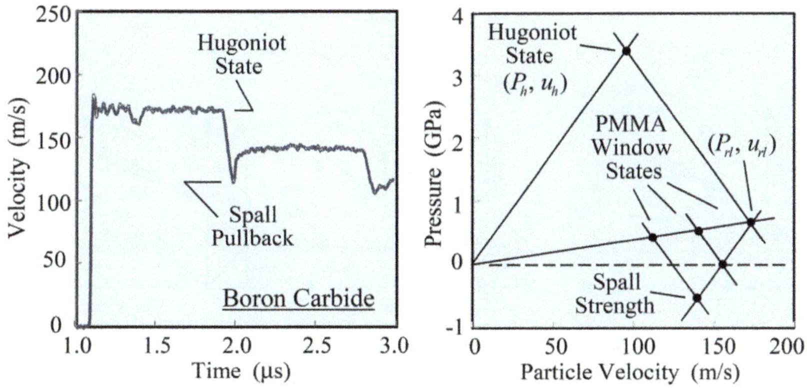

Figure 11.5. Velocity recording of spall failure in boron carbide ceramic (left) and pressure–particle-velocity analysis (right) carried out to determine spall strength from the measurement. Data from Grady and Moody (1996).

recorded by the measurement (~140 m s^{-1}). Further measurements of spall strength in boron carbide ceramics and some of the underlying physics of spall failure are discussed in chapter 1.

11.2.2 Distance–time plots in shock analysis

The use of distance–time plots has entered shock wave analysis in nearly all of the preceding chapters. Nonetheless, distance–time plots are fundamental to the analysis of shock wave experiments and data. A repeat of this integral tool of shock analysis is warranted. Three different representations of the distance–time plot are considered here and each has features that makes it useful under certain applications. First is the Eulerian distance–time plot in which the spatial position x and time t are independent coordinates. Second is the Lagrangian distance–time plot where in the initial reference frame the material particle position h and the time t are the independent coordinates. Lastly, is the inertial distance–time plot where coordinates are provided by encompassed material mass m and time t.

A representative illustration of each of the three distance–time plots is shown in figure 11.6. In each case the plots are illustrative of a step shock wave followed by a centered isentropic (dispersive) release wave that decompresses the material back to near the initial state. The function f of the independent variables in each case specifies any feature of the wave such as the particle velocity, pressure or shock velocity at the position and time.

Eulerian distance–time. The independent position coordinate is a point fixed in Euclidian space (at least for us concerned with the more ordinary issues of ballistic and shock wave engineering). The history of any feature of the wave motion at the spatial position x is then some functional form of the time. The pressure $p(x,t)$ or density $\rho(x,t)$, for example, or the Lagrangian position $h(x,t)$ are captured at the specified position x. The path of a wave is determined by the material velocity of the wave plus the motion of the medium through space. For example, in the plot on the left in figure 11.6 the initial shock wave propagates into material at rest and the

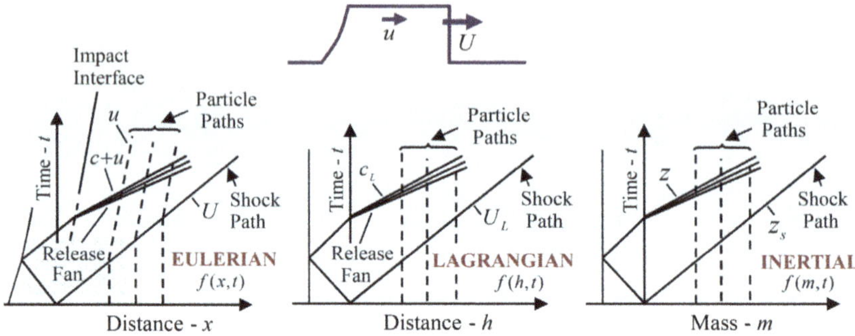

Figure 11.6. Distance–time plots of one-dimensional wave motion in the Eulerian, Lagrangian and inertial reference frames.

path is $x = Ut$, where U is the shock velocity. In contrast, release waves propagate into material in motion and the distance–time path reflects both the wave and material velocity, $dx/dt = c + u$. The spread of characteristics is a consequence of reduction in both the speed of the wave and the speed of the material as decompression proceeds. The paths of material elements in Eulerian space are correspondingly determined by the particle velocity $dx/dt = u$.

Lagrangian distance–time. Position in the Lagrangian distance–time plot is specified by the distance to a material point in an initial reference state of the participating media. The reference state is almost always the initial one-atmosphere condition where sample thicknesses are measured and distances to experimental recording planes are measured before shock experimentation proceeds. In the Lagrangian plot the time history of gage planes and other recording interfaces are vertical lines as illustrated in the center plot in figure 11.6. Wave speeds between interfaces are determined by the transit time and the initial (Lagrangian) distance of separation. Shock speeds in the Eulerian and Lagrangian plots for the example shown in figure 11.6 are the same $U = U_L$. If material ahead of the shock wave is in motion due to, for example, a preceding shock wave such as an elastic precursor wave or a phase-transformation precursor wave, then the Eulerian and Lagrangian shock speeds differ. The absolute Eulerian shock speed differs due to both the velocity and compressed state of the media ahead. The relative Eulerian and Lagrangian sound speed in the compressed state are related through the initial and compressed state density by the relation $\rho_o c_L = \rho c$.

Inertial distance–time. Lastly, illustrated on the right in figure 11.6, is the inertial distance–time plot. This plot, of course, relates closely to the Lagrangian plot in that position is identified with material points in the medium. There are, however, important differences. The inertial distance–time plot is not identified with any reference configuration of the system. Position is the encompassed mass m, the integrated density per unit area of the material in the direction of wave propagation. The inertial wave speed is the rate dm/dt at which mass is traversed irrespective of any spatial representation of the system. Inertial shock speed in the figure 11.6 inertial plot is $z_s = \rho_o U$, and is recognized as the shock impedance of the medium at the specified shock amplitude. The temporal rate at which the shock traverses mass is the inertial plot expression of the shock speed z_S. Sound speeds in the three distance time plots are related through $z = \rho_o c_L = \rho c$. The inertial sound speed follows from the isentropic derivative of the EOS pressure through $z^2 = (dp/dv)|_S$, where $Z = -(dp/dv)|_S$ is the isentropic inertial modulus of the material.

Application of the distance–time plot to experiment design and analysis is aptly illustrated in figure 11.7 by shock Hugoniot and EOS experiments conducted on a porous limestone rock (Murri and Grady 1973). In the experiments, in-contact explosives input a shock wave into the test material. The Hugoniot pressure induced is determined by shock impedance of the limestone and the selected explosive system. Planar slabs of the limestone are prepared of specified thickness and bonded with inductive particle-velocity gages between planes of the material and on the back free surface. The test design subjects the limestone first to shock compression to a near-constant Hugoniot state, and secondly to pressure decompression from the release wave emerging from the back free surface of the test material. (The post-Hugoniot state, prior to arrival of the release wave, experiences modest reduction in pressure because of the explosive Taylor wave.) Pulsed Helmholtz coils provide a uniform magnetic field in the proximity of the inductive gages over the duration of the shock measurement. Gage motion induced by the shock wave through the experiment duration provides time-resolved measurements of the particle velocity through a shock compression and subsequent decompression event.

The Lagrangian distance–time plot shown in figure 11.7 is used in experiment design, and in assessing the timing of instrument recording and inductive field application. Two inductive particle-velocity gages are placed within the test limestone at the positions shown in the distance–time plot. A third gage is placed on the back free surface of the limestone test system. Shock input is introduced at the explosive and test material contact interface identified at distance position zero and time zero. The Hugoniot shock wave traverses the test sample, emerges at the back

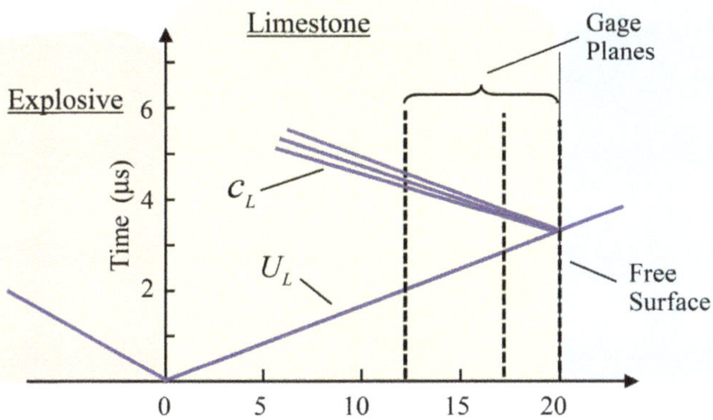

Figure 11.7. Distance–time plot in the Lagrangian representation for explosive-driven shock Hugoniot and release experiments on porous limestone.

free surface, and returns as a dispersing decompression wave. The experimental method is described further in section 11.8.4.

Particle-velocity histories measured in one such test are shown on the left in figure 11.8. Hugoniot particle velocity is provided directly be the interior gage measurements. Shock velocity is provided by transit times between the three gage records. The shock speed in the test shown is slightly in excess of 6 km s^{-1}. The shock velocity and measured particle velocity yield a Hugoniot pressure for the present test of 30.1 GPa.

The material tested in the study is a naturally occurring porous limestone identified as Indiana limestone, also referred to as Bedford limestone or Salem limestone. The density of the specific test material is 2.31 g cm^{-3}, as compared to the pycnometery-determined crystal density of 2.87 g cm^{-3}. The porosity is about 15%, and thus test data yield states on a slightly elevated temperature Hugoniot for limestone brought about by dissipation in the shock compaction process. A test series of four experiments with differing explosive systems provides a segment of Hugoniot over a pressure range of about 10–30 GPa. Hugoniot shock-velocity versus particle-velocity from the particle-velocity gage records and analysis of wave arrival time data are shown on the right in figure 11.8. Further EOS properties of the limestone are calculated from the decompression wave emerging from reflection of the shock at the back free surface. Decompression paths are assessed through appropriate Lagrangian analysis of the collective wave profiles as discussed in section 11.8. The wave speed of the first arrival of the decompression wave is plotted on the right in figure 11.8. This wave speed is identified as the bulk Lagrangian sonic velocity of the material at the Hugoniot pressure. The slope of the Hugoniot along with the bulk sonic wave speed determines the local pressure–energy–volume EOS of the material in the neighborhood of the Hugoniot state. The Gruneisen parameter at the Hugoniot state can be calculated from the local slope of the Hugoniot and the sonic wave speed by methods outlined in chapter 6. A Gruneisen gamma of approximately 1.6 is determined from the data in figure 11.8 (Murri and Grady,

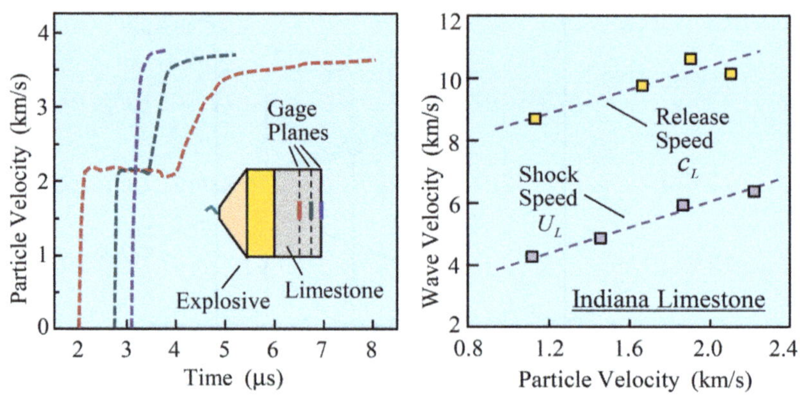

Figure 11.8. Shock Hugoniot and release wave experiments on porous Indiana limestone with inductive particle-velocity gages (Murri and Grady 1973).

1973), a reasonable measure of this thermodynamic property for limestone over the 10–30 GPa shock pressure range of the data.

11.3 Shock equilibration in a gage plane

The measurement of a structured shock wave is best accomplished through the placement of a diagnostic within the material under test. Experience has shown that better time resolution is achieved by the use of a high-fidelity diagnostic such as VISAR interferometry at an interface between the test material and a calibrated interferometer window material. The downside of an interface measurement is that dissimilar material properties lead to a wave structure that is distorted. Analysis necessary to recover the corresponding in-material wave measurement can be difficult, and not always possible.

The alternative measurement method of placing a structured shock wave diagnostic, such as a piezo-resistive stress gage, or an inductive particle-velocity gage, within the material is an attractive option, and has been pursued by many. Again, difficulties arise. The diagnostic of choice must be placed between plates of the test material. This is accomplished with a bonding agent, such as epoxy, that is commonly of lower mechanical impedance than the material being tested. The process of this lower impedance interface plane of finite thickness achieving stress equilibrium as the structured shock wave passes through is a wave interaction event. The stress equilibration process in a gage plane provides a useful application of the methods of shock wave analysis.

11.3.1 Stress wave analysis

The problem of interest is abstracted in figure 11.9. A step shock wave of amplitude σ_o is propagating in a medium with shock impedance z_1 (both materials are approximated as linear elastic) and is normal incident on the gage plane material of thickness a and shock impedance $z_2 < z_1$. The shock wave transmitted into the

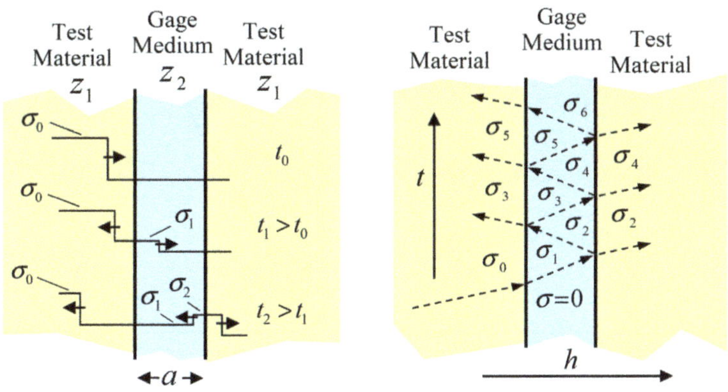

Figure 11.9. Left: A step shock wave propagating in the test material incident on a gage plane with a different (lower) wave impedance. Stress wave profiles at several later times after wave reflections are also illustrated. Right: A Lagrangian distance–time plot of the wave interaction in the gage plane, and the successive increasing stress levels.

gage medium reduces to lower stress amplitude σ_1 because the gage medium is a lower shock impedance material. The shock wave then echoes between the two interfaces within the test material as successively higher stress levels are achieved. The gage material equilibrates over time to the incident stress level σ_o. Of interest to the stress gage application is the reduction in measurement resolution brought about the time to stress equilibrium. At what time is measurement of stress in the gage plane a reasonable approximation to stress in the test medium? In alternative words, what frequency response can be assigned the gage measurement?

The pressure–particle-velocity diagram necessary to calculate the stress history within the gage plane is illustrated on the right in figure 11.10. The corresponding Lagrangian distance versus time plot is provided on the right in figure 11.9. Structure of the stress wave through the first several stress levels is shown on the left in figure 11.9. As the step stress wave impinges on the left-hand boundary of the gage plane a markedly reduced stress σ_1 is transmitted into the gage medium while a decompression wave is reflected back into the bounding test material. From the stress–particle diagram it is seen that the first stress level in the gage region is written as

$$\sigma_1 = z_2 u_1 = z_2(u_o + (u_1 - u_o)). \tag{11.5}$$

Next the relations

$$\sigma_o = z_1 u_o, \qquad \sigma_o - \sigma_1 = z_1(u_1 - u_o), \tag{11.6}$$

are used to eliminate particle velocity from equation (11.5), yielding

$$\sigma_1 = z_2 u_1 = z_2\left(\frac{\sigma_o}{z_1} + \left(\frac{\sigma_o - \sigma_1}{z_1}\right)\right). \tag{11.7}$$

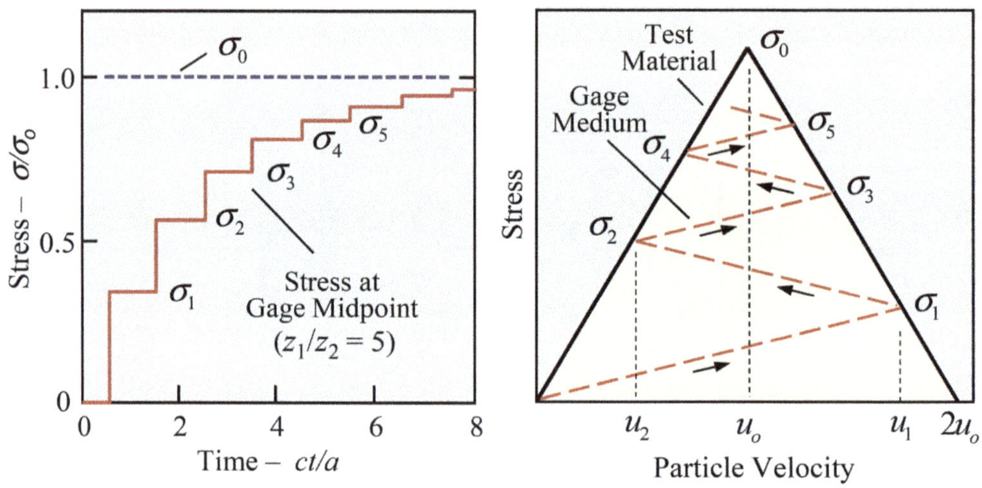

Figure 11.10. Wave interactions within the gage medium are shown in the stress–particle-velocity diagram on the right. The stress history of the midpoint of the gage region is shown on the left. An impedance ratio (z_1/z_2) of 5 is used for the calculation.

Equation (11.7) is solved for the first stress level in the gage zone and the stress amplitude of the reflected wave,

$$\sigma_1 = \frac{z_2}{z_1}\frac{2z_1}{z_2 + z_1}\sigma_o, \quad \sigma_1 - \sigma_o = \frac{z_2 - z_1}{z_2 + z_1}\sigma_o. \tag{11.8}$$

These relations are commonly written as

$$\sigma_1 = \frac{z_2}{z_1}T\sigma_o = \frac{z_2}{z_1}(1 - R)\sigma_o, \quad \sigma_1 - \sigma_o = R\sigma_o, \tag{11.9}$$

where T and R are, respectively, referred to as the transmission and reflection coefficients. Note that the sign of the reflection coefficient is chosen so that a negative R identifies a change in sign of the reflected wave in keeping with other areas of physics (e.g. optics, seismology). The reflection and transmission coefficients are related through $T + R = 1$.

Calculation of stress following wave interaction at the right-hand boundary of the gage plane proceeds by writing

$$\sigma_2 = \sigma_1 + z_2(u_1 - u_2) = z_2(2u_o - (2u_o - u_1) - u_2). \tag{11.10}$$

The respective terms for particle velocity on the right are replace by expressions for stress

$$\sigma_2 = \sigma_1 + z_2\left(\frac{2\sigma_o}{z_1} - \frac{\sigma_1}{z_1} - \frac{\sigma_2}{z_1}\right). \tag{11.11}$$

Solving for σ_2 yields

$$\sigma_2 = -\frac{z_2 - z_1}{z_2 + z_1}\sigma_1 + \frac{2z_2}{z_2 + z_1}\sigma_o. \tag{11.12}$$

Joining with equation (11.9) yields

$$\sigma_2 = (1 - R^2)\sigma_o. \tag{11.13}$$

A similar calculation for the next stress level σ_3 results in

$$\sigma_3 = (1 + R^3)\sigma_o, \tag{11.14}$$

and with modest effort one can show that subsequent stress levels result in the relation

$$\sigma_n = (1 - (-R)^n)\sigma_o. \tag{11.15}$$

Equation (11.15) for the stress at the midpoint of the gage plane is plotted on the left in figure 11.10. (Note that $R < 0$.) The stress history plotted in the figure for an impedance ratio $z_1/z_2 = 5$ ($R = -2/3$) illustrates the response and lack of fidelity of a gage region to a step stress wave. Gage equilibration time to the input stress level increases as the impedance ratio increases. The longitudinal impedance of aluminum oxide ceramic or copper metal is in the neighborhood of 4×10^7 kg s^{-1} m^{-2}, while that of an epoxy adhesive is perhaps 4×10^6 kg s^{-1} m^{-2}, or an impedance ratio of approximately 10.

The thickness of the gage region also plays a role in the time require to equilibrate to the input stress level. The number of reflections in the gage plane determines the value of n in equation (11.15) and the corresponding stress. If the thickness of the gage region in figure 11.9 is a, and the wave velocity in the gage region material is c, then a wave reflection and stress jump occurs at each time that ct/a achieves an integer value. If a number of steps are necessary to approach the input stress level, then it is reasonable to approximate equation (11.15) with $n = ct/a$, a continuous function of time. The power of the reflection coefficient in equation (11.15) then becomes

$$(-R)^{n(t)} = (-R)^{ct/a} = e^{(ct/a)\ln(-R)}, \tag{11.16}$$

and equation (11.15) becomes the exponential function

$$\sigma(t) = (1 - e^{-t/\tau})\sigma_o, \tag{11.17}$$

where

$$\tau = \frac{a}{-c \ln(-R)}, \tag{11.18}$$

provides a characteristic measure of the time required for the gage region to approach equilibrium.

An illustrative example is shown in figure 11.11 where stress wave measurements on shock compression of magnesium oxide (MgO) crystals are compared using a quartz piezo-electric pressure gage in one case and a Manganin piezo-resistive gage in another (Grady 1977). The quartz crystal is bonded to the MgO crystal with epoxy less than a few micrometers thick and results in very little degradation in the

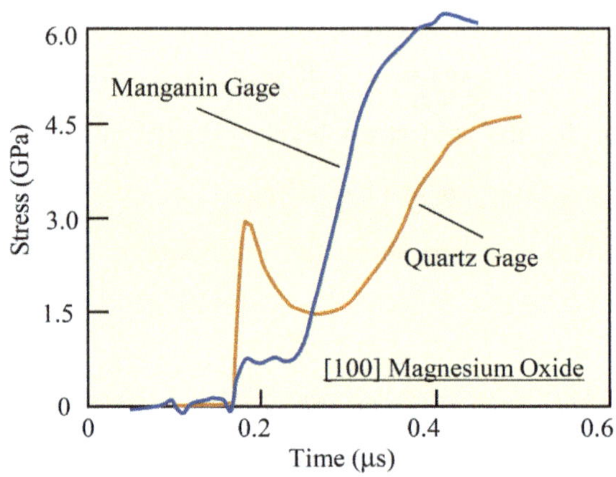

Figure 11.11. Stress wave measurements in 3 mm thickness of [100] oriented single-crystal magnesium oxide, comparing manganin piezo-resistive and quartz piezo-electric pressure gages (Grady 1977). Gage plane impedance mismatch is responsible for marked disparities in precursor wave structure and amplitude. Earlier second wave arrival reflects the higher shock pressure in the manganin gage test.

measured structure of the elastic precursor wave. (The resolution for the quartz gage is principally governed by the finite sensing area of the gage and small errors in the alignment of the planar stress wave incident on this area.) In contrast, the Manganin gage placed between crystals of MgO with a gage plane thickness of nearly 100 μm results in unacceptable degradation of both the wave shape and amplitude of the precursor wave.

11.3.2 Stress wave attenuation from planar defects

The previous analysis focuses on the equilibration of a stress wave passing through a dissimilar-impedance gage plane. The analysis also has application to a markedly different problem in shock wave interaction; namely, the passage of a stress wave through a heterogeneous layered medium. This subsection continues the analysis and examines the propagation and attenuation of a stress pulse through a medium of parallel planar defects. The subsequent section pursues application of the analysis to stress wave propagation in oil shale.

When a step stress wave undergoes normal incidence on a planar region of dissimilar acoustic impedance, such as illustrated in figure 11.9, a fraction of the energy of the wave is reflected back towards the direction from which the wave originates. If such planes are thin and sparsely distributed through the medium then they may be regarded as elastic planar defects and the analysis performed in the preceding section can be used to calculate the loss of energy and attenuation of waves due to such defects. Planar regions in which the acoustic impedance is either larger than, or smaller than, that of the principal medium, will exhibit similar behavior. Since the latter case was considered in the previous section, the case where $z_2 > z_1$ in figure 11.9 is considered here.

The stress–particle-velocity diagram shown on the right in figure 11.12, and the sequence of stress levels within the defect region, differs markedly from the previous

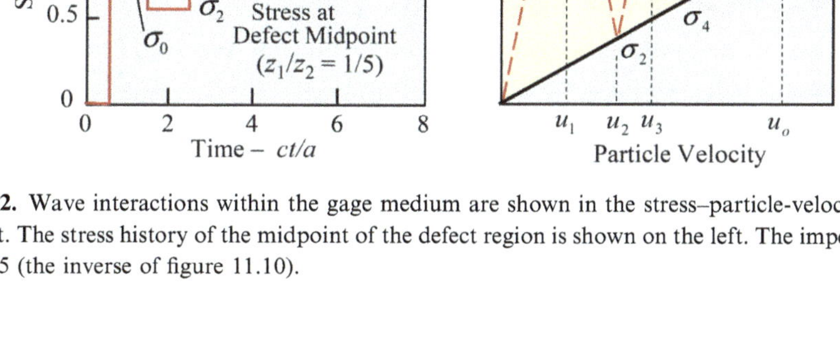

Figure 11.12. Wave interactions within the gage medium are shown in the stress–particle-velocity diagram on the right. The stress history of the midpoint of the defect region is shown on the left. The impedance ratio (z_1/z_2) is 1/5 (the inverse of figure 11.10).

calculation. Analysis, however, proceeds similarly. Writing for the first stress state in the defect zone

$$\sigma_1 = z_2 u_1 = z_2(u_o + (u_1 - u_o)), \tag{11.19}$$

and eliminating particle velocity from the relation using

$$\sigma_o = z_1 u_o, \quad \sigma_o - \sigma_1 = z_1(u_1 - u_o), \tag{11.20}$$

results in

$$\sigma_1 = \frac{z_2}{z_1} T\sigma_o, \quad \sigma_1 - \sigma = R\sigma_o, \tag{11.21}$$

and these are the same relations obtained for the previous case provided by equation (11.9). Continued analysis shows that the subsequent pressure steps in the defect zone are provided by

$$\sigma_n = (1 - (-R)^n)\sigma_o, \tag{11.22}$$

which is again the same relation as equation (11.15) for the previous case. The difference in the solution in this case is that the sign of the reflection coefficient R is positive, and highlights the fact that the defect medium impedance $z_2 > z_1$. The resulting stress history transmitted into the defect region is shown on the left in figure 11.12 and, for this case, reverberates between stress values above and below as equilibration to the input pressure level is sought.

The history of the reflected stress wave is also provided by equation (11.22) and is shown as the dashed curve in figure 11.12. For the reflected wave only the odd numbered stress levels constitute the history and equation (11.22) can be written as

$$\sigma_n = (1 + R^n)\sigma_o, \quad n = 1, 3, 5 \cdots. \tag{11.23}$$

Through the same arguments leading to equation (11.17), a continuous representation of the reflected stress history is provided by

$$\sigma(t) = (1 + \mathrm{e}^{-t/\tau})\sigma_o, \tag{11.24}$$

where in this case the characteristic equilibration time is

$$\tau = \frac{a}{-c\ln(R)}. \tag{11.25}$$

11.3.3 Application to multiple planar defects

The objective of the analysis is now to assess the energy removed from a stress pulse as it propagates through a collection of such parallel and planar defects. Consider a square pulse with stress amplitude σ and pulse duration T incident on one defect. The pulse duration is considered to be much longer than the characteristic defect time τ in equation (11.25). As the front of the square pulse passes over the defect, energy from the pulse is lost through reflection from the planar defect. This reflected energy is provided by

$$\varepsilon_r = \int \sigma \, dx = \int \sigma u \, dt = \frac{1}{z_1} \int \sigma^2(t) dt, \tag{11.26}$$

and finally

$$\varepsilon_r = \frac{1}{z_1} \int \sigma^2 \mathrm{e}^{-2t/\tau} \, dt. \tag{11.27}$$

The integrand contains the reflected portion of the stress from equation (11.24), and σ is now the constant amplitude of the pulse passing over the defect (identified as σ_o in equation (11.24)). Carrying out the integral and doubling this value (because an equal energy is also reflected when the back of the square pulse passing over the defect) provides

$$\varepsilon_r = \frac{1}{z_1} \sigma^2 \tau \tag{11.28}$$

for the back scattered energy from a single planar defect (again $\sigma = \sigma_o$, a constant).

If many such planar defects are parallel and distributed through the body with an average spacing b, then the energy loss from the pulse is

$$dE = -\varepsilon_r \, dN = -\left(\frac{1}{z_1} \sigma^2 \tau \right) \left(\frac{dh}{b} \right), \tag{11.29}$$

where $dN = dh/b$ is the number of planar defects that is passed over by the wave pulse in propagating a Lagrangian distance dh. The rate of attenuation of the energy of the square pulse is then provided by

$$\frac{dE}{dh} = -\frac{1}{bz_1} \sigma^2 \tau. \tag{11.30}$$

The energy of a planar wave pulse at any position h in the body can be defined by the work on that plane over the time history of the stress in the wave pulse, or

$$E(h) = \int \sigma \, dx = \int \sigma u \, dt = \frac{1}{z_1} \int \sigma^2 \, dt, \tag{11.31}$$

where $dx = u \, dt$ identifies the Eulerian displacement of the plane at Lagrangian position h. The latter step assumes an elastic relation between stress and particle velocity, $\sigma = z_1 u$. The loss of energy with propagation distance of the stress pulse is then

$$\frac{dE}{dh} = \int \sigma \frac{\partial u}{\partial h} dt + \int u \frac{\partial \sigma}{\partial h} dt = -\int \sigma \frac{\partial \varepsilon}{\partial t} dt - \rho_o \int u \frac{\partial u}{\partial t} dt. \tag{11.32}$$

The second step makes use of the Lagrangian conservation equations of mass and momentum described in more detail in chapter 1. Completing the differentials yields

$$\frac{dE}{dh} = -\int \sigma \, d\varepsilon - \frac{1}{2} \rho_o u_{\mathrm{res}}^2. \tag{11.33}$$

The velocity u_{res} on the right is the residual velocity of the material remaining after passage of the pulse. This velocity is zero for the elastic material considered here and is nominally zero for most materials that return to the original density after passage of the stress pulse. It would not be zero, for example, for a porous solid that undergoes a permanent densification during compaction in the stress wave. The first term on the right is the stress–strain hysteresis in the compression–decompression cycle applied in passage of the stress pulse.

The momentum of the stress pulse at the same position h is provided by the impulse applied on that plane

$$I(h) = \int \sigma \, dt. \tag{11.34}$$

The change of momentum with propagation distance is then

$$\frac{dI}{dh} = \int \frac{\partial \sigma}{\partial h} dt = -\rho_o \int \frac{\partial u}{\partial t} dt = -\rho_o u_{res}. \tag{11.35}$$

Again for the present elastic material and most materials that return to the original density after passage of the stress pulse, the residual velocity u_{res} is zero, or very nearly zero, and momentum of the pulse is conserved.

For the square stress pulse of amplitude σ and duration T at position h in the material the momentum from equation (11.34) is $I(h) = \sigma T$. Stress amplitude σ_o and T_o will determine the initial impulse applied to the material and define the input pulse momentum amplitude $I_o = \sigma_o T_o$ at $h = 0$. Conservation of momentum of the stress pulse requires $I_o = \sigma_o T_o = \sigma T$. The energy at any position h of the square pulse is then from equation (11.31),

$$E(h) = \frac{1}{z_1} \sigma^2 T = \frac{I_o}{z_1} \sigma(h), \tag{11.36}$$

where the second step follows from conservation of momentum in the stress pulse. Attenuation of energy and stress of the pulse wave are then proportional through

$$\frac{dE}{dh} = \frac{I_o}{z_1} \frac{d\sigma}{dh}, \tag{11.37}$$

and equation (11.30) for the present application of wave scattering in a body with a distribution of planar defects becomes

$$\frac{d\sigma}{dh} = -\frac{\tau}{b I_o} \sigma^2. \tag{11.38}$$

A corresponding differential equation for the pulse energy is obtain by introducing $E(h) = \sigma(h) I_o / z_1$.

A similar analysis of stress wave attenuation is undertaken in chapter 3 on dynamic fragmentation where an impulse imparted to a concrete slap by explosive detonation leads to stress pulse propagation and back surface spall.

Equation (11.38) is readily solved yielding

$$\sigma(h) = \frac{\sigma_o}{1 + h/h_o}, \tag{11.39}$$

where

$$h_o = \frac{I_o b}{\sigma_o \tau} = \frac{T_o}{\tau} b. \tag{11.40}$$

The characteristic attenuation length h_o scales with the nominal defect spacing. At propagation distances $h \gg h_o$ the stress pulse exhibits first-power power-law attenuation.

The attenuation length scale h_o from equation (11.40) is dependent on the density of defects through the characteristic spacing b and the material time constant τ. Equation (11.25) provides this time constant in terms of the relative impedance and the thickness of the scattering zone. When the impedance of the scattering zone is appreciably higher than that of the medium, that is $z_2 \gg z_1$, the logarithmic can be approximated by $\ln R \simeq -2z_1/z_2$ and equation (11.25) becomes

$$\tau = \frac{\rho_2 a}{2z_1} = \frac{m}{2z_1}, \tag{11.41}$$

where $m = \rho_2 a$ is the mass per unit area of the scattering zone. The scattered energy from equation (11.28) then becomes

$$\varepsilon_r = \frac{1}{2} m \frac{\sigma_o^2}{z_1^2} = \frac{1}{2} m u_o^2. \tag{11.42}$$

Using equation (11.41) in equation (11.40) and noting that $m/b = (m/a)(a/b) = \rho_2 \nu_2$, where ρ_2 and ν_2 are the density and volume fraction of the scattering material, respectively, the attenuation length scale due to scattering becomes

$$h_o = \frac{1}{2} \frac{\rho_1}{\rho_2 \nu_2} c_1 T_o = \frac{1}{2\nu_2} \frac{z_1}{z_2} c_2 T_o. \tag{11.43}$$

11.4 Application to stress pulse attenuation in oil shale

An instructive example of stress pulse attenuation is provided by wave profile measurements on samples of oil shale. The test material is oil shale from the US state of Wyoming region of the Green River Formation. This specific oil shale has a kerogen content of about 40 gallons ton^{-1}. The density and wave speed in the material are 2150 kg m^{-3} and 3.0 km s^{-1}, respectively. Kerogen is the organic component from which oil is extracted from the shale. Kerogen in the oil shale tested is not distributed uniformly but rather resides in thin and parallel planes separated by regions relatively lean in kerogen content. The matrix material is carbonate marlstone (principally dolomite). An image of the surface of a sample of the Green River Formation oil shale is illustrated in figure 11.13.

Figure 11.13. An image of oil shale surface (approximately 1 cm^2) illustrating laminar character kerogen planes. Stress wave propagation in the tests described is from left to right. The wavy looking feature is a scratch on the surface.

11.4.1 Elastic scattering and stress pulse attenuation

In the experiments performed, thin sheets of PMMA, commonly referred to as Plexiglas, are mounted on projectiles and caused to undergo planar impact on samples of oil shale (Grady 1983). Induced planar stress wave pulses propagate through the oil shale samples in the direction of the normal to the planes of kerogen. Impact velocities of about 70 m s^{-1} resulted in impulsive pressure loading of oil shale samples with a nearly square pressure pulse to approximately 150 MPa in stress amplitude and 2 μs in time duration. Earlier shock wave experiments on oil shale indicated a Hugoniot elastic limit (HEL) and measure of the dynamic strength of slightly in excess of 200 MPa (Munson 1977). Accordingly, impact at the somewhat reduced shock pressure amplitude is expected to be nominally elastic in the stress wave response. The velocity histories shown in figure 11.14 are measured with the

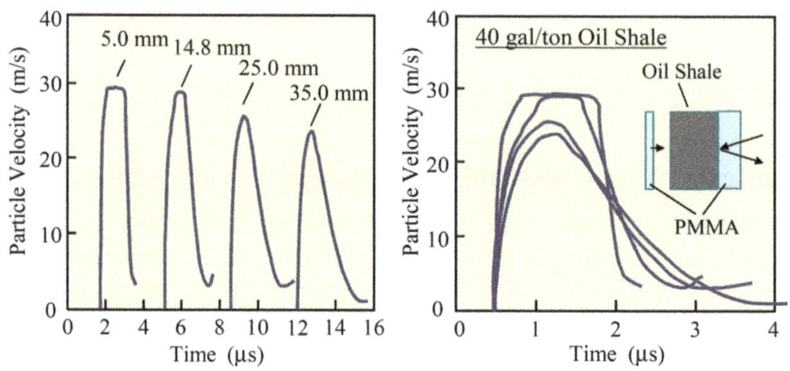

Figure 11.14. Measured particle-velocity histories resulting from stress pulse propagation through increasing thicknesses of 40 gal. ton^{-1} oil shale (Grady 1983). The stress pulse is produced by planar impact of a 2 mm PMMA plate at approximately 70 m s^{-1} velocity. Velocity is measured by interferometry methods within a PMMA window material directly behind the oil shale sample. The peak input pressure in the oil shale is approximately 150 MPa. The same velocity profiles overlay on the right-hand side to illustrate the change in wave structure with propagation distance.

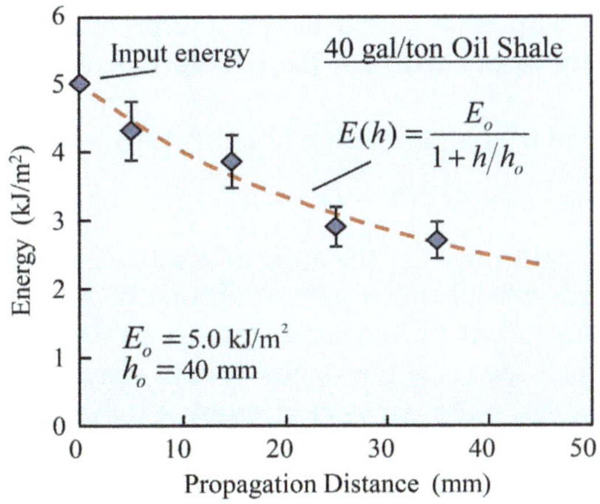

Figure 11.15. Energy of the stress pulse in 40 gal. ton^{-1} oil shale at increasing propagation distance (Grady 1983). Input energy of 5 kJ m^{-2} is calculated from the impact pressure amplitude and impulse duration.

velocity interferometer method (Barker and Hollenbach 1972) within PMMA window material backing increasing thicknesses of the oil shale test samples. Sample thicknesses and corresponding wave pulse propagation distances are indicated on the left in figure 11.14.

Present analysis of the stress wave data will focus on the energy of the shock pulse. The initial energy input of approximately 5 kJ m^{-2} is calculated from the impact pressure and duration of the input pulse. The subsequent energy of the wave pulse at increasing propagation distance is calculated from the integrated time history of the measured pulse accounting for the impedance mismatch between the oil shale sample and the PMMA velocity interferometer recording media. Energy at each propagation distance is calculated through equation (11.31) by relating stress to particle velocity through a constant wave impedance. The latter step is not precisely correct, although analysis shows that the inaccuracy is small.

Wave pulse energy at increasing propagation distance is plotted in figure 11.15 and compared with predicted attenuation based on wave scattering from equation (11.39). The experimental data and the theoretical attenuation relation based on scattering are in reasonable accord. The value of the attenuation length scale h_o of 40 mm represents a best fit to the rate of decay of the energy. Other mechanisms could of course be responsible for wave pulse attenuation at stress levels below onset of pervasive inelastic yield of oil shale. Viscoelastic behavior or modest plasticity of the highly organic material are also possible dissipation mechanisms.

11.4.2 Viscous attenuation of a stress pulse

Solid viscosity can also lead to the attenuation and dispersion of a shock stress pulse such as described in the previous subsection. An interesting question would be the form of a viscous constitutive relation that leads to the same first-power, power-law attenuation arrived at through an elastic scattering model that led to equation (11.39). An exploration of viscosity in the propagation of shock pulses exercises nonlinear wave analysis methods used in the interpretation of shock wave propagation. Solid viscosity associated with the structuring of shock waves is explored further in chapter 8.

Consider the elementary one-dimensional constitutive law of the form

$$\sigma(\varepsilon, \dot{\varepsilon}) = E\varepsilon + \eta\dot{\varepsilon}^m. \tag{11.44}$$

The first term on the right provides for a linear elastic response between stress and strain through an elastic modulus $E = \rho c^2$. Additionally, a viscous stress is provided by the second term dependent on the strain rate to a power m. A power of $m = 1$ would, for example, provide for a linear Newtonian viscosity.

Again, consider a stress pulse parameterized by a pulse stress amplitude σ and a pulse duration T. The analysis leading to the rate of pulse energy attenuation in equation (11.33) continues to apply and

$$\frac{dE}{dh} = -\int \sigma \, d\varepsilon = -\int (E\varepsilon + \eta\dot{\varepsilon}^m) d\varepsilon = -\int_0^T \eta\dot{\varepsilon}^m \, d\varepsilon = -\int_0^T \eta\dot{\varepsilon}^{m+1} \, dt. \tag{11.45}$$

The last steps show that over the duration of the pulse, elastic strain energy is recovered and only viscosity contributes to energy decay. Calculation of the integral in equation (11.45) requires knowledge of the time history of the strain through the stress pulse. For example, if the pulse is a half sine wave and $m = 1$, then the integral is readily calculated and leads to an average strain rate through the pulse of $\dot{\varepsilon} = (\pi/\sqrt{2})\varepsilon/T \simeq 2.2\varepsilon/T$, where ε is the maximum strain in the pulse and T the pulse duration. Other assumptions lead to comparable values for the average strain rate. To expedite the calculation the dimensionally correct assumption of $\dot{\varepsilon} = \varepsilon/T$ for the average strain rate is made. The integral then yields

$$\frac{dE}{dh} = -\eta\left(\frac{\varepsilon}{T}\right)^{m+1} T. \tag{11.46}$$

Again, momentum conservation in the stress pulse leads to a proportionality between the rates of energy and stress attenuation in equation (11.37) and provides the differential expression

$$\frac{I_o}{z}\frac{d\sigma}{dh} = -\eta\left(\frac{\varepsilon}{T}\right)^{m+1} T \tag{11.47}$$

for dependence of the amplitude of the stress pulse with propagation distance. The invariance of the impulse in the stress pulse $I_o = \sigma T$ reduces equation (11.47) to

$$\frac{d\sigma}{dh} = -\eta\frac{z}{(\rho c^2 I_o)^{m+1}}\sigma^{2m+1}. \tag{11.48}$$

Integration yields

$$\sigma(h) = \frac{\sigma_o}{(1 + h/h_o)^{1/2m}}. \tag{11.49}$$

The functional form of the elastic scattering pulse attenuation solution of equation (11.39) is replicated only for a viscous strain-rate power of $m = 1/2$. The length scale h_o in equation (11.49) for $m = 1/2$ is

$$h_o = \frac{(\rho c^2 I_o)^{3/2}}{\eta z \sigma_o}. \tag{11.50}$$

Equating equation (11.50) to the corresponding scattering attenuation length scale from equation (11.40) and solving for the viscosity coefficient yields

$$\eta = \sqrt{\rho c^2 I_o}\frac{c\tau}{b}. \tag{11.51}$$

Viscosity increases with the characteristic time τ of the scattering defects and decreases with the nominal defect spacing b as intuition expects. The calculated viscosity coefficient is not a property of the material, however, as the coefficient η depends on the magnitude of the initial impulse I_o. It is nonetheless interesting that a viscosity dependent on the square root of the strain rate also emerges from the fourth-power dependence of steady structured waves (see chapter 8).

11.5 Shock and release in concrete

An alternative study offers perspectives on extended applications of the analysis methods integral to shock EOS experimentation. The material is concrete. Because of the coarse heterogeneous microstructure and modest shock impedance of concrete, the more conventional impact compression and decompression experiment to assess the shock constitutive response is rendered difficult. An alternative shock and release experimental technique (Rosenberg and Ahrens 1966, Lysne *et al* 1969, Chhabildas and Grady 1984) is well suited to the dynamic testing of concrete and comparable materials. Analysis of such data brings forward issues in the interpretation of shock wave measurements.

11.5.1 Reverberation shock decompression experiment

The experimental design used to assess the shock Hugoniot and pressure release properties through the technique is illustrated in figure 11.16 (Grady 1996). Conventional gas gun technology is used to perform planar impact experiments over the impact pressure range of interest. The concrete sample is mounted on the projectile as shown. A thin metal plate (the witness plate) is placed in a stationary target assembly and back-surface motion of the witness plate is monitored with VISAR diagnostics throughout the shock compression and subsequent decompression of the test concrete material. Both copper and tantalum metal provide appropriate witness plate materials. EOS properties of the metal are necessary, and the shock impedance needs to be markedly higher than that of the concrete.

The pressure–particle-velocity and distance–time plots necessary to the design, interpretation and analysis of the experimental data are illustrated in figure 11.17. Planar impact of the concrete sample onto the stationary metal witness plate shocks

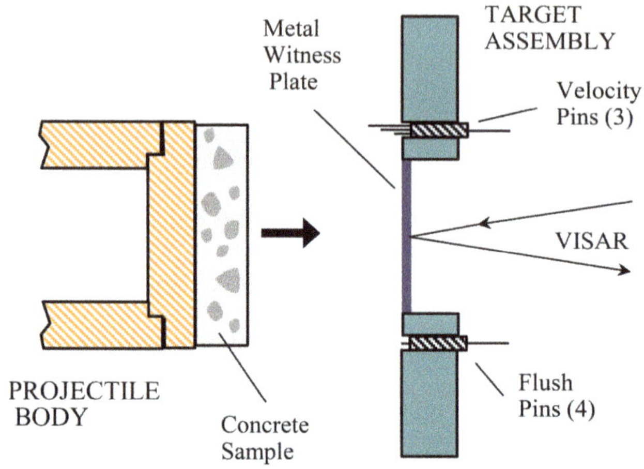

Figure 11.16. Experimental configuration for testing shock Hugoniot and decompression properties of concrete using the reverberating shock technique.

both concrete and metal to their respective Hugoniot states. Continuity of pressure and particle velocity is required at the impact interface. The sequence of pressure and particle-velocity states is illustrated with the plot on the left in figure 11.17. When the shock in the metal witness plate emerges at the recording surface, a reflected wave decompressing the metal to zero pressure is propagated back into the plate. The amplitude of the shock, along with the shock Hugoniot and release isentrope of the plate metal, determines the jump in particle velocity to the value u_1. The particle velocity u_1 is the amplitude of the first plateau in the VISAR interferometer measurement.

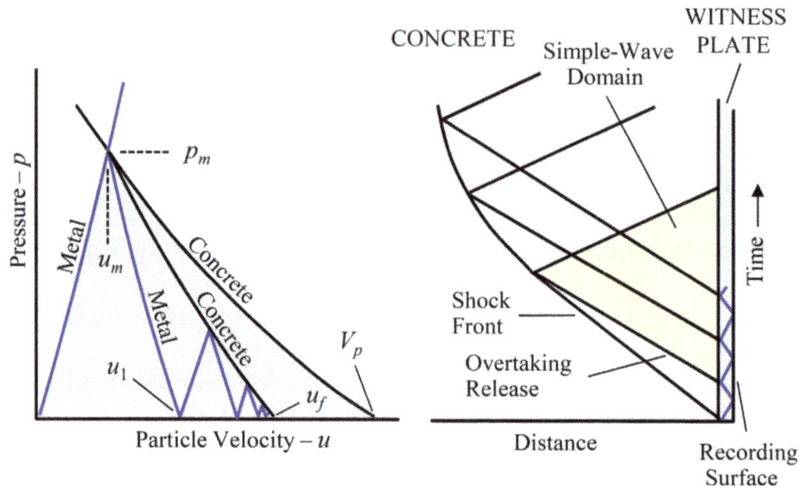

Figure 11.17. Pressure–particle-velocity (left) and distance–time plots (right) for interpretation and analysis of reverberating shock and release experiments on concrete.

Subsequent acceleration of the witness plate following the initial Hugoniot state plateau is governed by the dynamic decompression properties of the concrete from the Hugoniot state. The plot on the left in figure 11.17 reveals that the test material decompresses in distinct pressure intervals determined by the decompression isentrope of the witness plate metal and the unique release properties of the concrete. Decompression states in the test material are mirrored in the particle-velocity measurement through corresponding steps in the velocity history.

11.5.2 Shock decompression measurements for concrete

Experimental velocity histories using the test configuration of figure 11.16 are shown in figure 11.18. The test material is identified as SAC-5 concrete (Ehrgott 1993). Studies at the time revealed comparable shock wave constitutive response of Conventional Strength Portland Cement (CSPC). Copper metal provides the witness plate material. The series of tests span a Hugoniot pressure range in the concrete over about 10–25 GPa. The test records reveal the Hugoniot state plateau u_1 and several subsequent plateaus corresponding to decompression steps in the concrete before transitioning to more or less continuous acceleration late in time. Irregularities in the time-resolved particle-velocity data are a consequence of the coarse aggregated nature of the concrete material.

Experimental time-resolved measurements of figure 11.18 are used to extract shock Hugoniot and decompression EOS properties of the test concrete. Computational methods can be applied directly to benchmark computational models against the measured experimental data. Alternatively, analytic methods applying the tools outlined in the present chapter are pursued to assess model-independent EOS responses of the material.

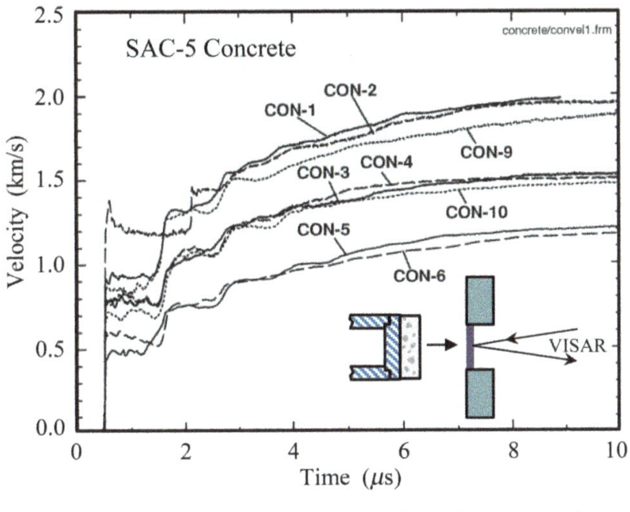

Figure 11.18. Experimental velocity histories for shock Hugoniot and decompression tests on concrete with copper witness plate (Grady 1996).

Hugoniot properties. Planar impact of the concrete sample on the stationary metal witness plate shocks both concrete and metal to the respective Hugoniot states. Emergence of the shock at the recording free surface decompresses the metal to zero pressure and accelerates the plate to velocity u_1 as illustrated in the pressure–particle-velocity plot in figure 11.17.

The pressure and particle-velocity Hugoniot states p_h and u_h in the test concrete are calculated from the measurement of u_1 and the EOS properties of the witness plate metal. Namely

$$p_h = p_m = \rho_m C_m u_1/2 + \rho_m S_m (u_1/2)^2 \tag{11.52}$$

and

$$u_h = V_p - u_m = V_p - u_1/2. \tag{11.53}$$

The projectile impact velocity is V_p, while the set of parameters (ρ_m, C_m, S_m) constitute a linear shock-velocity–particle-velocity Hugoniot description of the witness-plate metal. The sensible assumption is also made that the decompression isentrope of the metal mirrors the Hugoniot of the metal in pressure–particle-velocity space. Hugoniot shock velocity, specific volume and specific energy, completing the Hugoniot state description of the concrete, are calculated through the Hugoniot conservation relations.

Decompression properties. Subsequent acceleration of the witness plate following the first Hugoniot state plateau is governed by the dynamic decompression properties of the concrete from the Hugoniot state. These data reflect the thermo-elastic EOS properties of the test material through the span of pressures below the Hugoniot state. The data further reflect constitutive responses including void compaction, material strength and perhaps other thermo-mechanical issues, as the initial lattice strain is recovered.

Discrete decompression states in the test concrete are readily calculated from the subsequent particle-velocity plateaus in the particle-velocity measurements (e.g. Chhabildas and Grady 1984). Scatter in the data, a consequence of the coarse heterogeneity of the test material, correspondingly translates to significant scatter in such calculated states. A discrete analysis does not fully utilize the experimental data. Previous assessments of stress waves incident on in-material gage planes in section 11.3 reveal the advantage of continuous analytic descriptions of the stress equilibration process. Similar analysis is appropriate to the present witness-plate acceleration data. For times subsequent to initial Hugoniot compression, a continuous description of witness-plate acceleration is effected by treating the plate as a lumped mass per unit area accelerating under the time-dependent pressure at the concrete–metal interface. This approximation is acceptable when numerous wave reflections occur within the witness plate over the acceleration time.

The pressure–particle-velocity release behavior of the test concrete is determined through the solution of the equation of motion governing acceleration of the metal witness plate. In turn, the corresponding pressure–specific-volume response of the concrete on decompression is determined through solution of the appropriate

Riemann integral. This aspect of the analysis is dependent on the simple-wave behavior of the decompression wave propagating from the concrete–metal interface back into the concrete. Concepts of simple waves and Riemann analysis are address in chapter 1. Decompression wave propagation remains simple until the lead characteristic of the decompression wave overtakes the initial shock wave imparted to the concrete. The domain of simple wave propagation is identified in the distance versus time plot of the experiment in figure 11.17.

The relation governing rigid acceleration of the metal witness plate is

$$m\, du/dt = p(u), \tag{11.54}$$

where $m = \rho_m h$, while ρ_m and h are density of the metal and metal plate thickness, respectively. An analytic form for the decompression pressure–particle-velocity response of the concrete is assumed of the form

$$p(u) = \rho_o C(u_f - u) + \rho_o S(u_f - u)^2, \tag{11.55}$$

with ρ_o the initial density of the concrete, u_f the terminal plate velocity, while C and S are unspecified constants to be determined by the witness plate acceleration data. The relation provided by equation (11.55) corresponds to the decompression pressure–particle-velocity path in the concrete shown on the left in figure 11.17. The form of equation (11.55) is familiar to the shock wave compressibility of many materials.

The particular solution of equation (11.54) and equation (11.55) yields

$$\frac{u_f - u}{C + S(u_f - u)} = \frac{u_f - u_m}{C + S(u_f - u_m)} e^{-(\rho_o S/m)t}, \tag{11.56}$$

with u_m the Hugoniot state particle velocity in the witness-plate metal as identified in figure 11.17. Further analysis requires optimizing equation (11.56) to the experimental velocity measurements through adjustments of the parameters C and S, and, if not adequately revealed in the data, the terminal plate velocity u_f.

The experimental pressure–particle-velocity relations provided by equation (11.55) are reflected and translated as described in section 11.2 of the present chapter. Decompression paths are then compared with the primitive pressure–particle-velocity shock Hugoniot for concrete. This operation yields

$$p(u) = \rho_o C(u - u_{\text{res}}) + \rho_o S(u - u_{\text{res}})^2. \tag{11.57}$$

where $u_{\text{res}} = V_p - u_f$ is the residual particle velocity on decompression from the Hugoniot state. The residual particle velocity u_{res} is unique to each experimental record.

Compression strain on release from the Hugoniot state is calculated from equation (11.57) and solution of the Riemann invariant

$$d\varepsilon = du/c(u), \tag{11.58}$$

where $c(u) = dp/du$. Equations (11.57) and (11.58) yield an analytic solution for the volumetric strain on the decompression path from the Hugoniot state

$$\varepsilon(u) = \varepsilon_{res} + \frac{1}{2S} \ln\left(\frac{2S}{C}(u - u_{res}) + 1\right). \tag{11.59}$$

In equation (11.59) the volume strain is $\varepsilon(u) = 1 - \rho_o/\rho$ and ε_{res} is the residual volume strain on decompression to zero pressure. Again ε_{res} is unique to each experimental record. Eliminating particle velocity between equations (11.57) and (11.59) provides the pressure–volume–strain decompression relation from the Hugoniot state as assessed from the experimental measurements of figure 11.18.

Selected decompression curves from the principal Hugoniot of SAC-5 concrete determined from the data in figure 11.18 are shown on the left in figure 11.19. Decompression curves are referenced to the shock Hugoniot for the concrete as determined by an optimum fit of the Hugoniot data and augmented by the known Hugoniot EOS of quartz. (SAC-5 concrete, between aggregate and grout, is approximately 80% quartz by mass.) The scatter in the Hugoniot data is a consequence of the coarse aggregate composition of the concrete and illustrates the difficulties of testing such materials. Decompression from 4.6 GPa reveals irreversible compaction of the 10%–12% percent porosity of the concrete, consistent with static compression studies on the same concrete (Ehrgott 1993). Decompression from Hugoniot states of 10 and 24 GPa reveals residual strain recovery. There are several explanations for this behavior.

11.5.3 Decompression strain recovery

First, there will be a degree of thermal expansion on shock recovery due to the extensive shock heating during shock compression. As noted, silica aggregate and sand in the concrete mixture constitute about 80% by weight of the material. Thus the properties of quartz are appropriate in examining the consequences of thermal expansion in the shock release process. The thermal expansion properties of crystalline quartz over the temperature range of concern are strongly nonlinear primarily due to the α-quartz to β-quartz transformation in the range of 550 °C–600 °C. Estimates of volume recovery based strictly on calculations of the residual

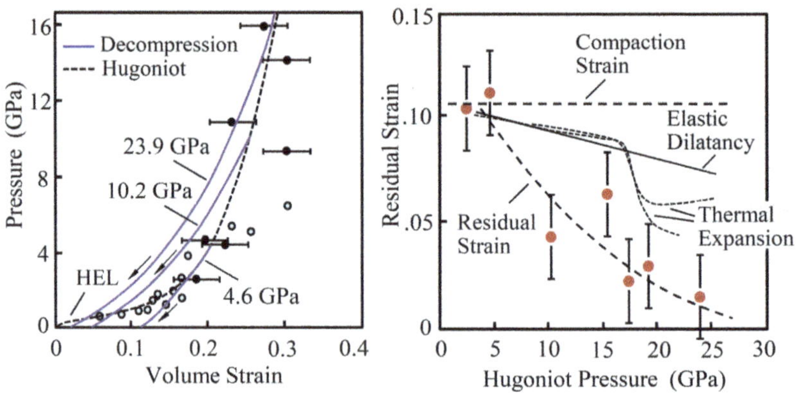

Figure 11.19. Left: Hugoniot and decompression path measurements for concrete. Right: Experimental residual strain assessed from shock and subsequent decompression data (Grady 1996).

temperature on decompression from the concrete Hugoniot and thermal expansion properties for quartz from Dmitriyev *et al* (1972) are shown on the right in figure 11.19. Residual shock temperature calculations are based on an assumption of uniform energy dissipation and concomitant temperature rise in the shock compression and release cycle. Predictions of volume recovery based strictly on thermal expansion due to shock heating can account for at best half of the measured residual strain. The heterogeneous nature of dissipation in the shock compression process, however, warrants further consideration.

There is a further mechanism for strain recovery on decompression that has broader implications to the shock compression constitutive response of high-strength compounds. The mechanism is traced to the anisotropic compliance of the crystalline material. Compaction and inelastic shear during shock compression eliminates all, or most, of the initial void volume, but leaves the granular media in a debonded, or weakly bonded, state at the Hugoniot pressure. The crystalline elastic compliance of quartz is anisotropic. The trigonal crystal symmetry of quartz yields a compliance with axial symmetry described by two elastic constants. Crystal decompression from the Hugoniot state will accommodate a shape change as well as a volume change. Random grain orientation will lead to incompatible compliance during decompression and the opening of void volume.

The compliance of crystalline quartz is characterized by two constants—the compliance β_p parallel to the symmetry axis and the compliance β_s perpendicular to this axis. Values for quartz are $\beta_p = 0.0067\ \text{GPa}^{-1}$ and $\beta_s = 0.0095\ \text{GPa}^{-1}$ (Nye 1957). As components of a second rank tensor, the invariant $\beta_s + 2\beta_p$ is responsible for the volume recovery, while shape change is determined by the deviator expression $|\beta_s - \beta_p|$. The ratio of the two provides a sensible measure of the ratio of the dilatant void strain to lattice volume strain recovery on decompression. Calculation of this athermal dilatant strain recovery based on the properties of quartz is shown on the right in figure 11.19 identified as the curve elastic dilatancy. Although neither thermal expansion nor athermal decompression dilatancy separately account for the experimental residual strain recovery, taken together they are of the right order to explain the trend of the data.

11.6 Shock velocity versus particle velocity

Measurement of the shock velocity and the particle velocity behind the shock wave is fundamental to characterization of the shock response of matter. Other state parameters such as the pressure, compression and temperature can be, and have been, measured. For the overwhelming preponderance of materials, however, shock wave properties and the shock wave EOS are established though measurement of shock velocity and particle velocity.

For many materials a series of such measurements leads to a single curve in shock-velocity–particle-velocity space (U–u space). This is the case for example in the LANL compendium of shock wave data (Marsh 1980), or the comparable compendium of Trunin *et al* (2001). For many of these materials the U–u data are well described by a linear curve defined by $U = C_o + Su$. For select materials an additional quadratic term Qu^2 is included. The parameters C_o, S and sometimes Q are fundamental properties of the Hugoniot response of the material. Other materials require a more complex curve to describe the data. Discussions of the details of the shock-velocity versus particle-velocity relations are addressed in chapters 1 and 10, while the uses of such relations pervade all of the chapters. Data from the LANL compendium for selected metals and compounds are shown in figure 11.20.

The data in figure 11.20 are, in particular, the shock-velocity–particle-velocity curves for the LANL impedance match standards (Carter *et al* 1971, Marsh 1980). Hugoniot curves for these standards have been establish with additional care. They provide known basis materials in subsequent tests, through impedance match techniques, on which a large portion of the LANL database is dependent. Span of the curves in figure 11.20 establish the experimentally determined range of the Hugoniot for each material. The parameters C_o and S, and for the case of iron Q, are provided in the figure caption.

With evolution in shock wave instrumentation, more sensitive measurements have revealed that compression to a shock state is not always achieved through a single shock wave. When materials exhibit cusps in the pressure versus dynamic compression response, the potential exists for a schism of the wave into two or more shock waves in achieving the final Hugoniot state. Cusps are a consequence of material elastic-plastic failure, polymorphic phase transformation, or other critical state phenomena encountered in the shock compression event.

In the 1980s and 1990s a shock wave diagnostic combining an inclined mirror with high-speed streak-camera technology (Ahrens *et al* 1968) was refined for purposes of resolving multi-wave structure in the shock response of solids. An exemplary example is displayed in figure 11.21 exhibiting three-wave structure in the shock compression of

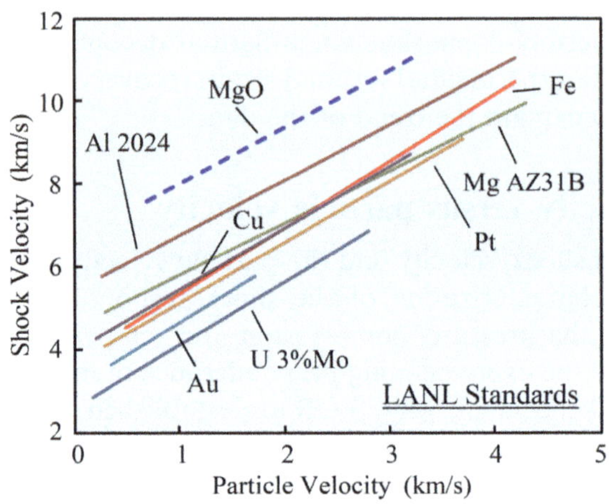

Figure 11.20. Shock-velocity–particle-velocity data for the LANL impedance match standards (Marsh 1980). Values for C_o and S, and for the case of iron Q, are provided by Al 2024 (5.33, 1.34), Au (3.06, 1.57), Cu (3.94, 1.49), Mg AZ31B (4.52,1.27), MgO (6.60, 1.37), Pt (3.63,1.47), U 3%Mo (2.57, 1.53) and Fe (3.57, 1.92, −0.068).

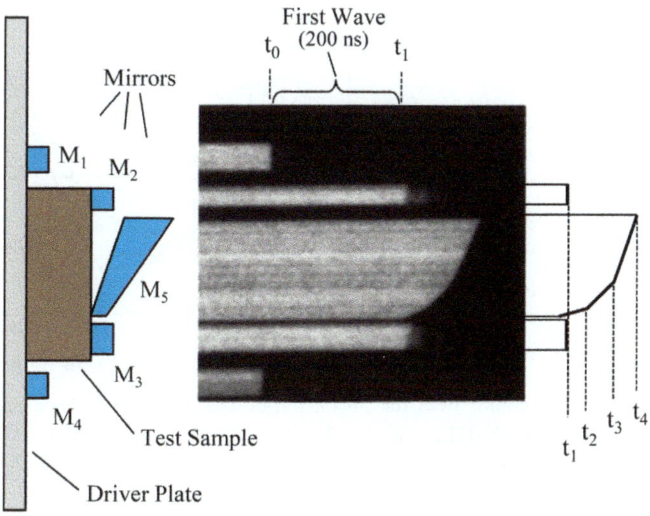

Figure 11.21. Shock EOS experiment with the inclined-mirror streak-camera technique on silicon nitride ceramic exhibiting three-wave structure . Reprinted figure with permission from Hongliang *et al* (2000). Copyright 2000 by the American Physical Society.

a silicon nitride ceramic with two-stage light-gas-gun techniques (Hongliang *et al* 2000). High-quality silicon nitride ceramic sustains a HEL strength of about 16 GPa. Further, silicon nitride initiates phase transformation on the Hugoniot at about 36 GPa to a higher density solid phase. Both HEL failure and onset of phase transformation are apparent in the streak-camera record and delineated by the change in slope in the streak record. Experimental shock velocity and free-surface velocity are determined from the arrival time and slope of the respective streak record segments in the measurement. The common assumption is made that the in-material Hugoniot particle velocity is one-half of the measured free surface velocity. As with earlier methods, basic data are presented as a display of shock velocity versus particle velocity. Hugoniot pressure versus density follows from the Hugoniot conservation relations applied to each of the several shock waves.

Both the basic data and resulting pressure–density Hugoniot for silicon nitride assessed from the data are shown in figure 11.22. Hugoniot data are compared with the reference isentrope for the higher pressure phase determined from both diamond anvil cell (DAC) and the present shock Hugoniot data (Hongliang *et al* 2000). The behavior on the Hugoniot above the transition pressure of 36 GPa to about 100 GPa is identified by the authors as a mixed-phase region. That transformation does not go to completion through the shock transition at these intermediate Hugoniot pressures is posed as an explanation.

Such apparent mixed-phase Hugoniot behavior has been reported for other high-strength ceramic compounds. Representative is the shock compression response of aluminum nitride (AlN) ceramic which has been experimentally investigated with both the inclined-mirror streak-camera method and with time-resolved velocity interferometry. One velocity interferometer recording shown on the left in figure 11.23 is illustrative of the intricate shock response of this compound. A first wave is

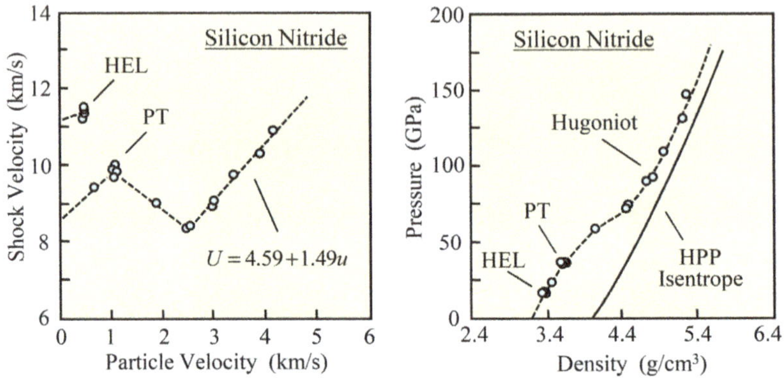

Figure 11.22. Experimental Hugoniot EOS for silicon nitride based on the inclined-mirror streak-camera method including Hugoniot strength (HEL) and shock-induced phase transformation (PT). Isentrope for the high-pressure phase (HPP) is also shown on the right. Data are from Hongliang *et al* (2000).

indicative of the elastic response of the ceramic traveling through the material with a wave velocity of about 11 km s^{-1}. The limiting elastic wave amplitude (the HEL) identifies the dynamic HEL strength of about 8 GPa supported by the ceramic before onset of failure. Separation into a second and third wave reveals a cusp in the compression at about 20 GPa that is a consequence of the onset of solid–solid crystallographic phase transformation and densification in the aluminum nitride lattice. A final Hugoniot state of about 40 GPa is achieved in this experiment. The additional complications in the measured wave structure seen in figure 11.23 and identified on the left are brought about by the wave interactions introduced at the sample–window interface due to differing sample and window shock impedance properties. Comparable wave measurements shown on the right further illustrate the complex shock response of this compound. The wave in the left-hand plot and the second highest amplitude wave in the right-hand plot are the same. Further discussions of the shock structure displayed and shock-induced phase transformation in AlN are provided in chapter 9.

In the shock compression experimental results shown, velocity interferometry or VISAR diagnostics monitor the transient velocity history at an interface between the test sample and a laser window material (optical grade lithium fluoride in the present case). This velocity history is brought about through the controlled impact of a projectile plate on the front surface of the test sample. The ensuing measured shock wave that evolves from the impact load contains constitutive property information on the dynamic-strength and shock EOS characteristics of the material.

As in other material property testing methods, the initial data do not contain independent material information, but are intimately linked to the ancillary testing apparatus and impact conditions. For example, in the present shock EOS test method the measured data also reflect window material characteristics. Further data analysis is required to extract test-independent constitutive data for the material.

Hugoniot states associated with the limiting strength (HEL), onset of phase transformation and the final Hugoniot state require the determination of shock

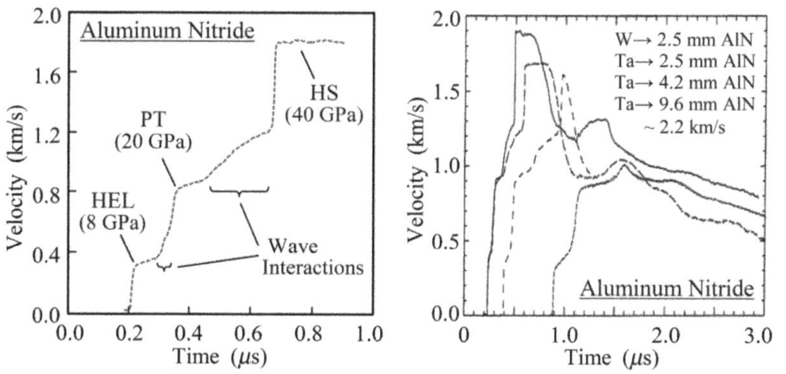

Figure 11.23. Velocity interferometer shock wave profile measurements for aluminum nitride at an interface between a test sample and a lithium fluoride window material. Wave interaction effects are emphasized on the left. Identified in the profile are the limiting dynamic strength (HEL), onset of phase transformation (PT) and the final Hugoniot state (HS). On the right are shown the impactor metal and ceramic sample thicknesses of lower peak particle-velocity wave profiles. The impact velocity was approximately 2.2 km s^{-1} in each test. Data from Grady and Moody (1996). See also Kipp and Grady (1994).

velocities and shock amplitudes of the respective shock waves. Shock EOS determination from the experimental data requires assumptions regarding both material response and test conditions. For example, the assumptions of discrete steady shock waves through the respective steps is made. Simplifying assumptions must be made concerning the complex reflection and interaction of waves at the recording interface between the test sample and window materials. The Hugoniot EOS for aluminum nitride is extracted from the interferometer wave profile measurements through the impedance matching methods in the pressure versus particle-velocity plane by methods described earlier in the chapter. Hugoniot data from the analysis are shown in the plot in figure 11.24.

Corresponding Hugoniot measurements have been made on the same aluminum nitride ceramic using the incline-mirror streak-camera technique discussed previously and as illustrated in figure 11.21 (Mashimo *et al* 1999). Hugoniot EOS data for aluminum nitride is extracted from measurements by the methods discussed for silicon nitride. These Hugoniot state data are compared with those assessed from the interferometer measurements in figure 11.24. Shown also in the plot is the compression isentrope for the high-density phase of aluminum nitride assess from DAC studies (Xia *et al* 1993).

Differences in Hugoniot properties assessed from the interferometer wave profile data and from the incline-mirror streak-camera technique are unsettling. Data from the latter suggest a mixed-phase transition on the Hugoniot from the 20 GPa transition onset to nearly 75 GPa Hugoniot pressure, comparable to the results reported for silicon nitride. In contrast, the Hugoniot response assessed from interferometer measurements suggests near complete phase transformation on the Hugoniot not higher than about 35 GPa.

Experimental difficulties with the interaction of multiple waves at a free surface or an impedance interface has been recognized as leading to inaccuracies in the

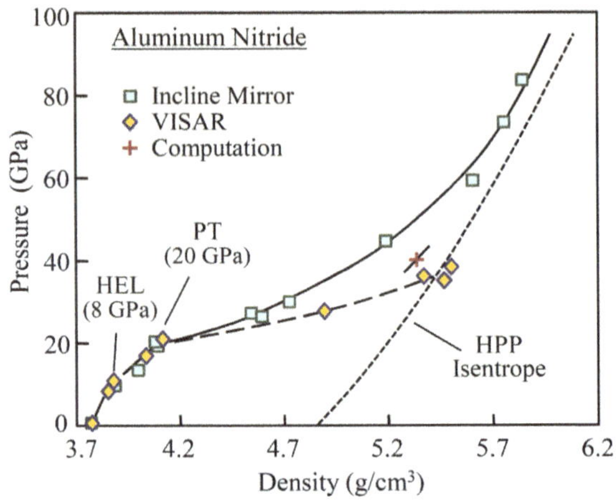

Figure 11.24. Shock Hugoniot states assessed from incline-mirror streak-camera techniques (Mashimo *et al* 1999) and from VISAR interferometer wave profile measurements (Kipp and Grady 1994). The model computation match to VISAR profile provided state is identified by the (+). High-pressure phase (HPP) isentrope assessed from DAC data of Xia *et al* (1993).

assessment of the Hugoniot EOS (e.g. Grady 1986, Grady and Kipp 1998). Generally it is argued that such effects are usually sufficiently small as to be ignorable. Present comparisons show that this is not always the case. Similar observations have been made concerning Hugoniot EOS measurements in high-strength ceramics (Alexander *et al* 2006).

To explore sources of inaccuracies in Hugoniot state interpretation from data such as the foregoing, computer wave code simulations of shock wave evolution and interface interaction have been pursued for aluminum nitride. For this purpose a one-dimensional variable modulus model was developed and calibrated to available interferometer data for aluminum nitride shown in part in figure 11.23 (Kipp and Grady 1994). This computer model was then used to explore the effects of wave interactions on data interpretation by comparing computational simulations of transmitted wave profiles with waves emerging at the sample free surface and at the interface between the sample and window material (Grady and Kipp 1998).

Comparisons of velocity profiles under the respective experimental conditions are shown in figure 11.25. A transmitted wave (no interface) at a point 4 mm from the impact surface reveals three distinct waves. The first wave is a consequence of compression softening at a HEL of about 8 GPa. The second wave results from a further compression softening at the approximately 20 GPa solid state phase transition. The third wave compresses the material to the final Hugoniot state of approximately 45 GPa.

A velocity history resulting from the same impact-produced shock wave emerging at a free surface 4 mm from the impact plane is also shown in figure 11.25. Although shock fronts and shock plateaus corresponding to the same three waves can be identified, substantial complexity resulting from wave reflections at the free surface are observed. Importantly, since first arrival, or perhaps some intermediate point,

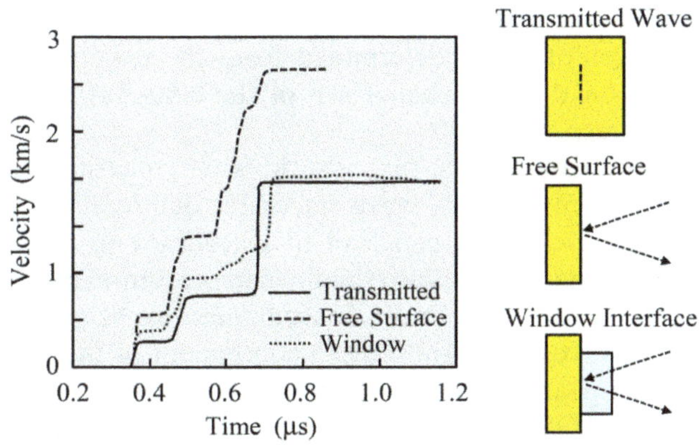

Figure 11.25. Computational simulations based on the model calibrated to VISAR wave profile measurements (Grady and Kipp 1998). The final pressure–density state is identified in the previous figure with Hugoniot data assessed from the incline-mirror and interferometer measurements.

would be interpreted as determining the shock velocity of this test, it is clear that significant inaccuracy will occur in the interpretation of the shock velocity and hence the Hugoniot state. Also important is that the final particle-velocity amplitude of the free surface motion is significantly less than twice the final Hugoniot particle velocity of the transmitted shock wave. That the free surface approximation is not appropriate in this application is a consequence of the significant density increase associated with the phase transformation and incorporated in the computational model. Inappropriate assumption of the free-surface approximation further contributes to error in the assessed Hugoniot states.

A velocity history at an interface between the aluminum nitride ceramic and a lithium fluoride window at 4 mm from the impact interface is also shown in figure 11.25. Again, the three waves are readily identified. Complexities due to wave interactions are also observed, although not as substantial as in the free-surface profile. The calculated interaction wave structure compares well with the measured wave profiles shown in figure 11.23. In this case, however, the arrival identified with the third shock wave occurs later than the transmitted wave arrival. This effect is a consequence of a reduction in propagation velocity as the shock wave enters the region of wave interaction with stress levels significantly reduced because of the lower impedance lithium fluoride window.

Computation solutions show the measured motion of an emerging shock wave at either a free surface or at a lithium fluoride interface, with wave interactions caused by a multiple wave structure can lead to errors in the inferred Hugoniot data. An alternative method of determining a Hugoniot state from the shock wave measurement is suggested by the present analysis. Namely, adjusting a computer model until optimal agreement with a measured wave profile is achieved and then accepting the calculated compression states as the Hugoniot state sought. This approach is attractive as the computational wave code in principal accounts for the complicating wave interactions. A Hugoniot state so determined is compared with other data interpretation methods in figure 11.24. The single calculated state differs from comparable pressure level states assessed by the other two methods. Of concern, however, is that Hugoniot states determined from the experimental data by this method may depend on the particular form of the constitutive model selected to describe the material response.

The present assessment shows that shock wave measurements made at a perturbing interface, when complex wave structures due to elastic-plastic or phase transition processes are involved, can lead to inaccuracy in the interpretation of shock EOS states. Such inaccuracies are small when precompression waves are small compared to the overall wave amplitude. Inaccuracies can be quite significant when such precompression waves are a substantial component of the wave structure.

There does not appear to be an optimal solution for extracting accurate technique-independent EOS data from wave measurements at a perturbing interface when complex wave structure is involved. It is perhaps best to recognize that alternative test techniques should probably be pursued. In-material diagnostic methods that measure profile structure with minimal perturbation of the propagating wave are an attractive approach. In-material methods, as discussed earlier in the chapter, have their own deleterious effects on wave profile resolution. An alternative experimental shock wave method impacts test samples onto a calibrated and monitored stationary plate, providing shock wave EOS measurements that do not rely on a measurement of shock velocity. The method provides unambiguous pressure–particle-velocity states. Inference of Hugoniot density relies on an interesting interpretation of the effective shock impedance when strength and/or phase transformation is expected to complicate the shock transition. The next section, pursuing EOS analysis of shock wave data on uranium alloys, explores further an effective shock impedance concept.

The lesson to learn is that a single diagnostic method cannot be counted on to provide optimal shock EOS properties over all materials and pressure ranges of interest. Different test geometries and diagnostic methods need to be exploited to achieve the desired quality of EOS data.

11.7 Elastic-plastic waves in uranium alloys

The present section provides a case study of applying shock analysis methods to the interpretation of the strength of metals subject to shock wave compression (Grady 2005). In particular, wave profile measurements obtained through velocity interferometer (VISAR) methods provide the experimental data of interest. The materials subjected to testing are alloys of uranium.

The shock profile data necessary to assess the EOS and dynamic-strength response of uranium metal subject to impact or explosive loading are available through several studies on unalloyed uranium (Grady 1986, Hixson *et al* 1998), and several engineering alloys of uranium (Dandekar *et al* 1980, Hixson *et al* 2000). The computational assessment of the response of metals in applications subjecting the material to a strong shock environment requires characterization of the Hugoniot EOS. This first-order approximation to the dynamic response of the material often adequately captures the important phenomenology. There are applications that require a more discerning description of the dynamic strength of the material during dynamic loading. Material response models ranging from a simple limiting von Mises stress to highly detailed strain-, strain-rate- and temperature-history-dependent models have been developed for such purposes.

The present analysis pursues available wave profile data for alpha-uranium and two uranium alloys, and assesses the dynamic strength of these materials from the wave profile data. The analysis is motivated by experiments on uranium 6% niobium (U6Nb), in which a clear precursor wave indicative of the material HEL and limiting dynamic shear strength were not observed (Zurek *et al* 2000). Detailed deformation models for U6Nb have been developed (e.g. Addessio *et al* 2003, Hayes *et al* 2003). These efforts leave unclear the dynamic-strength properties of U6Nb, including a sensible Hugoniot strength. The analysis through pressure–particle-velocity plots described earlier in the chapter provides understanding of the dynamic-strength response.

The available wave profile data on U6Nb also provides contrast with earlier comparable data on uranium metal. Wave profile measurements using quartz gage techniques have been performed on uranium 3/4% titanium (U3/4Ti) by Dandekar *et al* (1980) and unalloyed alpha-uranium has been tested with time-resolved VISAR (Grady 1986, Hixson *et al* 1998). Profiles measured in these studies show clear multi-wave shock profile structure associated with dynamic strength. These earlier data, with the U6Nb data, provide an opportunity to examine and contrast the respective wave profiles and dynamic-strength behaviors of the several uranium alloys through appropriate analysis methods applied to the measured wave profiles.

Briefly, analysis of the wave profile data reveals a dynamic strength associated with onset of pervasive yield in the neighborhood of $Y = 0.7$–1.0 GPa for the three uranium metals. The HEL for the U6Nb alloy is markedly higher (~ 2.0 GPa) than that of alpha-uranium and U3/4Ti—a consequence of the anomalously high Poisson's ratio of this alloy. Self-similar dispersive wave propagation (see chapter 1) at stress levels below the HEL indicates additional inelastic deformation processes. Dynamic features in the wave profiles suggest twinning as the underlying wave structuring mechanism. Analysis of the data support the proposition of Meyers *et al* (2001) that twinning and dislocation slip are distinct deformation mechanisms with independent activation stress levels.

11.7.1 Uranium 6% niobium (U6Nb)

Transmission shock profile data for U6Nb using time-resolved VISAR diagnostics are reported (Hixson *et al* 2000, Zurek *et al* 2000). These data reveal dynamic response characteristics essential to the EOS and dynamic-strength properties of U6Nb. U6Nb is a unique alloy of uranium that, when appropriately quenched from the stable γ^o phase, achieves a metastable α'' phase which has useful engineering properties (e.g. Addessio *et al* 2003). The α'' phase of U6Nb has marked ductility and corrosion resistance, but can decompose to a more brittle and less corrosion resistant material if heated or aged for an appreciable time. Under static load (tension or compression), inelastic deformation in U6Nb initiates through crystal reorientation or twinning, at several tenths of a percent strain, and at a stress of a few tenths of a GPa. Pervasive yield through dislocation slip proceeds in the range of about 0.6–1.0 GPa. In tension, dynamic fracture failure of this alloy occurs in the neighborhood of 1.0 GPa stress at an extension of approximately 30% (Grady and Olsen 2003). Dynamic tensile failure studies of U6Nb are discussed in chapter 2.

Anomalies in the compressive shock structure have been reported. Several of the papers have stated that U6Nb does not exhibit a HEL. This anomaly is more accurately characterized by Hayes *et al* (2003), who states that the measured shock profiles in U6Nb do not exhibiting a distinct elastic precursor wave. A compressive precursor wave is a common signature of the dynamic elastic limit, or HEL, of a material under shock compression. The compression wave profiles for U6Nb (Hixson *et al* 1998, Zurek *et al* 2000) shown in figure 11.26 illustrate the rather featureless shock profile structure measured. The data encompass a shock pressure range of about 2–12 GPa.

Clearly the U6Nb alloy will experience an elastic limit under some amplitude of shock compression. At issue is the stress at which this limit is achieved along with the nature of the governing yield mechanisms. The HEL for U6Nb is not quantified in the earlier studies. Analysis of the shock profile data through methods other than the elastic precursor wave structure, however, can be performed to assess the dynamic strength and HEL.

Before proceeding with the analysis, it is worthwhile defining the Hugoniot state and the HEL as is used here, as well as fairly commonly in the literature. Rigorously, a Hugoniot state is the stress and strain state achieved through passage of a steady structured shock wave. A steady wave is one that remains unchanged in structure as it propagates through the medium. The Hugoniot is the locus of such states achieved through successively higher amplitude shock waves. A cusp in the Hugoniot will lead to a two-wave structure if the strength of the shock loading exceeds the amplitude of the cusp. The first wave has an amplitude determined by the position of the cusp point on the Hugoniot. The second is a slower shock wave centered at the

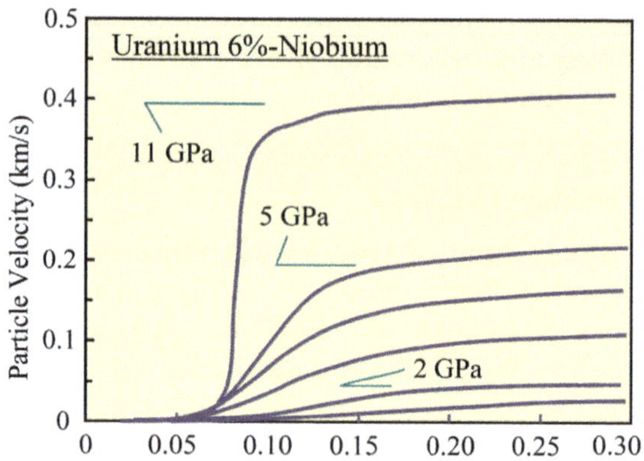

Figure 11.26. Particle-velocity wave profiles in U6Nb measured with time-resolved VISAR methods (Hixson *et al* 2000).

cusp point and with final amplitude determined by the intensity of the impact. Each of the two waves should separately be a steady structured wave.

Deformation yield at the elastic limit of the metal is one source for a cusp on the Hugoniot. This amplitude is commonly identified as the HEL. For various reasons the shock wave loading the material to the elastic limit, or the second wave loading the material to the final Hugoniot state, may not be steady. It is nonetheless common to identify such states as the HEL and the final Hugoniot state even though the more rigorous steady wave requirement is not strictly met. Usually there is reasonable wave separation between the first wave (the elastic precursor wave), and the second deformation shock wave, such that this lack of rigor is a sensible experimental expedient. Also, states determined through more detailed unsteady wave analysis are not usually appreciably different. We find that for the present U6Nb wave profile data that this relaxed definition is problematic.

Constitutive modeling and computational simulation studies of the dynamic response of U6Nb have been pursued. Addessio *et al* (2003) provide a multidimensional thermo-mechanical description of the deformation of U6Nb. Hayes *et al* (2003) provide a rate-dependent stress-relaxation model more focused on the structured wave profile data. Both studies undertake computational simulations of the shock response of U6Nb and quite adequately capture the reported compressive wave profile structure. Hayes *et al* (2003) attribute the lack of a distinct precursor to time-dependent shear-stress relaxation. Addessio *et al* (2003) conclude that crystal reorientation during compressive load accounts for the absence of the precursor. That outwardly different physics are brought into play by the two studies to describe the phenomenon suggests further investigation is warranted. Without detracting from these previous studies, the present discussion is a code- and model-independent assessment of the U6Nb wave profile data applying the analysis tools of the shock physicist.

In the analysis the dynamic strength and the HEL of U6Nb, along with other constitutive features, are directly extracted from the shock profile data. Analysis methods pertinent to the present chapter are applied. Reasonable quantitative bounds are placed on the HEL and dynamic strength of U6Nb. The dynamic-strength and Hugoniot states for U6Nb are compared with the strength and EOS properties of unalloyed alpha-uranium and U3/4Ti.

11.7.2 Impedance analysis of the uranium 6% niobium (U6Nb) wave profile data

In shock wave testing of the dynamic response of solids, measurement of the transit time of the structured compression shock wave provides a first-order indication of the dynamic-strength supporting property of the material. The velocity of the measured wave profile, when compared with known elastic properties of the material, reveals a pre-yield strength-supported longitudinal wave in contrast to a post-yield bulk or near-bulk wave velocity under the uniaxial strain constraint of shock wave loading.

Transit time measurements are not always made. Transit time measurements were not made for the wave profile data for U6Nb analyzed here. The transit time measurement is frustrated by the difficulty of correlating disparate diagnostics instrumentation over the very brief transit times (~1 μs), compounded with the added uncertainty in timing brought about by non-planarity of impact. Even when attempts are made to account for all of the factors contributing to a transit time measurement, resulting uncertainties can be unacceptably large.

An alternative method for directly assessing the material dynamic compressibility from the shock wave profile data is through the impedance analysis outlined earlier in the chapter. The method is more commonly associated with earlier shock EOS studies where, for direct explosive or flyer plate loading, a known standard material is introduced in the experimental assembly. Measurements of shock-induced particle velocity along with the appropriate continuity and Hugoniot conditions provided values for the shock impedance EOS properties of the material.

The ancillary quartz and sapphire materials used in the testing of U6Nb (Hixson *et al* 2000) are appropriately characterized over the stress range of interest. These are the standard materials that allow for impedance analyses of the U6Nb shock profile data. In most of the reported tests on U6Nb, z-cut quartz impactors are used (z-cut sapphire was used in one test) and the time-resolved motion of the free surface is monitored as the impact-induced shock wave emerges at the back free surface. The appropriate impedance analysis is illustrated in figure 11.27. Free surface velocity is measured with the VISAR diagnostic (Barker and Hollenbach 1972). From this measurement, a Hugoniot particle velocity in the material before emerging at the free surface can be inferred. The simplest assumption is the free surface approximation where $u_h \simeq u_{fs}/2$. If additional material response information (such as elastic-plastic strength) is available, either *a priori*, or through iterative analysis of the data, higher order free surface methods are possible (e.g. Grady 1986, Mesheyakov *et al* 2003).

With the experimentally determined Hugoniot particle velocity, the Hugoniot stress is calculated from the known compression relation (the σ–u compression curve shown in figure 11.27) of the impactor material. The dynamic impedance of the test

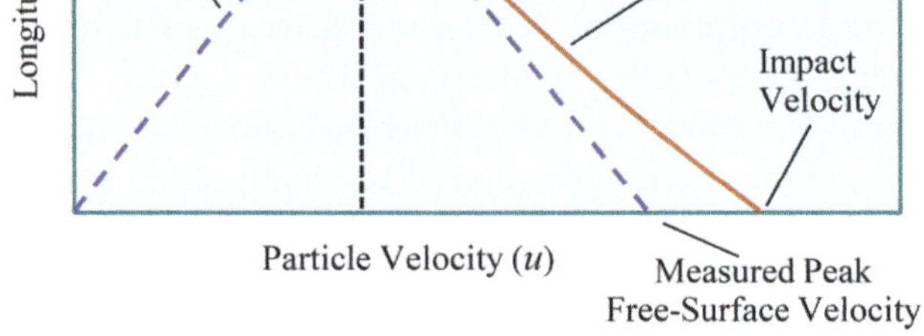

Figure 11.27. Hugoniot state and effective impedance analysis of U6Nb shock wave profile data.

material is in turn determined from the ratio of the two Hugoniot properties $z = \sigma_h/u_h$. Uncertainties in these experimental properties arise from uncertainties in projectile velocity, VISAR final state velocity, the EOS for the ancillary standard material and, importantly, assumptions relating the free-surface velocity to the *in situ* Hugoniot particle velocity.

If the dynamic response of the material is elastic, or hydrodynamic, with the modulus increasing with compression, the analysis method provides a sensible quantitative measure of the shock impedance of the test material. If, however, loading to the final state is through a structured unsteady wave, the impedance inferred must be regarded as an average, or effective, impedance associated with the final shock state. It has meaning only as an experimental expedient for examining the trend of the collective shock data. Features such as phase transformations are uncovered by such analysis and the resulting trends in the effective impedance data. Unusual shock impedance behavior is displayed by the available U6Nb data. Similar behavior results for example from analysis of the shock Hugoniot data of silicon carbide (Gust *et al* 1973) discussed in chapter 6, where a level of free silicon leads to an observed phase transition on the Hugoniot.

The effective shock impedance inferred from the shock wave data by the method discussed above can be normalized by the initial material density and plotted, with dimensions of wave velocity, against the corresponding Hugoniot particle velocity or Hugoniot stress. This plot for the U6Nb data is provided in figure 11.28. These data are compared with earlier shock Hugoniot data for U6Nb from Marsh (1980). Also shown in the figures are the reported values for the bulk and longitudinal elastic ultrasonic wave speeds for U6Nb (Zurek *et al* 2000, Marsh 1980).

Impedance amplitudes for the two lowest data points at 1.4 GPa and 1.7 GPa Hugoniot stress are sensibly consistent with expected longitudinal elastic precursor velocities for U6Nb. An elastic, or near-elastic, response of the U6Nb material under shock compression is suggested by the data to at least this stress amplitude.

Markedly lower effective impedance is determined from the analysis for the three successively higher amplitude U6Nb experiments. The trend of these data suggests

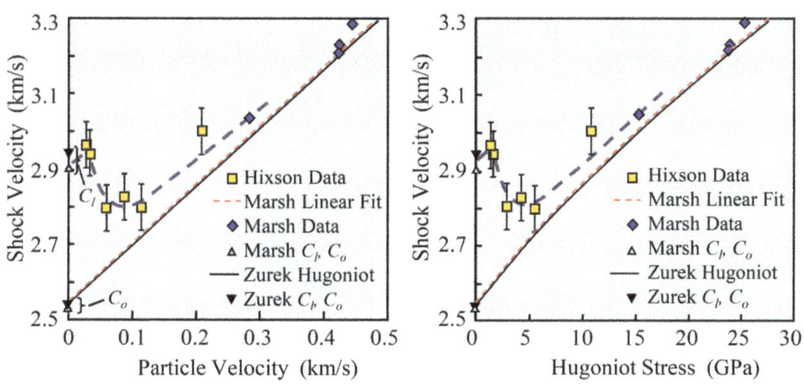

Figure 11.28. Shock velocity (effective shock impedance divided by reference density) versus Hugoniot particle velocity (left) and Hugoniot stress (right) for the Hixson U6Nb data. See text for references.

multi-wave, and probably unsteady wave, behavior for the higher stress amplitude waves. The highest stress data point is not inconsistent with the shock Hugoniot measurements of Marsh (1980), although application of the free surface approximation with $u_h \simeq u_{fs}/2$ is less appropriate because of the more complex wave behavior expected for the higher amplitude wave profile. Decreasing effective shock impedance with increasing shock amplitude can be compared with Hugoniot data for silicon carbide and silicon ceramic mixture in chapter 6 where similar behavior is attributed to phase transition on the Hugoniot.

For von Mises yield, the unconfined dynamic yield strength is related to the constrained uniaxial strain HEL through the relation

$$\sigma_{hel} = \frac{1 - \nu}{1 - 2\nu} Y, \tag{11.60}$$

assuming a constant Poisson's ratio to the HEL amplitude. For U6Nb, a Poisson's ratio of $\nu = 0.377$ yields for the HEL

$$\sigma_{\text{hel}} \cong 2.54 Y. \tag{11.61}$$

The relatively large ratio of the HEL to von Mises strength for this uranium alloy is a consequence of the modest increase of deviator stress (large Poisson's ratio) with the axial shock stress in comparison with the behavior of most metals. If the 1.7 GPa Hugoniot state is identified as a lower limit to the HEL (the highest experimental point indicative of elastic response), and the 2.9 GPa state as an upper limit (the lowest experimental point revealing failure on the Hugoniot), then the bounds for the dynamic von Mises strength of 0.7 to1.1 GPa are calculated for U6Nb (see table 11.1).

11.7.3 Alpha uranium

Unalloyed α-uranium also undergoes stress-induced inelastic deformation and plastic flow through twinning and dislocation glide. Over a variety of heat treatment

Table 11.1. Properties for alpha uranium and uranium alloys.

Property	Alpha uranium	U6Nb uranium	U3/4Ti uranium
Density (g cm^{-3})	18.94	17.44	18.62
C_l (km s^{-1})	3450	2950	3400
C_s (km s^{-1})	2130	1310	2000
C_o (km s^{-1})	2420	2530	2500
Poisson's ratio	0.192	0.377	0.235
G (GPa)	86.1	29.9	74.5
K(GPa)	110.9	111.9	116.0
HEL (GPa)	1.2–1.4	1.7–2.9	1.1–1.4
Y(GPa)	0.92–1.07	0.67–1.13	0.76–0.97
Y/HEL ratio	0.763	0.394	0.692

and rolling conditions, α-uranium at ambient conditions exhibits tensile inelastic deformation due to twinning in the range of 0.15–0.35 GPa and exhibits an ultimate tensile strength of 0.4–0.6 GPa at an elongation in the range of 5%–20% (Holden 1958). Onset of twinning is generally an increasing function of temperature (Tablin 1967). Under large-strain compressive deformation α-uranium exhibits high strain hardening with an unconfined true flow stress of approximately 1.2 GPa at 30% strain and approximately 2.5 GPa at 300% strain (Hockett and Sherby 1975, Hockett *et al* 1977). Grain size plays and important role in the deformation strength (Tablin 1967) and a marked dependence on strain rate in the $10^2\ s^{-1}$ to $10^3\ s^{-1}$ range has been reported (Huddart *et al* 1980).

Shock profile measurements are provided for unalloyed D-38 α-uranium (Grady 1986, Hixson *et al* 1998). Free surface velocity from the earlier study measured with VISAR at the back of 7.5 mm uranium discs impacted by sapphire discs provide the velocity profiles shown in figure 11.29. Peak (Hugoniot) stresses achieved in the series of tests ranged from about 3–15 GPa. Similar structure is observed in the α-uranium investigated by Hixson *et al* (1998).

In contrast to the previous U6Nb alloy, α-uranium exhibits a well-defined precursor wave, providing a signature of the dynamic strength. A clearly resolved rise to a distinct break at approximately 0.04 km s^{-1} is observed in the detailed precursor wave structure shown on the right in figure 11.29. All tests in this figure were performed on the same 7.5 mm thickness samples, so a steady wave versus unsteady wave behavior below this break cannot be determined from the data.

If the break at 0.04 km s^{-1} is identified as the HEL amplitude, then an HEL stress of 1.2 to 1.4 GPa is calculated for α-uranium. If a more central midpoint of the hardening plateau is selected (~0.05 km s^{-1}) the HEL amplitude is closer to 1.6 GPa. Based on a Poisson's ratio of 0.192 for α-uranium, a von Mises dynamic yield Y is 0.9–1.1 GPa based on an HEL stress of 1.2–1.4 GPa. Comparable dynamic-strength values are obtained from the data of Hixson *et al* (1998). Material and calculated properties for α-uranium are also provided in table 11.1.

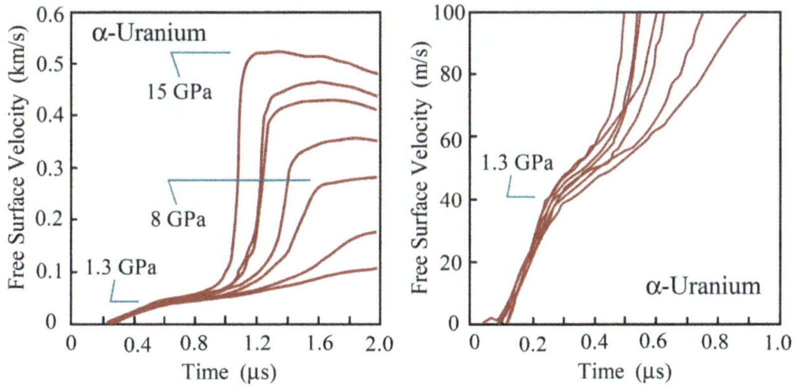

Figure 11.29. Free surface VISAR shock wave velocity profiles measurements for α-uranium (Grady 1986). Detailed images of the precursor wave are shown on the right. Wave structure evolves over 7.5 mm propagation distance.

11.7.4 Uranium 3/4% titanium (U3/4Ti)

Shock profile data for α-phase U3/4Ti (actually U0.78Ti) are provided by the study of Dandekar *et al* (1980). Planar impact experiments use x-cut quartz gages to monitor the shock wave stress profiles. Both direct impact stress and transmission stress histories are measured. Analysis here is focused on the transmission wave profile data. Peak impact stresses over the range of 2.8–9.6 GPa are achieved in the series of experiments. Transmission stress measurements are for test samples ranging from 2 mm to 9.2 mm in thickness.

Structured compressive shock profiles with reasonably distinct two-wave features are observed in all transmission tests. The authors identify a hesitation cusp at the top of the precursor wave as the HEL for U3/4Ti. The amplitude of the HEL is observed to decrease with test sample thickness. Stable HEL amplitudes are achieved within the 5–9 mm sample thickness range. The stable HEL amplitude for U3/4Ti is reported to be 1.1–1.4 GPa. A Poisson's ratio of 0.235 leads to a dynamic von Mises yield Y of 0.8–1.0 GPa. This compares with a reported static compressive yield of 0.75–1.04 GPa. A self-similar plot of the structured compressive wave profiles reveal unsteady simple centered wave propagation (see chapter 1) within the amplitude range of the observed precursor. This key experimental observation may have implications for both the α-uranium and U6Nb wave profile structure and the responsible underlying inelastic deformation mechanisms.

11.7.5 Comparative behavior of uranium and alloys

Analysis of the shock wave profile data for uranium and the several uranium alloys indicates similarities in their deformation response to impact loading. HELs are determined from the respective wave profile data. The dynamic shear strengths provided in table 11.1 are comparable in amplitude. These measures of dynamic strength are also in sensible agreement with the stress amplitude at onset of the pervasive dislocation glide deformation observed in static testing of the same materials. Precursor attenuation and the evolution of wave structure in the thinnest (~2–3 mm) test samples suggest rate sensitivity to the dislocation-governed deformation. Unsteady (centered wave) features of the precursor wave suggest some sub-HEL inelastic deformation. Shear-induced twinning observed under static deformation at comparable stress amplitudes provides the simplest explanation for this sub-HEL deformation in the shock wave. Further shock wave analysis considers these observations in the following subsections.

11.7.5.1 Hugoniot elastic limit

The strength data for α-uranium, U6Nb and U3/4Ti are compared in figure 11.30. The plot relates the difference in the axial (shock propagation direction) and the lateral stress on the ordinate to the axial stress on the abscissa. The elastic shock-loading path in this plot assumes a constant Poisson's ratio, and relates stress difference to axial stress through

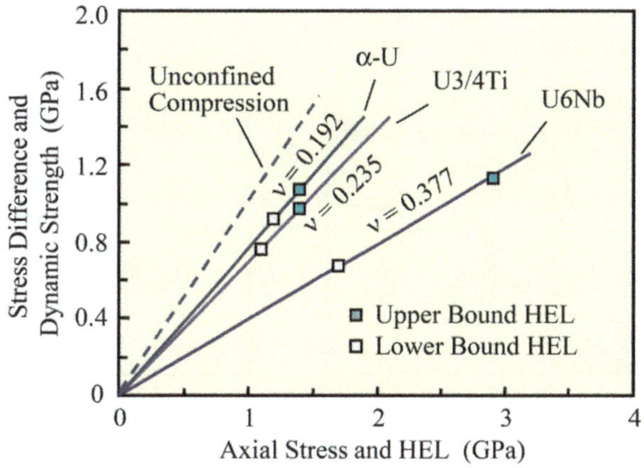

Figure 11.30. Upper- and lower-bound HEL and dynamic shear strength for α-uranium, U6Nb and U3/4Ti from shock wave profile analysis. The path in stress space corresponds to uniaxial strain shock compression.

$$\sigma_1 - \sigma_2 = \frac{1 - 2\nu}{1 - \nu}\sigma_1. \tag{11.62}$$

Poisson's ratio for the three uranium alloys is provided in table 11.1. A load path for unconfined uniaxial compression is also shown in the graph. In unconfined uniaxial compression $\sigma_2 = 0$ and the load path has a slope of 1 in the plot. The range of HEL values inferred from the wave profile data for each of the uranium alloys is plotted as the symbols on the respective load paths. Corresponding yield strength is provided on the ordinate.

The broader spread in the U6Nb dynamic strength reflects the absence of wave profile data in the critical range. Nonetheless, onset of pervasive inelastic shear deformation is nominally the same for α-uranium and the two uranium alloys, and is in sensible agreement with onset of pervasive dislocation plasticity observed in static unconfined compressive testing.

11.7.5.2 Precursor wave structure

The fully resolved finite risetime (approximately 0.2 μs) of the precursor wave measured on α-uranium in figure 11.29 is not commonly observed in other metals (Grady 1986). Metals (aluminum for example) more typically exhibit a near-discontinuous rise to the precursor amplitude associated with the HEL. Tests on differing thickness samples to determine whether this structure of the precursor wave for α-uranium is steady were not performed in the study. Wave profiles measured on U3/4Ti (Dandekar *et al* 1980) were obtained on samples ranging over 2–9 mm. A plot of the precursor wave history as a function of distance divided by time is self-similar, indicative of unsteady centered wave structure (see chapter 1).

A similar analysis of the U6Nb profiles (Hixson *et al* 1998) can be performed. Precursor waves shown for tests performed on samples that are 2.55, 3.81 and 5.11 mm in thickness are plotted on the left in figure 11.18. The VISAR velocity data were numerically differentiated to calculated the maximum acceleration (maximum

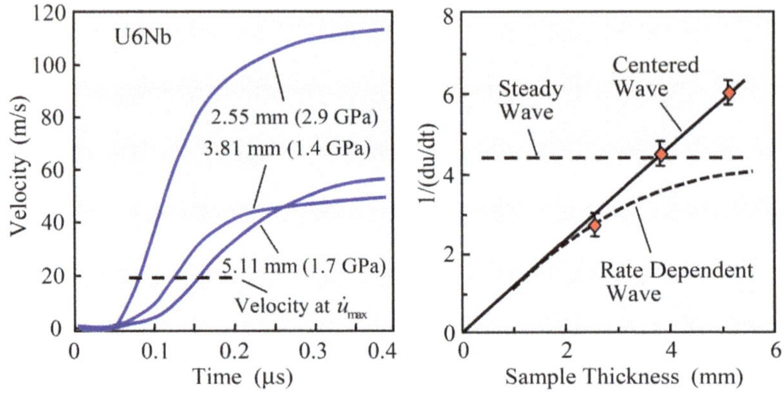

Figure 11.31. Left: Wave structure for U6Nb from different sample thickness experiments (Hixson *et al* 2000). Right: The reciprocal of the maximum acceleration for the U6Nb wave profiles as a function of sample thickness.

of $du/dt = \dot{u}_{max}$) in the rise of the precursor wave. This maximum occurs at about 0.02 m s^{-1} free surface velocity. The 1.7 GPa profile with a sapphire window was corrected to an equivalent free surface velocity.

The reciprocal of this maximum acceleration in the shock wave front is plotted as a function of sample thickness on the right in figure 11.31. Three curves are also shown in the plot. For a steady wave (unchanging in form) $\dot{u}_{max}$ or its reciprocal, should be independent of sample thickness as illustrated by the horizontal line in the figure. If an impact-induced Heaviside stress load applied to the sample front is in the process of relaxing to a steady wave through some rate mechanism over the range of experimental sample thickness, the data should qualitatively plot as the curve identified as rate-dependent in the figure. If, however, centered simple-wave propagation is exhibited by the selected profiles, as indicated by the data for U3/4Ti, the points should fall on a straight line emanating from the origin. The analysis shows that the sub-HEL wave profile data for U6Nb are most consistent with centered simple-wave propagation.

11.7.5.3 Deformation mechanisms

Several deformation mechanisms appear to participate in the shock compression of uranium and its alloys. The trends are common to the three uranium metals, although activation stress levels and relative contributions differ somewhat. First, the present analysis of the wave profile data has shown that a dynamic yield stress in the neighborhood of about 0.7–1.1 GPa is appropriate for the three uranium metals. The corresponding HEL depends on the elastic Poisson's ratio of the material, but a value in the range of 1.1–1.5 GPa would be appropriate for α-uranium and U3/4Ti, while a value closer to 2.0 GPa would apply for U6Nb.

Analysis of the precursor data also reveals that deformation below the HEL is not strictly elastic. The unsteady centered wave propagation observed in this stress range indicates a rate-independent softening of the longitudinal modulus. Although softening of the elastic shear modulus is not unknown (for example amorphous

SiO_2), considering the propensity for deformation twinning of uranium alloys under static loading in this same stress range, a twinning mode of inelastic deformation more likely explains the observed wave structure. Twinning is, in any case, a minor contributor to the deformation in this range of dynamic loading. The total strain in U6Nb approaching the HEL (say 1.5 GPa) is about 1%. An estimate of the twinning strain based on the rate of spread of the precursor wave is nearer to 0.1%, or about 10% of the total strain. The remainder is recoverable elastic strain. This quantitative estimate of the twinning strain is carried out by comparison of the stress–strain path for the U6Nb centered wave to 1.5 GPa with the corresponding linear elastic loading to the same stress.

Under static unconfined compression, twinning can accommodate about 2%–3% inelastic strain before saturation and deformation through pervasive dislocation plasticity takes over (Addessio *et al* 2003). It remains to be explained why twinning strain is markedly reduced under constrained dynamic loading.

Considering the minor contribution from twinning to the total shear deformation up to the 1.0–2.0 GPa axial stress level, it is sensible, at least to first-order, to identify onset of pervasive plastic yield, and the HEL, with the higher 1.0–2.0 GPa stress level for uranium. Above this amplitude plastic shear deformation dominates the Hugoniot strain.

This interpretation of the shock data for the several uranium metals is illustrated in figure 11.32. The relative contributions of elastic and inelastic strain below and above the HEL (as interpreted in the present analysis) are shown as a function of axial stress. Below the HEL, shear deformation is predominantly elastic with a small inelastic contribution due to deformation twinning accounting for the centered precursor wave propagation. The marked deformation transition identified as the HEL occurs in the 1.0–2.0 GPa range where pervasive dislocation plasticity dominates subsequent deformation.

The different stress activation levels and magnitudes of inelastic strain for deformation twinning and dislocation plasticity suggested by the wave profile data

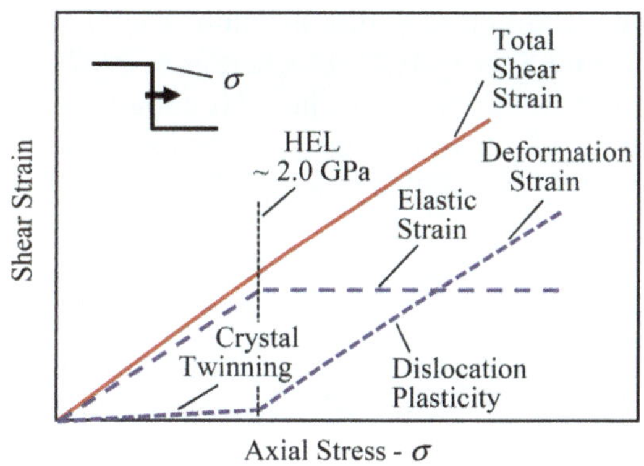

Figure 11.32. Qualitative description of deformation modes in the shock compression of uranium inferred from the analysis. The plot illustrates the relative contributions of elastic and deformation strain below and above the HEL.

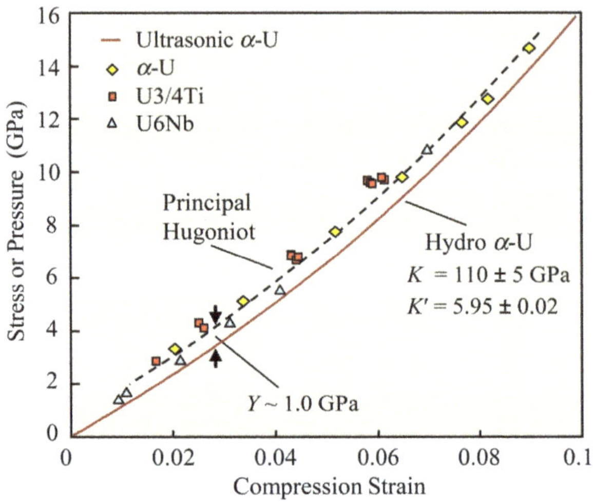

Figure 11.33. Shock Hugoniot states in uranium and uranium alloys (Dandekar *et al* 1980, Grady 1986, Hixson *et al* 2000), and comparison with extrapolated hydrostatic compression curve for alpha uranium (Abbey and Bonner 1975).

for the uranium metals provide support for the constitutive model of Meyers *et al* (2001) in which twinning and dislocation slip are distinct deformation mechanisms with independent activation stress levels. Further, the amplitude dispersive nature of the centered precursor wave propagation supports a distributed stress activation of twinning (possibly Weibull in character) also suggested by Meyers *et al* (2001).

The post-HEL strength of the uranium alloys is not well constrained by the available wave profile data. The higher resolution VISAR data for unalloyed α-uranium (Grady 1986, Hixson *et al* 1998), where a well-resolved precursor wave is observed, suggest a modest rate of continued hardening following initial yield at the HEL. This dynamic behavior is not surprising considering the marked deformation hardening character of this uranium metal observed in static large strain testing (Hockett and Sherby 1975, Hockett *et al* 1977). An indication of post-HEL strength is provided in figure 11.33 where Hugoniot data for the three uranium alloys are compared with a hydrostatic compression curve based on an extrapolation of the ultrasonic data of Abbey and Bonner (1975) for α-uranium.

11.8 Lagrangian analysis

The preponderance of shock EOS data accumulated on condensed matter over the past sixty plus years has been determined through measurement of the mechanical shock velocity and particle velocity induced by the imparted shock wave. The pressure–energy and pressure–density Hugoniot EOS is extracted from the data through application of the Hugoniot conservation relations of mass, momentum and energy.

In later years, increasing sophistication in experimental techniques for measuring transient stress and particle velocity in shock wave studies has shown that the assumptions of steady state wave propagation and the Hugoniot jump conditions are not always acceptable approximations. This is even more evident in spherical or cylindrical stress wave studies where geometric attenuation is always present. Considerable structure is commonly observed in wave profiles. Relating measured wave profile structure to the underlying physical processes accountable for the dynamic response, and posing material constitutive relations descriptive of the transient dynamic behavior, is usually the objective of such experimental research. One approach relies on the application of finite difference computer codes to match experimental profiles by varying parameters in analytic forms of theoretical or semi-empirical constitutive relations. Alternatively, explicit methods of determining constitutive response (stress, strain, strain rate, energy) directly from measured profiles obtained by in-material gages have been sought to supplement computer methods. This goal has served as motivation for the development of a formalism known as Lagrangian analysis (Fowles and Williams 1970, Cowperthwaite and Williams 1971). This analysis couples experimental wave profiles obtained from multiple in-material gages with the governing equations of dynamic flow to determine the actual stress, strain, strain-rate and energy paths experienced by material subjected to transient stress wave loading. Such analysis of experimental data provide an experiment-independent material dynamic response for purposes of constitute model development and validation.

Fowles and Williams (1970) first put forward the concept of Lagrangian analysis for purposes of extracting dynamic constitutive information from in-material measurements of structured waves of stress and particle velocity. The concept of phase velocities of constant stress and particle velocity was introduced. Cowperthwaite and Williams (1971) established conditions on the structure of the wave phase velocities and determined experimental measurement criteria for classes of material constitutive behavior. An alternative Lagrangian analysis appropriate to strongly attenuating waves was posed that exploited an optimal directional gradient analysis applied to multiple stress or particle-velocity measurements (Grady 1973). The latter method was extended by Seaman (1974) and conditions were determined for application. Applications of Lagrangian analysis methods to wave profile data are assessed in subsequent studies (Cagnoux *et al* 1984, Aidun and Gupta 1991). Although not a commonly used tool in the analysis of solid dynamics, when appropriate the methods continue to find application (e.g. Tao *et al* 2011).

The several Lagrangian analysis methods are discussed that are variations on the same scheme. One follows from the seminal study of Fowles and Williams (1970), where the concept of phase velocities is introduced. Second is a path-derivative method more appropriate to rapidly attenuating wave profile data. Lagrangian analysis is most readily performed on multiple stress or particle-velocity gage data conducted in uniaxial one-dimensional flow. With additional effort, the analysis has application to experiments conducted under cylindrical or spherical symmetry.

11.8.1 Basic equations

The conservation laws for continuous one-dimensional hydrodynamic flow in one-dimensional uniaxial strain compression are expressed in the following partial differential equations,

$$\left.\frac{\partial \varepsilon}{\partial t}\right|_h + \left.\frac{\partial u}{\partial h}\right|_t = 0, \tag{11.63}$$

$$\left.\frac{\partial u}{\partial t}\right|_h + \left.\frac{1}{\rho_o}\frac{\partial \sigma}{\partial h}\right|_t = 0, \tag{11.64}$$

$$\left.\frac{\partial E}{\partial t}\right|_h + \left.\frac{\sigma}{\rho_o}\frac{\partial u}{\partial h}\right|_t = 0. \tag{11.65}$$

In these relations $\varepsilon = 1 - \rho_o/\rho$ is a measure of the one-dimensional volumetric compression strain with ρ and ρ_o the current and initial density, respectively, u the particle velocity, σ the longitudinal stress and E the specific internal energy. Each field parameter is a function of the two variables h and t, which are the Lagrangian position and time, respectively. These equations are generally valid. The only restriction occurs in equation (11.65) where heat flow and radiation are neglected. Subject to these restrictions, the equations apply at any point in the wave profile and do not depend on assumptions concerning steady wave behavior or constitutive properties of the material.

For comparison, the corresponding Hugoniot conservation equations in Lagrangian form, which emerge from the equations above and apply when the flow is steady state or when discontinuities arise in the flow, are

$$U[\varepsilon] - [u] = 0, \tag{11.66}$$

$$U[u] - \frac{1}{\rho_o}[\sigma] = 0, \tag{11.67}$$

$$U[E] - \frac{\sigma + \sigma_o}{2\rho_o}[u] = 0. \tag{11.68}$$

The bracket signifies a jump in the property through a structured steady wave or discontinuous shock wave and U is the shock velocity.

The same differential equations of motion can be written in the integral form

$$\varepsilon(h,t) = \varepsilon(h, t_o) - \int_{t_o}^{t} \left.\frac{\partial u}{\partial h}\right|_t dt, \tag{11.69}$$

$$u(h,t) = u(h,\, t_o) - \frac{1}{\rho_o}\int_{t_o}^{t} \left.\frac{\partial\sigma}{\partial h}\right|_t dt, \tag{11.70}$$

$$E(h,t) = E(h,\, t_o) - \frac{1}{\rho_o}\int_{t_o}^{t} \sigma\left.\frac{\partial u}{\partial h}\right|_t dt. \tag{11.71}$$

The time integration in each expression is along a path of constant h. Lagrangian analysis capitalizes on this representation of the conservation relations. Experimental gages that measure the time history of stress or particle velocity are located at interfaces in the test material that are Lagrangian positions of constant h.

11.8.2 Phase-velocity analysis

The original development of Fowles and Williams (1970) follows directly from equations (11.69)–(11.71). Using identities valid for continuous functions of two variables,

$$\left.\frac{\partial u}{\partial h}\right|_t = -\frac{\partial u/\partial t|_h}{\partial h/\partial t|_u} \tag{11.72}$$

and

$$\left.\frac{\partial \sigma}{\partial h}\right|_t = -\frac{\partial \sigma/\partial t|_h}{\partial h/\partial t|_\sigma}, \tag{11.73}$$

define the phase velocities as $c_u = \partial h/\partial t|_u$ and $c_\sigma = \partial h/\partial t|_\sigma$. Equations (11.69)–(11.71) then take the form

$$\varepsilon(h,t) = \varepsilon(h,\, t_o) + \int_{u_o}^{u} \frac{du}{c_u}, \tag{11.74}$$

$$u(h,t) = u(h,\, t_o) + \frac{1}{\rho_o}\int_{\sigma_o}^{\sigma} \frac{d\sigma}{c_\sigma}, \tag{11.75}$$

$$E(h,t) = E(h,\, t_o) + \frac{1}{\rho_o}\int_{u_o}^{u} \frac{\sigma\, du}{c_u}. \tag{11.76}$$

In this form the phase velocities are attributed some physical significance. They represent propagation velocities of points of constant stress or constant particle velocity as the wave traverses the medium. They may or may not equate to acoustic velocities of the material at corresponding stress or particle velocity amplitudes. In the experimental analysis the phase velocities are determined as a function of stress or particle velocity amplitude from multiple experimental profiles and used in equations (11.74)–(11.76) to determine the remaining flow parameters.

The integrals as displayed in equations (11.69)–(11.71) are of course equivalent to those in equations (11.74)–(11.76). The differences arise in how the two sets of integrals are applied to experimental data in extracting dynamic constitutive properties of the test material from experimental wave profiles.

11.8.3 Path-derivative analysis

An alternative approach to Lagrangian analysis identified as a path-derivative analysis (Grady 1973, Seaman 1974) makes optimal use of multiple stress or particle-velocity wave profile measurements to extract the unmeasured strain and energy histories. The approach described is applicable in particular to attenuating wave profiles that are problematic for analysis through the phase-velocity formulation.

In figure 11.34, the stress field $\sigma(h,t)$ or particle-velocity field $u(h,t)$ is shown as a surface in σ–h–t space or in u–h–t space. The experimental data occur in the form of multiple stress and/or particle-velocity profiles. These profiles are recorded through judicious placement of the in-material gages. The profiles appear in the figure as traces on the surface along paths of constant h. With these profiles, and the equations of flow, in the form of equations (11.69)–(11.71), the objective is to determine the entire flow field $\sigma(h,t)$, $u(h,t)$, $\varepsilon(h,t)$ and, when EOS issues are of interest, $E(h,t)$. In practice it is convenient, although not necessary, to determine the flow properties along paths at the gage locations. For the example outlined, reference will be to gage, or path, number two (the middle profile) in figure 11.34. Also, in the interest of brevity, discussion will focus on stress gage analysis. The corresponding analysis is appropriate to multiple particle-velocity gages.

Gage number two in the figure provides a continuous stress–time profile. This stress profile alone is not sufficient to determine the corresponding particle-velocity history along the same gage path. Knowledge of the outer derivative $\partial\sigma/\partial h|_t$ along the gage path in conjunction with equation (11.70) provides the particle-velocity history at the

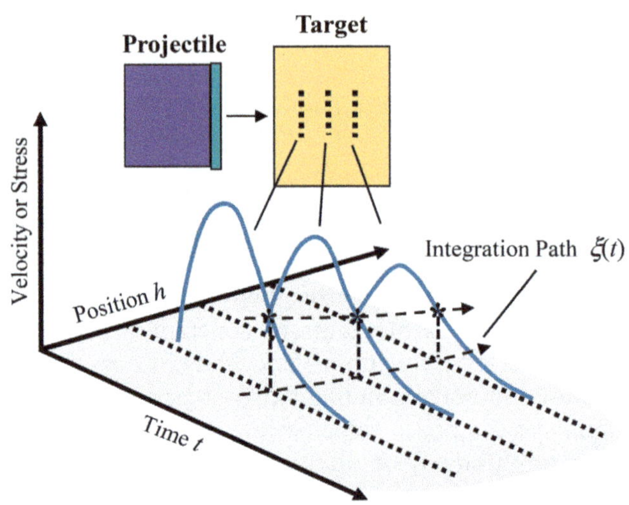

Figure 11.34. Multiple stress or particle-velocity profiles that are traces on a stress or particle-velocity surface that in turn is approximated through the path-line intergral Lagrangian analysis methods.

gage location. The remaining stress-gage profiles (gages number one and number three) are used to provide an estimate of this derivative along the gage path.

There are various ways that the gage data can be used to estimate the derivative $\partial\sigma/\partial h|_t$ as a function of time along the gage path. The following approach optimizes the available multiple wave profile data in determining the derivative. Note the three stress points, denoted by asterisks in the figure, which lie vertically above the curve $\xi(t)$ in the h–t plane. The directional derivative along this $\xi(t)$ path is provided by

$$\frac{d\sigma}{dt} = \left.\frac{\partial\sigma}{\partial t}\right|_h + \left.\frac{\partial\sigma}{\partial h}\right|_t \dot{\xi}(t). \tag{11.77}$$

This equation is solved for the outer derivative

$$\left.\frac{\partial\sigma}{\partial h}\right|_t = \frac{1}{\dot{\xi}(t)}\left[\frac{d\sigma}{dt} - \left.\frac{\partial\sigma}{\partial t}\right|_h\right]. \tag{11.78}$$

An equivalent relation holds for the particle velocity,

$$\left.\frac{\partial u}{\partial h}\right|_t = \frac{1}{\dot{\xi}(t)}\left[\frac{du}{dt} - \left.\frac{\partial u}{\partial t}\right|_h\right]. \tag{11.79}$$

The quantity on the right in either equation is estimated from the data. The path $\xi(t)$ is arbitrary and is chosen to optimize the available data. It can be chosen, for example, to connect HEL points on the multiple measured wave profiles. For attenuating waves, it is chosen to connect points of equivalent form. When reflected waves occur it can be chosen to connect the reflected wave front as it appears in each profile.

This alternative method of Lagrangian analysis differs in detail but not in principle from the original development of Fowles and Williams (1970). Theorems concerning the necessary information needed for given gage configurations are discussed by Cowperthwaite and Williams (1971). These theorems fully carry over to the path-derivative analysis. The technique is explored in detail for multiple attenuating wave profiles in Seaman (1974). Applications to spherically divergent multiple wave profile measurements have also been pursued (Grady 1973).

In the path-derivative analysis, the phase velocities of Fowles and Williams are not explicitly determined from the data. Phase velocities can, however, be useful indicators of special types of flow (Cowperthwaite and Williams 1971). Phase velocities are determined in the path-derivative Lagrangian analysis through

$$c_\sigma = \frac{\dot{\xi}(t)}{1 - \dfrac{d\sigma/dt}{\partial\sigma/\partial t|_h}} \tag{11.80}$$

and

$$c_u = \frac{\dot{\xi}(t)}{1 - \dfrac{du/dt}{\partial u/\partial t|_h}}. \tag{11.81}$$

11.8.4 Application of the phase-velocity analysis

The several methods of Lagrangian analysis have not achieved a significant following in the assessment of structured stress wave measurements in solid materials. This is probably a consequence of the expedience of high-speed computing in comparing material models with measured data. There are, nonetheless, cases in which direct analysis through Lagrangian methods can be informative, and usefully complement other analysis techniques.

Here the phase-velocity method of Lagrangian analysis as put forth by Fowles and Williams (1970) is applied to time-resolved measurements of in-material stress or particle-velocity records. Although illustrative, the example has application to data of interest in the shock wave study of solids. The example has either application to stress or particle-velocity measurements. Consider two closely spaced particle-velocity measurements as illustrated on the left in figure 11.35. The wave is characterized by shock compression to a Hugoniot state followed by a linear decrease in particle velocity. The unloading slope of the wave is quantified by the constant a (negative if a decreasing slope). The shock propagates at a speed U and with peak Hugoniot particle velocity decaying with propagation distance according to a slope of magnitude b (also negative for a decreasing slope).

The waves are representative, for example, of the initial stages of elastic precursor decay, attenuating waves in porous solids, or when waves of such structure are input to the materials. Thus, the analysis is focused on a small patch of the flow surface associated with the initial stages of decompression subsequent to the shock state.

The particle-velocity field as a function of time t and Lagrangian position h is explicitly written

$$u(h,t) = u_o\left(1 + b\frac{h}{U} + a\left(t - \frac{h}{U}\right)\right), \tag{11.82}$$

where a is the rate of decrease of particle velocity in the unloading wave and b is the rate of decay of the peak Hugoniot state. The Hugoniot particle velocity is u_o at the

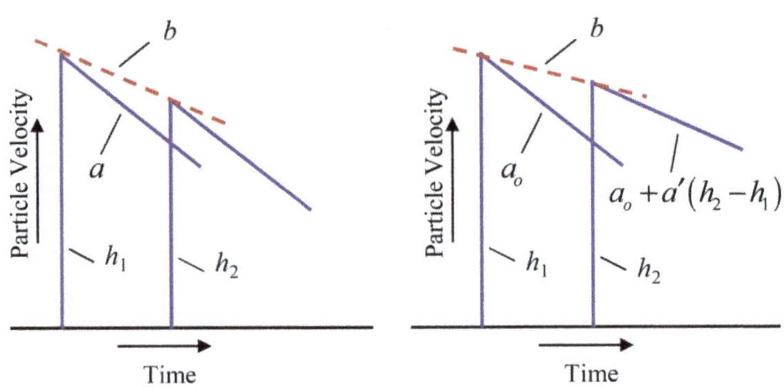

Figure 11.35. Particle-velocity histories at two closely space Lagrangian positions h_1 and h_2. The slope of the release wave and slope of the peak decay are denoted by a and b, respectively. On the right the slope a changes linearly with position h.

first gage at position $h = h_1 = 0$ and at time $t = 0$. Consider first the wave profiles on the left, where the slope a remains constant with propagation distance.

The differential of equation (11.82) yields the phase velocity

$$c_u = \left.\frac{\partial h}{\partial t}\right|_u = \frac{a}{a - b}U. \tag{11.83}$$

Equation (11.64) provides for the stress field the relation

$$\sigma(h,t) = \sigma_h(t) + \rho_o \int_{u_h(t)}^{u} c_u \, du. \tag{11.84}$$

Integration from the Hugoniot state $u_h(t)$ yields

$$\sigma(h,t) = \sigma_h(t) + \rho_o c_u(u(h,t) - u_h(t)) \tag{11.85}$$

or

$$\sigma(h,t) = \rho_o U u_o\left(1 + b\frac{h}{U} + (a + b)\left(t - \frac{h}{U}\right)\right). \tag{11.86}$$

Alternatively, a pair of experimental stress gage measurements exhibiting the wave profile structure characteristic sketched on the left in figure 11.35 would, through similar analysis, yield a particle-velocity field of the form described by equation (11.82).

The differential of equation (11.86) yields the constant stress phase velocity

$$c_\sigma = \left.\frac{\partial h}{\partial t}\right|_\sigma = \frac{a + b}{a}U. \tag{11.87}$$

Equations (11.63) and (11.64) join to provide

$$\left.\frac{\partial \varepsilon}{\partial t}\right|_h - \frac{1}{\rho_o c_u c_\sigma}\left.\frac{\partial \sigma}{\partial t}\right|_h = 0. \tag{11.88}$$

The same relation can be expressed as

$$\frac{d\sigma}{dv} = -\rho_o \left.\frac{\partial \sigma}{\partial \varepsilon}\right|_h = -\rho_o^2 c_u c_\sigma = -\rho_o^2 U^2 \frac{a + b}{a - b}. \tag{11.89}$$

Phase velocities assessed from the experimental wave profile data determine the stress versus specific volume path of a material element on unloading from the Hugoniot state. This assessment of the initial decompression modulus of the stress–volume path is independent of any model of the material response. No thermodynamic condition is implied. Rate- or time-dependent processes may be active as decompression ensues.

The particle velocity expressed by equation (11.82) additionally provides the strain field in the release wave. This is achieved through the corresponding relation in equation (11.74). Alternatively, and equivalently, equation (11.69) provides the strain field where the partial derivative of the particle velocity is calculated from the equation (11.82). The two calculations are not equivalent when the multiple gages are used to provide the phase velocities, or alternatively, the partial derivatives through approximate differencing methods. Depending on the wave structure, approximation of these alternative local measures of the differential properties can have differing sensitivities. These approximation difficulties in application of Lagrangian analysis methods are explored in depth by Seaman (1974) and by Aidun and Gupta (1991). Application of equation (11.74) yields

$$\varepsilon(h,t) = \varepsilon(h,\, t_o) - \int_{t_o}^{t} \left.\frac{\partial u}{\partial h}\right|_t dt = \frac{u_o}{U}\left(1 + b\frac{h}{U} + (a - b)\left(t - \frac{h}{U}\right)\right). \qquad (11.90)$$

The ratio of the time derivative of the stress field in equation (11.86) and the strain field in equation (11.90) provides the modulus

$$K = \left.\frac{\partial \sigma}{\partial \varepsilon}\right|_h = \rho_o U^2 \frac{a + b}{a - b}, \qquad (11.91)$$

and the same expression arrived at in equation (11.89). Depending on the relative magnitudes of a and b, the decompression path can have a positive or negative slope.

In accessing the structure of the two closely placed wave profile measurements, there may be a discernable difference between the slope a of the two waves as illustrated on the right in figure 11.35. This added parameter is captured through a dependence of the slope on position $a(h) = a_o + a'h$ and analysis remains tractable. The field assessed through the two particle-velocity measurements remains as expressed in equation (11.82). Stress is calculated through integration of the partial time derivative of the particle velocity from the Hugoniot state along a path at constant time,

$$\sigma(h,t) = \rho_o u_o U\left(1 + b\frac{h}{U} + (a + b)\left(t - \frac{h}{U}\right) + \frac{1}{2}a'U\left(t - \frac{h}{U}\right)^2\right). \qquad (11.92)$$

Strain is calculated from integration of the partial position derivative of the particle velocity from the Hugoniot state along a path of constant position,

$$\varepsilon(h,t) = \frac{u_o}{U}\left(1 + b\frac{h}{U} + (a - b)\left(t - \frac{h}{U}\right) - \frac{1}{2}a'U\left(t - \frac{h}{U}\right)^2\right). \qquad (11.93)$$

Calculation of the decompression modulus from the Hugoniot state yields again to first-order the relation in equation (11.91). The added parameter a' associated with the change in slope between the two measurements yields a measure of the change in the modulus with stress on decompression identified as K'. To first-order in the ratio b/a_o

$$K' = \left.\frac{d}{d\sigma}\frac{\partial \sigma}{\partial \varepsilon}\right|_h = 2\frac{U^2}{u_o}\frac{a'}{a_o^2}\left(1 + \frac{b}{a_o}\right). \qquad (11.94)$$

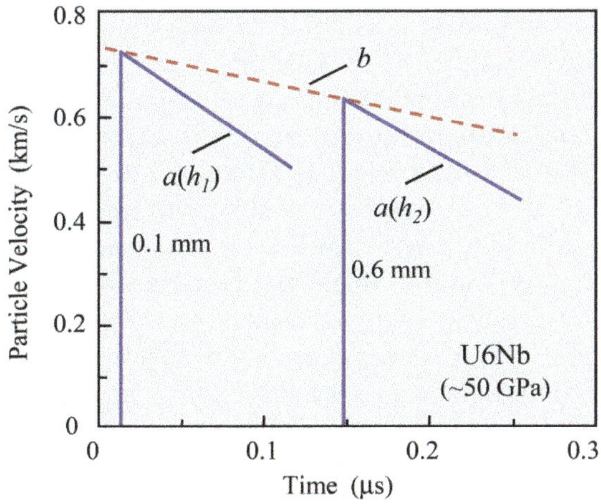

Figure 11.36. Evolution of wave structure in U6Nb of an initial 50 GPa shock and subsequent decompression. Rate of decay of the peak pressure and change in decompression slope is determined by the bulk modulus and pressure derivative at the Hugoniot state.

Equation (11.94) is also a useful relation when b equals zero and U is the wave velocity at the start of the decompression wave.

An application of this limited form of Lagrangian analysis is warranted. Consider an approximately 50 GPa Hugoniot shock compression (approximately 20% compression strain) in uranium 6% niobium. Appropriate Hugoniot EOS properties are provided in chapter 7 and in table 11.1 of the present chapter. Shock compression followed by decompression is input to the sample. Particle velocity is monitored at $h_1 = 0.1$ mm and $h_2 = 0.6$ mm into the sample. Wave profiles are shown in figure 11.36.

The rate of unloading (the initial value of a) is an experimental parameter adjusted in the test. Here an initial input value of the decompression slope of $a = -3\ \mu s^{-1}$ is assumed. An equilibrium thermodynamic response is further assumed. Consequently, the profile parameters b and a' are constrained by equations (11.91) and (11.94), respectively. At 50 GPa Hugoniot pressure in U6Nb, K and K' are approximately 440 GPa and 4.0, respectively. The calculated values for b and a' are $-0.9\ \mu s^{-1}$ and 0.8, respectively. The changes in profile structure over the 0.5 mm propagation distance are illustrated in figure 11.36. Analysis of the data as illustrated in the figure would provide a reasonable measure of the bulk modulus at pressure. The modest change in the slope a is not readily discernable over the half-millimeter propagation distance, and consequently does not readily constrain the pressure derivative of the bulk modulus.

11.8.4.1 Experiments on quartz rock

Certain shock wave data are amenable to assessment through the present abbreviated Lagrangian analysis methods. One such set of experimental data is provided in figure 11.37. Shown are in-material particle-velocity profiles obtained from shock wave experiments on Arkansas novaculite, a very fine-grained, near-full-density (2.63 g cc^{-1}) quartzite (Grady *et al* 1974, Murri *et al* 1975).

Velocity measurements are made using an electromagnetic transducer technique whereby a moving conductor in a stationary magnetic field generates an electromotive signal proportional to the velocity (Dremin and Adadurov 1964, Dremin *et al* 1965, Petersen *et al* 1970). A magnetic field is produced with a pulsed Helmholtz coil. Measurement particle-velocity transducers are placed between plates of the quartz material nominally 3 mm in thickness. One transducer is placed on the back free surface of the test assembly. A planar shock wave is introduced into the quartz with an explosive lens system. The Taylor wave developed in the explosive when coupled into the test material leads to the subsequent ramp unloading following the initial shock compression observed in the figure 11.37 wave profile measurements. A subsequent acceleration wave originates from the back free surface of the quartzite target on arrival of the shock wave and propagation of the returning decompression wave through gage positions. Construction and application of the test and the measurement method is illustrated in figure 11.38.

The initial shock wave and subsequent gentle decompression portion of the wave is a consequence of an approximately 30 GPa explosive shock compression and following Taylor wave. This portion of the particle-velocity wave history can be approximated by the wave structure depicted on the left in figure 11.35 and values estimated for the slopes *a* and *b*.

The unloading slope of the several waves is reasonably captured with a value of $a = 0.05\ \mu s^{-1}$. The decay *b* of the wave peaks is sensibly the same as the slope *a*. (All values of *a* and *b* are negative.) The slope *b* can be varied to assess the sensitivity of the decompression modulus from equation (11.91). When $a = b$, the unloading modulus is infinite (a vertical stress–volume path). The acceleration wave identified in the figure associated with initial decompression originating from shock arrival at the back free surface provides a sonic Lagrangian wave speed of 16.5 km s^{-1} and a

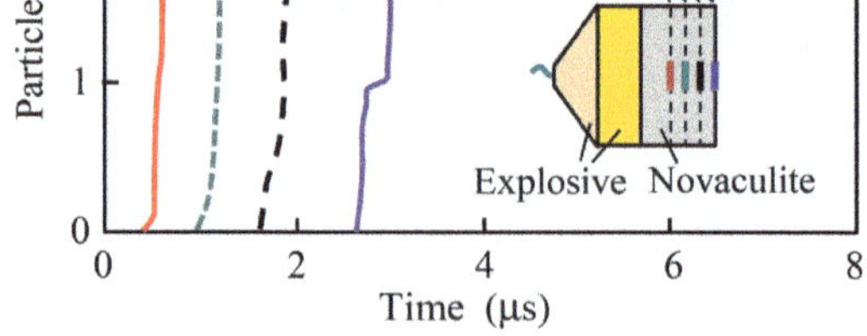

Figure 11.37. Particle-velocity measurements in quartz rock subject to approximately 30 GPa shock compression (Grady *et al* 1974).

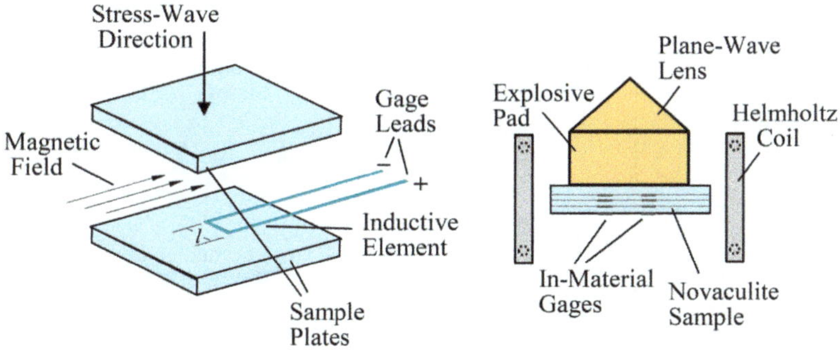

Figure 11.38. Configuration and construction of explosion-induced planar shock wave experiment with multiple in-material particle-velocity measurements (left). The explosive, electromotive transducer, coil assembly and sample test system are illustrated on the right.

Lagrangian acoustic modulus of about 720 GPa. A modulus of this magnitude from the analysis through equation (11.91) requires a value of the slope b = 0.04 μs^{-1}. Attenuating waves approximating the data with a = 0.05 μs^{-1} and b = 0.04 μs^{-1} are illustrated on the left in figure 11.39.

Quartz initially in the alpha phase undergoes a four-fold to six-fold coordination change to the stishovite phase under shock compression within the Hugoniot pressure range of approximately 20–40 GPa. In this range on the Hugoniot, quartz appeared to reside as a mixture of the low-pressure and high-pressure phases. The present test and a comparable test at lower pressure were performed to assess whether the quartz material upon achieving the Hugoniot state remained frozen at the mixture achieved, or whether continued transformation to the high-pressure phase proceeded with time. Explosive shock compression followed by ramp unloading was performed to assess such phase transformation kinetics. Hugoniot states achieved in the experiments are shown on the right in figure 11.39. The downward-directed cones emerging from the Hugoniot states represent the range of decompression paths that are sensibly consistent with the measure experimental profiles shown in figure 11.37 considering the experimental uncertainty in the data. The dashed line identifying a slope of b = 0.04 μs^{-1} in the plot on the left in figure 11.39 corresponds to the modulus assessed from the speed of the acceleration wave emerging from the back free surface. The latter modulus is presumably a reasonable measure of the frozen mixed-phase compressibility of the material.

In summary, the shock and ramp-unloading experiment, although a valiant effort, was not sufficiently sensitive to observe continued time-dependent phase transformation if indeed such transformation occurs. Within the accuracy of the data, the ramp unloading and the acceleration wave speed indicate nominally the same modulus of the material at the Hugoniot state. In later studies on quartz a more definitive transformation pressure on the Hugoniot in the neighborhood of 22 GPa has been accepted. Further, concerns have been raised regarding accuracy of the quartz Hugoniot measurements in that era over the extended transition range. Further discussions of the shock response of quartz through the α-quartz to stishovite transformation are pursued in chapters 7 and 9.

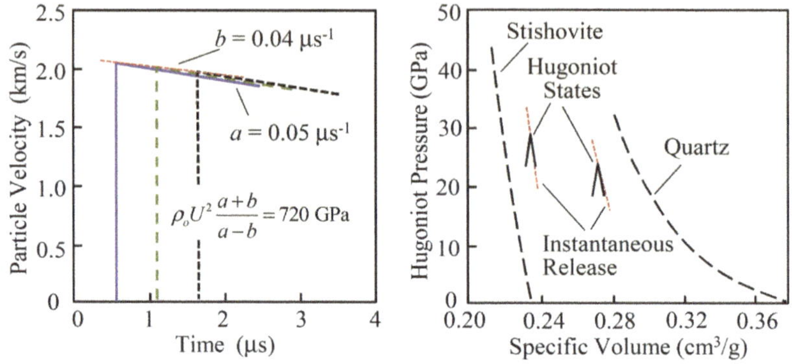

Figure 11.39. Wave structure based on acceleration wave speed modulus of 720 GPa. The experimental Hugoniot states for Arkansas novaculite in the mixed-phase region and the decompression paths are assessed through Lagrangian analysis of ramp unloading and from the acceleration speed modulus.

11.8.4.2 Experiments on dolomite rock

Attenuating wave profiles of the form illustrated in figure 11.35 can also result from material conditions in the shock wave event. Experimental stress wave profiles shown in figure 11.40 and figure 11.41 are a consequence of a time-dependent phase transformation and stress relaxation occurring in dolomite rock subjected to a strong shock wave (Grady *et al* 1976, Grady 2015). More is said on phase transformation in dolomite rock later. Features of the material response in the neighborhood of the attenuating peak are assessed with the methods of Lagrangian analysis of attenuation waves. An attenuating stress wave approximated by the expression in equation (11.92) will, through the same integration analysis procedures performed previously, yield particle-velocity and volume strain fields of the forms provided in equations (11.82) and (11.93). Correspondingly, the decompression pressure–volume strain path in the neighborhood of the peak is given by equation (11.91) and, if a dependence of the initial slope with Lagrangian distance is assessed from the data, the pressure dependence of this modulus comes from equation (11.94).

Wave structure near the peak yields an absolute magnitude for b substantially less than a. Consequently, the decompression modulus from equation (11.91) is positive

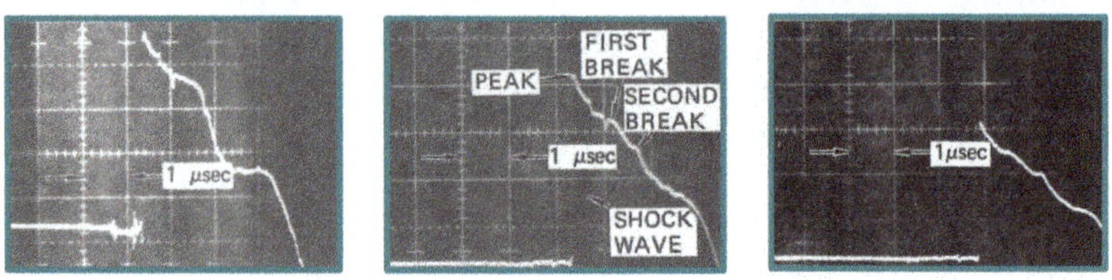

a) 1.58 mm b) 4.80 mm c) 8.17 mm

Figure 11.40. Manganin pressure gage measurements in dolomite rock subject to explosive-driven aluminum (0.25 in.) flyer plate shock loading to approximately 42 GPa Hugoniot pressure (Grady *et al* 1976). The distances shown are from impact plane to gage location. From Grady *et al* 1976. Copyright © 1976 by John Wiley Sons, Inc. Reprinted by permission of John Wiley & Sons, Inc.

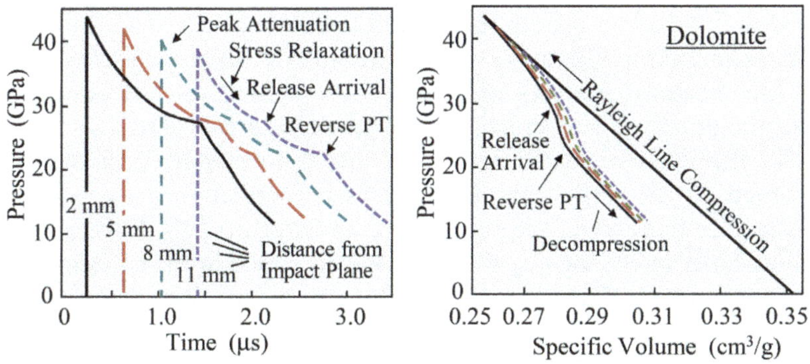

Figure 11.41. Experimental shock compression and release stress wave profiles in Blair dolomite (Grady *et al* 1976). Wave profiles are statistically smoothed from seven data records, for which three are shown in figure 11.40. Shock compression and decompression pressure versus volume paths assessed from path-derivative Lagrangian analysis are shown on the right.

and found to be approximately 40% larger than the Hugoniot modulus of about $\rho_o U^2 = 160$ GPa. Further, inspection of profiles in figure 11.41 suggests that the slope a is increasing in absolute magnitude with propagation distance indicating from equations (11.91) or (11.94) that the decompression path is steepening with decreasing pressure. A more detailed Lagrangian analysis of the data more completely reveals the response of the material to this shock-induced time-dependent phase transformation.

11.8.5 Application of the path-derivative analysis

The path-derivative Lagrangian analysis owes its genesis to the seminal phase-velocity analysis of Fowles and Williams (1970). The method's practicality in application becomes apparent in addressing multiple waves of complex form. Tests of the method using waveforms of known analytic solutions (e.g. Grady 1973) demonstrate the reliability of the path-derivative analysis (see the subsequent subsections of this chapter). Uncertainties in application arise principally from experimental measurement errors.

11.8.5.1 Application to dolomite rock

A unique demonstration of the path-derivative Lagrangian analysis is shown for the wave profiles in figures 11.40 and 11.41 (Grady *et al* 1976). The wave measurements are made using in-material Manganin stress gages. The test material is dolomite rock. Specifically the material is Blair dolomite from a Martinsburg, WV, quarry. Blair dolomite is a fine-grain dolomite (>98% $CaMg(CO_3)_2$) with a density of 2840 kg m^{-3}.

A planar shock wave is introduced into the material with an explosive driven aluminum flyer plate. The maximum pressure achieved in this test is approximately 42 GPa. In figure 11.40 are shown oscilloscope records of three of seven successful gage recordings located over four Lagrangian positions. The experimental configuration mirrors that of figure 11.38 with the exception that particle-velocity gages are replaced with Manganin stress gages and the explosive system launches a metal flyer plate. The seven stress wave measurements were statistically smoothed providing the reconstructed set of stress wave profiles at the four Lagrangian positions in the dolomite test shown on the left in figure 11.41.

Details in the decompression inferred from the attenuating wave profiles are uniquely suited to the path-derivative analysis. Derivative paths are selected to encompass similar features in the closely spaced wave profiles. Particle velocity and specific volume are determined by the methods discussed in the earlier subsection on path-derivative analysis. The pressure–specific-volume histories determined from the analysis are shown on the right in figure 11.41 at four Lagrangian positions comparable to the gage locations.

Shock compression followed by immediate decompression relaxation of the pressure in the dolomite material was an unexpected result. The thickness dimension of the impacting flyer plate and location of the recording pressure gages suggested a flat-topped pressure wave of several microseconds duration. Such flat-top pressure wave behavior was observed in comparable experiments in dolomite below 30 GPa (Grady *et al* 1976) and also for earlier experiments on quartz rock at pressures in excess of 40 GPa (Grady *et al* 1974). Additional tests were performed to assure that the Manganin piezo-resistive gage assembly was not compromised at this excessively high shock pressure in dolomite rock. The use of Manganin pressure gages at such high pressures was new technology at the time.

Consideration of the shock physics responsible for the observed experimental measurements led to the supposition that dolomite entered the field of a new high-pressure phase at pressures less than the 42 GPa Hugoniot pressure of the present test. Stress relaxation is a consequence of time-dependent kinetics associated with transformation from the low-pressure to the high-pressure phase. Inflection points in the measured decompression wave arise, as best illustrated in the figure 11.41 reconstruction, first from arrival of the release wave from the back of the impacting flyer plate and second from the supposition of reversion transformation back to the low-pressure phase at approximately 22 GPa. At the time of this experimental shock physics study on dolomite, no other high-pressure data were available to support a shock-induced phase transformation interpretation of the shock data on dolomite.

11.8.5.2 Shock wave and DAC compression of dolomite rock

It is instructive to examine the structured wave-profile measurements and analysis relative to the experimental Hugoniot for dolomite. Hugoniot data for dolomite to nearly 180 GPa are provided in figure 11.42. The data from four independent studies reasonably characterize the Hugoniot response of dolomite rock as assessed from standard shock-velocity and particle-velocity measurements used at the time.

In the linear fit to the shock versus particle-velocity data, the intercept of C_o = 5.29 km s^{-1} is less than the ultrasonic bulk sound speed of 5.74 km s^{-1} (Heard *et al* 1973). Although disparity between the ultrasonic bulk sound speed and the shock-velocity intercept has been suggested as a possible indicator of a phase transition on the Hugoniot, there are other possible explanations. The Hugoniot data provide no clear indication for a phase transition in the dolomite mineral over the pressure

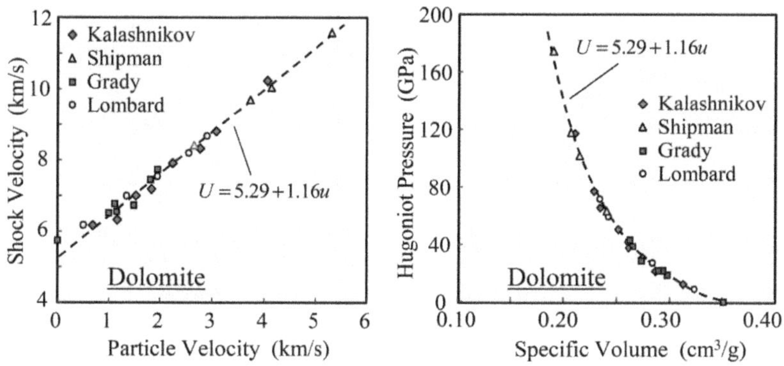

Figure 11.42. Shock Hugoniot data for near-full-density (2.82–2.85 g cm^{-3}) dolomite (Lombard 1961, Shipman *et al* 1969, Kalashnikov *et al* 1973, Grady *et al* 1976).

range tested. Static high-pressure studies of dolomite using DAC technology offer a stark contrast with EOS properties inferred from shock Hugoniot measurements.

Santillán *et al* (2003), using DAC and infrared spectroscopy methods, identified a first-order transition to a phase identified as dolomite II at 20 GPa transition pressure and with a 4% volume change. Confirmation of the high-pressure phase transition in dolomite by Mao *et al* (2011) placed the transformation pressure closer to 17 GPa with a markedly larger 15% volume change. Additionally, a further first-order transition to dolomite III with an additional 8% volume change is observed at a transformation pressure of 36 GPa. The phases are stable under increasing temperature conditions and the dolomite III phase is stable to the 83 GPa test pressure limit. Dolomite III persists on decompression to 26 GPa. The dolomite tested contained approximately 8% iron (siderite). Merlini *et al* (2012), report further DAC investigations on iron-bearing dolomite. They report that the dolomite II transformation at 17 GPa exhibits negligible volume change more in line with the displacive calcite I–II transformation. The dominant volume change occurs in dolomite III transition to densities in agreement with the study of Mao *et al* (2011).

DAC studies on dolomite confirm a first-order high-pressure transformation of dolomite. Two high-pressure phases (dolomite II and III) are observed. Uncertainties in transition density within the range of 17–36 GPa are noted among the several studies.

Overlay of shock and DAC compression of dolomite in figure 11.43 illustrates the stark contrast in EOS response inferred from the two test methods. The diamond anvil compressibility and transformation pressures associated with dolomite I, II and III is that of Mao *et al* (2011) on the mild iron-bearing dolomite. There are differences among the three DAC studies concerning the location of the phase boundaries and to which phase to ascribe the transformation volume. The general trend, however, is representative of all three static studies. Phase transformation on decompression inferred in the shock wave profile experiments (Grady *et al* 1976) falls within the pressure range reported in the several static studies. The decompression path lies below the Hugoniot, indicative of an EOS anomaly; most likely phase transformation in lieu of the more recent DAC data. Compression volumes

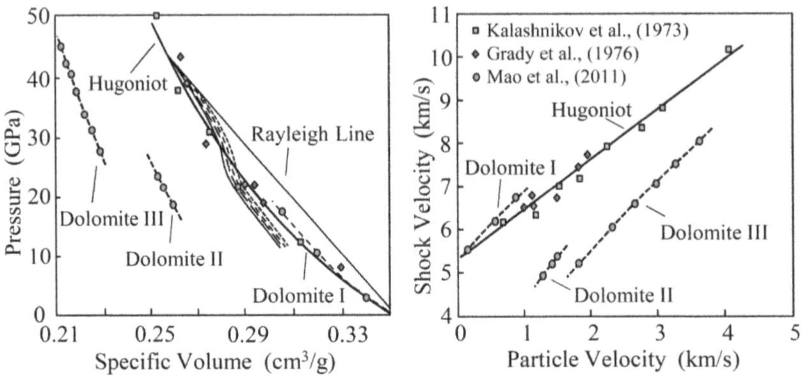

Figure 11.43. DAC and shock wave compression of dolomite. Static experimental points on the left are mapped into the shock-velocity versus particle-velocity plane on the right through application of the Hugoniot conservation relations. References identified on the right apply to both plots.

achieved at comparable pressure are, however, markedly less than reported transformation volumes in the static studies.

For further perspective, the static pressure versus volume points are mapped into the shock-velocity versus particle-velocity plane through application of the Hugoniot mass and momentum conservation relations. Regions of two-wave structure and markedly slower shock velocities would be expected if reported static EOS conditions were achieved in the shock compression event.

The disparities between the compression of dolomite as assessed through shock Hugoniot measurements and that determined through DAC experiments are considerable. These differences should be disconcerting to those who rely on shock Hugoniot data to constrain EOS models for purposes of ground shock calculations in which time scales significantly exceed that of the Hugoniot state measurement.

Differences in mineralogy may play a role. Significant iron content is noted in most of the dolomite subject to DAC testing. Composition is not generally available for the materials subjected to shock Hugoniot measurement. The dolomite subjected to time-resolved pressure measurements (Blair dolomite) contained negligible iron content. Composition differences would seem an unlikely explanation for the substantial compressibility differences, however.

DAC compression studies are performed on dolomite crystals (1μ–5μ grain size). Shock Hugoniot measurements are conducted on dolomite rock (dolostone). Competent dolostone are commonly micritic (2μ–4μ grain size) to microsparitic (5μ–20μ grain size) in grain structure. Hugoniot compression entails heterogeneous deformation and temperatures, non-hydrodynamic stress states and grain boundary effects among others. Thus, initial material conditions differ markedly in the two test methods. Nonetheless, no obvious mechanism is suggested to account for the differences.

A more likely explanation is provided by the time-resolved pressure measurements illustrated by the results shown in figure 11.43. Namely, the dolomite transformation exhibits time-dependence when examined with instrumentation on the microsecond time scale. Thermally activated kinetics associated with the dolomite I–II–III transitions may play a role. Alternatively, the athermal kinetics associated with localized shear (and/or temperature) activated transformation and subsequent time-dependent transformation boundary propagation can be involved. One experiment conducted by Santillán *et al* (2003), under elevated temperature conditions, at pressures comparable to the shock wave experiment of figure 11.41, suggests thermally activated kinetics associated with the dolomite phase transformation.

Experimental concerns here lean sharply towards the shock Hugoniot experiment. The one structured wave experiment discussed above suggests that dolomite may be an unusual material exhibiting rate-dependent EOS transitions that deceived the Hugoniot measurement techniques of the time. Such techniques rely principally on determination of shock velocity through shock transit-time measurement, and particle velocity through analysis of measured shock velocity and impact projectile velocity. A modestly decaying shock wave and post-shock stress relaxation would not be observed with such techniques. Stress gages in the structured wave experiment of figures 11.40 and 11.41 reveal a brief period of stress relaxation before arrival of the decompression wave originating from the back of the impact plate. With the recent DAC studies, further shock experiments on dolomite are warranted. Impact experiments maintaining the shock state for longer periods of time, combined with higher resolution time-resolved measurements (VISAR for example) are clearly needed.

11.8.6 Lagrangian analysis in spherical symmetry

The path-derivative Lagrangian analysis is particularly appropriate in the evaluation of experimental measurements in symmetric curvilinear geometries (Grady 1973). In particular, consider the system of partial differential equations governing spherically symmetric wave motion. The equations follow from the conservation laws of mass and momentum, and the relation for the material velocity

$$\frac{\rho_o}{\rho} = \frac{r^2}{h^2}\left(\frac{\partial r}{\partial h}\right)_t, \tag{11.95}$$

$$\rho\left(\frac{\partial u}{\partial t}\right)_h + \left(\frac{\partial \sigma}{\partial r}\right)_t = 2\frac{\varphi}{r}, \tag{11.96}$$

$$u = \left(\frac{\partial r}{\partial t}\right)_h. \tag{11.97}$$

The Lagrangian and Eulerian radial position are h and r, respectively; σ and u are the radial stress and velocity, respectively; and φ is the difference between the radial and tangential stress. The system of equations assumes nothing about the properties of the material and is generally applicable throughout the flow field except where discontinuities occur. At discontinuities, the jump conditions in the Lagrangian coordinates apply.

The system of three equations in the five unknowns (r, ρ, u, σ, φ) is incomplete, and knowledge of the constitutive properties of the material is necessary for closure. In Lagrangian analysis in spherical coordinates, closure of the system is achieved through multiple gage measurements of the history of radial stress and particle velocity. The system of equations is solved in integral form for the unknown flow variables in terms of the measured stress and particle velocity. The Eulerian position is obtained by a time integration of the particle velocity along a path of constant h,

$$r(h,t) = h + \int_{t_o}^{t} u\, dt. \tag{11.98}$$

The volume strain $\varepsilon = 1 - \rho_o/\rho$ follows from the continuity equation

$$\varepsilon(h,t) = \varepsilon(h,\, t_o) - \frac{1}{h^2}\int_{t_o}^{t} \left(\frac{\partial r^2 u}{\partial h}\right)_t dt. \tag{11.99}$$

Again, only particle-velocity measurements are required. The stress difference is determined from the momentum relation and is written

$$2\varphi = \rho r\left[\left(\frac{\partial u}{\partial t}\right)_h + \frac{1}{\rho_o}\frac{r^2}{h^2}\left(\frac{\partial \sigma}{\partial h}\right)_t\right]. \tag{11.100}$$

Efficacy of the Lagrangian analysis method is difficult to quantify and depends on the approach pursued as demonstrated by Aidun and Gupta (1991). Accuracy is nonetheless determined to first-order by the accuracy and consistency of the experimental data. As discussed previously, when the outer derivative of stress or particle velocity is required, a path-derivative relation,

$$\left(\frac{\partial f}{\partial h}\right)_t = \frac{1}{\dot{\xi}(t)}\left[\frac{df}{dt} - \left(\frac{\partial f}{\partial t}\right)_h\right], \tag{11.101}$$

provides the needed relations in equations (11.98)–(11.100). The differential df/dt is the directional derivative along the path $\xi(t)$ in the h–t plane illustrated previously in figure 11.34. The partial time derivative in equation (11.101) is determined from the gage records. The path $\xi(t)$ is chosen to span the multiple gage profiles making optimum use of the gage data in approximating the differential df/dt.

11.8.6.1 Application to an elastic analytic solution

The analysis method in spherical curvilinear geometry is readily tested through known analytic solutions. For a linear elastic material the wave equation is generally solved through the elastic potential

$$\psi(H,T) = f(H - T)/H, \tag{11.102}$$

where $f(H - T)$ is an arbitrary function with $H = h/a$ and $T = ct/a$ non-dimensional expressions for the radial Lagrangian position and time. The length scale a is the radius of an internal cavity at which an impulsive load initiates outwardly directed spherical wave propagation, and c is the elastic wave speed. The solution for the elastic wave motion begins through specification of the displacement time history at $h = a$ through

$$U(1,T) = g(1 - T), \quad T \geqslant 1, \tag{11.103}$$

and subject to the condition of zero displacement at the wave front,

$$U(T,T) = 0, \quad T \geqslant 1. \tag{11.104}$$

The function $g(y)$ is an arbitrary but appropriate representation of the initial motion history provided by the internal load, while the form of the argument y in equation (11.103) is for later analytic convenience.

Displacement throughout the field is prescribed by the derivative of the elastic potential $U = \partial\psi/\partial H$ yielding the differential relation

$$f'(y) - f(y) = g(y),\ y \leqslant 0, \tag{11.105}$$

where $y = 1 - T$ and has the solution

$$f(y) = \mathrm{e}^{y} \int_{0}^{y} \mathrm{e}^{-x} g(x) dx. \tag{11.106}$$

The elastic potential from equation (11.102) then allows the calculation of selected stress and velocity histories as representative experimental data for testing of the Lagrangian analysis technique. In particular the radial particle velocity follows from

$$V(H,T) = \frac{\partial^2 \psi}{\partial T \partial H} = -\frac{1}{H} f'' + \frac{1}{H^2} f'. \tag{11.107}$$

The radial and tangential strains are accordingly

$$\varepsilon_r = \frac{\partial U}{\partial H} = \frac{1}{H} f'' - \frac{2}{H^2} f' + \frac{2}{H^3} f, \tag{11.108}$$

$$\varepsilon_t = \frac{U}{H} = \frac{1}{H^2} f' - \frac{1}{H^3} f. \tag{11.109}$$

Radial and tangential stresses follow from the stress–strain relations in spherical coordinates.

An appropriate and convenient functional form for the cavity boundary displacement that generates wave structures that are representative of measurements resulting from explosion-induced spherical stress waves in solid media is given by the relation

$$g(y) = u_o y^2 \mathrm{e}^{y}, \quad y \leqslant 0. \tag{11.110}$$

A measure of the maximum displacement strain at the cavity boundary is provided by the coefficient u_o. Radial stress and particle-velocity histories at selected Lagrangian radial positions calculated from the analytic solution are shown in figure 11.44. A value of $u_o = 0.02$ is used in equation (11.110). Values for the stress plotted in the figure are normalized by the bulk modulus of the material. Particle velocity is normalized by the elastic sound speed. A Poisson's ratio of 0.25 is assumed for the material. Stress and velocity histories at selected Lagrangian radii are representative of multiple experimental recordings that are subjected to the path-derivative Lagrangian analysis as a test of the method.

In undertaking Lagrangian analysis of the test experimental records depicted in figure 11.44, volume strain and strain difference are first calculated from the particle-velocity data and the equation of continuity. The stress difference and mean stress are then determined using both the particle velocity and radial stress data in conjunction with equation (11.100). Time is then eliminated as a parameter yielding the relations for pressure versus volume strain and stress difference versus pressure determined from the path-derivative Lagrangian analysis method. Plots resulting from this test study are provided in figure 11.45. Dashed curves illustrate results of the Lagrangian analysis and are compared with the solid curves for the exact paths from the analytic solution. The test analysis illustrates that, at least for this example, when the gage profile data are exemplary, the Lagrangian analysis method can yield quite satisfactory results.

The Lagrangian analysis yields a near-linear dependence between mean stress (pressure) and volume strain, and between stress difference and strain difference,

$$P = \alpha\varepsilon_v, \qquad \varphi = 2\beta\varepsilon_d. \tag{11.111}$$

Values for the constants from the analysis of $\alpha = 0.555$ and $\beta = 0.332$ compare well with the analytic bulk modulus of 5/9 and shear modulus of 1/3.

11.8.6.2 Application to experimental data for Westerly granite

Application of the Lagrangian analysis method to experimental data more effectively illustrates sensitivities to gage calibration errors and frequency response, in addition to errors in time and position correlation. A truly unique set of data is provided by the painstaking study of Swift (1971) in which blocks of Westerly

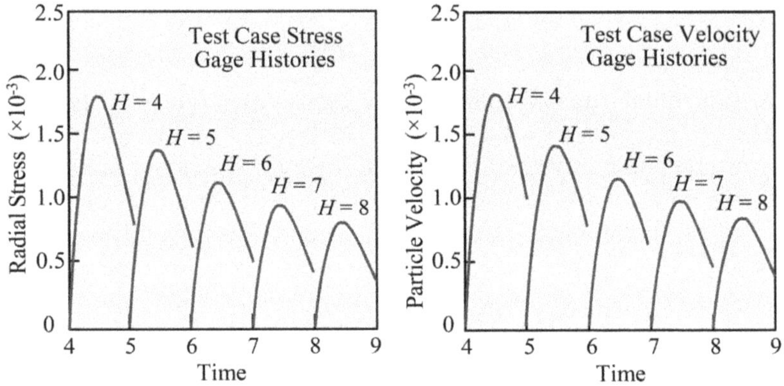

Figure 11.44. Radial stress (left) and particle-velocity (right) profiles (non-dimensional) calculated in the linear elastic analysis and assumed as experimental stress and velocity measurements for purposes of testing the path-derivative Lagrangian analysis methods.

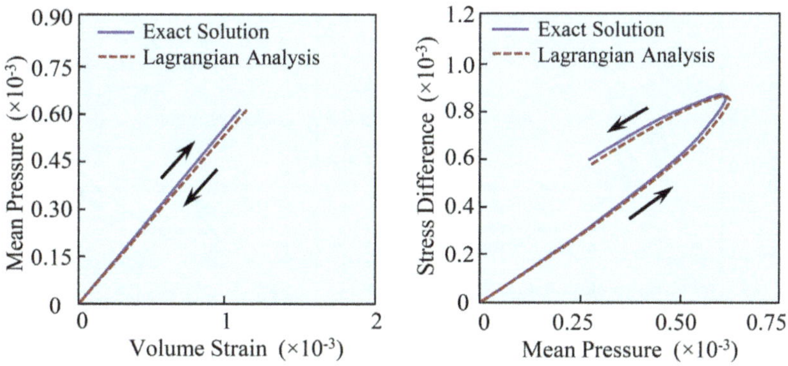

Figure 11.45. Solid curves are exact solutions from the elastic analysis at the intermediate Lagrangian position ($H = 6$) for the mean pressure versus volume strain (left) and stress difference versus mean pressure (right) through the compression and decompression wave history. Dashed curves are the same paths determined through the path-derivative Lagrangian analysis.

granite were subjected to spherical one-dimensional stress wave loading resulting from detonation of explosive placed in a one-inch radius cavity within the block. Radial stress and particle velocity was measured at multiple radial positions. Particle velocity was measured using the electromagnetic gage technique (Dremin and Shvedov 1964, Edwards *et al* 1970). Radial stress was measured with piezo-resistive Manganin stress gages (Keough 1968).

Reported experimental radial stress and particle-velocity records achieved in the test on Westerly granite are shown in figure 11.46. Sources of error are discussed, and a nominal error of 12% for the particle-velocity records and 25% for the radial stress records are recommended. The data of Swift are subjected to the same path-derivative Lagrangian analysis applied to the previous elastic analytic solution (Grady 1973). The stress and velocity data in figure 11.46 span radial positions in the granite block from about four to eight times the radius (one inch) of the explosive cavity. Analysis is identified with a Lagrangian position at six times the radius,

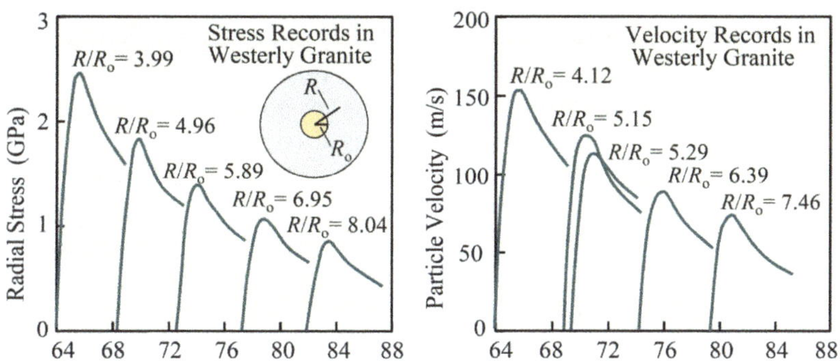

Figure 11.46. Radial stress (left) and particle-velocity (right) measurements in a spherical symmetry explosive experiment on Westerly granite (Swift 1971). Cavity radius R_o is equal to one inch.

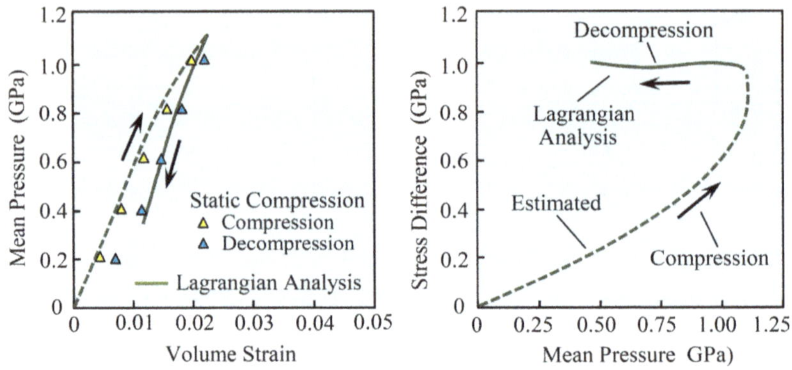

Figure 11.47. Path-derivative Lagrangian analysis of radial stress and particle-velocity measurements of spherical explosive experiment on Westerly granite (Swift 1971). Static compression data from La Mori (1968).

intermediate to the experimental data in estimating outer derivatives necessary to calculation of the dynamic material response of the granite.

Strain histories including the volumetric and difference strain are obtained directly from the particle-velocity data alone. Volume strain at the analysis position and the corresponding peak radial stress are found to be in sensible agreement with independent Hugoniot measurements on Westerly granite (Rosenberg 1971). The implication is that material response to the stress wave loading, at least to the maximum stress, is not markedly effected by the spherical geometry of the motion. Correspondingly, Lagrangian analysis of the wave before and up to the peak is not effective in calculating the deviator stress from equation (11.100). Estimated derivatives in this region are many times larger than the stress difference calculated, and measurement errors negate analysis results in this region of the wave motion.

This difficulty is not the case in the unloading portion of the wave motion. Spherical flow significantly influences this later-time material response and the stress difference of approximately 1 GPa is readily calculated from the path-derivative estimates through equation (11.100). Stress and strain responses derived from Lagrangian analysis of the multiple radial stress and particle-velocity gage data are illustrated in the two plots in figure 11.47. Dashed curves in the plots are estimated responses in the loading portion of the wave where spherical Lagrangian analysis is ineffective in determining stress difference. Solid curves correspond to calculated response from the unloading wave data. A calculated stress difference of about 1 GPa shown in the right-hand plot is used to determine the mean stress on decompression shown on the left with $P = \sigma - 2\varphi/3$. The analysis results are compared with static compression and decompression measurements performed on Westerly granite (La Mori 1968). Further, the analysis yields a shear modulus of 18.5 GPa and a Poisson's ratio of about 1/3, in sensible agreement with ultrasonic properties for Westerly granite (Rosenberg 1971).

11.8.7 Lagrangian analysis methods

The techniques of Lagrangian analysis emerged in the early 1970s as a natural response to the evolution of the time-resolved instrumentation under development.

These emerging time-resolved measurement techniques revealed the complex unsteady large-amplitude shock wave structure supported by many natural and engineering materials of concern. New analysis methods were necessary to uncover the material response features essential to the development of constitutive models. The emergence of Lagrangian analysis methods was the response.

Like earlier Hugoniot analysis, Lagrangian analysis is an effort to extract model-independent and test-independent strength, rheologic and EOS properties relevant to the intense explosive and impact conditions to which materials are subjected. Results have been mixed. Lagrangian analysis is painstaking and uncertainties are difficult to quantify. Today's computational power, however, offers avenues for more thorough and exacting applications of Lagrangian analysis.

The alternative, of course, is the development of material constitutive models based on other insights into material response, and then validation and calibration of such models to the same time-resolved stress wave data. This latter approach has been the more popular. Perhaps a melding of both approaches offers insights into the ever-increasing detail in the dynamic material behavior of condensed matter relevant to the expanding applications of interest.

References

Abbey A E and Bonner B P 1975 Elastic constants of alpha-uranium *J. Appl. Phys.* **46** 1427–8

Addessio F L, Zuo Q H and Mason T A 2003 Model for high-strain-rate deformation of uranium–niobium alloys *J. Appl. Phys.* **93** 9644–54

Ahrens T J, Gust W H and Royce E B 1968 Material strength effects in the shock compression of alumina *J. Appl. Phys.* **39** 4610–6

Aidun J B and Gupta Y M 1991 Analysis of Lagrangian gauge measurements of simple and nonsimple plane waves *J. Appl. Phys.* **69** 6998–7014

Alexander C S, Vogler T J, Reinhart W D, Grady D E, Kipp M E and Chhabildas L C 2006 Influence of Shock Wave Measurement Technique on the Determination of Hugoniot States *Shock Compression of Condensed Matter-2005* ed S Schmidt and W Tao 1229–32

Barker L M and Hollenbach R E 1972 Velocity interferometer for measuring the velocity of any reflecting surface *J. Appl. Phys.* **43** 4669–80

Cagnoux J, Chartagnac P, Hereil P and Perez M 1984 Lagrangian analysis: modern tool of the dynamics of solids *Ann. Phys* **12** 451–524

Carter W J, Marsh S P, Fritz J N and McQueen R G 1971 The Equation of State of Selected Materials for High-Pressure References *Symposium on Accurate Characterization of the High-Pressure Environment*, National Bureau of Standards, Gaithersburg, MD, Oct. 14-18, 1968, (NBS, Washington, DC, Special Publication 326)

Chen P J, Graham R A and Davison L 1972 Analysis of unsteady waves in solids *J. Appl. Phys.* **43** 5021–7

Chhabildas L C and Grady D E 1984 Dynamic material response of quartz at high strain rates *Mater. Res. Soc. Symp. Proc.* **22** 147–150

Cowperthwaite M and Williams R F 1971 Determination of constitutive relationships with multiple gauges in nondivergent waves *J. Appl. Phys.* **42** 456–62

Dandekar D P, Martin A G and Kelly J V 1980 Deformation of depleted uranium—0.78 Ti under shock compression to 11.0 GPa at Room Temperature *J. Appl. Phys.* **51** 4784–9

Dmitriyev A P, Kuzyayev L S, Protasov Y I and Yamshchikov V S 1972 Physical properties of rocks at high temperature, *NASA Technical Translation*, NASA TT F-684, June.
Dremin A N and Adadurov G A 1964 The behavior of glass under dynamic loading *Soviet Phys. Solid State* **6** 1379–84
Dremin A N and Shvedov K K 1964 The determination of Chapman–Jouquet pressure and of the duration of reaction in the detonation wave at high pressure *Zh. Prikl. Mekh. Tekh. Fiz* **2** 144–55
Dremin A N, Pershin S V and Pogorelov V F 1965 Structure of shock waves in KCl and KBr under dynamic compression *Combust. Explos. Shock Waves* **1** 1–4
Edwards D J, Erkman J O and Jacobs S J 1970 The electromagnetic velocity gage and applications to the measurement of particle velocity in PMMA *Naval Ordnance Lab White Oak, MD* 70–9
Ehrgott J 1993 SAC-5 concrete *Waterways Experimental Station Memorandum for Record* CEWES-SP-O March
Fowles R and Williams R F 1970 Plane stress wave propagation in solids *J. Appl. Phys.* **41** 460
Grady D E 1973 Experimental analysis of spherical wave propagation *J. Geophys. Res.* **78** 1299–307
Grady D E 1977 Processes occurring in shock wave compression of rocks and minerals *High Pressure Research Applications in Geophysics* (New York: Academic) pp 389–438
Grady D E 1983 Stress wave studies on selected rocks and minerals *Sandia National Laboratories Technical Report* SAND83- 1898, September
Grady D E 1986 Steady-wave rise-time and spall measurements on uranium (3–15 GPa) *Metallurgical Applications of Shock-Wave and High-Strain-Rate Phenomena* ed L E Murr, K P Staudhammer and M A Meyers (New York: Marcel Dekker) p 763
Grady D E 1996 Shock equation of state properties of concrete *Structures under Shock and Impact IV* ed N Jones, C A Brebbia and A J Watson (Southampton: Computational Mechanics Publications) pp 405–14
Grady D E 2005 Investigation of explosively driven fragmentation of metals—metal breach, enhanced ductility, adiabatic shear, shock structure and strength *Applied Research Associates Technical Report* September
Grady D E 2015 Equation of state of dolomite from shock Hugoniot and static compression studies *Proc. APS Topical Group on Shock Compression of Condensed Matter, (Tampa, FL, 15–19 June)*
Grady D E, Murri W J and Fowles G R 1974 Quartz to stishovite: wave propagation in the mixed phase region *J. Geophys. Res.* **79** 332–8
Grady D E, Murri W J and Mahrer K D 1976 Shock compression of dolomite *J. Geophys. Res.* **81** 889–93
Grady D E and Moody R L 1996 Shock compression profiles in ceramics, *Sandia National Laboratories Technical Report.*, SAND96-0551, March
Grady D E and Kipp M E 1998 Discrepancies in the interpretation of Hugoniot states from inclined-mirror and VISAR data, VISAR data *Applied Research Associates Report* March
Grady D E and Olsen M L 2003 A statistical and energy based theory of dynamic fragmentation *Int. J. Impact Eng.* **29** 293–306
Gust W H, Holt A C and Royce E B 1973 Dynamic yield, compression, and elastic parameters for several lightweight intermetallic compounds *J. Appl. Phys.* **44** 550–60
Hayes D B, Hixson R S, Gray G T, Zurek A K, Vorthman J E and Anderson W W 2003 Precursor suppression by shear stress relaxation in U-Nb(6-wt%) *Shock Compression of*

Condensed Matter—2003 ed M D Furnish (New York: American Institute of Physics) pp 573–6

Heard H C, Duba A, Abey A E and Schock R N 1973 Mechanical properties of Blair dolomite, *Lawrence Livermore Laboratory Report* UCRL-51465

Hixson R S, Vorthman J E, Gustavsen R L, Zurek A K, Thissel W R and Tonks D L 1998 Spall wave-profile and shock recovery experiments on depleted uranium *Shock Compression of Condensed Matter—1997* ed A Schmidt, D Dandekar and J Forbes (New York: American Institute of Physics) pp 479–82

Hixson R S, Vorthman J E, Gustavsen R L, Zurek A K, Anderson D L and Tonks D L 2000 Spall response of U-NB (6%) alloy *Shock Compression of Condensed Matter—1997* ed M D Furnish, L C Chhabildas, L C Chhabildas and R S Hixson (New York: American Institute of Physics) pp 489–492

Hockett J E and Sherby O D 1975 Large strain deformation of polycrystalline metals at low homologous temperatures *J. Mech. Phys. Solids* **23** 87–98

Hockett J E, Gilman P S and Sherby O D 1977 Compressive deformation of polycrystalline uranium at low temperatures *J. Nuclear Materials* **64** 231–40

Holden A N 1958 *Physical Metallurgy of Uranium* (Reading, MA: Addison-Wesley)

Hongliang H, Sekine T, Kobayashi T, Hirosaki Hm and Suzuki T 2000 Shock-induced phase transition of *b*-Si3N4 to *c*-Si3N4 *Phys. Rev.* B **62** 11412

Huddart J, Harding J and Bleasdale P A 1980 The effect of strain rate on the tensile flow and fracture of alpha-uranium *J. Nuclear Mater* **89** 316–30

Kalashnikov N G, Pavlovsky M N, Simakov G V and Trunin R F 1973 Dynamic compressibility of calcite group minerals *Izv. Earth Phys* **2** 23–9

Keough D D 1968 Procedure for fabrication and operation of Manganin shock pressure gage, *Stanford Research Institute Technical Report* AFWL-TR-68-57

Kipp M E and Grady D E 1994 Shock phase transformation and release properties of aluminum nitride *J. Physique* 249–56

La Mori P N 1968 Compressibility of three rocks *Battelle Memorial Institute Tech Report* DASA 2151

Lombard D B 1961 The Hugoniot equation of state of rocks *University of California Radiation Laboratory* UCRL-6311, February

Lysne P C, Boade R R, Percival C M and Jones O E 1969 Determination of release adiabats and recentered Hugoniot curves by shock reverberation techniques *J. Appl. Phys.* **40** 3786

Mao Z *et al* 2011 Dolomite III: a new candidate lower mantle carbonate *Geophys. Res. Lett.* **38** L22303

Marsh S P 1980 *LASL Shock Hugoniot Data* ed S P Marsh (Berkeley, CA: University of California Press)

Mashimo T, Uchino M, Nakamurac A, Kobayashi T, Takasawa E, Sekine T, Noguchi Y, Hikosaka H and Syono Y 1999 Yield properties, phase transition, and equation of state of aluminum nitride under shock compression up to 150 GPa *J. Appl. Phys.* **86** 6710–6

Merlini M, Crichton W A, Hanfland M, Gemmid M, Müllerb H, Kupenkoe I and Dubrovinsky L 2012 Structures of dolomite at ultrahigh pressure and their influence on the deep carbon cycle *Proc. Natl Acad. Sci.* **109** 13509–14

Mesheyakov Y I, Divakov A K, Petrov Y A and Cline C F 2003 On the dynamic plasticity and strength of polycrystalline beryllium *Int. J. Impact Eng.* **30** 17–30

Meyers M A, Vohringer O and Lubarda V A 2001 The onset of twinning in metals: a constitutive description *Acta Mater.* **49** 4025–39

Munson D E 1977 Analysis of high pressure oil shale Hugoniots *Sandia National Laboratories Technical Report* SAND77-0118, February

Murri W J and Grady D E 1973 Equation of state of rocks *Stanford Research Institute Report* AT (04-3)-115, April

Murri W J, Grady D E and Mahrer K D 1975 Equation of state of rock *Stanford Research Institute Report* PYU 1883, AT(04-3)–115 July

Nye J F 1957 *Physical Properties of Crystals* (Oxford: Clarendon)

Petersen C F, Murri W J and Cowperthwaite M 1970 Hugoniot and release wave measurements for selected geological materials *J. Geophys. Res.* **75** 2063–72

Rosenberg J T 1971 Dynamic shear strength of shock-loaded granite and polycrystalline quartz *Stanford Research Institute Technical Report* DASA 2718

Rosenberg J T and Ahrens T J 1966 ASME paper No 66-WA/PT-7 *American Society of Engineering* December

Santillán J, Williams Q and Knittle E 2003 Dolomite-II: A High-Pressure Polymorph of CaMg $(CO_3)_2$ *Geophys. Res. Lett.* **30** 1054

Seaman L 1974 Lagrangian analysis for multiple stress or particle velocity gages in attenuating waves *J. Appl. Phys.* **45** 43034314

Shipman F M, Isbell W M and Jones A H 1969 High pressure Hugoniot measurements for several Nevada Test Site rocks, DTRA-AD854912, March

Swift R P 1971 The dynamic response of Westerly granite to spherical stress waves *Physics International Co. Technical Report* DASA 2712, San Leandro CA

Tablin D M R 1967 Tensile properties and fracture of uranium between –200 C and +900 C *J. Austral. Inst Metals* **12** 32–44

Tao W, Huan S, Tan X and Jiang G 2011 Lagrangian analysis method with least square cubic B-spline *Appl. Mech. Materials* **94** 1681–4

Trunin R F, Gudarenko L F, Zhernokletov M V and Simakov G V 2001 Experimental Data on Shock Compression and Adiabatic Expansion of Condensed Matter, Sarov: RFNC-VNIIF; ISBN 5-85165-624-7.

Xia Q, Xia H and Ruoff A L 1993 Pressure-induced rocksalt phase of aluminum nitride: a metastable structure at ambient condition *J. Appl. Phys.* **73** 8198

Zurek A K, Hixson R S, Anderson W W, Vorthman J E, Gray G T and Tonks D L 2000 Elastic response of U–Nb (6%) alloy *Proc. 6th Int. Conf. on Mechanical and Physical Behaviour of Materials under Dynamic Loading (25–29 September)* (EDP Sciences) pp 677–82

www.ingramcontent.com/pod-product-compliance
Ingram Content Group UK Ltd.
Pitfield, Milton Keynes, MK11 3LW, UK
UKHW051042120926
8556IPUK00030B/829

9 780750 312585